Cereal Grains

Ever since the beginnings of agriculture, cereals have provided unlimited health benefits to mankind as a staple food in our diet. Cereals are rich in complex carbohydrates that provide us ample energy and help to prevent many diseases such as constipation, colon disorders, and high blood sugar levels. They enrich our overall health with abundant proteins, fats, lipids, minerals, vitamins, and enzymes. In every part of the world cereals are consumed for breakfast, lunch or dinner. *Cereal Grains: Composition, Nutritional Attributes, and Potential Applications* provides an overview of cereals, including their properties, chemical composition, applications, post-harvest losses, storage, and quality.

Various well-versed researchers across the globe share their knowledge and experience covering cereal's role in food security, allergens in grains, phytochemical profile, industrial applications, health benefits, global standard of cereals, and recent advances in cereal processing.

KEY FEATURES

- Contains comprehensive information on general composition and properties of cereals
- Discusses the recent advances in cereal technology
- Provides knowledge on bioactive characterization of cereal grains
- Contains information on future aspects of grain quality and allergens in cereal grains

This handbook is a valuable resource for students, researchers, and industrial practitioners who wish to enhance their knowledge and insights on cereal science. Researchers, scientists, and other professionals working in various cereal processing industries and horticultural departments will also find the comprehensive information relevant to their work.

Cereal Grains

Composition, Nutritional Attributes and Potential Applications

Edited by
Gulzar Ahmad Nayik, Tabussam Tufail,
Faqir Muhammad Anjum, and Mohammad Javed Ansari

CRC Press
Taylor & Francis Group
Boca Raton London

CRC Press is an imprint of the
Taylor & Francis Group, an **informa** business

First edition published 2023
by CRC Press
6000 Broken Sound Parkway NW, Suite 300, Boca Raton, FL 33487–2742

and by CRC Press
4 Park Square, Milton Park, Abingdon, Oxon, OX14 4RN

CRC Press is an imprint of Taylor & Francis Group, LLC

Library of Congress Cataloging-in-Publication Data
Names: Nayik, Gulzar Ahmad, editor. | Tufail, Tabussam, editor. | Anjum,
    Faqir Muhammad, editor. | Ansari, Mohammad Javed, editor.
Title: Cereal grains : composition, nutritional attributes, and potential
    applications / edited by Gulzar Ahmad Nayik, Tabussam Tufail, Faqir
    Muhammad Anjum & Mohammad Javed Ansari.
Description: First edition. | Boca Raton, FL : CRC Press, 2023. | Includes
    bibliographical references and index.
Identifiers: LCCN 2022037739 (print) | LCCN 2022037740 (ebook) |
    ISBN 9781032156637 (hbk) | ISBN 9781032171593 (pbk) |
    ISBN 9781003252023 (ebk)
Subjects: LCSH: Grain. | Cereal products. | Cereals as food—Nutrition.
Classification: LCC SB189 .C396 2023 (print) | LCC SB189 (ebook) |
    DDC 633.1—dc23/eng/20220928
LC record available at https://lccn.loc.gov/2022037739
LC ebook record available at https://lccn.loc.gov/2022037740

ISBN: 978-1-032-15663-7 (hbk)
ISBN: 978-1-032-17159-3 (pbk)
ISBN: 978-1-003-25202-3 (ebk)

DOI: 10.1201/9781003252023

Typeset in Times
by Apex CoVantage, LLC

I dedicate this book to my beloved daughter

Eidain Hoor Binti Gulzar

Contents

Preface

Cereals and legumes are an important part of diets and contribute substantially to the nutrient intake of human beings. They are significant sources of energy, protein, dietary fiber, vitamins, minerals, and phytochemicals. However, cereals are prone to fungal infection, which produces mycotoxins during storage in addition to adulteration, especially in cereal and cereal products, which cannot be easily detected, posing significant risks to human health. It is therefore important to screen and assess the safety and quality attributes of cereal foods to ensure food safety and quality.

Approximately one-third of the food produced (about 1.3 billion tons), worth about US $1 trillion, is lost globally during post-harvest operations every year. The solutions to reduce post-harvest losses require relatively modest investment and can result in high returns compared to increasing the crop production to meet the food demand. Post-harvest loss includes food loss across the food supply chain from harvesting of crops until their consumption. The losses can be broadly categorized as weight loss due to spoilage, quality loss, nutritional loss, seed viability loss, and commercial loss. The magnitude of post-harvest losses in the food supply chain varies greatly among different crops, areas, and economies. In developing countries, people try to make the best use of the food produced; however, a significant amount of produce is lost in post-harvest operations due to a lack of knowledge, inadequate technology, and/or poor storage infrastructure.

Although some books are available on major cereals of the world, they lack general knowledge on cereals, so the main aim of this book is to provide a general view of cereals, including their properties, chemical composition, applications, phytochemical profiles, post-harvest losses, storage, quality, and so on.

We would like to thank and acknowledge all the contributors for their fruitful contributions and their dedication to the editorial guidelines and timeline. We are fortunate to have the opportunity to collaborate with many international experts. We would like to thank our colleagues from the production team of Taylor & Francis Group for their constant help during the editing and production process. Finally, we as editors have a message to all readers that this book may contain minor errors or gaps. Suggestions, criticisms, and comments are always welcome, so please do not hesitate to contact us for any relevant issue.

Dr. Gulzar Ahmad Nayik
Dr. Tabussam Tufail
Prof. Faqir Muhammad Anjum
Dr. Mohammad Javed Ansari

Editors

Dr. Gulzar Ahmad Nayik completed his master's degree in food technology from the Islamic University of Science & Technology, Awantipora, Jammu and Kashmir, India, and his PhD from Sant Longowal Institute of Engineering & Technology, Sangrur, Punjab, India. He has published more than 75 peer-reviewed research and review papers and more than 35 book chapters. Dr, Nayik has also edited 11 books with Springer, Elsevier, and Taylor & Francis. Dr. Nayik has also published a textbook on food chemistry and nutrition and has delivered several presentations at various national and international conferences, seminars, workshops, and webinars. Dr. Nayik was shortlisted twice for the prestigious Inspire-Faculty Award in 2017 and 2018 from Indian National Science Academy, New Delhi, India. He was nominated for India's prestigious National Award (Indian National Science Academy Medal for Young Scientists-2019–20). Dr. Nayik also fills the roles of editor, associate editor, assistant editor, and reviewer for many food science and technology journals. He has received many awards, appreciations, and recognitions and holds membership in various international societies and organizations. Dr. Nayik is currently editing several book projects with Elsevier, Taylor & Francis, Springer Nature, Royal Society of Chemistry, and others.

Dr. Tabussam Tufail is currently working as assistant professor in the University Institute of Diet & Nutritional Sciences, Faculty of Allied Health Sciences, University of Lahore, Lahore, Punjab, Pakistan. He completed his PhD in food science and technology and MS in food science and technology from Government College University Faisalabad. He has completed 25 international and national training sessions as well as courses from different organizations. He has published 70 peer-reviewed research/review papers, 14 book chapters, and 1 book and presented papers at 25 international and national conferences as well as attending a number of conferences, seminars, workshops, and webinars. Dr. Tufail serves as an editorial board member for Acta Scientific Publications. He is a frequent reviewer of several reputed journals in the area of food and nutrition as well as food science and technology. He is a lifetime member of the Scientific and Technical Research Association (STRA), lifetime member of the Healthcare & Biological Sciences Research Association (HBSRA), lifetime member of the Teaching and Education Research Association (TERA), lifetime member of the Social Science and Humanities Research Association (SSHRA), and lifetime member of the Pakistan Society of Food Scientists & Technologists (PSFST).

Prof. Faqir Muhammad Anjum is a renowned cereal technologist of Pakistan who is currently Chief Executive of Ifanca Pakistan Halal Apex Private Ltd. He has completed his PhD in grain science from Kansas State University, USA; PhD in food technology from the University of Agriculture, Faisalabad, Pakistan; and postdoc from the University of Reading, UK. Prof. Anjum has served as Vice Chancellor, University of the Gambia, The Gambia; Vice Chairman of the Association of West African Universities; and Founding Director General, National Institute of Food Science & Technology, University of Agriculture, Faisalabad. He has vast experience of 39 years of teaching and research. He is the recipient of Tamgha-i-Imtiaz, the fourth-highest award of Pakistan. Prof. Anjum has supervised 37 students at the doctorate level and 116 at the master's level. Prof. Anjum has published more than 290 national and international research/review papers. He has also published 15 books and is the author/co-author of many book chapters. Prof. Anjum has completed 26 R & D projects.

Dr. Mohammad Javed Ansari is currently working as assistant professor in the Department of Botany, Hindu College Moradabad, UP, India. Dr. Ansari also worked as assistant professor in the Department of Plant Protection, College of Food & Agricultural Sciences, King Saud University Riyadh, Saudi Arabia, from 2010 to 2017. He earned his PhD in biotechnology from the Indian Institute of Technology Roorkee, India, and MSc in botany from Ch. Charan Singh University Meerut, India. Dr. Ansari has published more than 150 research/review articles in high impact factor journals. He has worked as Principal Investigator in many funded projects at King Saud University Riyadh, Saudi Arabia. Dr. Ansari has participated in many national and international conferences (Abu Dhabi, Belgium, Indonesia, Turkey, United States, Italy, Romania, Germany, Spain, Cuba, Australia, Egypt, Ukraine, Ethiopia, Portugal, etc.). He is an editorial board member for many renowned journals and also a member of many international scientific associations and societies. Dr. Ansari has supervised many students at the master's and doctorate level.

Contributors

Muhammad Afzaal
Department of Food Sciences
Government College University
Faisalabad-Pakistan

Asif Ahmad
Institute of Food and Nutritional Sciences
PMAS-Arid Agriculture University
Rawalpindi, Pakistan

Taslima Ahmed
Department of Applied Food Science and
    Nutrition
Faculty of Food Science and Technology
Chattogram Veterinary and Animal Sciences
    University (CVASU)
Chattogram Bangladesh

Samreen Ahsan
Department of Food Science
    and Technology
Khwaja Fareed University of Engineering and
    Information Technology
Rahim Yar Khan, Punjab, Pakistan

Ayesha Ali
Department of Food Science and Technology
Khwaja Fareed University of Engineering
    and Information Technology
Rahim Yar Khan, Punjab, Pakistan

Muhammad Adil Farooq
Department of Food Science
    and Technology
Khwaja Fareed University of Engineering
    and Information Technology
Rahim Yar Khan, Punjab, Pakistan

Abha Anand
Department of Food Science, Nutrition and
    Technology
College of Community Science
Palampur, Himachal Pradesh, India

Panthavur Nairveetil Anjali
Department of Food Science and Technology
Pondicherry University
Puducherry, India

Mohammad Javed Ansari
Department of Botany
Hindu College Moradabad (Mahatma Jyotiba
    Phule Rohilkhand University Bareilly)
India

Misbah Aslam
Department of Food Sciences
Government College University
Faisalabad-Pakistan

Menithen Beber Rodrigues
Faculty of Agrarian Sciences
Federal University of Amazonas—UFAM,
Manaus, AM, Brazil

Zarina Begum
Shaheed Udham Singh College of Engineering
    and Technology
Tangori, Mohali, Punjab, India

Abida Bhat
Department of Immunology and Molecular
    Medicine
Sher-e-Kashmir Institute of Medical Sciences
Kashmir, J & K-India

Farhan M Bhat
Department of Food Science Nutrition and
    Technology
College of Community Science
Palampur, Himachal Pradesh, India

Sunanda Biswas
Department of Food and Nutrition
Acharya Prafulla Chandra College
    New Barrackpore
Kolkata, West Bengal, India

Tridip Boruah
PG Department of Botany
Madhab Choudhury College
Barpeta, Assam, India

Sowriappan John Don Bosco
Department of Food Science and Technology
Pondicherry University
Puducherry, India

Nilakshi Chauhan
Department of Food Science Nutrition
    and Technology
College of Community Science
Palampur, Himachal Pradesh, India

Twinkle Chetia
Department of Botany
Cotton University
Guwahati. Assam, India

Kamini Choubey
PG Department of Botany
Madhab Choudhury College
Barpeta, Assam, India

Muhammad Farhan Jahangir Chughai
Department of Food Science
    and Technology
Khwaja Fareed University of Engineering and
    Information Technology
Rahim Yar Khan, Punjab, Pakistan

Tanuva Das
Department of Food Engineering
    and Technology
Tezpur University
Assam, India

Barsha Devi
Plant Physiology and Biochemistry
    Laboratory Department of Botany
Gauhati University
Guwahati, Assam, India

Y.S. Dhaliwal
BR Ambedkar Institute of Medical
    Sciences
Mohali, Punjab, India

Subhamoy Dhua
Department of Food Engineering and
    Technology
Tezpur University
Assam, India

Shoaib Fayyaz
Department of Food Science and Technology
Khwaja Fareed University of Engineering and
    Information Technology
Rahim Yar Khan, Punjab, Pakistan

Arun Kumar Gupta
Department of Life Sciences, Graphic Era
    (Deemed to Be) University
Dehradun, India

Muzzamal Hussain
Department of Food Sciences Government
    College University
Faisalabad-Pakistan

Ali Ikram
University institute of Food science
    and Technology
University of Lahore
Pakistan

Rukhsana Iqbal
Department of Chemistry, Khwaja Fareed
    University of Engineering & Information
    Technology
Rahim Yar Khan, Punjab, Pakistan

Syed Junaid-ur-Rehman
Department of Food Science and Technology
Khwaja Fareed University of Engineering and
    Information Technology
Rahim Yar Khan, Punjab, Pakistan

Ghulam Mustafa Kamal
Department of Chemistry, Khwaja Fareed
University of Engineering & Information
    Technology
Rahim Yar Khan, Punjab, Pakistan

Kulwinder Kaur
Department of Processing and Food
    Engineering
Punjab Agricultural University
Ludhiana, India

Preetinder Kaur
Department of Processing and Food
    Engineering
Punjab Agricultural University
Ludhiana, India

Muhammad Khalid
Department of Chemistry
Khwaja Fareed University of Engineering &
    Information Technology
Rahim Yar Khan, Punjab, Pakistan

Adnan Khaliq
Department of Food Science and Technology
Khwaja Fareed University of Engineering and
    Information Technology
Rahim Yar Khan, Punjab, Pakistan

Mohammed Ayub Khan
Defence Food Research Laboratory
Siddharthanagar, Mysuru
Karnataka, India

Arfa Liaquat
Department of Chemistry
Khwaja Fareed University of Engineering &
    Information Technology
Rahim Yar Khan, Punjab, Pakistan

Atif Liaqat
Department of Food Science and Technology
Khwaja Fareed University of Engineering
    and Information Technology
Rahim Yar Khan, Punjab, Pakistan

Chamman Liaqat
Institute of Food and Nutritional Sciences
PMAS-Arid Agriculture University
Rawalpindi, Pakistan

Maristela Martins Pereira
Faculty of Agrarian Sciences
Federal University of Amazonas—UFAM,
    Manaus, AM
Brazil

Tariq Mehmood
Department of Food Science
    and Technology
Khwaja Fareed University of Engineering
    and Information Technology
Rahim Yar Khan, Punjab, Pakistan

Poonam Mishra
Department of Food Engineering and
    Technology
Tezpur University
Assam, India

Bharti Mittu
National Institute of Pharmaceutical Education
    and Research (NIPER)
S.A.S. Nagar, Mohali
Punjab, India

M Pal Murugan
Defence Food Research Laboratory
Siddharthanagar, Mysuru
Karnataka, India

Muhammed Navaf
Department of Food Science
    and Technology
Pondicherry University
Puducherry, India

Gulzar Ahmad Nayik
Department of Food Science &
    Technology
Govt. Degree College Shopian
J & K-India

Bushra Niaz
Department of Food Sciences
Government College University
Faisalabad-Pakistan

Ayesha Noreen
Department of Chemistry
Khwaja Fareed University of Engineering &
    Information Technology
Rahim Yar Khan, Punjab, Pakistan

Nahidur Rahman
Department of Food Processing
    and Engineering
Faculty of Food Science
    and Technology
Chattogram Veterinary and Animal
    Sciences University
Chattogram, Bangladesh

Amara Rasheed
Department of Food Sciences
Government College University
Faisalabad-Pakistan

Muhammad Ahtisham Raza
Department of Food Sciences
Government College University
Faisalabad-Pakistan

Jagbir Rehal
Department of Food Science
    and Technology
Punjab Agricultural University
Ludhiana, India

Asma Sabir
Department of Chemistry
Khwaja Fareed University of Engineering &
    Information Technology
Rahim Yar Khan, Punjab, Pakistan

Aqib Saeed
Department of Food Science and Technology
Khwaja Fareed University of Engineering and
    Information Technology
Rahim Yar Khan, Punjab, Pakistan

Farhan Saeed
Department of Food Sciences
Government College University
Faisalabad-Pakistan

Kanza Saeed
Department of Food Science
    and Technology
Khwaja Fareed University of Engineering
    and Information Technology
Rahim Yar Khan, Punjab, Pakistan

Nimra Sameed
Department of Food Science and Technology
Khwaja Fareed University of Engineering and
    Information Technology
Rahim Yar Khan, Punjab, Pakistan

Muhammad Saqib
Department of Chemistry
Khwaja Fareed University of Engineering &
    Information Technology
Rahim Yar Khan, Punjab, Pakistan

Nazmul Sarwar
Department of Food Processing and
    Engineering
Faculty of Food Science and Technology
Chattogram Veterinary and Animal Sciences
    University (CVASU)
Chattogram, Bangladesh

Anil Dutt Semwal
Defence Food Research Laboratory
Siddharthanagar, Mysuru
Karnataka, India

Sakshi Sharma
Defence Food Research Laboratory
Siddharthanagar, Mysuru
Karnataka, India

Ayesha Siddiqa
Department of Food Science and Technology
Khwaja Fareed University of Engineering and
    Information Technology
Rahim Yar Khan, Punjab, Pakistan

Kappat Valiyapeediyekkal Sunooj
Department of Food Science
    and Technology
Pondicherry University
Puducherry, India

Barbara Elisabeth Teixeira-Costa
Faculty of Agrarian Sciences
Federal University of Amazonas—UFAM
Manaus, AM, Brazil

Karabi Talukdar
PG Department of Botany
Madhab Choudhury College
Barpeta, Assam, India

Rahul Thakur
Department of Food Process
    Engineering
National Institute of Technology
Rourkela, India

Tabussam Tufail
University Institute of Diet & Nutritional
    Sciences
The University of Lahore-Pakistan

Huma Bader Ul Ain
University Institute of Diet & Nutritional
    Sciences
The University of Lahore-Pakistan

Maryam Umar
Department of Food Sciences
Government College University
Faisalabad-Pakistan

Ranjana Verma
Department of Food Science
Nutrition and Technology, College of
    Community Science
Palampur, Himachal Pradesh, India

Dadasaheb Wadikar
Defence Food Research Laboratory
Siddharthanagar, Mysuru
Karnataka, India

Amir Sasan Mozaffari Nejad
School of Medicine
Jiroft University of Medical Sciences
Jiroft, Iran

Umar Farooq
Department of Botany
Govt. Degree College Shopian
J & K-India

Saeme Asgari
Department of Biochemistry and Biophysics,
    Tehran Medical Sciences
Islamic Azad University
Tehran, Iran

CHAPTER 1

Cereals
An Overview

Tabussam Tufail, Huma Bader Ul Ain, Muzzamal Hussain, Muhammad Adil Farooq, Gulzar Ahmad Nayik and Mohammad Javed Ansari

CONTENTS

DOI: 10.1201/9781003252023-1

1.1 INTRODUCTION

In the family of *Gramineae*, cereals are known to be edible seeds or grains (Tufail et al., 2021). In many nations, rye, oats, barley, maize, triticale, and millet are farmed among the grains. The most common vegetation in the world is wheat and rice, accounting for more than half of all grain output. The hereditary material for another plant (embryo) and the endosperm, which is loaded with starch grains, are structurally equal in all cereals. Different kinds of grain may be found all over the world, and their production is influenced by genetic and environmental variables. Cereals are referred to as grains because of their very nutritious seeds. Since the dawn of civilization, a few grains have been staple foods for both humans and animals. The most significant sources of nutrition are the cereals, and they provide significant quantity as meals based on protein calories, B vitamins, and minerals. They are generally inexpensive to grow, store, and transport and don't fall apart quickly when kept dry (Bader Ul Ain et al., 2019).

Grains are formed from blooms, and while the architecture of many cereals varies, there a few fundamental features. The germ (or growing life) might be a thin-walled structure that houses the growing plant. The endosperm is set apart from the majority of the grain by the scutellum (included in preparation of the seed's nutrition stores in the course of formation). The endosperm is made up of starch-filled cells with thin walls. On the off chance that the grain sprouts, the seedling utilizes endosperm-enhanced supplements until the improvement of green starts, allowing photosynthesis to start. The aleurone surrounds the endosperm. The pericarp (defined by ovary of bloom) encompasses the seed coat on the outside of the grain. Bran is framed by thick-walled structures on the outside (Saeed et al., 2021).

1.2 DIFFERENT CEREALS

1.2.1 Wheat

Wheat has the potential to be a prime cereal crop in many ways. The tritium family has thousands of varieties (Bader Ul Ain et al., 2019); wheat belongs to it, and it is the most economically predominant (Saeed et al., 2016). It is created in colder-season and seedtime grain, as well as being used in various countries all over the world due to the variety of available species and combinations, as well as its adaptability. The United States, China, and Russia are among the world's top wheat makers; widespread wheat farming occurs in Pakistan, India, Canada, the European Union, and Australia. In 2003, it was estimated that 556.4 million tons of wheat were planted, constituting 30% of global wheat creation. Spikelets make up an ear or spike of grain. Each spikelet has a wheat grain encapsulated between the lemma and palea. The grain can be round, oval, or in form, and the brush hairs can be short or long. Wheat has fusiform spikes, can be awned (unshaven) or awnless, and is successfully sifted in most modern varieties (Bader Ul Ain et al., 2019).

1.2.2 Rice

Rice is a significant yield, filling in as staple nourishment for an enormous part of the total populace, most significantly in Asia. Rice is generally developed for human use, including morning grains, and it is likewise used for this purpose in Japan (Tufail et al., 2021). There is a massive number of rice assortments (north of 100,000); however, a couple have broadly evolved (for example, arrangements of the advanced semi-bantam plant sort with leaves). Rice is made up of two parts: an exterior supportive layer (known as body) and edible rice caryopsis. The pericarp (which includes shade), seed coat, fetal, and endosperm are the exterior layers of brown rice (Tufail et al., 2018).

1.2.3 Maize

Zea mays L., also called maize, began on the western hemisphere (Calder, 2013). It very well may be a minimal-expense source of starch and an enormous source of sustenance for creatures (Hussain et al., 2021). Regardless there being many particular combinations, the four essential classifications of commercial importance are flint maize, dent, sweet corn, and popcorn. The four basic elements of the maize portion (the plant's regenerating seed) are the endosperm, germ, tip cap, and pericarp. US production dwarfs that of other countries (Raza et al., 2020), and the maize genome has been extensively researched to date in the United States.

1.2.4 Barley

Barley is a tough flora that can withstand a broad range of environments and originated around 15000 BC (Hussain et al., 2022). *Hordeum vulgare*, a developed grain, is largely used for animal nutrition, particularly for pigs, as well as malting and brewing in the production of lager and refining in the production of whiskey. For sustenance, a small amount of grain is utilized. In the United Kingdom and the Middle East, pearled grain is utilized in soups and stews; the grain is likewise utilized in bread and squashed into porridge in a some nations. Spikelets are connected to the rachis in a rotating example to shape the grain head or spike. The husk completely covers the grain, the testa or seed coat, the pericarp (the part wherein the husk is safely located in many species), and aleurone, the outside layers of the grain segment.

1.2.5 Oats

Oats have largely been produced for nourishment and may be produced effectively in deficient ground and in chilly, wet areas. A small amount is used for humans—oatcakes and porridge as cereals, porridge as rolled oats, and for baby oat flour food. In a variety of applications (not food based), such as in cement and makeup products, oats are used (Saeed et al., 2021). There are a few distinct animal groups, with *A. sativa* L. (normal spring or white oat) being the main cultured type. It's conceivable that red-oat varieties have adjusted to warmer locales, where it is developed as an oat for colder times of the year. Its pieces make up an oat spikelet. Each portion is wrapped in a body (consisting of two coatings—a lemma and a palea) that is slightly connected to the groat. The groat accounts for 65–85% of a portion and is surrounded by bran layers (Lundin et al., 2003).

1.2.6 Rye

Rye is a sturdy flora that grows best in chilly climates where other cereals are unable to grow. It may likewise develop at high altitudes and in semi-bone-dry conditions. It's a crop for colder times of year that is cultivated in the early reap season and collected in the late spring. The plant might develop from 30 cm to multiple meters tall. It is the principal bread grain in Russia, Poland, Germany, and the Scandinavian nations; in this manner it could be a major crop. Rye is likewise used to make crisp bread and whiskey, as well as being utilized as animal feed. It is yellowish-grey in type. It is longer and looser than wheat. The debilitated endosperm, the pericarp, and the testa make up the cereal, whereas microorganisms make up the rest (Vinkx & Delcour, 1996).

1.2.7 Millet

Millet is a genus belonging to small-grained yearly cereal grasses, according to Saleh et al. (2013). The most prevalent sort is pearl millet. Small millets incorporate finger, proso, and foxtail, but they represent under 1% of grains for human use, making them less significant as far as world food supply. On the other hand, they are crucial in some areas of Asia and Africa, where principal grains are unable to provide long-term yields. Growth period duration, size, grain consistency, and flavor vary based on the variety, climate, and soil conditions.

1.2.8 Sorghum

Sorghum is a summer crop that is susceptible to cold temperatures but has some pest and disease resistance. In numerous nations such as African and Central Eastern Europe, it is a typical dish in numerous cuisines. It is composed of a bare caryopsis, which comprises a pericarp, endosperm, and microorganism. In light of their actual properties, sorghums are ordered into four types: grain sorghum, sweet sorghum, sudangrass sorghum, and Sudan sorghums and broomcorn (Awika & Rooney, 2004; Awika et al., 2003).

They are classified based on the following features:

- Shape of the pericarp
- Thickness of the pericarp
- Endosperm type
- Shade of the endosperm

1.2.9 Triticale

By crossing rye and wheat, a primary human-made cereal is triticale (Triticosecale). It has the colder-season durability of rye as well as the baking limits of wheat. It is essentially utilized as a grub crop yet might likewise be ground into flour for making bread; however, because it does not have the same amount of gluten as wheat, recipe modifications are required (Gupta & Priyadarshan, 1982).

1.2.10 Other Grains

Beside the cereals referenced previously, there are a couple of others that, while not as significant on worldwide, might be pivotal in certain regions of the planet. Buckwheat (otherwise called Saracen corn) is a cooked grain, porridge, or hotcake ingredient from the plant named *Fagopyrum esculentum*. In Chile and Peru, quinoa, a grain from the South American plant *Chenopodium collection*, is utilized to make bread (Cordain, 1999).

1.3 NUTRITIONAL VALUE

People have been eating oats for millennia. Oats are a staple food in both developed and emerging nations, and they are rich in supplements. Grains and cereal items are high in calories, carbs, protein, fiber, micronutrients like vitamin E and B, and minerals such as magnesium and zinc. Cereals can support the utilization of nutrients and minerals, but how many micronutrients depends on how much of the microorganism, grain, and endosperm are present (Nasir et al., 2020). In spite of the fact that thiamin, niacin, calcium, and press are added to wheat flour in the United Kingdom, the pericarp, microbe, and aleurone layer are plentiful in nutrients, and handled grain items lose a portion of these supplements (with the exception of whole meal).

1.3.1 Carbohydrates

Grains are regularly classified as carb-rich food sources since they contain roughly 75% carbohydrates. The endosperm contains starch granules, which are the grain's essential part. Rice granules have a width of just 5 m, though wheat granules may be 25–40 m (either large, focal point–molded granules or small, round granules). The amylose-to-amylopectin proportion in starch granules changes depending upon the grain and type. Amylose makes up 25–27% of the starch in typical cereal variations, but amylopectin makes up the majority of the starch in waxy varieties (such as rice and maize). In cereals, however, a part of this starch Isn't processed and is retained in the small intestine (Tufail et al., 2022).

1.3.2 Lipids

Lipid content in cereals ranges from 1 to 3% in barley, rye, rice, and wheat to 5 to 9% in maize and 5 to 10% in oats in dry-matter measurement, in spite the microorganism being the best source of lipids, as shown in Table 1.1 (Ahmed et al., 2022).

1.3.3 Protein

Protein content in cereals ranges from 6–15%. Wheat has gliadins and glutenins, whereas rice and maize have glutelin (oryzenin) and prolamin (zein), respectively; oats have albumins and globulins, barley has hordeins and glutelins. Notwithstanding the way that cereals contain a different scope of amino acids, which are the blocks for building proteins, a few are just present in the following amount. Food must include amino acids, which are essential, and the human body can synthesize additional (non-essential) amino acids from them, as shown in Table 1.2. The fundamental amino acid with the lowest supply in relation to the need is known as the limiting amino acid. In grains, lysine is the limiting amino acid (Tufail et al., 2021).

The grain seed is a mind-boggling structure with various parts and physical perspectives. The three essential physical parts of the bit, or caryopsis, are the wheat, incipient organism, and endosperm. The structure and size of the initial two grains change significantly, requiring the use of particular strategies to eliminate them. Grain is composed of the pericarp, testa, nucellar layer, and aleuronic layer. It is high in fiber, debris, compounds, nutrients, and globulin-stockpiling proteins. Twenty-five percent of the protein, 18% of the carbs; 16% of the lipids; and 5% of the debris, tocopherol, and complex water-dissolvable nutrients, as well as numerous compounds, are present in raw grain (Koehler & Wieser, 2013).

Table 1.1   Nutritional Characterization of Cereals

Fat & Fatty Acids (g/100 g food)	Barley, Pearl, Raw	Oatmeal, Quick Cook, Raw	Wheat Flour, White	Rye Flour	Rice, Brown, Raw	Rice, White, Raw
Total fat	1.7	9.2	1.2	2.0	2.8	3.6
Saturated fatty acids	0.29	1.61	0.16	0.27	0.74	0.85
Cis-monounsaturated fatty acids	0.14	3.34	0.13	0.21	0.66	0.91
Polyunsaturated fatty acids						
Total cis	0.77	3.71	0.51	0.95	0.98	1.29
n-6 (as 18:2)	0.70	3.52	0.48	0.82	0.94	1.26
n-3	0.07	0.19	0.03	0.13	0.04	0.03

Table 1.2 Amino Acid Contents in Different Cereals

Amino acid (g/100 g protein)	Wheat (hard)	Rice		Maize		Barley	Oats	Rye	Millet (average of 7 types)	Sorghum	
		B	M	N	HL					N	HL
Phenylalanine	4.6	5.2	5.2	4.8	4.3	5.2	5.4	5.0	5.5	5.1	4.9
Histidine	2.0	2.5	2.5	2.9	3.8	2.1	2.4	2.4	2.0	2.1	2.3
Isoleucine	3.0	4.1	4.5	3.6	3.4	3.6	4.2	3.7	3.8	4.1	3.9
Leucine	6.3	8.6	8.1	12.4	9.0	6.6	7.5	6.4	10.9	14.2	12.3
Lysine	2.3	4.1	3.9	2.7	4.3	3.5	4.2	3.5	2.7	2.1	3.0
Methionine	1.2	2.4	1.7	1.9	2.1	2.2	2.3	1.6	2.5	1.0	1.6
Threonine	2.4	4.0	3.7	3.9	3.9	3.2	3.3	3.1	3.7	3.3	3.3
Tryptophan	2.4	1.4	1.3	0.5	0.9	1.5	–	0.8	1.3	1.0	0.9
Valine	3.6	5.8	6.7	4.9	5.6	5.0	5.8	4.9	5.5	5.4	5.1

Oats small outer layer, and undeveloped organism make up 10% of cereal dry weight, and they are generally taken out before use by processing (wheat), cleaning, pearling (grain), or decorticating (rice). The embryo of maize makes up about 10 to 11% of the grain as well as being vital for optimum nutrition due to its high protein and oil content. During germination, the endosperm, which makes up the bulk of the seed, provides the nutrients essential for embryo development. Stocks are hydrolyzed by the enzymes coming from the aleurone layer and embryo and make them accessible to the seed, releasing and activating the nutrients. In endosperm stocks, starch granules are scattered in a framework of capacity proteins. The centralizations of starch grow as the separation from the fringe does, but the groupings of protein decline as separation from the outskirts to the middle develops (Evers and Millar, 2002).

Brown rice has 6.6 to 7.3% protein, milled rice contains 6.2 to 6.9% protein, and basmati rice contains 8.2 to 8.4% protein. Because these elements are largely found in the kernel's outer layers, protein content diminishes linearly as the degree of polish increases. Furthermore, amylose from starch granules binds to 0.7% of protein, as depicted in Table 1.3 (Keenan et al., 2006).

Protein content varies by variety, environment, and laboratory extraction; these numbers are crucial in identifying grain species.

1.3.4 Vitamins

Cereals lack vitamins C and B12, as well as vitamin A and beta-carotene, with the exception of yellow maize (Cordain, 1999). Cereals, on the other hand, provide an excellent source of thiamin, riboflavin, and niacin, among other B vitamins (Germain et al., 2003). Grains also have a large amount of nutrients, such as vitamin E.

Table 1.3 Protein Content in Different Cereals

Cereals	Protein Content
Wheat	8.0–17.5
Maize	8.8–11.9
Rice	6.6–8.4
Barley	7.0–14.6
Oats	8.7–16.0
Rye	8.0–17.7
Sorghum	7.0–15.0
Millet	8.3–13.3
Buckwheat	21.6–25.3
Quinoa	13.0–14.0

1.3.5 Minerals

Grains, like other plant foods, have a high potassium content level and a low salt level. Whole grain cereals also include significant levels of magnesium, iron, and zinc, as well as a lower quantity of a variety of trace minerals, including selenium. Rice has the most selenium of any cereal grain, ranging from 10 to 13 g per 100 g. The amount of selenium in a cereal depends on the amount of selenium in the soil. Cereals such as wheat developed in North America have a higher selenium content than wheat grown in Europe, and the recent change to European wheat has been faulted for a drop in selenium utilization in the United Kingdom (Shehzad et al., 2014).

1.3.6 Non-Starch Polysaccharides

Non-starch polysaccharides (NSPs) occur in huge amounts in all grains. NSPs come in two types: insoluble and dissolvable, and keeping in mind that both may assist with weight loss (by deferring the clearing of food from the stomach), their impacts on the body are unique. Although most oats have comparable insoluble NSP content, the water-solvent NSP present changes. In wheat, rye, and grain, arabinoxylans are the most plentiful water-dissolvable NSP, yet beta-glucans are the most bountiful NSP in oats. Wheat has more beta-glucans and arabinoxylans than grain, oats, and rye (for dry weight, 3–11%, 3–7%, 1–2%, and 1%, individually). (Berthon et al., 2013).

1.3.7 Phytochemicals

Grains include phytochemicals, also known as plant bioactive compounds, which may have health-promoting qualities. Other antioxidants such as tocotrienols, tocopherols, and carotenoids are also present in grains; however, flavonoids are only detected in minimal amounts. In lab trials, breakfast with whole grain cereals was shown to contain an amount of antioxidants equivalent to vegetables and fruits, and one study found that bound phytochemicals constitute the greatest contributor to the overall activities of antioxidants. A kind of phytoestrogen, lignans, are found in grain, and while the quantity is small (in comparison to linseed), grains may be a significant source due to the large quantities ingested daily (Adom & Li, 2002).

1.3.8 Anti-Nutrients

In dry weighing, corn contains 0.89% phytate, wheat has 1.13%, brown rice has 0.89%, grain has 0.99%, and oats have 0.77%. The aleurone layer and, less significantly, the top of most cereals are high in phytate. This suggests that processing modifies the phytate level of most oats; white flour, for instance, has very little phytate. Phytate might bind minerals, including calcium, iron and zinc, and there is also little proof that phytate represses retention of these components (e.g., when phytate was added to white bread, McCance and Widdowson noticed a decrease in calcium assimilation in people) (Harris & Shearer, 2014). This impact may be especially critical for individuals who eat diets with a very low amount of kcal (Kaur et al., 2015).

Tannins, which can be found in brown sorghum, bind and hasten protein, making it more challenging to process. The dietary benefit of sorghum is expanded by germination and treatment with potassium carbonate, calcium oxide, ammonium, or sodium bicarbonate. A goitrogen is found in the wheat and endosperm of pearl millet grain (thioamide). Trypsin inhibitors have likewise been recognized in pearl millet and rye, which can restrict protein absorbability, but they are generally ineffective at warm-environment. Other anti-nutrients present in rye affect nourishment yet are of little worry to people since they are taken out or destroyed during handling or baking (Ertop & Bektaş, 2018).

1.4 CEREALS IN HEALTH AND DISEASE

The relevance of certain dietary products, such as cereals, and their influence on health is a current issue. Whole grain cereals have been the subject of some study, and the results suggest that those who eat whole grains have better nutritional profiles (Chavan et al., 1989). Grains might have numerous of advantages related to health, which will be explained in the following.

1.4.1 Balancing Energy

Grains have lower energy density, and a whole grain cereal–rich diet (Holt et al., 2008) may help alleviate hunger by being bulky. Cereals may likewise impact body weight by influencing hormonal variables. By zeroing in on expanding grain utilization, a decrease in the use of other food as well as a decrease in fat intake is conceivable. For instance, an investigation of adults found that everyday consumption of 60 grams of breakfast cereal with low-fat milk decreased normal fat utilization by 5.4% while helping carb intake by a similar sum.

A high-fiber, sugar-rich breakfast was the most filling meal, and it was related to less food usage close to the start of the day and at lunch. Hunger returned at a slower rate after this lunch than after a low-fiber, sugar-rich meal. Altogether fat- and sugar-rich morning meals tasted better, they were less filling, and a high NSP content might be a deterrent element against obesity and overweight.

1.4.2 Glycemic Index

Carb-rich dinners are classified utilizing the glycemic index (GI) scale. It's characterized as

> the steady region under the blood glucose bend subsequent to eating 50 grams of starch from a test food, partitioned by the region under the bend in the wake of eating a comparable measure of sugar from a control food (commonly white bread or glucose).
>
> *(Pereira & Ludwig, 2001)*

The glycemic load (GL) is determined by multiplying the dietary GI by the aggregate sum of carb in the meal (Jenkins et al., 2002). It computes a meal's total glycemic impact. In a new WHO/FAO study on nutrition and chronic conditions, lower–glycemic index dinners were related to a general betterment in glycemic control in people with diabetes mellitus.

1.4.3 Heart-Related Health Benefits

Individuals who eat a great deal of whole grains have a lower risk of coronary disease and stroke, per many significant partner studies conducted in the United States, Finland, and Norway. Numerous partner studies demonstrated an inverse relationship between whole grain diet and the risk of cardiovascular disease (CVD) (Hu, 2003). Also, a wellbeing investigation by doctors discovered that men who ate one portion of whole grains for breakfast every day had a 20% lower chance of dying from cardiovascular disease than those who ate the lowest amount of whole grains (Liu et al., 2003). There was a statistically insignificant relationship between total or purified grain consumption as well as death rate from CVD. Despite the fact that there is presently not enough proof to demonstrate an immediate relationship between cardiovascular wellbeing and whole grain meals, the WHO/FAO inferred that NSP and whole grain oats bring down the frequency of CVD (Baum et al., 2012).

1.4.4 Diabetes

A potential role of fiber in diabetes counteraction has long been conjectured, and a high intake of grain fiber has always been connected to a lower chance of diabetes (Willett et al., 2002). The frequency of type 2 diabetes was demonstrated to be inversely related to whole grain consumption, with a rate ratio (RR) of 0.65 between the highest and lowest quartiles of whole grain consumption, meaning a 42% risk decrease. Oat fiber has likewise been connected with a lower risk of type 2 diabetes (RR 0.39). Information from seven planned partner studies was combined to get a gauge of a 30% decrease in risk (counting the Montonen research) (Liu, 2002).

1.4.5 Digestive Health

Insoluble fiber, which can be found in many meals, together with cereals, is known to work on gastrointestinal well-being. Insoluble fiber retains fluid, bringing about a weight gain in the stool. It additionally advances stomach microscopic organism improvement and movement. When contrasted with a diet low in fiber, McIntosh et al. (2003) found that high-fiber wheat and rye food sources helped develop different stomach microbiota, such as waste beta-glucoronidase, auxiliary bile acids, and para-cresol fixations, as well as lower waste pH.

1.4.6 Other Cancers

Fiber consumption has likewise been connected with a lower risk of pancreatic and breast cancer. Specialists noticed a converse relation between multigrain meal consumption and the occurrence of upper gastrointestinal, bladder, and kidney malignant growths in a progression of case-control studies conducted in Italy. Grains might safeguard against chemical-related tumors due to their high lignan content. Stomach microscopic organisms modify the construction of lignans, a sort of phytoestrogen, to make them more like mammalian lignans (Tufail et al., 2020).

1.4.7 Hypertension

Hypertension is a key risk component for cardiovascular disease as well as kidney infection. Salt consumption has been shown to influence heart rate in more established individuals, as well as those with high blood pressure and diabetes, yet a food-based system has only recently been researched. The different studies expected to build consumption of a wide scope of food varieties, including whole grain oats but with a focus on foods grown in the ground, as well as low-fat dairy items. Among hypertensive people (n = 133), the Scramble diet impacted hypertension, with a drop in systolic pulse (SBP) of 11.4 mmHg and a decrease in diastolic circulatory strain (DBP) of 5.5 mmHg.

1.4.8 Other Health Benefits

In human and animal research, inhibitory impacts and satiety have been shown by proteins present in oat grain. A review looked at the satiety impacts of protein present in milk products and wheat in rodents and found that fibrin had no effect on the general hunger decrease detailed in all rodents, proposing that proteins have physiological properties free of their traits or resources. Proteins in wheat were found to be a decent trigger of a cholecystokinin and glucagon-like peptide 1 (GLP-1) blend in human duodenal tissue, suggesting that they might diminish hunger. Protein in wheat is part of a diet for weight reduction. Dipeptidyl peptidase-IV (DPP-IV) inhibitor peptides were found in oats, grain, and wheat in a programmatic trial (Lacroix & Li-Chan, 2012). DPP-VI has been shown to obstruct wheat development in vitro by Velarde-Salcedo and associates (2013). Oat proteins were found to display pro-inhibitory exercises in an in silico and in vivo study (Cheung et al., 2010).

Grain showed an assortment of inhibitory impacts when utilized as wheat flour in bread. The impacts of grain protein and calcium caseinate on numerous wellbeing indicators in men with hypercholesterolemia and postmenopausal women were studied in 2010. (Jenkins et al.). Altogether, protein sources had amino acids, as well as ramifications for serum lipids, cell support response, and heartbeat, toward the end of the treatment time span, showing that grain protein would be a good plant of protein in the diet. Even without the effects of gluten, rice prolamins have been found to have a higher in cereal and also work against leukemia influence than wheat glutenins and gliadins (Chen et al., 2010).

Lunasin, a chemopreventive and anticancer bioactive peptide, was recognized in nine different grain cultivars (Jeong et al., 2010). These lunasin sources from grain were shown to be bioactive and bioavailable in both in vivo and in vitro assessments (Jeong et al., 2010). A lunasin-like peptide found in amaranth grain reduced histone acetylation in the center of NIH-3T3 cells (Maldonado-Cervantes et al., 2010).

Amaranth peptides have been shown to have a high DPP-IV inhibitory effect and antidiabetic potential (Velarde-Salcedo et al., 2012), especially when presented to simulated gastrointestinal assimilation (Velarde-Salcedo et al., 2013). This similarly underlines the chance of clinical benefits from bioactive peptides through a standard diet rather than through a medication. ACE-inhibitory peptides (ALEP, VIKP) and anti-hypertensive peptides (IKP, LEP) were found in amaranth 11S globulin after an in silico study using 3D models and mirrored protein docking development (Vecchi & Añon, 2009).

It's not unexpected that eating a diet high in whole grains decreases the risk of chronic disease, yet it's less clear how these cases work and whether different supplements, like proteins, have a synergistic impact that can bring medical advantages credited to other grain parts, like fiber and cell reinforcements (Jensen et al., 2007). There is evidence that shows that eating cereals, particularly whole grains, on a regular basis may help prevent chronic diseases. Although cause and effect cannot be proven, those who consume whole grain–rich diets tend to have a lower risk of developing a variety of chronic diseases. It's unclear if whole grains are a cause-and-effect link or simply a marker of a healthy lifestyle.

Biological activities of wheat binding proteins are LMW glutenin were high in pro-inhibitors, general inhibitors like DPP-IV, and celiac harmful peptides. LMW glutenin incorporated a follow measure of narcotics ($A = 0.003$), which may be utilized as analgesics or cell reinforcement peptides ($A = 0.0206$). Of all the wheat proteins considered, HMW glutenin had the most elevated occurrence of cell reinforcement peptides, as well as ACE inhibitors, narcotic, antithrombotic, and DPP-IV inhibitor peptides. HMW glutenin ($A = 0.006$) likewise included anticancer peptide sequences.

Alpha-gliadin was high in ACE inhibitors, as well as DPP-IV and Enthusiasm inhibitors. It was additionally one of a handful of proteins with neuropeptide action in our examination. Both gamma- and omega-gliadin contained high amounts of ACE inhibitor, DPP-IV-inhibitor, and cancer prevention agent peptides; however, omega-gliadin had more celiac-unsafe peptides, while gamma-gliadin had some hypotensive (rennin-inhibitor) and DPP-IV-inhibitor peptides. Generally, wheat capacity proteins contained a different range of bioactive peptides with assorted organic functions, including anticancer and cell reinforcement qualities, as well as inhibitors of DPP-IV.

We investigated the most basic oat binding proteins. They contained 11 S globulin, 12 S globulin, and Avenin N9. The two globulins included similar degrees of ACE inhibitors, anti-amnestics (Kick inhibitors), DPP-IV inhibitors, cell reinforcements, and hypotensive peptides. The two globulins and bacterial permease ligand action had antithrombotic peptides, distinguishing them from other protein molecules explored. Avenin N9 had a lower recurrence of ACE inhibitor and cell reinforcement peptides than globulins, in spite of the fact that it showed narcotic and celiac harmful impacts. DPP-IV inhibitor peptides were likewise embedded in similar amounts. The discovery of celiac disease–causing peptides in this protein was startling, considering oat is generally utilized as a wheat substitute for those with celiac disease.

The most fundamental stockpiling proteins in grain are B-hordein forerunner, C-hordein, D-hordein, and globulin. The B-hordein forerunner was rich in ACE inhibitors, as well as Enthusiasm and DPP-IV inhibitor peptides. This protein likewise incorporated a limited quantity of celiac hurtful peptides and had minor antibacterial and antithrombotic impacts. C-hordein has an equilibrium of ACE inhibitors, DPP-IV inhibitors, and celiac poisonous peptides, as well as some neuropeptide (anxiolytic) action. D-hordein had the most ACE inhibitor peptides ($A = 0.511$), as well as narcotic inhibitor, DPP-IV-inhibitor, and DPP-IV-inhibitor action (Tufail et al., 2020).

Cell reinforcement action was more common in grain hordeins than in the other stockpiling proteins explored. Grain globulin, similar to C-hordein, showed ACE and DPP-IV inhibitor peptide action, yet had the most uneven cell reinforcement action ($A = 0.009$) of the grain proteins. The cell reinforcement action of grain hordeins was demonstrated to be more prominent than that of the other stockpiling proteins tried. The typical cell reinforcement event recurrence of wheat proteins was 0.0321, oat proteins 0.0274, rice 0.0325, and grain proteins 0.0344, with the hordeins' average being 0.0428. HMW glutenin, on the other hand, had the most noteworthy single protein event recurrence of cell reinforcement action at 0.0497.

1.5 CONCLUSION

Around 60 to 70% of daily energy requirements worldwide are met by cereals. Cereals can be eaten whole, partially processed, or processed. Several functional bioactive ingredients, such as polyphenols, tocopherol, and antioxidants, can be found in cereals, particularly varieties of rice, wheat, maize, and some millets. They minimize the risk of cancer, type 2 diabetes, hypertension, high blood pressure, and the glycemic index while preventing or controlling several diseases. In some regions, they may be ingested fermented, whereas in others, they may not. The B complex vitamins and certain other nutrients are increased in fermented grains, which also improves nutrient availability and digestion.

REFERENCES

Adom K.K., & Liu R.H. (2002). Antioxidant activity of grains. *Journal of Agricultural and Food Chemistry*, 50, 6182–6187.

Ahmed, H.M., Amiri-Ardekani, E., & Ebadi, S. (2022). Phytotoxicity of natural molecules derived from cereal crops as a means to increase yield productivity. *International Journal of Agronomy*, 2022.

Awika, J.M., & Rooney, L.W. (2004). Sorghum phytochemicals and their potential impact on human health. *Phytochemistry*, 65(9), 1199–1221.

Awika, J.M., Rooney, L.W., Wu, X., Prior, R.L., & Cisneros-Zevallos, L. (2003). Screening methods to measure antioxidant activity of sorghum (Sorghum bicolor) and sorghum products. *Journal of agricultural and food chemistry*, 51(23), 6657–6662.

Bader Ul Ain, H., Saeed, F., Ahmed, A., Asif Khan, M., Niaz, B., & Tufail, T. (2019). Improving the physicochemical properties of partially enhanced soluble dietary fiber through innovative techniques: A coherent review. *Journal of Food Processing and Preservation*, 43(4), e13917.

Bader Ul Ain, H., Saeed, F., Khan, M.A., Niaz, B., Khan, S.G., Anjum, F.M., . . . & Hussain, S. (2019). Comparative study of chemical treatments in combination with extrusion for the partial conversion of wheat and sorghum insoluble fiber into soluble. *Food Science & Nutrition*, 7(6), 2059–2067.

Bader Ul Ain, H., Saeed, F., Khan, M.A., Niaz, B., Rohi, M., Nasir, M.A., . . . & Anjum, F.M. (2019). Modification of barley dietary fiber through thermal treatments. *Food Science & Nutrition*, 7(5), 1816–1820.

Baum, S.J., Kris-Etherton, P.M., Willett, W.C., Lichtenstein, A.H., Rudel, L.L., Maki, K.C., . . . & Block, R.C. (2012). Fatty acids in cardiovascular health and disease: A comprehensive update. *Journal of Clinical Lipidology*, 6(3), 216–234.

Berthon, B.S., Macdonald-Wicks, L.K., Gibson, P.G., & Wood, L.G. (2013). Investigation of the association between dietary intake, disease severity and airway inflammation in asthma. *Respirology*, 18, 447–454.

Calder, P.C. (2013). Omega-3 polyunsaturated fatty acids and inflammatory processes: nutrition or pharmacology?. *British Journal of Clinical Pharmacology*, 75(3), 645–662.

Chavan, J.K., Kadam, S.S., & Beuchat, L.R. (1989). Nutritional improvement of cereals by fermentation. *Critical Reviews in Food Science & Nutrition*, 28(5), 349–400.

Chen, X., Penman, L., Wan, A., & Cheng, P. (2010). Virulence races of Puccinia striiformis f. sp. tritici in 2006 and 2007 and development of wheat stripe rust and distributions, dynamics, and evolutionary relationships of races from 2000 to 2007 in the United States. *Canadian Journal of Plant Pathology*, 32(3), 315–333.

Cheung, W.Y., Di Giorgio, L., & Åhman, I. (2010). Mapping resistance to the bird cherry-oat aphid (Rhopalosiphum padi) in barley. *Plant Breeding*, 129(6), 637–646.

Cordain, L. (1999). Cereal grains: Humanity's double-edged sword. *World Review of Nutrition and Dietetics*, 84, 19–19.

Ertop, M.H., & Bektaş, M. (2018). Enhancement of bioavailable micronutrients and reduction of antinutrients in foods with some processes. *Food and Health*, 4(3), 159–165.

Evers, T., & Millar, S. (2002). Cereal grain structure and development: Some implications for quality. *Journal of Cereal Science*, 36(3), 261–284.

Germain, E., Bonnet, P., Aubourg, L., Grangeponte, M.C., Chajès, V., & Bougnoux, P. (2003). Anthracycline-induced cardiac toxicity is not increased by dietary omega-3 fatty acids. *Pharmacological Research*, 47, 111–117.

Gupta, P.K., & Priyadarshan, P.M. (1982). Triticale: Present status and future prospects. *Advances in Genetics*, 21, 255–345.

Harris, W.S., & Shearer, G.C. (2014). Omega-6 fatty acids and cardiovascular disease: Friend, not foe? *Circulation*, 130(18), 1562–1564.

Holt, A.K., Baevre, A.B., Rodbotten, M., Berg H., & Knutsen, S.H. (2008). Antioxidant properties and sensory profiles of breads containing barley flour. *Food Chemistry*, 110, 414–421.

Hu, C., Zawistowski, J., Ling, W., & Kitts, D.D. (2003). Black rice (*Oryza sativa* L. indica) pigmented fraction suppresses both reactive oxygen species and nitric oxide in chemical and biological model systems. *Journal of Agricultural and Food Chemistry*, 51(18), 5271–5277.

Hussain, M., Saeed, F., Niaz, B., Afzaal, M., Ikram, A., Hussain, S., . . . & Anjum, F.M. (2021). Biochemical and nutritional profile of maize bran-enriched flour in relation to its end-use quality. *Food Science & Nutrition*, 9(6), 3336–3345.

Hussain, M., Tufail, T., Saeed, F., Ain, H.B.U., Shahzadi, M., & Suleria, H.A.R. (2022). Functional and nutraceutical properties of cereal bran industrial waste: An overview. *Bioactive Compounds from Multifarious Natural Foods for Human Health*, 43–63.

Jenkins, D.J., Kendall, C.W., Vuksan, V., Vidgen, E., Parker, T., Faulkner, D., . . . & Corey, P.N. (2002). Soluble fiber intake at a dose approved by the US Food and Drug Administration for a claim of health benefits: Serum lipid risk factors for cardiovascular disease assessed in a randomized controlled crossover trial. *The American Journal of Clinical Nutrition*, 75(5), 834–839.

Jensen, E.S., Ambus, P., Bellostas, N., Boisen, S., Brisson, N., Corre-Hellou, G., . . . & Pristeri, A. (2007). Intercropping of cereals and grain legumes for increased production, weed control, improved product quality and prevention of N-losses in European organic farming systems. In *1st European joint organic congress* (pp. 180–181). Danish Research Centre for Organic Food and Farming, DARCOF.

Jeong, J.S., Kim, Y.S., Baek, K.H., Jung, H., Ha, S.H., Do Choi, Y., . . . & Kim, J.K. (2010). Root-specific expression of OsNAC10 improves drought tolerance and grain yield in rice under field drought conditions. *Plant Physiology*, 153(1), 185–197.

Kaur, S., Sharma, S., Singh, B., & Dar, B.N. (2015). Effect of extrusion variables (temperature, moisture) on the antinutrient components of cereal brans. *Journal of Food Science and Technology*, 52(3), 1670–1676.

Keenan, M.J., Zhou, J., Mccutcheon, K.L., Raggio, A.M, Bateman, H.G., Todd, E., Jones, C.K., Tulley, R.T., Melton, S., Martin, R.J., & Hegsted, M. (2006). Effects of resistant starch, a non-digestible fermentable fiber, on reducing body fat. *Obesity*, 14, 1523–1534.

Koehler, P., & Wieser, H. (2013). Chemistry of cereal grains. In *Handbook on sourdough biotechnology* (pp. 11–45). Boston, MA: Springer.

Lacroix, I. M., & Li-Chan, E. C. (2012). Evaluation of the potential of dietary proteins as precursors of dipeptidyl peptidase (DPP)-IV inhibitors by an in silico approach. *Journal of Functional Foods*, 4(2), 403–422.

Liu, S. (2002). Intake of refined carbohydrates and whole grain foods in relation to risk of type 2 diabetes mellitus and coronary heart disease. *Journal of the American College of Nutrition*, 21, 298–306

Liu, S., Sesso, H.D., Manson, J.E., Willett, W.C., & Buring, J.E. (2003). Is intake of breakfast cereals related to total and cause-specific mortality in men?. *The American Journal of Clinical Nutrition*, 77(3), 594–599.

Lundin, K.EA., Nilsen, E.M., Scott, H.G., Løberg, E.M., Gjøen, A., Bratlie, J., . . . & Kett, K. (2003). Oats induced villous atrophy in coeliac disease. *Gut*, 52(11), 1649–1652.

Maldonado-Cervantes, E., Jeong, H. J., León-Galván, F., Barrera-Pacheco, A., De León-Rodríguez, A., De Mejia, E. G., . . . & De La Rosa, A. P. B. (2010). Amaranth lunasin-like peptide internalizes into the cell nucleus and inhibits chemical carcinogen-induced transformation of NIH-3T3 cells. *Peptides*, 31(9), 1635–1642.

McIntosh, G.H., Noakes, M., Royle, P.J., & Foster, P.R. (2003). Whole-grain rye and wheat foods and markers of bowel health in overweight middle-aged men. *The American Journal of Clinical Nutrition*, 77(4), 967–974.

Nasir, M., Jabbar, M.A., Ayaz, M., Ali, M.A., Imran, M., Gondal, T.A., . . . & Sharifi-Rad, J. (2020). Bio-therapeutics effects of probiotic strain on the gastrointestinal health of severely acute malnourished children. *Cellular and Molecular Biology*, 66(4), 65–72.

Pereira, M.A., & Ludwig, D.S. (2001). Dietary fiber and body-weight regulation: Observations and mechanisms. *Pediatric Clinics of North America*, 48(4), 969–980.

Raza, M.A., van der Werf, W., Ahmed, M., & Yang, W. (2020). Removing top leaves increases yield and nutrient uptake in maize plants. *Nutrient Cycling in Agroecosystems*, 118(1), 57–73.

Saeed, F., Afzaal, M., Hussain, M., & Tufail, T. (2021). Advances in assessing product quality. In *Food losses, sustainable postharvest and food technologies* (pp. 191–218). London: Academic Press.

Saeed, F., Ahmad, N., Nadeem, M.T., Qamar, A., Khan, A.U., & Tufail, T. (2016). Effect of arabinoxylan on rheological attributes and bread quality of spring wheats. *Journal of Food Processing and Preservation*, 40(6), 1164–1170.

Saeed, F., Hussain, M., Arshad, M.S., Afzaal, M., Munir, H., Imran, M., . . . & Anjum, F.M. (2021). Functional and nutraceutical properties of maize bran cell wall non-starch polysaccharides. *International Journal of Food Properties*, 24(1), 233–248.

Saleh, A.S., Zhang, Q., Chen, J., & Shen, Q. (2013). Millet grains: Nutritional quality, processing, and potential health benefits. *Comprehensive Reviews in Food Science and Food Safety*, 12(3), 281–295.

Shahzad, Z., Rouached, H., & Rakha, A. (2014). Combating mineral malnutrition through iron and zinc biofortification of cereals. *Comprehensive Reviews in Food Science and Food Safety*, 13(3), 329–346.

Tufail, A., Li, H., Naeem, A., & Li, T. X. (2018). Leaf cell membrane stability-based mechanisms of zinc nutrition in mitigating salinity stress in rice. *Plant Biology*, 20(2), 338–345.

Tufail, T., Fatima, A., Arshad, M.U., Ain, H.B.U., Imran, M., Siddiqui, H., . . . & Masood, S. (2021). Dietary guidelines to combat complications during pregnancy. *Bioscience Research*, 29, 10–20.

Tufail, T., Kour, J., Saeed, F., Ain, H.B.U., Ikram, A., Khalid, W., . . . & Chopra, H. (2022). An overview of food toxins. In *Handbook of plant and animal toxins in food* (pp. 1–26). April 13, Boca Raton, Florida: CRC Press.

Tufail, T., Riaz, M., Arshad, M.U., Gilani, S.A., Ain, H.B.U., Khursheed, T., . . . & Saqib, A. (2020). Functional and nutraceutical scenario of flaxseed and sesame. *International Journal of Biological Sciences*, 17, 173–190.

Tufail, T., Saeed, F., Afzaal, M., Ain, H.B.U., Gilani, S.A., Hussain, M., & Anjum, F.M. (2021). Wheat straw: A natural remedy against different maladies. *Food Science & Nutrition*, 9(4), 2335–2344.

Tufail, T., Saeed, F., Arshad, M.U., Afzaal, M., Rasheed, R., Bader Ul Ain, H., . . . & Shahid, M.Z. (2020). Exploring the effect of cereal bran cell wall on rheological properties of wheat flour. *Journal of Food Processing and Preservation*, 44(3), e14345.

Vecchi, B., & Añon, M.C. (2009). ACE inhibitory tetrapeptides from Amaranthus hypochondriacus 11S globulin. *Phytochemistry*, 70(7), 864–870.

Velarde-Salcedo, A.J., Barrera-Pacheco, A., Lara-González, S., Montero-Morán, G.M., Díaz-Gois, A., De Mejia, E.G., & De La Rosa, A.P.B. (2013). In vitro inhibition of dipeptidyl peptidase IV by peptides derived from the hydrolysis of amaranth (Amaranthus hypochondriacus L.) proteins. *Food Chemistry*, 136(2), 758–764.

Velarde-Salcedo, A.J., González de Mejía, E., & Barba de la Rosa, A.P. (2012). In vitro evaluation of the antidiabetic and antiadipogenic potential of amaranth protein hydrolysates. In *Hispanic foods: Chemistry and bioactive compounds* (pp. 189–198). Washington, DC: American Chemical Society.

Vinkx, C.J.A., & Delcour, J.A. (1996). Rye (*Secale cereale* L.) arabinoxylans: A critical review. *Journal of Cereal Science*, 24(1), 1–14.

Willett, W., Manson, J., & Liu, S. (2002). Glycemic index, glycemic load, and risk of type 2 diabetes. *The American Journal of Clinical Nutrition*, 76(1), 274S–280S.

CHAPTER 2

Role of Cereals in Food Security

Ranjana Verma, Nilakshi Chauhan, Farhan M Bhat, Abha Anand and Y.S. Dhaliwal

CONTENTS

2.1 INTRODUCTION

The concept of food security is diverse and can be examined at the individual, domestic, state, geographical, cultural, and international levels when people from all over the globe have available and accessible appropriate, secure, and nutritionally rich food to satisfy their daily intake requirements and food selection according to the standard of living (FAO, 1996). Food insecurity seems to be the most fundamental form of human privation, meaning an inability to obtain adequate amounts and quality of food to meet one's basic nutritional demands (Park et al., 2012). The challenges faced in the 21st century include global hunger. The major causes of global insecurities and hunger are food insecurity, malnutrition, and poverty (IFPRI, 2016).

In the 1970s, the term "food security" was coined. Food security includes ensuring availability of food to meet increasing demands, avoiding volatility in production of agricultural produce, and maintaining market stability by the Food and Agriculture Organization (Capone et al., 2014). Later, the Food and Agriculture Organization also studied the availability of food and recognized the significance of supply and demand as important parameters in balancing food security. Half the world's population is living below the poverty line (Berry et al., 2015). Regardless of this, the global food supply chain has doubled in the last 30 years, but approximately 800 million people around the globe are suffering from severe hunger/starvation (FAO, 2014). Malnourishment affects 22,000 adolescents per day, according to UNICEF. There are around 750 million individuals who do not have clean water to drink, and more than 2 billion individuals are suffering from many diseases due to deficiencies of micronutrients. (FAO, 2014).

DOI: 10.1201/9781003252023-2

Furthermore, food security is intrinsically linked to long-term development, indicating a varied set of strategies ranging from agriculture to food handling in social, ecological, geographical, and socioeconomical contexts that all are immensely complicated (Fusco et al., 2020). Some authors have their own point of view on food insecurity, even if people, families, and local communities are affected by food insecurity, but the reasons may be more multifaceted, involving decisions related to political and economical as well as agricultural produce systems. Therefore, it is necessary to do in-depth research and take additional action in this context (Allen, 2013).

Over the last half-century, cereal grains like maize, rice, and wheat have made significant contributions to ensuring the world's food security, mostly by increasing crop production and making crops highly resistant to drought, flood, pests, and diseases. However, there is still much work to be done, with over 800 million people suffering from severe starvation and others suffering from insufficient nutrition. The main challenges in food security are the changing climate, extensive deterioration of the ecosystems that support agricultural production, increasing exponential population growth, and uneven availability of resources. These challenges are essential key factors for improved livelihoods. The aim of this chapter is to discuss food security at the world and national level and the role of cereals in food security.

2.2 WORLDWIDE STATUS OF FOOD SECURITY

The economies of countries all around the world are slowing and receding. No country is exempt, and as is customary, the world's poorest nations and most vulnerable people bear the brunt of the consequences. According to the authors and data compiled on the state of food insecurity in the world since 1970, approximately 1.2 billion people are hungry or undernourished around the world—more than 1 billion than last year and roughly one-sixth of humanity. The current situation is unique in history, with numerous variables combining to harm those who are in danger of food poverty. This will put people in a vulnerable position. Different experiences in achieving food security in different nations and timeframes are emphasized in scientific research.

In European countries, for instance, the overall food supply chain isn't the only factor to consider; individuals' purchasing capacity and the nutritional quality of meals are also important (Fusco et al., 2020). As a result, half a million people in European countries cannot afford an adequate amount of food, and over 20 million households cannot even afford good-quality food on a daily basis (Eurostat, 2013). Beginning with the Industrial Revolution, the United Kingdom grew increasingly reliant on food imports, prompting the government to implement a number of measures aimed at ensuring food security. These policies must take into account the rising restrictions on food availability in the world; climate variability; and accessibility of natural resources such as soil, fresh water sources, energy sources, and bio-fuels in the twenty-first century (Kirwan & Maye, 2013).

Food security may appear to be an issue for underdeveloped and developing countries, given the fact that scarcity of resources can lead to difficulties and flaws associated with potential food insecurity concerns. However, according to the experts, the characteristics of food security can be seen in developing and developed countries due to several economical concerns such as price inaccessibility, debts, monetary obligations, food patterns, and the busy schedule of people's everyday lives (Fusco et al., 2020). For example, studies have shown that a balanced diet is very difficult for Irish people who come from low-class backgrounds and live on subsistence wages. This is also evident in Great Britain, where people on unemployment funds or on government pensions have a hard time meeting their fundamental needs for a healthy lifestyle. As a result, the issue of food security in various countries is tackled from a human rights perspective, ensuring each individual's right to food and health (Dowler et al., 2012).

The FAO predicted global cereal production in the 21st century would be reduced by 2.10 million ton, and currently, production of cereal grains is approximately 2.791 million ton, up 0.71% (19.21 million ton) from the previous year's total production (FA0, 2021). The decrease in use of coarse grains was mostly due to a forecast somewhat weaker export of coarse grain output and reduced production of coarse grains like barley and sorghum, which more than offset an upward revision to the world's production of maize crops that was better than expected (Agah et al., 2017).

Nevertheless, coarse grain global production is approximately 1504 million ton in Russia and the United States, which is 1.4% more than the previous year. In the case of wheat, recent data from Argentina and the United Kingdom and Republic of Ireland reveal that lower yields than projected have resulted in a significantly reduced worldwide output. Current estimated production is approximately 769.7 million ton, which means a 1% yearly reduction.

In the case of rice, the government of the Islamic Republic of Pakistan has estimated the records of cereal crops harvested in 2020–21 and observed a minor decline in the output of rice, mainly due to water restrictions in several areas. Lower-than-expected rice production was recorded in Thailand owing to floods in September and October 2021, and in Bangladesh due to lower-than-expected main-crop yields, according to data compiled from around the world. As a result, rice output is expected to reach 518–520 million ton (milled basis) in 2021, representing a 0.9% increase over the previous year and a new high. World cereal usage in 2021–22 is expected to increase by 1.7% over 2020–21, reaching 2810 million ton. Since November, the wheat utilization prediction has been reduced by 1.8 million ton to 779 million ton.

2.3 NATIONWIDE STATUS OF FOOD SECURITY

In India, food security is one of the major concerns; more than one-third of Indian people are considered impoverished, and 50% of all children are undernourished in some fashion. In India's food security environment, there have been many new concerns in the last 20 years: (i) The influence of economic liberalization in the 1990s on agricultural produce and food availability. (ii) The World Trade Organization (WTO) was constituted and, under it, the Agriculture Agreement. (iii) The triple F crises: food price, fuel prices, and financial crises, and issues related to change in climatic conditions. (iv) The phenomenon of hunger in the midst of abundance, that is, inventories in the initial periods of this century and in 2008–2009. (v) In the 1990s, the Public Distribution System (PDS) was the first to implement the distribution of food grain at affordable prices as a part of government policy for managing economy of country. This policy targeted around 1775 blocks of the country including tribal, deserts, and drought prone areas. (vi) The judiciary of India wants to implement the Mid-Day Meal scheme all over the country through its movement program "Right to Food." The main aim of this program is to ensure food and nutrition security in the country. (vii) The National Food Security Law (Right to Food) has been proposed. (viii) Monitorable target. It envisages an average growth rate of agricultural GDP to 4% per year. Over the previous 20 years, these changes have generated both possibilities and challenges for the country's food and nutrition security by ensuring reduction in anemia among children and women in the productive age group by one third. It is also targeted to reduce anemia by 50% along with 30% reduction in low birth weight by 2025.

Aside from food production, availability, and need, food security involves a wide range of elements. According to the Food and Agriculture Organization, food security occurs when all individuals have the right to access adequate, secure, and nutrient-dense food to meet their food requirements in an adequate quantity and quality (FAO). Food security has three components: accessibility, affordability, and consumption (nutrition).

At the national level, food security refers to the country's ability to supply domestic consumption by stockpiling enough cereal grains, whether from local production or imports. The effectiveness and strategies linked to food accessibility are examined in this section. The country's cereal grain production surged from roughly 50 million ton in 1951 to 287.17 in 2018–2019. The annual growth rate of cereal grains was roughly 3.6% in 2013–14, whereas oil-containing seeds, sugarcane, fruits and vegetables, and dairy production output have all increased significantly.

India is the world's second-largest producer of cereal grains such as rice, wheat, and maize. The growing demand for cereal grains on the worldwide market is providing a good climate for Indian cereal grain exports. To address food demands, India prohibited the sale of rice, wheat, and other grains in 2008 to different countries in the world. Because of the high demand in the worldwide market and the nation's surplus productivity, the restriction has been abolished, although only a restricted number of food commodities exports are permitted. The permitted minimum amount of exported cereal grains had no discernible effect on economic prices or conditions utilized for storage of cereal grains.

Wheat, rice, coarse millet, barley, and corn are the most valuable food crops in terms of food security. According to the Ministry of Agriculture's preliminary estimate for the year 2019–20, total production of cereal grains such as paddy, corn, and other grains is at 100.36 million ton, 19.88 million ton, and 8.28 million ton, respectively (Radhakrishna & Venkata Reddy, 2020).

2.4 YEARLY COST ESTIMATION OF CEREALS

Of the Asian countries, India is the largest producer of cereal grains such as rice, wheat, and maize and is also the biggest exporter of cereal goods in the world's market. In the financial year 2020–21, India's grain exports totaled Rs. 74,490.834/$10,064.044 million. Major cereal grains of India, such as rice, which includes basmati and non-basmati rice, have a major share in India's cereal exportation, accounting for 87.65% in 2020–21. Wheat, on the other hand, accounts for only about 12.38% of total Indian cereal exports during this time period (APEDA, 2021).

An increase in the price of cereal significantly affects the energy/calorific consumption of the people below the poverty line. Increasing trends were observed in the actual price of grains in the early 90s, which was mainly associated with a decrease in poverty alleviation. According to the authors, the reduction was shown to be more significant in remote areas, mainly due to advancements in infrastructural facilities in rural regions, which increased the availability of various other food products to farming families (Rao, 2000). The authors further observe that a decline in the consumption of cereal grains was not associated with the wellbeing of humankind.

2.5 NUTRITIONAL IMPORTANCE OF CEREAL GRAINS

Cereals contribute approximately 11,000–18,000 kJ/Kg of energy, that is, about 20 times more than fruits and vegetables, and are considered an important vehicle for enhancing nutrition. Cereals are regarded as important sources of nutrition for both humans and animals due to their contribution in providing protein, carbohydrates, fiber, vitamin E, iron, vitamin B complex, niacin, riboflavin, thiamine, and minerals. The high calorific value in food grains is mainly due to the presence of starch in them, but the fat and protein contents also contribute to energy generation. The starch composition in cereals varies from 60 to 70% of their overall weight, proteins vary from 7 to 13% and fat from 1 to 4%, along with moisture content, cellulose, trace elements (minerals), and vitamins. The presence of bran in grains is essential for reducing the risk of developing cardiovascular disease by lowering blood cholesterol levels.

Cereals are also rich sources of bioactive compounds, polyunsaturated fats like omega-3, linolenic acid, soluble and insoluble fiber, and resistant starches. The presence of phytochemicals in cereals makes them ideal for the formulation of functional foods and nutraceuticals due to their antioxidative, antimutagenic, and anticarcinogenic activities (Serafini et al., 2002). The production of cereals is much greater than other types of crops throughout the world, and they are considered a staple food crop (Sarwar et al., 2013). The nutritional elements, in addition to chemical components that are present in small quantities, usually in the bran and germ part of cereal grains, are called bioactive compounds. These have beneficial effects on health, like antioxidant properties and anti-carcinogenic, anti-inflammatory, anti-allergenic, anti-atherogenic, and anti-proliferative activity, but are not regarded as essential for the human body (Kris-Etherton et al., 2002).

2.6 ROLE OF CEREALS IN FOOD SECURITY

2.6.1 Wheat and Maize

Wheat production throughout the world varies between continents and accounts for an annual average of about 750 million tons (FAO, 2019), with Europe and North America being the major producers, while the Asian countries exceed this in wheat consumption, with China and India being the major producers. The yield per hectare of wheat has been estimated to be around 3.5 tons globally, in which the East Asian and the European countries account for an average yield of 4.3–5.3 t/ha. However the yield is lowest in the countries of South Asia and Africa. The difference in production of cereal crops is mainly due to differences in crop management practices (e.g., fertilizer use in crops, irrigation facilities) and agro-ecological conditions (e.g., rain distribution in the region, quality aspects of soil). The production of wheat in India was recorded by USDA in 2017–2018 around 97.0 million metric tons. This increase in production was accounted due to enhanced flooded irrigation among other parameters at most crucial stages of plant growth like crown root development, tillering, flowering, ripening, and maturity stages (Dixon et al., 2009). Higher yields, income generation, and technical assistance encourage farmers, mostly in Asian countries, to increase the area of wheat production as compared to other cereal crop (Hazell, 2009). The total global wheat production accounted to 760 million tons, out of which two-third of the world's wheat production, amounting to 65%, is used for human consumption, and the remaining one-third is utilized as animal feed (IDRC, 2010). The annual per capita consumption of wheat has been found to vary between27 and 170 kg in eastern and southern Africa and central Asia, respectively (Shiferaw et al., 2013), whereas in Asian countries such as China and India, each country consumes approximately 18% of world's wheat (RaboResearch, 2017). However, a large segment of the population relies on wheat to meet their dietary requirements of energy and nutrition. The estimation of energy per capita by wheat has been found to be approximately 450 to 500 kcal in a single day (Dixon et al., 2009).

The world production of maize yearly was estimated at 1,127 million tons (FAO, 2019). The per-hectare production of maize depends upon the various types of agro-ecological condition, including diversification in temperature range, different geographical conditions such as differences in altitude and latitude, terrain types and soil composition. The major producer of maize in the world has been found to be North America, followed by East Asia. Larger production of maize is due to a wide range of maize cultivars that have been subjected to different agricultural techniques.

The main aim of these cereal crops to provide an adequate amount of nutrition in terms of calories and proteins to the people who consume these crops in their diets at an affordable price. Whole grain cereals, including wheat and maize, provide a broad spectrum of vital nutrients that can help to address malnourishment, trace element deficiency, and chronic malnutrition (obesity and noncommunicable illnesses) (Poole et al., 2020). However, the major cereal food sources are not regarded as a solution for food security, and diversified diets are required for reciprocal efforts in diverse types of foods, food supply chain restructuring, and overall growth. Sorghum and millets, being nutrient-dense coarse cereals, are considered good examples in enhancing food security because of their ability to anticipate disastrous effects of climate (Rodríguez et al., 2020; Arbex et al., 2018). Nelson et al. (2018) emphasized the importance of nutritious foods, which provide a high concentration of micronutrients while also improving macronutrient accessibility. A number of scientific organizations throughout the world are working for food system transformation and recommending worldwide objectives for nutritious diets and sustainable production of food.

2.6.2 Other Cereals

Barley (*Hordeum vulgare* L.) is the cereal grain with world's fourth-largest production after wheat, rice, and corn and could be classified into hulled barley and hull-less barley according to husk content on the grain (Mayer, 2012). Approximately three-quarters of the worldwide barley production is used for animal feed, while 20% is malted for use in alcoholic and non-alcoholic beverages, and 5% is used as a component in the production of many food products (Blake et al., 2011). Barley (*Hordeum vulgare* L.) is one of the first domesticated cereals of the world, and in India it is considered a sacred grain from ancient times. Barley and oats are two cereals that have higher content of soluble fibers called beta-glucans as compared to other cereal grains (Shah et al., 2017). It is available in different varieties such as hulled, hull-less, normal, and waxy barley, as well as low and high β-glucan barley. In India, its utilization as a food crop is restricted to the tribal areas of hills. Barley products like sattu have been used in ayurvedic medicine for treatment of various ailments and are consumed in the summer because of their cooling effects on the human body (Verma et al., 2011). The low incidence of hyperlipidemia and diabetes in the Tibetan population has been associated with the consumption of hull-less barley. The complex carbohydrates present in barley ensure slow release of glucose upon digestion into the bloodstream, thereby resulting in a lower glycemic index, and thus it is essential for patients suffering from diabetes (Foster-Powell et al., 2002). It has been shown that consumption of barley and oats leads to better colon health. The consumption of oats in breakfast has greatly increased in last few years due to its nutritional and health-promoting abilities. Barley to be used for food purposes, either raw, roasted, or malted, should have a desirably higher protein content. Barley has a relatively low glycemic index as compared to several other cereals.

Food barley could be an important area of research and development in India from the perspective of farmers, industry, and consumers. If consumption of food barley increases, farmers may get better prices with lower inputs as compared to other same-season crops. However, industry has to come forward to introduce barley-based products in the market with aggressive information on its health-benefitting properties. There is an urgent need to develop hull-less or naked barley varieties with comparable yield to hulled barley and better quality traits. Standardization of barley-based products and popularization of its health benefits are some of the ways to enhance the economic returns of cultivators and provide healthy foods to people.

Sorghum (*Sorghum bicolor*) is the fifth most-grown cereal and is a staple source of food consumed mainly in the arid and semi-arid regions of world with nutritional composition similar to that of maize (Venkateswaran et al., 2014). The United States, India, Nigeria, Mexico, and Argentina are the leading producers of sorghum in the world, accounting for about 57.6 million tons annually. Sorghum is mainly used for feeding the population in African countries and for livestock feeding, along with ethanol production in America and Australia (Hariprasanna & Patil, 2015). Sorghum is used as malt for the preparation of alcoholic and non-alcoholic beverages. Being gluten free, sorghum is blended with wheat flour to formulate a variety of bakery items suitable for patients with celiac diseases. It is also processed in various types of specialty and functional foods due to its ability to reduce rises in blood glucose level and its high fiber, mineral, and protein content and is also an important source of manufacturing weaning food. Sorghum can used for preparations of ready-to-eat breakfast cereals in the form of popped, puffed, shredded, and flaked cereals (Awika & Rooney, 2004).

Rye (*Secale cereale*) is a cereal belonging to grass family *Gramineae*, cultivated after wheat, barley, and oats but recorded as far back as over 2000 years ago. Rye is mainly consumed for human food in Europe, and its qualities resemble those of wheat in bread-making formulations. Rye is well known as flour used for baking bread, with different lines possessing different viscosities (Andersson et al., 2011). Rye, being hull less and having a higher amount of starch-degrading enzymes, is regarded best for the brewing of whisky, retaining characteristically pungent and a hard-edged, spicy, grain-like flavor. Rye grains possess higher content of soluble fiber pentosans and β-glucan that aids in the formulation of a diet suited for cardiovascular patients, severely diabetic patients, and obese persons (Andersson et al., 2009).

Oats (*Avena sativa*) is a cereal grain belonging to the *Poaceae* grass family and is valued for its nutritional and health properties. The presence of soluble fiber β-glucan in oats retains the function characteristic of binding high water content and cholesterol-rich bile acids. This property makes oats the best choice for obese and cardiovascular patients due to its slow digesting capability, increase in satiety, and suppressing of appetite. The β-glucan also acts a prebiotic in the human body due to its ability to be fermented in the intestines by the residing bacteria and thus enhances gut health and reduces colorectal cancer. The presence of phytoestrogens in oats reduces the prevalence of chronic inflammation in cardiovascular and diabetic patients. Oats contain polyphenols and avenanthramides that enhance the production of nitric oxide and thus help maintain lower blood pressure along with reduction of inflammation in patients suffering from heart problems.

2.6.3 Small Millets

Small millets, generally known as coarse cereals, consist of eleven species, finger millet (*Eleusine coracana*), barnyard millet (*Echinochloa crus-galli*), foxtail millet (*Setaria italica*), proso millet (*Panicum miliaceum*), little millet (*Panicum sumatrense*), kodo millet (*Paspalum scrobiculatum*), teff (*Eragrostis tef*), fonio (*Digitaria exilis*), Job's tears (Coix lacryma-jobi), guinea millet (*Brachiariadeflexa*), and browntop millet (*Urochloa ramosa*) (Muthamilarasan et al., 2019; Goron & Raizada, 2015). All of these species are known by different vernacular names in various parts of the country. Coarse millets are mainly grown as conventional food crops in semi-arid regions and areas around the world, but their farming is limited to specific areas due to the advent of cash crops (Weber, 1998). The morphological properties of small millets resemble those of rice, and they belong to the family of major cereals, *Poaceae*. In terms of sustainable farming features, nutritional properties, and the ability to meet demand for food security, these coarse cereals are preferable in comparison to cereal grains. Small millets are dry land cereals. These crops do not require much water for their growth and can survive under water-limiting conditions. The plants of millet efficiently use nitrogen from the soil (Fuller et al., 2004); for example, small millet, especially finger millet, requires approximately 250 ml of water to produce 1 g of dry matter, while in the case of major cereals, wheat and corn consume 450 to 500 ml of water, respectively (Feldman et al., 2018). According to the authors, finger millet requires 20–60 kg of N_2 fertilizer per hectare to increase productivity when compared to major cereal crops (Li & Brutnell, 2011).

In little millets, nutritional compounds such as mineral content; protein content, especially essential amino acids; dietary fiber; and resistant starch are all abundant. Finger millet, for example, contains a lot of calcium, iron, and potassium, while little and barnyard millet contain a higher amount of iron (10–20 mg per 100 g). These small millets contain a high amount of protein, more than 10 to 12%, and are also a good source of crude fiber, ranging from 7 to 13%. Furthermore, these millets are devoid of gluten protein and are utilized for the development of products suited for patients with celiac disease (Muthamilarasan, 2016; Taylor & Emmambux, 2008).

2.7 CONCLUSION

Food heterogeneity across the world is a serious vulnerability for global food security. If grain prices are subject to supply and demand, with the authorities functioning as a facilitator, the cultivated areas of major cereal crops such as paddy and rice would be shifted to fulfil the increased demand for other food crops such as oilseeds, fruits, and vegetables, due to dietary diversification. The long-term goals of food security require research and innovation in other food crop production as well as technological availability and skill development for underprivileged farmers. The proposed policies developed by the governments of each particular region of the globe must include the quality of food products as well as minimizing the prices of the products associated with market demand. The dietary intake of the world's people can be reduced by increasing the production of agricultural produce, especially in rain-fed regions, and increasing the availability of the food at reasonable rates through the PDS system, as well as other activities that are related to alleviating poverty. Programs related to food and nutritional security should be based on the nutritional needs of individuals.

REFERENCES

Agah, S., Kim, H., Mertens-Talcott, S. U., and Awika, J. M. 2017. Complementary cereals and legumes for health: Synergistic interaction of sorghum flavones and cowpea flavonols against LPS-induced inflammation in colonic myofibroblasts. *Molecular Nutritional Food Research*, 61(7). doi: 10.1002/mnfr.201600625

Agricultural and Processed Food Products Export Development Authority. 2021. *Field crop survey report*. Visakhapatnam: Ministry of Commerce and Industry, Government of India

Allen, P. 2013. Facing food security. *Journal of Rural. Studies*, 29, 135–138

Andersson, R., Fransson, G., Tietjen, M., and Aman, P. 2009. Content and molecular-weight distribution of dietary fiber components in whole-grain rye flour and bread. *Journal of Agricultural and Food Chemistry*, 57(5), 2004–2008.

Andersson, U., Dey, E. S., Holm, C., and Degerman, E. 2011. Rye bran alkyl resorcinols suppress adipocyte lipolysis and hormone-sensitive lipase activity. *Molecular Nutrition and Food Research*, 55, 290–293

Arbex, P. M., Castro Moreira, M. E., Lopes Toledo, R. C., Morais Cardoso, L., Pinheiro-Santana, H. M., Anjos Benjamin, L., Licursi, L., Carvalho, C. W. P., Vieira Queiroz, V. A., and Duarte Martino, H. 2018. Extruded sorghum flour (*Sorghum bicolor* L.) modulates adiposity and inflammation in high fat diet-induced obese rats. *Journal of Functional Foods*, 42, 342–346.

Awika, J. M., and Rooney, L. W. 2004. Sorghum phytochemicals and their potential impact on human health. *Phytochemistry*, 65, 1193–1199

Berry, E., Dernini, S., Burlingame, B., Meybeck, A., and Conforti, P. 2015. Food security and sustainability: Can one exist without the other? *Public Health Nutrition*, Cambridge UK: Cambridge University Press, 18, 2293–2302.

Blake, T., Blake, V., Bowman, J., and Abdel-Haleem, H. 2011. *Barley: Production, improvement and uses*. Wiley-Blackwell, 522–531.

Capone, R., El Bilali, H., Debs, P., Cardone, G., and Driouech, N. 2014. Food system sustainability and food security: Connecting the dots. *Journal of Food Security*, 2, 13–22.

Dixon, J., Braun, H. J., Kosina, P., and Crouch, J. 2009. *Wheat facts and futures 2009*. Mexico: CIMMYT. https://repository.cimmyt.org/handle/10883/1265.

Dowler, E.A., and Connor, D. 2012. Rights-based approaches to addressing food poverty and food insecurity in Ireland and UK. *Social Science and Medicine Journal*, 74, 44–51

Eurostat. 2013. *Income and living conditions in Europe (EU-SILC)*. Luxembourg: European Union.

FAO. 1974. *Report of the world food conference*. Rome: FAO.

FAO. 1983. *World food security: A reappraisal of the concepts and approaches*. Rome: FAO

FAO. 1996. Rome Declaration on World Food Security and World Food Summit Plan of Action. World Food Summit 13–17 November 1996. Rome.

FAO. 2021. Record cereal production seen keeping markets adequately supplied in 2021/22. Rome: FAO.

FAO, IFAD, UNICEF, WFP and WHO. 2019. The State of Food Security and Nutrition in the World 2019. Safeguarding against economic slowdowns and downturns. Rome: FAO

FAO, IFAD and WFP, 2014. The state of food insecurity in the World 2014: Strengthening the enabling environment for food security and nutrition. Rome: FAO.

Foster-Powell, K., Holt, S.H., and Brand-Miller, J.C. 2002. International table of glycemic index and glycemic load values. *The American Journal of Clinical Nutrition*, 76, 5–56.

Fuller, D.Q., Korisettar, R., Vankatasubbaiah, P.C., Jones and M.K. 2004. Early plant domestications in southern India: some preliminary archaeobotanical results. *Vegetation History and Archaeobotany*, 13, 115–129.

Fusco, G., Coluccia, B., and De Leo, F. 2020. Effect of trade openness on food security in the EU: A dynamic panel analysis. *International Journal of Environment Research Public Health*, 17, 4311.

Goron, T. L., and Raizada, M.N. 2015. Genetic diversity and genomic resources available for the small millet crops to accelerate a new green revolution. *Frontiers in Plant Science*, 6, 157. doi: 10.3389/fpls.2015.00157

Hariprasanna, K., and Patil, J.V. 2015. Sorghum: Origin, Classification, Biology and Improvement. In: Sorghum Molecular Breeding. Madhusudhana, R., Rajendrakumar, P., Patil, J. (eds) pp. 3–20, Springer, New Delhi. https://doi.org/10.1007/978-81-322-2422-8_1

Hazell, P. B. R. 2009. *The Asian green revolution. IFPRI Discussion Paper 00911.* Washington, DC: International Food Policy Research Institute.

IDRC. 2010. *Facts and figures on food and biodiversity.* IDRC Communications, International Development Research Centre. www.idrc.ca/en/research-in-action/facts-figures-food-and-biodiversity

IFPRI. 2016. *2016 Global food policy report, IFPRI.* Washington, DC: IFPRI.

Khorsandi, P. 2020. WFP chief warns of 'hunger pandemic' as global food crises report launched. New York: World Food Programme Insight

Kirwan, J., and Maye, D. 2013. Food security framings within the UK and the integration of local food systems. *Journal of Rural Studies*, 29, 91–100

Kris-Etherton, P.M., Hecker, K.D., Bonanome, A., Coval, S.M., and Binkoski, A.E. 2002. Bioactive compounds in foods: Their role in the prevention of cardiovascular disease and cancer. *American Journal of Medicine*, 113(Suppl. 9B): 71S–88S.

Li, P., and Brutnell, T.P. 2011.*Setaria viridis* and *Setaria italica*, model genetic systems for the panicoid grasses. *Journal of Experimental Botany*, 62, 3031–3037

Mayer, K.X.F. 2012. The international barley genome sequencing consortium. A physical, genetic and functional sequence assembly of the barley genome. *Nature*, 491, 711–716.

Muthamilarasan, M. et al., 2016. Exploration of millet models for developing nutrient rich graminaceous crops. *Plant Science*, 542, 89–97

Muthamilarasan, M. et al., 2019. Multi-omics approaches for strategic improvement of stress tolerance in underutilized crop species: A climate change perspective. *Advances in Genetics*, 103, 1–38

Nelson, G., Bogard, J., Lividini, K., Arsenault, J., Riley, M., Sulser, T. B., et al., 2018. Income growth and climate change effects on global nutrition security to mid-century. *Natural Sustainability*, 1, 773–781.

Park, C.Y., Hwa-Son., H., and A.D.B. San-Andres. 2012. *Food security and poverty in Asia and the Pacific: Key challenges and policy issues.* Mandaluyong: Asian Development Bank

Poole, N., Donovan, J., and Erenstein, O. 2020. Agri-nutrition research: Revisiting the contribution of maize and wheat to human nutrition and health. *Food Policy*, 101976. doi: 10.1016/j.foodpol.2020.101976

RaboResearch 2017. *Global wheat consumption.* Utrecht: Rabobank. https://research.rabobank.com/far/en/sectors/grainsoilseeds/global_wheat_demand_article_1.html

Radhakrishna, R., and Venkata Reddy, K. 2020. Food security and nutrition: Vision 2020. http://planningcommission.nic.in/reports/genrep/bkpap2020/16-bg2020.pdf

Rao, C. H. H. 2000. Declining demand for foodgrains in rural India: Causes and implications. *Economic and Political Weekly*, 201–206.

Rodríguez, J. P., Rahman, H., Thushar, S., and Singh, R. K. 2020. Healthy and resilient cereals and pseudocereals for marginal agriculture: Molecular advances for improving nutrient bioavailability. *Frontiers in Genetics*, 11, 49. https://doi.org/10.3389/fgene.2020.00049

Sarwar, M.H., Sarwar, M.F., Sarwar, M., Qadri, N.A., and Moghal, S. 2013.The importance of cereals (*Poaceae: Gramineae*) nutrition in human health: A review. *Journal of Cereals Oilseeds*, 4, 32–35.

Serafini, M., Bellocco, R., Wolk, A., and Ekström, A. M. 2002. Total antioxidant potential of fruit and vegetables and risk of gastric cancer. *Gastroenterology*, 123, 985–991.

Shah, A., Gani, A., Masoodi, F. A, Shoib, M. W., and Bilal, A. A. 2017. Structural, rheological and nutraceutical potential of β-glucan from barley and oat. *Bioactive Carbohydrates and Dietary Fibre*, 10, 10–16

Shiferaw, B., Smale, M., Braun, H. J., Duveiller, E., Reynolds, M., and Muricho, G. 2013. Crops that feed the world 10. Past successes and future challenges to the role played by wheat in global food security. *Food Security*, 5, 291–317.

Taylor, J.R.N., and Emmambux, M.N. 2008. Gluten-free foods and beverages from millets. In Gallagher, E. (ed.), *Gluten-free cereal products and beverages* (pp. 1–27). London: Elsevier

Venkateswaran, K., Mauraya, M., Dwivedi, S.L., and Upadhyaya, H.D., 2014. Wild Sorghums - Their Potential Use in Crop Improvement. In: *Genetics, Genomics and Breeding of Sorghum*. Wang, Y-H., Upadhyaya, H.D., Kole C.R, (Eds). Taylor & Francis Group (CRC-Press), Boca Raton, Florida, USA.

Verma, R.P.S., Kharub, A.S., Kumar, D., Sarkar, B., Selvakumar, R., Singh, R., Malik, R., Kumar, R., and Sharma, I. 2011. *Fifty years of coordinated barley research in India*. Directorate of Wheat Research, Karnal-132001. Research Bulletin No. 27: 46.

Weber, S. 1998. Out of Africa: The initial impact of millets in South Asia. *Current Anthropology*, 39, 267–274

Wynne, B., and Zhang, Y. 2012. *European association for the study of obesity (EASO)*. London: SAGE Publications.

CHAPTER 3

Physical Properties of Cereal Grains

Panthavur Nairveetil Anjali, Sowriappan John Don Bosco, Muhammed Navaf
and Kappat Valiyapeediyekkal Sunooj

CONTENTS

3.1 INTRODUCTION

Cereal grains are the foremost significant plants that can be considered the backbone of agriculture and have had equal importance in all centuries. The term "cereal" is derived from "Ceres," the ancient Roman goddess of sustenance. It belongs to the Monocotyledonous family *Gramineae*, commonly known as grasses (Serna-Saldivar, 2016). Cereals can be broadly grouped into four categories, Oryzeae (rice), Triticeae (wheat and barley), Andropogoneae (maize/corn), and Aveneae (oats). Among them, wheat and rice constitute more than 50% of cereal production worldwide. Apart from wheat and rice,

DOI: 10.1201/9781003252023-3

maize, sorghum, barley, millets, rye, and oats constitute the cereal crops. According to FAOSTAT, India is one of the significant grain-producing countries. As one of the world's largest populated countries, India requires cereal commodities to provide its people nutritional security (2017).

Whole-grain foods are irreplaceable for their nutritional benefits (Hemdane et al., 2016). Studies have reported that whole-grain foods can reduce the risk of heart disease, obesity, and type 2 diabetes. A suitable combination of antioxidants, dietary fiber, vitamins, and minerals accounts for these health benefits (Cho et al., 2013). Wheat is considered one of the most prolific food sources that provide energy supply in the human diet. The origin of native wheat grasses is from the Kashmir region to Syria and then to parts of Africa. *Triticum aestivum* is the major wheat variety cultivated, which can be utilized for the production of cake, bread, or different ready-to-eat food products. Another variety commonly produced is *Triticum durum*, widely used to make pasta (Kulp & Ponte, 2000). Rice is believed to be one of the most sacred plants in Asia. Studies show that *Oryza sativa* is the one species consumed worldwide, even though approximately 20 species of the genus *Oryza* are present (Serna-Saldivar, 2016). The ecogeographical classification of *Oryza sativa* includes *indica* (tropical), *japonica* (temperate), and *javanica* (semitropical) (Perdon & Holopainen-Mantila, 2020).

Corn (*Zea mays* L.) is the only grain crop and dominant staple food of native Americans, known as "mahiz." There are different varieties of corn, namely dent, waxy dent, flint, pod, flour/soft, sweet/sugar, popcorn, and so on. The color of the pericarp varies from the most common color, yellow, to orange, red, blue, or white, according to the variety. Dent corn is mostly preferred for cereals like corn flakes (Perdon & Holopainen-Mantila, 2020). Sorghum and millets are essential crops of most regions of Asia and Africa, cultivated in less time compared to wheat. Sorghum, millet, and barley can tolerate severe salinity stress and water shortage conditions. They can be cultivated in regions susceptible to drought conditions, maintaining sustainable agriculture (Manjunatha et al., 2007). Barley is most commonly consumed in the form of malt and is widely used in the brewing and bakery industry. The structure is similar to wheat, but the proportion of bioactive compounds like β-glucan, phenolic compounds, and antioxidants varies, enhancing its health benefits (Kaur & Gill, 2021).

The most prominent advantage of cereals is their extended shelf life. The matured grains can be stored and used for food production for quite a long time without expiring, and they can also be preserved as seeds for sowing in the future. Another benefit of cereals is their capacity to store calories and other nutrients in a concentrated form as relatively smaller units while being the largest supplier of calories among the major food groups. Moreover, except for phytic acid, antinutritional factors are not present in cereal grains (Serna-Saldivar, 2016).

3.2 PHYSICAL PROPERTIES OF CEREAL GRAINS

The physical properties of cereal grains during the design, optimization, and improvement of technology associated with different unit operations after the crop harvesting process are important (Tarighi et al., 2011). Several types of equipment and machinery have been designed and developed according to grain characteristics for appropriate cleaning, sorting, grading, threshing, transporting, ventilating, drying, storing, and processing. The dimensions of bulk storage tanks and holding bins with a given capacity are calculated based on grain kernel properties. Proper knowledge regarding the physical and engineering properties of grains help to resolve any problems (Irtwange, 2000). Moreover, accurate size distribution and characteristic dimensions of kernels allow us to calculate the grain's total surface area, sphericity, and volume (Al-Mahasneh & Rababah, 2007). Several studies have reported the use of accurate characteristic dimensions in determining grain size (Sologubik et al., 2013). Table 3.1 shows the physical properties of some major cereal grains. Physical properties like bulk density help us find the grain storage and transport capacity, whereas the true density helps in designing the appropriate separation equipment. Likewise, another physical property, the porosity of grain, allows us to calculate the airflow resistance during the drying process. Moreover, different types of storage structures and containers are designed according to cereal grain dimensions by analyzing certain properties like the coefficient of friction or the angle of repose (Vilche et al., 2003). Hence, proper knowledge

Table 3.1 Physical Properties of Some Major Cereal Grains

Physical Properties	Rice	Wheat	Maize	Millet
Length (mm)	8.45	7.08	9.87	3.118
Width (mm)	2.60	3.27	7.41	1.936
Thickness (mm)	1.86	2.98	3.25	1.702
Equivalent diameter (mm)	3.44	4.10	6.19	1.720
Sphericity (%)	40.76	57.93	62.76	0.9374
Aspect ratio	0.31	0.46	0.75	1.067
Surface area (mm^2)	20.27	31.2	120.02	12.547
Volume (mm^3)	30.76	40.45	90.30	3.795
Bulk density (kg/m^3)	471	650	765	866.1
True density (kg/m^3)	1193	1150	1315	1578
Porosity (%)	60.52	43.48	41.83	45.1
Angle of repose (deg.)	37.5	36.2	30.2	25
Thousand kernel weight (g)	21.64	30	220	5.985

of geometric properties like the length, breadth, thickness (linear dimensions), sphericity, roundness, surface area, volume, aspect ratio, hardness, color value; gravimetric properties like thousand kernel weight, true density, bulk density, porosity; and frictional properties like the angle of repose and coefficient of friction of grains are required in the cereal industry.

3.2.1 Geometric Properties

3.2.1.1 Grain Dimensions (mm)

The three principal linear dimensions, length (L), width (W), and thickness (T), of grains have a vital role in designing and selecting sieve separators for milling operations and also in calculating the power required for the process. They are determined with the help of a vernier caliper, precise to 0.01 mm (Mir et al., 2013). The shape of cereal grains is one of the characteristics that differentiate grain kernels from one another. It has a significant role in cleaning, sorting, and grading grains and, moreover, influences their bulk behavior, such as the angle of repose. The shape of cereal grains can be described in three ways; 1) by comparing the grain kernel with a geometric figure, 2) by using the shape factor, and 3) with the help of virtual models. The most common method of determining the shape of a grain is by comparing it to geometric figures. These include spherical, pyramidal, lenticular, polyhedral, and ellipsoid (Kaliniewicz et al., 2015). Different categories like short, medium, and long grains are classified according to the size and shape of the grains. A classification of rice kernels based on grain dimensions in the United States is given in Table 3.2 (Perdon & Holopainen-Mantila, 2020). Considering the cereal grain a prolate spheroid in shape, the equivalent diameter (D_e) of the kernel can be evaluated using the following expression (Mir et al., 2013):

$$D_e = \left[L \frac{(W+T)^2}{4} \right]^{1/3}$$

Table 3.2 Classification of Rice Kernels Based on the Grain Dimensions

Grain Dimensions	Short Grain	Medium Grain	Long Grain
Length (mm)	5.2–5.4	5.4–6	6.7–7
Width (mm)	2.7–3.1	2.3–2.7	1.9–2
Thickness (mm)	1.9–2	1.7–1.8	1.5–1.7
L/W ratio	Less than 1.9	2–2.9	Greater than 3

Studies have observed a linear increment in all the dimensions of millets with respect to the moisture present in the grain. It indicates the expansion in length, width, thickness, and equivalent diameter of millet kernels on water absorption within a range of 5–22.5%. The percentage of moisture content along the dimensions varies depending on the difference in the cell arrangement of the grain structure (Baryeh, 2002). A similar linear dependence of grain dimensions on moisture content was observed in corn seeds and barley (Tarighi et al., 2011; Sologubik et al., 2013).

Another study reported that the size and dimensions of barley and wheat grains augmented with an increment in the moisture percentage, from 14 to 22%. The length of barley and wheat grains increased from 9.87 to 11.13 mm and 5.98 to 6.86 mm, width from 3.28 to 3.81 mm and 3.12 to 3.75 mm, and thickness from 2.21 to 2.96 mm and 2.80 to 3.22 mm, respectively. Moreover, the arithmetic, geometric, and equivalent mean diameter of all three axes increased with the increment of moisture content. The results showed that barley grains have the capacity to absorb more water to increase their size than wheat grains. In addition, the study also reported that grain dimensions determine the cylinder-concave clearance in threshing machinery. If the clearance is small, a high amount of grain breakage loss happens due to the crushing of grains, and if the clearance is large, un-threshed grains will pass, along with straw (Belay & Fetene, 2021).

3.2.1.2 Sphericity (%)

The sphericity (Ø) of a grain is the index of its roundness. It is defined as the ratio of the surface area of the sphere having the same volume as that of the grain kernel to the surface area of the kernel (Jain & Bal, 1997). It is measured using the following equation (Mir et al., 2013):

$$\emptyset = \frac{\left(LWT\right)^{1/3}}{L}$$

Researchers have reported that most agricultural materials will have a sphericity value within the range of 0.32 to 1. The results obtained for the sphericity of raw rough rice showed similar range (Varnamkhasti et al., 2008). Lower sphericity indicates elongated grains with less resemblance to a sphere (Hamdani et al., 2014). The studies reported by Mir and co-workers also showed similar properties for paddy grains and brown rice. The sphericity of paddy grains with pointed tips along the length axis was observed to be lower than brown rice, which has a comparatively more rounded edge (Mir et al., 2013). It is noticed that the amount of moisture present in the kernels also affects the sphericity value. A study on millet varieties revealed a linear increase from 5–18%, after which a sharp increment in sphericity is observed with the augmentation of percentage of moisture content (Baryeh, 2002).

Studies have reported that when the amount of moisture in barley rose from 13.15 to 35.29%, the sphericity also increased from 51.43 to 53.14%. A further increase in moisture content up to 45.82% reduced the sphericity value to 52.26% (Sologubik et al., 2013). The reason for the initial increment in sphericity might be the proportional increase in length, width, and thickness of the grain. In contrast, moisture content above 35.29% might have caused a significant increase only in length, resulting in lower sphericity. Similar results of an initial increment followed by a reduction in sphericity were observed for okra seeds (Sahoo & Srivastava, 2002).

Another study reported an increase in barley and wheat grain sphericity with reference to an increment in the percentage of moisture from 14 to 22%. The sphericity value rose from 42 to 45% and 62.5 to 63.5% for barley and wheat grains, respectively. The study suggested that as the percentage of moisture rises, the shape of the grain approaches a sphere. This is due to changes in dimensions occurring in the grain when moisture absorption takes place (Belay & Fetene, 2021). Shah and his co-workers reported that the sphericity of three varieties of oat cultivars, Sabzaar, SKO20, and SKO90, were observed to be 39.813, 38.543, and 34.58%, which varied significantly ($p \leq 0.05$).

The result indicates the SKO90 variety is comparatively elongated, and none of the three varieties resemble a sphere (Shah et al., 2016).

3.2.1.3 Roundness

Roundness indicates the sharpness of edges and corners of a grain. It is described as the ratio of the average radius of curvature of the edges and corners to the radius of curvature of the biggest inscribed circle (Sahay & Singh, 1996). The roundness of the grains was determined by tracing the shadow graphs of grains with similar moisture content using a microfilm reader with an appropriate magnification. The circumscribing and inscribing circles obtained from the projected view were analyzed for their area and diameter (Dutta et al., 1988). The most widely accepted method to determine the roundness of any particle is given as follows (Sahay & Singh, 1996):

$$Roundness = \frac{A_p}{A_c}$$

where A_p = Largest projected area of the grain when it is in the rest position
A_c = Area of the smallest circumscribing circle.

$$Roundness\ ratio = \frac{r}{R}$$

where r = Radius of curvature of the sharpest edge
R = Mean radius of the grain.

Comparing grains of unknown roundness with standard images of known values is another method to determine the roundness of any particles. According to this method, there are six categories of roundness: very angular (corners will be sharp), angular, sub-angular, sub-rounded, rounded, and well rounded (edged will be perfectly rounded) (Dutta et al., 1988).

3.2.1.4 Volume and Surface Area

The principal linear dimensions were used to calculate the volume (V; mm^3) and surface area (S; mm^2) of cereal grains with the help of the following equations:

$$V = 0.25\left[\left(\frac{\pi}{6}\right)L\left(W+T\right)^2\right]$$

$$S = \frac{\pi BL^2}{2L - B}$$

where $B = \sqrt{WT}$.

Studies reveal that the volume and surface area of grains increase with an increment in the percentage of grain moisture. Baryeh (2002) noticed a 50% augmentation in the surface area of millet when the percentage of moisture varied from 5–22.5%. A study on a different type of pearl millet variety indicated the grain surface area was 16.38 mm^2 at 7.4% moisture. The same variety at a similar percentage of moisture showed a higher grain surface area of 24.02 mm^2 because of the higher grain dimensions. Similarly, a linear increment in volume was also noticed for the millet variety from 8.43 to 15.23 mm^3 at 5% to 22.5% grain moisture content (Baryeh, 2002). Another study reported an increase in the surface area of corn seeds from 155.38 to 166.38 mm^2 and volume from 162.84 to 179.024 mm^3 when the grain percentage of moisture rose from 5.15 to 24.07% (Tarighi et al., 2011).

Sologubik and co-workers revealed a significant linear increment in the surface area of barley grains from 54.60 to 63.79 mm^2, with an augmentation in the grain percentage moisture from 13.15 to 45.86% (2013). The jatropha and rice varieties studied by several other researchers also showed similar results (Garnayak et al., 2008; Zareiforoush et al., 2009). With an augmentation in the percentage of grain moisture, a remarkable linear increment in surface area was observed in Nosrat barley grains compared to Scarlett grains. It might be due to relatively minor expansion occurring in the length of the Scarlett grains (Tavakoli et al., 2009).

Another study revealing the physical properties of rough rice grain observed a grain volume of 20.27 and 21.06 mm^3 for two different varieties, Sorkheh and Sazandegi. Their surface area was observed to be 31.76 and 32.58 mm^2, respectively, which showed a significant difference between the varieties. The cereal grain particles separate in the air stream depending on the air velocity, terminal velocity, and number of particles entrapped per unit volume of the airflow. The surface area is significant in evaluating the rate of heat transfer to the particle. Rapid heat transfer is possible with a smaller volume of particles on a unit surface area. Thus, the surface area and particle size affect the drying rate, represented by the surface-to-volume ratio (Varnamkhasti et al., 2008). If the drying rate is affected by any diffusion of water content, larger particles dry slowly compared to smaller particles of a similar shape. The drying time and the energy required also depend on the surface area to volume ratio (Markowski et al., 2013).

3.2.1.5 Aspect Ratio (R_a)

It is the ratio of width to length or the ratio between shorter and longer sides. It is very important to measure the aspect ratio distribution of grain kernels to classify and analyze the off-size grains in market grade. The aspect ratio is calculated using the following equation (Mir et al., 2013):

$$R_a = \frac{W}{L}$$

The aspect ratio is an indication of the tendency of cereal grain towards an oblong shape. A study on the physical properties of barley and oats observed a relatively low aspect ratio value, which generally represents the difficulty of grains to roll properly. Hence there is a chance for barley and oats grains to slide over a flat surface. Designing hoppers and other processing equipment requires knowledge of these characteristics. The aspect ratio values obtained for hulled barley, hull-less barley, Sabzaar oats, and the SKO-20 variety were 43.52 ± 4.4, 39.13 ± 1.6, 21.03 ± 1.3, and 23.92 ± 1.8%, respectively. It is clear that the oats variety has a comparatively low aspect ratio value, indicating its lower rolling tendency (Hamdani et al., 2014).

Another study reported a significant difference in aspect ratio at a 1% probability level for rough rice grain varieties. The values obtained for Sorkheh and Sazandegi rice cultivars were 0.28 and 0.29, respectively (Varnamkhasti et al., 2008). Similarly, the aspect ratios of paddy and brown rice kernels of the Pusa-3 cultivar were 0.19 and 0.24, the lowest value among all the samples used for the study. In contrast, the highest values obtained were 0.44 and 0.61 for paddy and brown rice of the Koshar variety. There was a significant difference ($p \leq 0.05$) in the aspect ratio between the paddy and brown rice cultivars (Mir et al., 2013).

Ramashia and his co-workers reported the physical and functional properties of finger millets from Africa. It was noticed that the milky cream sample had the highest percentage of mean results for aspect ratio (92.21 ± 0.83%) and the lowest value was obtained for the black cultivar (73.55 ±

0.23) (Ramashia et al., 2018). Another study reported that millet grains had an aspect ratio value of 59.62% at a 10% grain moisture content (Adebowale et al., 2012). In addition, a 47.4% aspect ratio was observed at 9.95% grain moisture content (Markowski et al., 2013).

3.2.1.6 Grain Hardness

The hardness of grain (N) indicates the texture of kernels, whether the endosperm is physically hard or soft. Generally, metal hardness is their resistance to permanent deformation. Similarly, the test for hardness or firmness in food materials also tends to bring about permanent deformation in the tested material. The hardness of cereals is related to the stiffness of grains, while the cellular objects are in terms of density. The material strength is the force at which it breaks, whereas toughness is the degree of cracking (Evers & Millar, 2002).

Wrigley and Batey (2012) assessed grain quality for bread making and reported that grain hardness is a varietal characteristic. It is one of the basic quality attributes that determine the grade of the grain in international trade, which leads to finding the suitability of the grain flour for different production purposes. Hard grain is used for bread making and soft grain for products like cookies, biscuits, cake, and pastries. Hardness is generally evident as a horny texture for the grain, whereas soft grain is opaque. The significance of grain hardness is that the starch content of soft grain falls apart during milling, leaving the intact granules. However, fractures occur in the starch particles of hard grains during milling, creating greater starch damage in the flour of hard grains. It is desirable for bread making, providing access to amylase to the starch and promoting water absorption (Wrigley & Batey, 2012).

Studies reported that hardness is a significant property of corn grains, affecting the power requirements for grinding, processability, bulk density, dust formation, nutritional properties, and milling yield. Corn hardness refers to the amount of hard endosperm present relative to the amount of soft floury endosperm in the grain. It is an inherited feature developed by different conditions and influences the damage caused by post-harvest handling of kernels. The different types of corn, flint and popcorn, are hard, having a high percentage of the vitreous endosperm. In contrast, flour and opaque corn with the lowest proportion of endosperm are soft, and the dent variety of corn is intermediate (Paulsen et al., 2018).

Grain hardness can be measured using a texture analyzer. The grain kernel is kept individually in the resting position on the platform of the analyzer, and a suitable load is applied until the grain starts being crushed. Then it is repeated five times with different sample grains to obtain a mean value. The texture analyzer will automatically record the maximum force required to break the grain (Balasubramanian & Viswanathan, 2010). Mwithiga and Sifuna (2006) studied the effect of moisture content on the physical properties of three different sorghum seeds. Rupture strengths of 48.66, 59.64, and 90.84 N were observed for Serena, Seredo, and Kari-mtama varieties, respectively, at a moisture content of 13.64%. It is evident from the report that the rupture strength lowered gradually with an increment in the percentage of moisture up to 19.05% (Mwithiga & Sifuna, 2006).

Another study showed a decreasing trend for the hardness of minor millets with an increment in the percentage of moisture. Higher hardness was noticed for common millet, followed by kodo millet, little millet, and finger millet. However, the hardness of foxtail millet and barnyard millet were almost equal at every moisture content. The need for a lower rupture strength at a greater percentage of moisture might be because the grain becomes softer and more sensitive to breakage at an increased moisture content. It means a higher rupture force is required to break grains with a reduced percentage of moisture (Balasubramanian & Viswanathan, 2010). Similar results were observed for the study of dried pomegranate seeds, in which the hardness reduced linearly from 87 to 50 N for an increment in the percentage of moisture from 6 to 18.13% (Kingsly et al., 2006).

3.2.1.7 Color Characteristics

Color is an important quality attribute in the food industry. Consumers make a decision on the overall appearance of food, including the color, aroma, and texture. For example, if a cereal product looks dark, the chance of a burnt taste is high. If the product's appearance is a light color, it will be mostly underdone. The ingredient color or the manufacturing process affects the color of the final product. Therefore, from the beginning of ingredient procurement until the dispatch of a product, the color must be monitored and maintained according to the requirements.

A color scale is related to how we perceive color, which is able to quantify the color difference and is linear throughout the color space. The food industry's most widely used color scales are the three-dimensional HunterLab L*, a*, b* and the modified CIE system called the CIELAB color scale. It works based on the opponent-color theory, which states that the human eye cone responses to red, green, and blue colors are mixed into black-white, red-green, and yellow-blue colors when they move up the optic nerve to the brain (Pathare et al., 2013). The L axis indicates lightness from 0 (black) to 100 (white); the a-axis indicates red to green, in which positive values are red, negative values are green, and 0 is neutral. The b-axis indicates yellow to blue, in which positive values are yellow, negative values are blue, and 0 is neutral. Every color that can be visually seen can be analyzed using an L, a, b scale. Moreover, the difference between a sample and a standard can also be determined by it (Granato & Masson, 2010).

- Chroma ($C*$)

This is considered the quantitative attribute of colorfulness, which is used to measure the difference in the degree of hue compared to a gray color with similar lightness. The intensity of a sample color perceived by humans will be higher when there is a higher chroma value. The expression used to calculate the chroma value is (Pathare et al., 2013):

$$C^* = \sqrt{a^{*2} + b^{*2}}$$

- Hue angle ($h°$)

Hue angle ($h°$) is the qualitative attribute of color, according to which the colors are conventionally termed reddish, greenish, bluish, and so on. It evaluates the difference between a specific color and a gray color with similar lightness. Hue angle is related to the fact that the absorbance of color will differ at different wavelengths. A lower hue angle represents a greater yellow character for the sample. The angles 0° and 360° indicate a red hue, whereas 90°, 180°, and 270° indicate a yellow, green, and blue hue, respectively. The color characteristics of a wide variety of food products have been evaluated with these parameters. The expression to determine the hue angle is given as (Pathare et al., 2013):

$$h° = \tan^{-1}\left(\frac{b^*}{a^*}\right)$$

Mir and his co-workers studied the color characteristics of various brown rice varieties, and a significant difference was observed between L*, a*, and b* color parameters. The lowest L* value was observed for the darkest Pusa-3 variety (L* = 55.99), followed by SR-1 (58.10). In contrast, the highest L* value (67.19) was observed in the lightest variety, Jehlum (Mir et al., 2013). The color difference in grain kernels might be due to genetic traits, the presence of pigmentation, or flour composition. The antioxidant content will also be high in colored rice (Sompong et al., 2011). Polishing and processing will be more challenging if the grain kernel has a deeper color (Shittu et al., 2012).

The highest a* value was observed for the SR-1 cultivar (7.73) and the lowest for the Jehlum variety (4.23). Another variety, Koshar, was reported to have the highest b* value (26.29), followed by the Pusa-3 (25.33) cultivar (Mir et al., 2013).

Another study reported the color characteristics of different varieties of finger millet grains and flour. The L* value of milky cream millet flour was higher than both black and brown grain/flour samples. The L* value of black grain to milky white grain samples ranged from 19.23 ± 0.42 to 52.97 ± 1.76. Positive values were obtained for the a* and b* coordinates, which indicates the presence of red and yellow pigmentation of grains in different concentrations. The a* value for grain was observed to be 18.28 ± 0.81. Similarly, the mean b* value of grain was 19.38 ± 0.15. The highest H° was observed for the milky cream grain, whereas the lowest was for the black cultivar. It varied from a range of 35.73° ± 1.06 to 68.63° ± 0.06 in grain. The C* values observed for the grain cultivars were in a range of 10.1 ± 3.99 to 29.1 ± 2.03. The increase in pigment concentration increases the chroma value, and similar H° and C* values can be observed, but they can be distinguished by their L* values only. The color intensity of samples discerned by humans will be high if there is a higher C* value (Ramashia et al., 2018).

Kaur and Gill (2021) studied the color characteristics of four germinated cereal grains, wheat, barley, sorghum, and millet. It was observed that as the germination time progressed, the a* and b* values of cereals increased, while the L* value decreased. This might be due to the action of enzymes like polyphenol oxidase or peroxidase, which results in a browning reaction, leading to an increase in red and yellow of the grain color. Moreover, there is a chance of producing a higher amount of starch and other products from the hydrolysis of protein during the germination period. It will result in a Maillard reaction during the drying process (Kaur & Gill, 2021). Similar results were observed for germinated oats (Tian et al., 2010).

3.2.2 Gravimetric Properties

3.2.2.1 Thousand Kernel Weight

Thousand kernel weight (g) is one of the indices determining the quality of cereal commodities. It indicates grain ripeness and affects grains' sowing potential (Warechowska M. et al., 2013). It is determined by counting 100 cereal grains and weighing them using a digital electronic balance with 0.001 g accuracy. Then the obtained value is multiplied by 10 to give the mass of a thousand kernels (Karimi et al., 2009). Mir and co-workers studied the physical properties of paddy and brown rice varieties, in which the significance of measuring a thousand kernel weight is adequately covered. It is an index of the outturn obtained after the milling of cereal grains. The total proportion of weight constituted by the husk can be determined by measuring the thousand kernel weight of paddy and milled rice. Moreover, the relative proportion of dockage present in paddy and the proportion of shriveled or immature kernels are also obtained. It was observed that the thousand kernel weight of paddy ranged from 22.23 to 28.63 g, whereas, for brown rice, it varied from 18.81 to 22.92 g. A decrease in thousand kernel weight is observed according to the stage of processing from rough rice to brown rice (Mir et al., 2013).

Several studies have reported a linear increase in the thousand kernel weight according to the increase in grain moisture content. A linear increase from 23.2 to 39.7 g was observed in wheat when the grain moisture content increased from 0 to 22% (Tabatabaeefar, 2003). Similarly, for different varieties of sorghum, namely Serena, Seredo, and Kari-mtama, the thousand kernel weight determined was 19.66, 20.89, and 33.91 g, respectively, at 13.64% grain moisture content, and it increased linearly with an increase in the percentage of moisture (Awika & Rooney, 2004). Another study also reported a linear increase in the thousand kernel weight of corn seeds from 267.7 to 305.8 g when the grain moisture content increased (Tarighi et al., 2011).

Belay & Fetene, 2021 reported that the average thousand kernel weight of three cereal grains, barley, wheat, and tef, increased with an increase in moisture content from 14 to 22%. The values obtained for barley, wheat, and tef were 31.00 to 34.5 g, 33.00 to 37.5 g, and 0.28 to 0.29 g, respectively. The increase in thousand kernel weight is due to the increase in the amount of water content in the individual grain.

The mass of individual grains of barley and wheat increased from 0.03 to 0.04 g, but the individual mass of tef grains cannot be determined because of their smaller size (Belay & Fetene, 2021).

3.2.2.2 Bulk Density

Bulk density (ρb; kg/m^3) is an important physical property indicating the grade and test weight of cereal grains during drying, storage, and processing (Adebowale et al., 2012). At earlier times, different approaches were introduced to analyze the volume of persisting grains in the combined hopper. Bulk density helps in determining the weight of grains in the hopper. Therefore, knowledge about bulk density is required in designing silos and hoppers for the proper handling of grains (Nalladurai et al., 2002). A study reported the mean bulk density value of two different rough rice cultivars, Sorkheh and Sazandegi, to be 544.34 and 471.21 kg/m^3, respectively. Since the bulk density of the Sazandegi variety is comparatively low, the Sorkheh variety requires a larger silo (Varnamkhasti et al., 2008). The results obtained by analyzing the bulk density denote whether the grains are similar in size and shape, which is a good indicator of grain quality that enhances better cereal flour production (Ramashia et al., 2018).

The mass/volume ratio is used to determine the bulk density by filling an empty cylinder of pre-determined volume and taring to measure the weight of grains continuously poured from a constant height until it reaches the top level and then striking off the excess grains to weigh accurately (Mir et al., 2013). The expression used for measuring bulk density is:

$$Bulk\ density = \frac{Mass}{Volume}$$

Several studies reported a negative relationship between bulk density and moisture content. Karimi and co-workers studied the effect of moisture content on the physical properties of wheat (Karimi et al., 2009). The bulk density varied from 0.72 to 0.66 g/cm^3 at different moisture levels, showing a decrease in bulk density with respect to an increase in moisture content with significant variations. This negative relation might be due to the higher volumetric expansion of grains when there is an increase in the moisture gain of grain samples. The overall increase in mass of grains is low compared to the volumetric expansion (Tabatabaeefar, 2003).

Another study revealed the bulk density of sorghum seeds was 757.6, 686.3, and 588.5 kg/m^3 for the Serena, Seredo, and Kari-mtama varieties, respectively, at 13.64% moisture content. These bulk density values decreased with an increase in the moisture content (Mwithiga & Sifuna, 2006). Other research also reported a negative relationship between the bulk density and grain moisture content. Some of the studies include the bulk density of finger millet varieties ranging from 684.99 to 777.50 kg/m^3 (Balasubramanian & Viswanathan, 2010) and bulk density of quinoa seeds decreasing from 747 to 667 kg/m^3 with an increase in grain moisture content (Vilche et al., 2003). The bulk density value obtained for hulled and hull-less barley was 690 ± 0.5 and 530 ± 0.1 kg/m^3, respectively, and for oats varieties Sabzaar and SKO-20, it was 410 ± 0.1 and 399 ± 0.2 kg/m^3, respectively (Hamdani et al., 2014). Studies have also reported that the long rice varieties have bulk density ranges from 541 to 579 kg/m^3. However, the bulk density of long rice was found to be 582 kg/m^3 at 12% grain moisture content (Mir et al., 2013).

Belay and Fetene (2021) reported a similar decrease in the bulk density of barley and wheat grains with respect to an increase in moisture content from 14 to 22%. The values obtained for barley and wheat grains varied from 497.83 to 444.14 kg/m^3 and 932.50 to 887.64 kg/m^3, respectively. The authors suggested that the negative relation of bulk density to moisture content might be due to the increase in the volume of the bulk grains than the increase in their mass. Therefore the volumetric expansion of grains was found to be comparatively high, resulting in a reduction in bulk density (Belay & Fetene, 2021).

3.2.2.3 True Density

True density (ρ_t; kg/m^3) is the ratio of the weight of a sample to its pure volume. The toluene displacement method is generally used to determine the true density. The liquid displacement method using distilled water can also be used, but toluene is less absorbed by the grains, so it is more efficient to measure the true density. It is calculated by the ratio of sample mass to the volume of toluene displaced by the grain mass (Hamdani et al., 2014).

$$\rho_t = \frac{M}{V_t}$$

where V_t = Displaced volume of toluene.

Studies have reported the true density of rough rice cultivars Sorkheh and Sazandegi was 1269.1 and 1193.4 kg/m^3, respectively. A remarkable difference in true density is observed in seeds with various impurities like field mustard, ryegrass, centaurea, and wild oats from the cereal crop seeds. A solution or suspension is used to determine the true density of different grain mixtures (Varnamkhasti et al., 2008). Hamdani and co-workers studied the physical properties of barley and oats. The true density value obtained for hulled and hull-less barley was 1112 ± 0.1 and 1333 ± 0.2 kg/m^3, respectively, and for Sabzaar and SKO-20 oats, it was 1112 ± 0.3 and 1112 ± 0.2 kg/m^3, respectively (Hamdani et al., 2014).

Reports show that true density increases linearly with an increase in the moisture content. A study on the physical properties of millet showed an increase of true density from 1550 to 1712 kg/m^3 at a grain moisture range of 5 to 22.5% (Baryeh, 2002). At a moisture content of 7.4%, it showed a true density value of 1585 kg/m^3, which is very similar to another variety of pearl millet that has 1590 kg/m^3 at the same grain moisture content (Jain & Bal, 1997). This indicates that despite the difference in the size and dimensions of various millet varieties, the true density value maintains a similar range. The higher increase in the grain mass compared to the increase in volume might be the reason for an overall linear increase when the percentage of moisture content increases. However, in contrast to the results observed in these studies, some researchers have reported a negative relationship between true density and grain moisture content for samples like soybean, neem nut, pumpkin seeds, and so on. It would have been because of a lower increase in total mass compared to the increase in the volume of the seeds according to the increase in moisture content (Baryeh, 2002).

The true density of quinoa seeds showed a linear increase with respect to the moisture content of grain. It varied from 928 to 1188 kg/m^3, showing an increase of 28% from the range of moisture content evaluated (Vilche et al., 2003). Similar results were obtained for the true density analysis of rye seeds. The true density of rye seeds increased linearly from 922.43 to 991.56 kg/m^3, with an increase in moisture content in a range of 9 to 13%. The total mass of grain might be higher than the volumetric expansion of grains (Jouki et al., 2012).

3.2.2.4 Porosity

Porosity (ε; %) is defined as the ratio of intergranular void space volume to that of bulk grain volume (Mir et al., 2013). It is the void space in the bulk grain mass that is not occupied by the grain (Saleh et al., 2013). Porosity is calculated using the expression:

$$\varepsilon = \left(\frac{\rho_t - \rho_b}{\rho_t} \right) \times 100$$

where ρ_t = True density value
ρ_b = Bulk density value of the grain sample.

Studies have reported that the porosity values observed for paddy and brown rice varieties were in the range of 52.70 to 59.45% and 41.06 to 46.70%, respectively. The highest porosity value was obtained for the variety Pusa-3, which might be due to its long grain length (Corrêa et al., 2007). A significant difference ($p \leq 0.05$) was observed in the porosity values of paddy and brown rice cultivars (Mir et al., 2013). The porosity value obtained based on the true and bulk densities of grain depends on the difference in the intrinsic characteristics of various rice cultivars (Varnamkhasti et al., 2008).

Another study on the effect of moisture content on the physical properties of barley reported a significant ($p < 0.01$) increase in porosity from 42.69 to 44.44% when the grain moisture content increased from 13.15 to 45.82% (Sologubik et al., 2013). Similarly, another study compared the porosity of barley variety Scarlett with different grains at varying moisture content. It was observed that the porosity of the Scarlett variety is lower than green wheat and canola (Al-Mahasneh & Rababah, 2007; Çalişir et al., 2005).

Solomon and Zewdu (2009) reported that when moisture is absorbed by the grains, there is an increase in the total volume of grain. The length axis elongates more quickly than breadth and thickness. This results in a decrease in the number of grains occupying a unit volume, thus lowering the bulk density value of the grain. In other words, the bulk volume of grain increases when there is an increase in the moisture percentage of the total grain mass. This addition of moisture content to the grain structure results in an increase in the true density. However, a greater increase in the bulk density value is observed than true density, depending on the moisture intake of grains, altering the characteristic dimensions of grain, mainly the length. This will lead to an increase in the porosity of grains (Solomon & Zewdu, 2009).

In contrast to this, several researchers have observed linear and non-linear decreases in the porosity value depending on the increase in moisture content. A study revealed a lowering of the porosity value of a millet variety from 48.5 to 48% at 7.5% grain moisture content, followed by a non-linear increase to 68% at 22.5% moisture (Baryeh, 2002). Porosity depends entirely on the true density and bulk density of grain samples. Therefore, the intrinsic characteristics of seed or grain affect densities and hence the porosity value.

The effect of moisture content on the porosity of barley and wheat revealed a linear decrease in porosity with respect to an increase in the moisture content from 14 to 22%. The porosity values were evaluated based on the bulk density and true density values obtained for barley and wheat grains, and it was found to vary from 46.75 to 42.80% and 33.01 to 26.12%, respectively. The variation in the two types of crops might be due to the difference in cell structure, mass characteristics, and volume during the increase in moisture content. The porosity value decreased due to the increase in bulk volume of grains than their mass, resulting in a decrease in the number of grains in a fixed volume and leading to a decrease in the porosity value (Belay & Fetene, 2021).

3.2.3 Frictional Properties

3.2.3.1 Angle of Repose

The angle of repose (θ) is the angle formed with the horizontal base at which the grain sample will heap up when piled. Figure 3.1 represents the apparatus for measuring the angle of repose. It can be evaluated by different techniques like the tilting box method, the fixed funnel method, and the revolving cylinder method.

3.2.3.1.1 Tilting Box Method

The tilting box method is appropriate for finely grained, non-cohesive materials generally having an individual particle size of less than 10 mm. It consists of a plywood box, two plates, a fixed plate, and an adjustable plate. The grain sample can be placed in the box, and the adjustable plate is slowly inclined or tilted to allow the grains to form a natural slope. Then the angle of repose is calculated

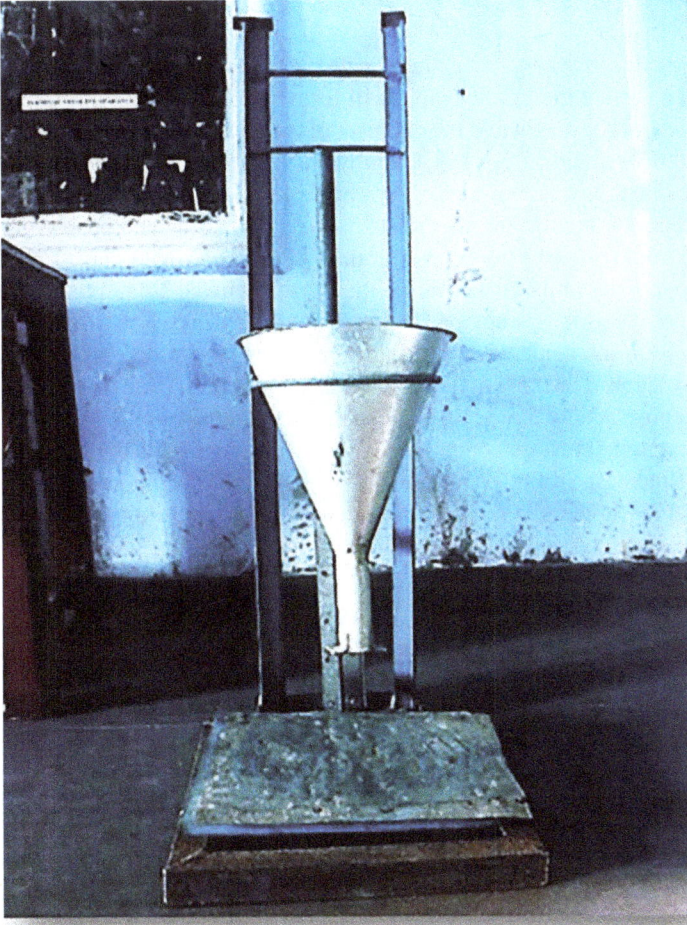

Figure 3.1 Apparatus for measuring the angle of repose.

using the measurements of sample-free surface depths at the end of the box, the midpoint of the sloped surface, and the horizontal distance from the end of the box to the midpoint (Baryeh, 2002).

3.2.3.1.2 Fixed Funnel Method

The fixed funnel method is used to analyze the angle of repose of different samples. The material to be evaluated is poured using a funnel to produce a cone, and the extreme end of the funnel should be nearest to the grain heap. When the grain heap reaches the already set point of height or width, stop pouring the sample into the funnel. The angle formed between the horizontal base and the grain heap has to be noted to determine the angle of repose. However, instead of direct measurement, dividing the height of corn by half the width of the base, followed by taking the inverse tangent ratio, gives the value of the angle of repose.

$$\theta = \tan^{-1}\left(\frac{2H}{D}\right)$$

where H = Height of the heap
D = Diameter of the heap (Mir et al., 2013).

3.2.3.1.3 Revolving Cylinder Method

The sample to be analyzed is kept in a cylinder with one of its ends transparent, and it is set to rotate at a constant speed so that the grain sample will form a certain angle with respect to the circular motion in the cylinder. The height and the diameter of the cone formed are applied in the equation to obtain the value for the angle of repose. Generally, the revolving cylinder method is recommended for evaluating the dynamic angle of repose (Mir et al., 2013).

The angle of repose is significant in creating hoppers for storing and processing cereal grains. It identifies the maximum angle of the heap of grain with the horizontal plane so that the inclination angle of the hopper wall can be made higher than that to ensure a continuous flow of samples in the hopper (Mir et al., 2013). A study on the physical properties of paddy and brown rice varieties reported a decrease in the angle of repose during each stage of processing. It was found that the milled rice variety showed the lowest value for the angle of repose (Razavi & Farahmandfar, 2008).

The angle of repose of sorghum seeds, Serena, Seredo, and Kari-mtama, was 20.11°, 24.41°, and 30.43°, respectively, at a moisture content of 13.64%. A linear increase in the angle of repose was observed with an increase in the grain moisture content of sorghum seeds (Mwithiga & Sifuna, 2006). Many other researchers also reported a linear increase in the angle of repose of grains with an increase in moisture content. Sologubik and co-workers reported the effect of moisture content on the physical properties of barley. The angle of repose values increased significantly ($p < 0.01$) from 18.18° to 27.31° with an increase in the moisture content from 13.15 to 45.82%. The authors suggested the increase in the angle of repose values might be due to the bounding of the surface layer of grains containing higher moisture content because of the surface tension produced. It is essential to determine the angle of repose of grains for designing the openings of hoppers, side walls, and storage structures. Therefore, grain moisture content should also be considered while designing such structures or equipment (Sologubik et al., 2013).

Belay and Fetene (2021) noticed a similar increase in the values of angle of repose of barley, wheat, and teff grains when the moisture content increased from 14 to 22%. The observed increase was in a range of 28.50° to 37.85°, 25.75° to 34.15°, and 35.20° to 40.25° for barley, wheat, and teff, respectively. This might be due to the surface tension formed on the outer layer surrounding the particle, which holds the whole grain mass together (Zareiforoush et al., 2009). Studying the angle of repose is necessary to design equipment and machinery for mass flow and storage structures. Similar observations were made in barley, maize, soybean, and oilseeds (Belay & Fetene, 2021).

3.2.3.2 Static Coefficient of Friction

The static coefficient of friction (μ) determines the amount of friction between two surfaces at rest. It is commonly measured against corrugated boards, fiberglass, wood, galvanized iron sheets, aluminum, plywood surfaces, and so on. The apparatus is designed in such a way that the test surface can face an adjustable tilting plate on a wooden table. Figure 3.2 represents the apparatus used for measuring the static coefficient of friction. The grain samples to be tested will be kept on an adjustable tilting plate, and the test surface will gradually be inclined so that about 75% of the grain samples slide. The tilted angle can be measured using the following formula (Hamdani et al., 2014):

$$\mu = \tan \alpha$$

Hamdani and co-workers studied the static coefficient of friction of different barley and oats varieties. The test materials used for the study were sun-mica, glass, and corrugated board. It was noticed that the static coefficient of friction of barley had significant variations, whereas oat varieties were similar for sun-mica, almost the same for glass, and slightly different for the corrugated board. The highest coefficient of friction was observed for corrugated boards, followed by sun-mica and glass.

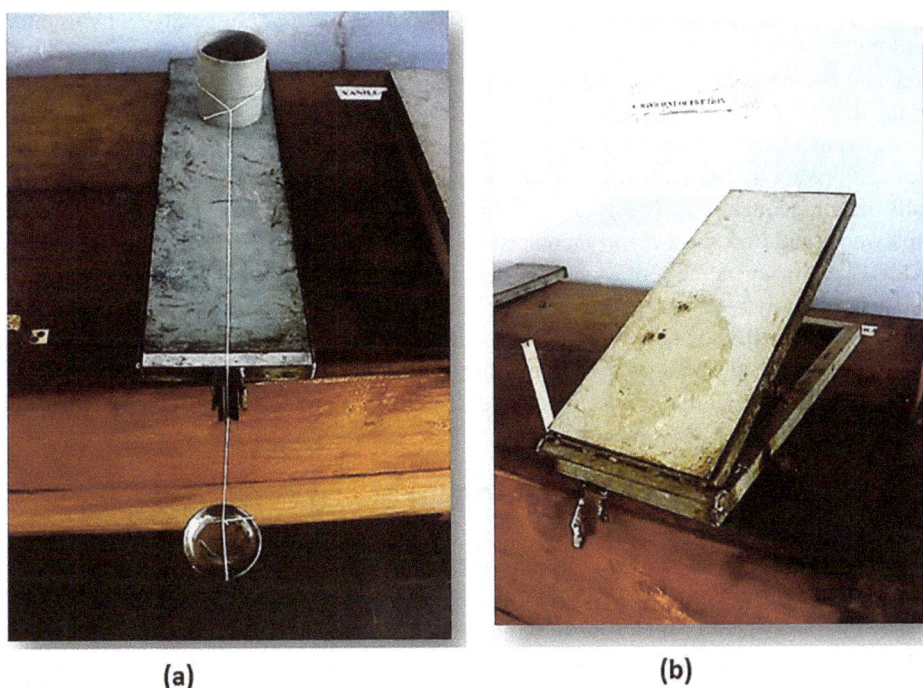

(a)   **(b)**

Figure 3.2 Apparatus for measuring the coefficient of friction.

It might be due to the increase in surface area and the adhesive force between the grain sample and the testing material (Hamdani et al., 2014).

Another study reported the frictional properties of barley varieties using plywood, galvanized steel, and aluminum test materials. A significant increase ($p < 0.01$) was observed with an increase in the moisture content from 13.15 to 45.86%. An increase in the coefficient of friction of 59.35, 38.41, and 60.17% was observed for aluminum, galvanized steel, and plywood surface, respectively. It might be because the high amount of moisture percentage increases the grain water content, and the test sample will have a higher frictional force to the surface in contact. The maximum frictional force was seen for galvanized steel, followed by aluminum and plywood (Tavakoli et al., 2009; Sologubik et al., 2013).

Tarighi and co-workers analyzed the physical properties of corn grains and observed the static coefficient of friction increased linearly with an increase in moisture content from 5.15 to 22% on all surfaces ($p < 0.05$). The test materials used for the study were galvanized iron, plywood, and compressed plastic. A greater value of the static coefficient of friction was observed on the compressed plastic surface (0.67%) and the lowest value on the galvanized iron sheet (0.57%) (Tarighi et al., 2011).

A study on the physical properties of quinoa seeds reported its static coefficient of friction using wood and galvanized iron surfaces. The highest value was obtained for the wooden surface than the iron sheet. According to the increase in moisture content, the static coefficient of friction on both surfaces also increased. Moreover, a higher variation of coefficient of friction with moisture content was observed in a galvanized iron sheet (65.3%) than in wood (25.6%) (Vilche et al., 2003).

Another study reported the frictional properties of different rough rice cultivars, namely the Sorkheh and Sazandegi varieties. There was a significant difference at a 1% probability between the rice varieties and the surfaces used for the study, including the glass sheet, galvanized iron sheet, and plywood. It was observed that the static coefficient of friction of the Sorkheh variety was 0.2899, 0.3185, and 0.4349% on the surface of the glass sheet, galvanized iron sheet, and plywood, while for the Sazandegi variety, it was 0.2186, 0.3153, and 0.4279%, respectively (Varnamkhasti et al., 2008).

Belay and Fetene (2021) conducted research to minimize the production loss by grain damage during the threshing of cereal crops. Three selected cereal crops, wheat, barley, and teff, at different moisture percentages of 14, 18, and 22%, were studied. The coefficient of friction on different surfaces can predict the movement of materials in various equipment and the pressure exerted on the walls of storage materials. It was noticed that the static coefficient of friction of grains on surfaces, plywood, and iron sheet metal increased with an increase in the moisture content in a range of 14–22%. The linear increase observed for barley, wheat, and teff was 36.10 to 62.85%, 35.41 to 60.92%, and 44.94 to 67.91%, respectively. The increase in friction upon the increase in moisture content might be due to the higher amount of water in the grain, which produces a cohesion on the contact surfaces. The sample surface becomes stickier, and the water content in it will try to adhere to the surface across which the grains are moved (Zareiforoush et al., 2010; Belay & Fetene, 2021).

3.3 CEREAL GRAIN QUALITY

Cereal grain quality is defined by several factors, including the physical properties, microbiological safety, compositional characteristics, and post-harvest management of grains. The physical properties include the grain dimensions, volume, surface area, aspect ratio, true and bulk densities, porosity, angle of repose, color value, hardness, static coefficient of friction, moisture content, percentage of damaged/broken kernels, presence of foreign materials, and so on. Microbiological safety is determined by any presence of fungal infection, mycotoxins, pests, mites, or insects and their fragments and odor. The compositional characteristics evaluate the milling yield, viability, starch composition, percentage of carbohydrate, protein, fat, fiber, vitamins, or mineral content (Jayas & Singh, 2012).

The post-harvest management of cereal grains requires three stages, storage, milling, and processing, before reaching the consumer. Most cereal grains are harvested in a specific time period every year, so they have to be stored in bulk quantity for future use. They are also acquired in large quantities if they have to be imported from other countries or states. These bulk amounts of grains have to be stored appropriately until needed, but on an average basis, world losses of cereal grains are 15% during storage. Moreover, about 50% of grain loss happens in tropical and subtropical areas due to insects and rodent infestation. Cereal grain storage is one of the major areas to be improved in the food sector (Serna-Saldivar, 2016).

Milling and further processing of different products also determine the grain quality. The time, temperature, pressure, humidity, performance of machinery, good manufacturing and hygiene practices followed by workers, and so on come under this category. Cereal grain quality highly depends on the nature of the grain and its end use. It also depends on the genetic traits, growth period, harvest time, machinery handling, system involved in the drying process, storage facilities and practices, and transportation procedures (Ratnavathi & Komala, 2016).

Another aspect of cereal grain quality is the requirement of a unique set of specifications for processing cereal-based products. A particular trait of grain or characteristic quality should be maintained to ensure a sustainable business. So the miller must choose grains that consistently produce flour of specified quality attributes. Hence, the opportunity for analyzing the quality of grains begins at the harvest stage. Moreover, the cereals can be segregated to separate storage areas based on the grade and quality of grain kernels (Wrigley & Batey, 2012).

The quality of cereal grain is usually categorized as (Krishnan et al., 2011):

- The efficiency in milling: generally, the head rice yield.
- Appearance and shape of the grain: the length, width, and thickness of grain before and after cooking.

- Cooking characteristics: the percentage of amylase present in the endosperm, aroma, and gelatinization temperature.
- Nutritional quality: carbohydrate, protein, fiber, fat, vitamin, and mineral content.

These four categories are subject to the consumer's choice and the products' intended end use. The quality of grain can also be classified into two other categories, genetic grain quality and acquired grain quality. Genetic grain quality is evaluated by various physicochemical parameters, such as grain shape and size, aroma, bulk density, gel consistency, gelatinization temperature, thermal conductivity, and equilibrium moisture content. Acquired quality includes the color, purity, moisture content, chalkiness, damage, immaturity, crack, whiteness, milling degree, and head grain recovery.

In the case of a rice grain, there are different seed-to-seed cycles at various stages. High-temperature stress at each stage can affect kernels' quality attributes (Krishnan et al., 2011).

3.3.1 Physical Defects and Impurities in Cereal Grains

Many types of post-harvest cereal grain contaminants like mycotoxins, fungi, rodents, insect pests, and so on create a huge loss of more than 421 million tons worldwide every year during storage (Mesterházy et al., 2020). Post-harvest contamination generally occurs during harvest and transportation, but it can also begin during the storage period of grains. Any unfavorable storage conditions like the presence of moisture content or higher temperature favor biological organisms like fungal pathogens growing and proliferating in stored cereal grains (Mannaa & Kim, 2017; Mason & McDonough, 2012).

Post-harvest cereal grains are biologically active, so the grain respiration process can elevate unfavorable conditions like higher temperature or moisture content to create "hot spots." Fungal spores germinate and grow to generate molds in the optimum conditions provided. These molds produce more heat and moisture, creating a self-generating cycle of spoilage (Januarius et al., 2011). Molds affect cereal grains by causing discoloration, protein and lipid degradation, loss of dry matter, production of mycotoxins and volatiles, reduction in germination, and so on. Some other damage and impurities in cereal grain are (Kaur et al., 2020):

1. Broken grains:
   a. Generally due to aggressive handling of grains.
   b. The endosperm of grains will be in an expanded form.
   c. Provides chances for mold infections.
   d. Broken grains are removed during the cleaning process.
   e. The yield of grains in milling is reduced.
2. Sprouted grains:
   a. Germinated grains as a result of wet harvesting.
   b. Alpha-amylase content will be higher.
   c. Even a few sprouted grains can lower the Hagberg falling number to an unacceptable value.
   d. It results in the rejection of grains.
3. Heat damage (burnt grains):
   a. Unacceptable; may arise from "hot spots" during the storage period.
   b. It might be due to excessive temperature during drying conditions.
   c. The color of grains varies from bronze to dark brown (charred).
   d. Overdried grains will alter their physicochemical properties, like irreversible gluten damage in wheat.
4. Screenings:
   a. Some unwanted materials like stones, straw, sticks, chaff, dust, mud balls, and so on have no value to the miller or the end-use consumer.
   b. These should be removed before milling.
   c. Materials like stones and metals can destruct the efficiency of machinery and may cause a spark.

    d. Large screenings include unthreshed grains, straws, beans, stones, and sticks, whereas small screenings include broken grains, weed seeds, shriveled grains, small straw pieces, and chaff.

    e. Mud balls and stones are very common screenings in wet harvested grains.

    f. When harvesting conditions are challenging, stones are picked up during combining.

    g. Dust, chaff, and fine soil may be present in grains. These grain dust particles are harmful and create respiratory problems if inhaled.

5. Smell:

    a. Any unusual smell might be a sign of an insect pest attack or the presence of any other chemicals or pesticides.

    b. A sweet or minty smell indicates an attack of mites.

    c. A fishy, musty smell indicates a mold attack.

    d. The smell of chemicals may also be present if any cleaning liquids or the presence of diesel occur in grain.

6. Moldy grains:

    a. Moldy grains indicate poor harvesting conditions.

    b. Dull-looking, weathered grains.

    c. It may impair the quality, such as wholemeal color.

    d. Mold occurs due to spores or mold attack, which is unacceptable because of the chances of mycotoxin formation.

7. Blackpoint:

    a. Blackpoint is the result of the response from plants to infection.

    b. The plant produces chemicals in the bran area, which occur in brown to black color.

    c. It is often associated with *Alternaria* infection, but it is not only the cause of blackpoint.

    d. The flour quality will be affected due to the dark bran specks produced.

8. Insect damage:

    a. Evidence of insects mainly indicates poor storage conditions and a chance for local "hot spots."

    b. The presence of active live insects hinders the processing of grains.

    c. Weevil damage is an example. Eggs are laid inside grains.

    d. Inside the kernels, the larvae might have eaten the endosperm of the grain.

9. Orange blossom ridge:

    a. Midges lay eggs in the empty florets.

    b. It infests crops at the flowering stage.

    c. Larvae attack the immature grains.

    d. It penetrates the bran and injects enzymes into the grain.

    e. It results in water ingress and the lowering of Hagberg's falling number.

    f. Other fungus infections can also occur along with orange blossom ridge.

10. Rodent droppings:

    a. Rodents like mice, rats, squirrels, and others urinate on grains, causing a food safety risk.

    b. Rodent dropping–contaminated grains are unacceptable.

    c. It directly damages the bulk grain and carries infection.

11. Splitting of grains:

    a. Excessive expansion or mechanical weakness results in cracks on the outer grain.

    b. Splitting mainly occurs through the ventral sides, but it can also occur on the dorsal and lateral sides of the grain.

    c. Splitting may lead to an infestation, commonly a mold attack.

    d. Excess water uptake occurs while processing, leading to a mushy steep with leached starch in the water.

12. Gape:

    a. Gape is generally a function of variety and environment.

    b. It becomes a defect only when it is associated with lateral splitting.

    c. Excessive expansion or poor development results in the development of a gap between the husk tissues.

    d. The endosperm remains intact.

13. Lost embryos:

    a. Commonly caused by mechanical damages.

    b. Grain will not germinate; hence it cannot be used for processes like malting grains.

14. Skinning:
    a. It is the separation and loss of lemma and palea.
    b. Skinning is caused by weather conditions, developmental factors, rough harvesting, and post-harvest handling.
    c. Loss of husk leads to filtration problems.
    d. The efficiency of the malting process will be reduced.
    e. Chances for dust problems may arise during the handling of grains.
    f. It is mostly seen in spring varieties.

3.3.2 Grade and Class of Cereal Grains

The quality parameters associated with the grain determine the grade of any lot of grain. Specific standard tables are used mainly by the grade assignation systems for the inspected grain. Different factors like the test weight, damaged kernels, broken kernels, shrunken kernels, and the presence of foreign materials are specified in these tables. Even though there are several kinds of classification systems around the world, the United States grain classification system is the most recognized one because the United States is one of the leading grain exporters in the world. The grading aims to check the grains' condition, health, and soundness. Consumers prefer insect-free, dry, sound grain that will be stored well for a long time and processors grain that will yield good-quality finished products. There are different tests for analyzing the grade of grains, which include mainly moisture, test weight, amount of foreign materials, percentage of damaged and broken kernels, and so on. After analyzing the tests, a specific grade is assigned to each grain. The maximum grade value depends on the minimum individual test value (Serna-Saldivar, 2012).

The United States Grain Standards Act defined corn as grain that consists of 50% or more whole kernels of shelled dent or flint corn and not more than 10% of other grains. There are three classes, white, yellow, and mixed. Yellow corn consists of yellow kernels and not more than 5% of other colored corn. However, partially reddish corns are also considered yellow corn. White corn consists of not more than 2% of other colored corn, but light pale straw or pink-colored corns are categorized as white. Mixed corn includes yellow corn with white-capped kernels (Paulsen et al., 2018).

The United States Department of Agriculture (USDA) established different grades and grade requirements for all cereal grains in which special grades of corn, flint, flint-dent, infested, and waxy corn, persist. Flint corn consists of 95% or more flint kernels, flint dent consists of a mixture of less than 95% flint corn, infested corn consists of living insects or weevils that cause damage to the stored corn, and waxy corn consists of 95% or higher percentage of waxy corn according to Federal Grain Inspection Service (FGIS) procedures. The corn grade varies from U.S. Grade number 1 to the lowest quality of corn, termed the sample grade (Paulsen et al., 2018). Table 3.3 shows the grades and criteria for corn grading established by the USDA (2020). Similar grades and grade

Table 3.3 Grades and Criteria for Corn Grading

U.S. Grade	Minimum Test Weight (lbs/bu)	Maximum Heat-Damaged Grain (%)	Maximum Total Damaged Grain (%)	Maximum Broken Grains and Foreign Matter (%)
1	56	0.1	3	2
2	54	0.2	5	3
3	52	0.5	7	4
4	49	1	10	5
5	46	3	15	7
Sample grade*				

Data source: United States Standards for Grain (USDA-GIPSA, 2020).

*Sample grade is distinctly the lowest quality grain, which does not meet any requirements for grades 1 to 5.

requirements are established for all other cereal grains like barley, wheat, rye, sorghum, and so on (USDA-GIPSA, 2020).

Rice is the only cereal grain graded as rough, brown, or white polished rice. The factors considered for grading are the percentage of chalky kernels, damaged kernels, and grain color. The grading system for rice is much more elaborate because the quality, yield, and characteristics during cooking depend on different properties. Rice marketing in the United States is based on size, form, and condition. Here, the test weight is not considered for the grading system. However, test weight is related to the milling yield, and it is lower when there is an increase in the percentage of dockage, immature, and empty kernels. The short, medium, and long rough rice varieties have test weights of 60, 58.5, and 56 kg/hL, respectively (Serna-Saldivar, 2012).

The percentage of kernels with chalky endosperm is a criterion of rice grading, but it is more susceptible to breakage during handling and milling. These kernels will be very soft textured, and there is a greater tendency for overcooking rice. Color is another criterion for rice grading. It ranges from a desirable white color to an undesirable dark grey color. The most commonly seen off-color is pink or rosy, which occurs mainly as contamination from red wild rice. The pigmentation in the aleurone layer contaminates white kernels to produce a pink coloration. According to color, parboiled rice can be classified as light or dark-colored rice, which is enhanced by hydrothermal processes (Serna-Saldivar, 2012).

The classification of oats includes white or yellow, red, black, gray, and mixed oats. Husked caryopsis oats are graded based on the test weight, percentage of dockage, and damaged kernels. The primary consideration in oat grading is wild oat contamination (*Avenafatua*), which mainly contaminates commercial plantations during harvesting. Commercial oats are different from wild oats in the absence of a twisted awn and pubescence in the germinal part.

The United States Federal Grain Inspection Service has not assigned any classes or subclasses to rye. The quality of rye entirely depends on the assigned grading system. The grading system of rye in the United States includes four grades and one other grade that does not meet the requirements. The Canadian Grain Commission also grades rye into four classes, but two more specific categories exist. The most important concern in the case of rye is the identification of ergot (*Claviceps purpurea*)-contaminated kernels. It produces a toxin that is highly toxic to humans (Serna-Saldivar, 2012).

3.4 CONCLUSION

It is essential to establish the physical properties of all cereal grains worldwide, since the knowledge of these physical and mechanical properties is very important in the food industry. One of the primary concerns regarding designing different equipment and machinery for properly handling and processing grains is their physical properties. It was observed that various kinds of cereal grains have distinct physical properties and are a function of the grain's moisture content. Therefore, the utmost care should be taken from the very beginning of harvest, followed by post-harvest management and processing of cereal grains until they reach the consumer. Moreover, several types of physical damage and impurities can be caused to grains if not adequately handled, stored, and processed.

REFERENCES

Adebowale AA, Fetuga GO, Apata CB, Sanni LO (2012) Effect of variety and initial moisture content on physical properties of improved millet grains. *Nigerian Food Journal*, 30, 5–10. https://doi.org/10.1016/s0189-7241(15)30007-2

Al-Mahasneh MA, Rababah TM (2007) Effect of moisture content on some physical properties of green wheat. *Journal of Food Engineering*, 79, 1467–1473. https://doi.org/10.1016/j.jfoodeng.2006.04.045

Awika JM, Rooney LW (2004) Sorghum phytochemicals and their potential impact on human health. *Phytochemistry*, 65, 1199–1221. https://doi.org/10.1016/j.phytochem.2004.04.001

Balasubramanian S, Viswanathan R (2010) Influence of moisture content on physical properties of minor millets. *Journal of Food Science and Technology*, 47, 279–284. https://doi.org/10.1007/s13197-010-0043-z

Baryeh EA (2002) Physical properties of millet. *Journal of Food Engineering*, 51, 39–46. https://doi.org/10.1016/S0260-8774(01)00035-8

Belay D, Fetene M (2021) The effect of moisture content on the performance of melkassa multicrop thresher in some cereal crops. *Bioprocess Engineering*, 5, 1–10. https://doi.org/10.11648/j.be.20210501.11

Çalişir S, Marakoğlu T, Öğüt H, Öztürk Ö (2005) Physical properties of rapeseed (*Brassica napus oleifera* L.). *Journal of Food Engineering*, 69, 61–66. https://doi.org/10.1016/j.jfoodeng.2004.07.010

Cho SS, Qi L, Fahey GC, Klurfeld DM (2013) Consumption of cereal fiber, mixtures of whole grains and bran, and whole grains and risk reduction in type 2 diabetes, obesity, and cardiovascular disease. *The American Journal of Clinical Nutrition* 98, 594–619. https://doi.org/10.3945/ajcn.113.067629

Corrêa PC, da Silva FS, Jaren C, et al (2007) Physical and mechanical properties in rice processing. *Journal of Food Engineering*, 79, 137–142. https://doi.org/10.1016/j.jfoodeng.2006.01.037

Dutta SK, Nema VK, Bhardwaj RK (1988) Physical properties of gram. *Journal of Agricultural Engineering Research*, 39, 259–268. https://doi.org/10.1016/0021-8634(88)90147-3

Evers T, Millar S (2002) Cereal grain structure and development: Some implications for quality. *Journal of Cereal Science*, 36, 261–284. https://doi.org/10.1006/jcrs.2002.0435

FAOSTAT (2017) Database of food and agricultural organization. www.fao.org/faostat/en/#search/Cereals

Garnayak DK, Pradhan RC, Naik SN, Bhatnagar N (2008) Moisture-dependent physical properties of jatropha seed (*Jatropha curcas* L.). *Industrial Crops and Products*, 27, 123–129. https://doi.org/10.1016/j.indcrop.2007.09.001

Granato D, Masson ML (2010) Instrumental color and sensory acceptance of soy-based emulsions: A response surface approach. *Ciencia y tecnología de alimentos*, 30, 1090–1096. https://doi.org/10.1590/s0101-20612010000400039

Hamdani A, Rather SA, Shah A, et al (2014) Physical properties of barley and oats cultivars grown in high altitude Himalayan regions of India. *Journal of Food Measurement and Characterization*, 8, 296–304. https://doi.org/10.1007/s11694-014-9188-1

Hemdane S, Jacobs PJ, Dornez E, et al (2016) Wheat (*Triticum aestivum* L.) bran in bread making: A critical review. *Comprehensive Reviews in Food Science and Food Safety*, 15, 28–42. https://doi.org/10.1111/1541-4337.12176

Irtwange SV (2000) The effect of accession on some physical and engineering properties of African yam bean. Unpubl PhD Thesis, Dep Agric Eng Univ Ibadan, Niger

Jain RK, Bal S (1997) Properties of pearl millet. *Journal of Agricultural Engineering Research*, 66, 85–91. https://doi.org/10.1006/jaer.1996.0119

Januarius OA, Lawrence, OG, Geoffrey, CM, Chepete, HJ (2011) *Rural structures in the tropics. Design and development*. Rome: FAO.

Jayas DS, Singh CB. (2012) *Grain quality evaluation by computer vision*. Sawston: Woodhead Publishing Limited

Jouki M, Emam-Djomeh Z, Khazaei N (2012) Physical properties of whole rye seed (secale cereal). *International Journal of Food Engineering*, 8, 1–14. https://doi.org/10.1515/1556-3758.2054

Kaliniewicz Z, Biedulska J, Jadwisieńczak B (2015) Assessment of cereal seed shape with the use of sphericity factors. *Technology Science*, 18, 237–246

Karimi M, Kheiralipour K, Tabatabaeefar A, et al (2009) The effect of moisture content on physical properties of wheat. *Pakistan Journal of Nutrition*, 8, 90–95. https://doi.org/10.3923/pjn.2009.90.95

Kaur H, Gill BS (2021) Changes in physicochemical, nutritional characteristics and ATR–FTIR molecular interactions of cereal grains during germination. *Journal of Food Science and Technology*, 58, 2313–2324. https://doi.org/10.1007/s13197-020-04742-6

Kaur M, Hüberli D, Bayliss KL (2020) Cold plasma: Exploring a new option for management of postharvest fungal pathogens, mycotoxins and insect pests in Australian stored cereal grain. *Crop & Pasture Science*, 71, 715–724. https://doi.org/10.1071/CP20078

Kingsly ARP, Singh DB, Manikantan MR, Jain RK (2006) Moisture dependent physical properties of dried pomegranate seeds (Anardana). *Journal of Food Engineering*, 75, 492–496. https://doi.org/10.1016/j.jfoodeng.2005.04.033

Krishnan P, Ramakrishnan B, Reddy KR, Reddy VR (2011) *High-temperature effects on rice growth, yield, and grain quality.* 1st ed. London: Elsevier Inc.

Kulp K, Ponte JG (2000) *Handbook of Cereal science and technology,* Revised and Expanded. London: Elsevier Inc

Manjunatha T, Bisht IS, Bhat KV, Singh BP (2007) Genetic diversity in barley (*Hordeum vulgare* L. ssp. vulgare) landraces from Uttaranchal Himalaya of India. *Genetic Resources and Crop Evolution*, 54, 55–65. https://doi.org/10.1007/s10722-005-1884-6

Mannaa M, Kim KD (2017) Influence of temperature and water activity on deleterious fungi and mycotoxin production during grain storage. *Mycobiology*, 45, 240–254. https://doi.org/10.5941/MYCO.2017.45.4.240

Markowski M, zuk-Gołaszewska K, Kwiatkowski D (2013) Influence of variety on selected physical and mechanical properties of wheat. *Industrial Crops and Products*, 47, 113–117. https://doi.org/10.1016/j.indcrop.2013.02.024

Mason LJ, McDonough M (2012) *Biology, behavior, and ecology of stored grain and legume insects.* New York: Springer

Mesterházy Á, Oláh J, Popp J (2020) Losses in the grain supply chain: Causes and solutions. *Sustain*, 12, 1–18. https://doi.org/10.3390/su12062342

Mir SA, Bosco SJD, Sunooj K V. (2013) Evaluation of physical properties of rice cultivars grown in the temperate region of India. *International Food Research Journal*, 20, 1521–1527

Mwithiga G, Sifuna MM (2006) Effect of moisture content on the physical properties of three varieties of sorghum seeds. *Journal of Food Engineering*, 75, 480–486. https://doi.org/10.1016/j.jfoodeng.2005.04.053

Nalladurai K, Alagusundaram K, Gayathri P (2002) Airflow resistance of paddy and its byproducts. *Biosystems Engineering*, 83, 67–75. https://doi.org/10.1006/bioe.2002.0091

Pathare PB, Opara UL, Al-Said FAJ (2013) Colour measurement and analysis in fresh and processed foods: A review. *Food and Bioprocess Technology*, 6, 36–60. https://doi.org/10.1007/s11947-012-0867-9

Paulsen MR, Singh M, Singh V (2018) *Measurement and maintenance of corn quality.* 3rd ed. London: Elsevier Inc.

Perdon AA, Holopainen-Mantila U (2020) Cereal grains and other ingredients. In: *Breakfast cereals and how they are made* (pp. 73–96). London: Elsevier Inc.

Ramashia SE, Gwata ET, Meddows-Taylor S, et al (2018) Some physical and functional properties of finger millet (*Eleusine coracana*) obtained in sub-Saharan Africa. *Food Research International*, 104, 110–118. https://doi.org/10.1016/j.foodres.2017.09.065

Ratnavathi CV, Komala VV (2016) Sorghum Grain Quality. In: Ratnavathi CV, Patil JV, Chavan UD (eds), Sorghum biochemistry: An Industrial Perspective. Academic Press, London, (pp. 1–61)

Razavi SMA, Farahmandfar R (2008) Effect of hulling and milling on the physical properties of rice grains. *International Agrophysics*, 22, 353–359

Sahay KM, Singh KK (1996) *Unit Operation Of Agricurtural Processing.* Vikas Publishing House Pvt. Ltd., Noida, UP

Sahoo PK, Srivastava AP (2002) Physical properties of okra seed. *Biosystems Engineering*, 83, 441–448. https://doi.org/10.1006/bioe.2002.0129

Saleh ASM, Zhang Q, Chen J, Shen Q (2013) Millet grains: Nutritional quality, processing, and potential health benefits. *Comprehensive Reviews in Food Science and Food Safety*, 12, 281–295. https://doi.org/10.1111/1541-4337.12012

Serna-Saldivar SO (2012) *Cereal Grains: Laboratory Reference and Procedures Manual.* CRC press, Boca Raton, Florida

Serna-Saldivar SO (2016) *Cereal grains: Properties, processing, and nutritional attributes.* London: CRC press

Shah A, Masoodi FA, Gani A, Ashwar BA (2016) Geometrical, functional, thermal, and structural properties of oat varieties from temperate region of India. *Journal of Food Science and Technology*, 53, 1856–1866. https://doi.org/10.1007/s13197-015-2119-2

Shittu TA, Olaniyi MB, Oyekanmi AA, Okeleye KA (2012) Physical and water absorption characteristics of some improved rice varieties. *Food and Bioprocess Technology*, 5, 298–309. https://doi.org/10.1007/s11947-009-0288-6

Sologubik CA, Campañone LA, Pagano AM, Gely MC (2013) Effect of moisture content on some physical properties of barley. *Industrial Crops and Products*, 43, 762–767. https://doi.org/10.1016/j.indcrop.2012.08.019

Solomon WK, Zewdu AD (2009) Moisture-dependent physical properties of niger (*Guizotia abyssinica* Cass.) seed. *Industrial Crops and Products,* 29, 165–170. https://doi.org/10.1016/j.indcrop.2008.04.018

Sompong R, Siebenhandl-Ehn S, Linsberger-Martin G, Berghofer E (2011) Physicochemical and antioxidative properties of red and black rice varieties from Thailand, China and Sri Lanka. *Food Chemistry,* 124, 132–140. https://doi.org/10.1016/j.foodchem.2010.05.115

Tabatabaeefar A (2003) Moisture-dependent physical properties of wheat. *International Agrophysics,* 17, 207–211

Tarighi J, Mahmoudi A, Alavi N (2011) Some mechanical and physical properties of corn seed (Var. DCC 370). *African Journal of Agricultural Research,* 6, 3691–3699. https://doi.org/10.5897/AJAR10.521

Tavakoli M, Tavakoli H, Rajabipour A, et al (2009) Moisture-dependent physical properties of barley grains. *International Journal of Agricultural and Biological Engineering,* 2, 84–91. https://doi.org/10.3965/j.issn.1934-6344.2009.04.084-091

Tian B, Xie B, Shi J, et al (2010) Physicochemical changes of oat seeds during germination. *Food Chemistry,* 119, 1195–1200. https://doi.org/10.1016/j.foodchem.2009.08.035

USDA-GIPSA (2020) *Grain inspection handbook. Book II. Grain grading procedures.* London: Federal Grain Inspection Service

Varnamkhasti MG, Mobli H, Jafari A, et al (2008) Some physical properties of rough rice (*Oryza sativa* L.) grain. *Journal of Cereal Science,* 47, 496–501. https://doi.org/10.1016/j.jcs.2007.05.014

Vilche C, Gely M, Santalla E (2003) Physical properties of quinoa seeds. *Biosystems Engineering,* 86, 59–65. https://doi.org/10.1016/S1537-5110(03)00114-4

Warechowska M., Warechowski J., Markowska A. (2013) Interrelations between selected physical and technological properties of wheat grain. *Technology Science,* 16, 281–290

Wrigley CW, Batey IL (2012) *Assessing grain quality.* 2nd ed. London: Woodhead Publishing Limited

Zareiforoush H, Komarizadeh MH, Alizadeh MR (2009) Effect of moisture content on some physical properties of paddy grains. *Research Journal of Applied Sciences, Engineering* 1:132–139

Zareiforoush H, Komarizadeh MH, Alizadeh MR (2010) Effects of crop-machine variables on paddy grain damage during handling with an inclined screw auger. *Biosystems Engineering,* 106, 234–242. https://doi.org/10.1016/j.biosystemseng.2010.02.008

CHAPTER 4

Chemical Composition of Cereal Grains

Asif Ahmad, Chamman Liaqat, Gulzar Ahmad Nayik and Umar Farooq

CONTENTS

4.1 INTRODUCTION

Cereal grains are an important source of food for human beings as well as a feed source for animals. These are considered the main agricultural products and staple foods all over the world. The milling process converts these grains into respective flours that are commonly used for the production of bakery items like cookies, cakes, pastries, and bread. Some other products that can be generated

DOI: 10.1201/9781003252023-4

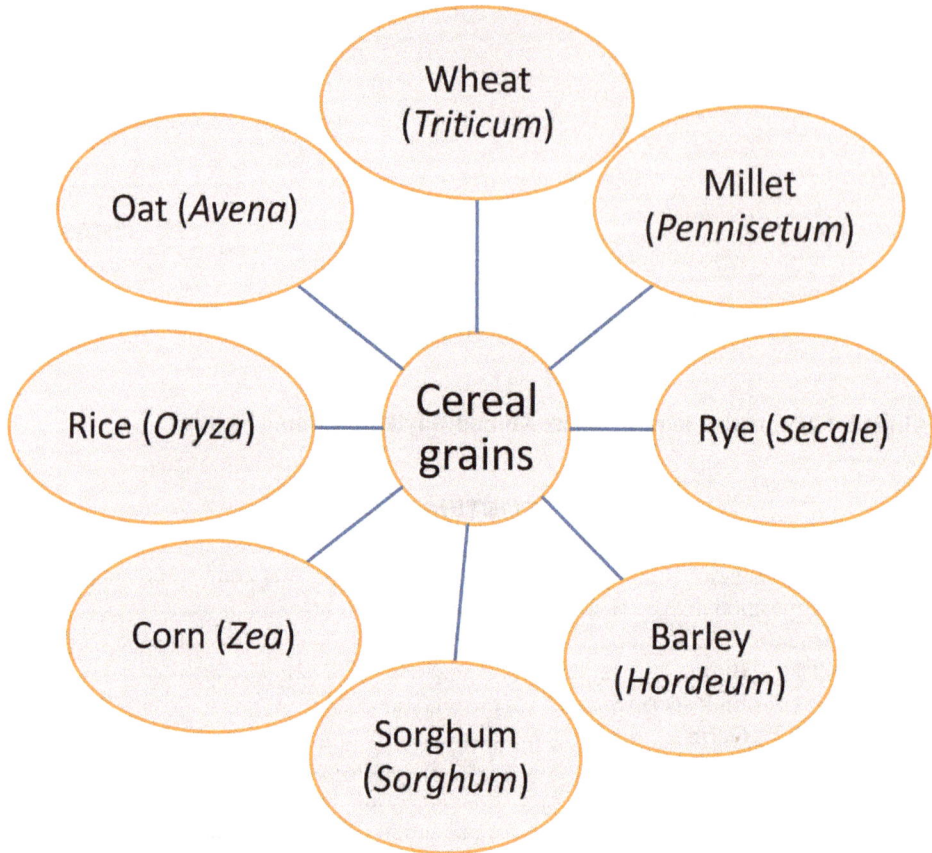

Figure 4.1 Commonly cultivated cereal grains.

from these grains flours are spaghetti, noodles, baby food, and confectionery products, and they are used as staple food products in various regions over the globe (Zareef et al., 2021). Among all the food groups, cereal grains are regularly consumed, and 60% of agricultural land is used for their cultivation. Moreover, they provide almost 50% of calories and proteins for the population all over the world (Los et al., 2018; Yu & Tian, 2018). By 2050, a nearly 70–100% increase in the production of cereal grains is required for the predicted population of 9.8 billion (Wang et al., 2018).

The term "cereals" represents the *Gramineae* family, and the main species are shown in Figure 4.1.

Cereal grains from different crops vary in chemical composition, shape, structure, and size. Structurally, all cereal grains mainly consist of three parts: bran, endosperm, and germ. Bran is the outer covering and constitutes 10–14% of the grain part. It is rich in fiber, bioactive substances, and minerals that impart numerous health benefits. Next to bran is the endosperm that makes up 80–85% of the grain and has several nutrients embedded in it. It possesses a body of endosperm cells that are covered by aleurone cells along with transfer cells and embryo-surrounding cells near the embryo. The third part is the germ (2.5–3%) that grows into a new plant in suitable conditions (Klerks et al., 2019; Olsen, 2020). Chemically, these consist of different nutrients like carbohydrates, proteins, lipids, dietary fiber, sugars, vitamins, and minerals. In all cereal grains, the major source of energy resides in carbohydrates, mainly starches, followed by proteins. Lipids also exist in appreciable amounts and contribute to organoleptic properties in addition to the provision of calories. Some cereals, like barley and oats, are rich in dietary fiber that nutritionally represents an undigested fraction and provides multiple health benefits for humans. Furthermore, sugars, minerals, and vitamins are also present in low quantities but fulfill the essential nutrient requirements in consumers (Tacer-Caba et al., 2015).

4.2 CEREAL COMPOSITION

All cereal grains, like wheat, rice, corn, sorghum, oat, and millet, have nearly the same basic structure but different nutritional composition. Structurally, all cereal grains are single-seeded fruits that are known as caryopses. A caryopsis consists of pericarp and testa that cover the endosperm and embryo (both represent the main storage tissues). The endosperm consists of two modified cell types: the aleurone cells that make up the outer layer and the central endosperm cells composed of gluten proteins and starch (Tosi et al., 2018). For example, wheat germ (2–3%) consists of the scutellum and embryo, and the aleurone layer separates them from the endosperm. The endosperm is mainly composed of proteins and starch and makes up 80–85% of the wheat kernel. Radicle and plumule are present at the embryonic axis, while wheat bran (13–17%) is composed of an outer covering that consists of three parts: pericarp, aleurone layers, and testa. During milling, bran is separated from the other parts of the grain (Boukid et al., 2018; Merali et al., 2015). Similarly, the grain of rice is harvested as the completely covered grain, in which a siliceous husk encloses brown rice (caryopsis). The husk is mainly composed of two distinct leaves called the larger lemma and palea that are composed of brittle and lignified cells. This portion contributes 20% of the average rice weight, with values ranging from 16–28% (Juliano & Tuaño, 2019), while the caryopsis is the single-seeded fruit in which the pericarp is bound to the seed (Figure 4.2).

The nutritional composition of all cereal grains differs. All cereal grains chemically comprise carbohydrates, proteins, lipids, dietary fiber, minerals, vitamins, and bioactive substances, but their percentages in all cereals vary accordingly. Carbohydrates are the important nutritional

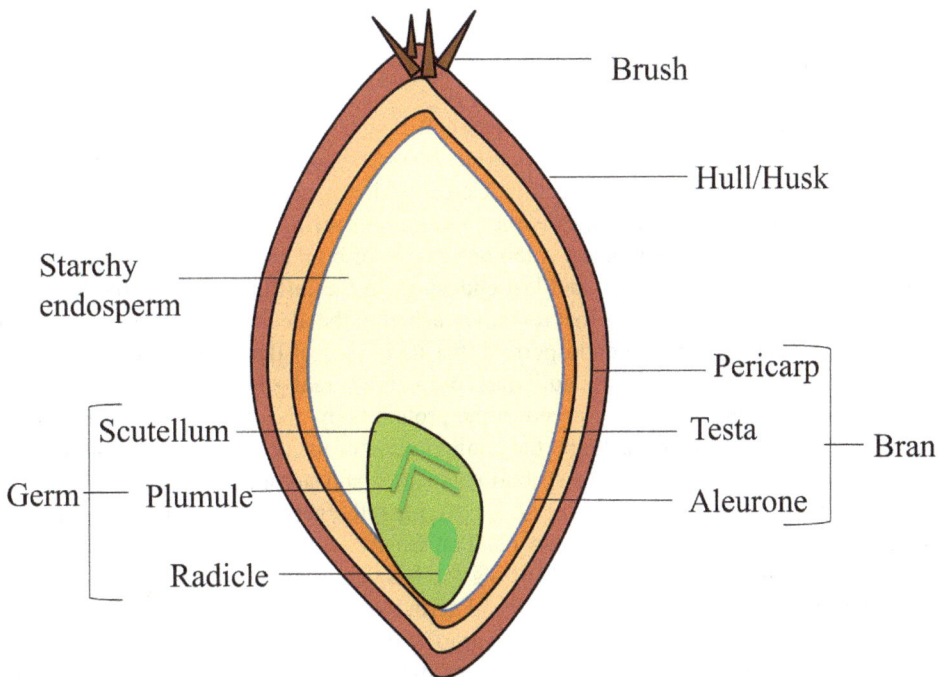

General Structure of Cereal Grain

Figure 4.2 General structure of cereal grain.

component required by the body and composed of monosaccharides like fructose and glucose; disaccharides like sucrose; oligosaccharides like raffinose; polysaccharides like fructans, cellulose, and starch; pentosans; and sugar alcohols like sorbitol, mannitol, and inositol (Quentin et al., 2015). Proteins are also considered a very important nutritional component. Amino acids are the basic unit of cereal proteins and play a significant role in the physiological functions of the human body (Hou et al., 2015). Similarly, cereal lipids that usually reside in the germ also play a great physiological role in the body. They are required by the body for different biological functions, and their deficiency may cause different diseases. Fatty acids are the main constructing units of lipids (De Carvalho & Caramujo, 2018). The presence of vitamins and minerals, like vitamins A, B, C, and E, from cereal sources plays a vital role in the human body. Similarly, cereal minerals like iron, magnesium, potassium, zinc, copper, sodium, manganese, chloride, fluoride, and chromium are also required by the body for different functions. Although these minerals and vitamins are required in small quantities, they play a significant role in the body (Gharibzahedi & Jafari, 2017; Maqbool et al., 2018). The chemical composition of cereal grains varies considerably, depends upon varieties and environmental factors, and has been reported by various researchers, but on average, wheat is composed of 78.1% carbohydrates, 14.7% protein, 2.1% minerals, 2.1% fat, and vitamins like B and thiamine (Ocheme et al., 2018). Moreover, the typical composition of corn based on the dry matter is 9.5% protein, 2.6% sugar, 71.7% starch, 4.3% oil, and 1.4% ash content (Ignjatovic-Micic et al., 2015). Similarly, rice contains almost 80% carbohydrates, 3% fat, approximately 7–8% protein, and 3% fiber content (Chaudhari et al., 2018). Moreover, sorghum comprises carbohydrates (73%), water (11%), protein (11.6%), energy (340 K/Cal), and fat (3%) by weight (Fano, 2017). Furthermore, oat cereal comprises carbohydrates (72.1%), protein (9.9%), moisture (11.7%), lipids (4.6%), ash (1.6%), crude fiber (14.6%), and starch (40.8%) (Anthero et al., 2019). Among the nondigestible carbohydrates, β-glucan (with 1→3, 1→4 glycosidic linkages) exists as a major dietary fiber of oat with characteristic health benefits. Health professionals recommend this dietary fiber for diabetic patients, weight watchers, and hypercholesterolemic patients (Ahmad et al., 2009; Ahmad & Kaleem, 2018).

4.2.1 Structure

All cereal grains, like corn (*Zea mays*), wheat (*Triticum*), rice (*Oryza sativa*), oat (*Avena sativa*), millet (*Pennisetum glaucum*), and sorghum (*Sorghum bicolor*) (Ahmad & Anjum, 2010; Papageorgiou & Skendi, 2018), have almost the same structure. Cereal grains structurally have three main components, bran, germ, and endosperm. Bran is the outermost covering of grain that covers the grain endosperm and germ part. The pericarp is the outer layer of bran and is composed of three structural layers: epicarp, then mesocarp, and last endocarp. In the milling industry, the term bran is collectively used for the tissue layers that remain attached to the aleurone after the milling process (Deroover et al., 2020). Then, the endosperm is the main part of the cereal grain and stores the nutritional components. It occupies the bulk mass of mature cereal grain. It mainly consists of protein and starch, which are separately stored in the protein body and starch granule. The quantity, shape, and size of the endosperm cell and the configuration of the protein body and starch granule affect the weight of cereal grains. The distribution of starch and storage proteins greatly vary in the different parts of the endosperm (Zhao et al., 2016). Last, the germ is the most important part of any cereal grain because it grows into a new plant on the availability of suitable conditions like air, water, and sunlight. It is the by-product of milling and has gained attention due to its functional and nutritional advantages in the food industry. Cereal germ has a high amount of polyunsaturated fatty acids (43.9–64.9%). It is also a great source of amino acids (24 to 36 g/100 g), vitamins (56 to 2700 mg/kg), unsaturated fatty acids (43–64 g/100 g), and dietary fiber (1 to 15 g/100 g). It also contains phytochemicals like sterols (1–6 g/100 g) and flavonoids (0.35 rutin equivalent/100 g) (Wang et al., 2021). The structural layers and chemical composition are represented in Figure 4.3.

Cereal grains

Bran (10-14%)	Endosperm (80-85%)	Germ (2.5-3%)

	Bran (10-14%)	Endosperm (80-85%)	Germ (2.5-3%)
Structural layers	• Pericarp • Testa • Nucellar epidermis • Aleurone	• Sub-aleurone cells • Central cells • Prismatic cells	• Scutellum • Embryonic axis • Radicle • Plumule
Chemical composition	• Dietary fiber • Minerals • Vitamins • Different bioactive compounds	• Starch • Proteins • Gluten (Gliadin & Glutenin) • Small quantities of vitamins and minerals	• Fatty acids • Minerals • Phytosterols • Amino acids (threonine, lysine, and methionine)

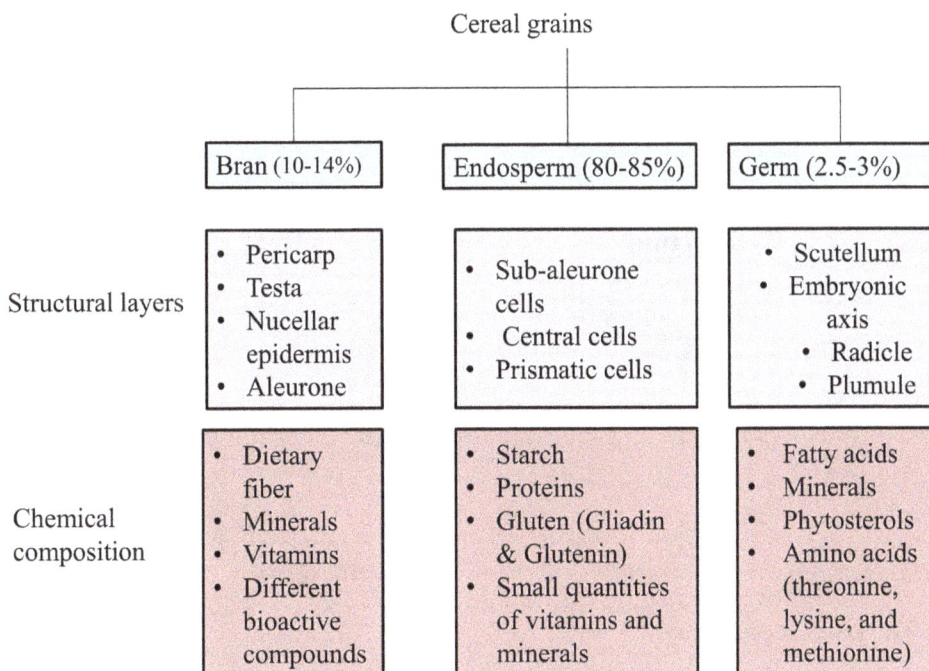

Figure 4.3 Structural layers and chemical composition of cereal grains.

4.2.1.1 Bran

Bran is the outermost part of every cereal grain that is mostly removed in the milling process, but it is a great source of dietary fiber, minerals and vitamins, and different bioactive compounds that can provide beneficial effects to consumers (Coda et al., 2015). The structure of bran for all cereal grains is almost the same, but the chemical composition of bran for each cereal grain is different. Structurally, bran consists of different cellular layers that surround the germ and endosperm part of cereal grain. From the outside of the cereal grain, one observes the pericarp that further consists of the epidermis layer, hypodermis, tubular and cross cells, the testa (seed coat), then the nucellar epidermis, and finally the aleurone layer, which is tightly bound to the seed coat (Deroover et al., 2020). Various layers of cereal bran are depicted in Figure 4.4.

The bran of all cereal grains slightly varies in composition. For example, wheat bran consists of approximately 19% starch, 18% protein, 19% cellulose, and 38% polysaccharides (non-starch) (Merali et al., 2015), while stabilized rice bran consists of protein (17.50 g/100 g), fat (13.10 g/100 g), total dietary fiber (TDF) (23.3 g/100 g), soluble dietary fiber (SDF) (2.1 g/100 g), insoluble dietary fiber (IDF) (21.17 g/100 g), carbohydrate (52.33 g/100 g), calcium (52.10 mg/100 g), zinc (6.02 mg/100 g), iron (28.10 mg/100 g), phosphorus (1185.20 mg/100 g), and energy (398 Kcal/100 g) (Bhosale & Vijayalakshmi, 2015). Moreover, on the basis of granulometric measurement (0.125 mm), corn bran consists of moisture (8.19 g/100 g), lipids (12.81 g/100 g), proteins (10.33 g/100 g), carbohydrates (56.97 g/100 g), ash (3.11 g/100 g), total dietary fiber (8.60 g/100 g), and energy (418.91 Kcal/100 g) (de Sousa et al., 2019). Similarly, extract of oat bran contains carbohydrate (39.45%), including glucose (0.61%), fructose (0.76%), lactose (<0.1%), sucrose (38.08%), maltose (<0.1%), moisture (9.8%), proteins (13.1%), minerals (3.2%), saponins (4.6%), β-glucans (3.1%), total pentosans (3.3%), crude fibers (1.5%), starch (0.3%), total dietary fibers (1.1%), and lipids (<1.7%). The mineral content

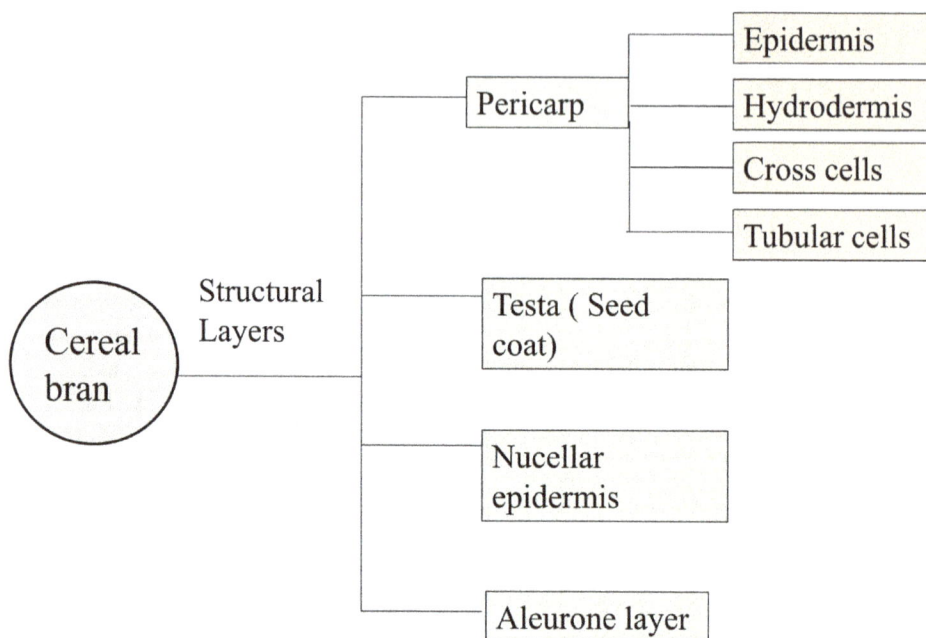

Figure 4.4 Layers of cereal bran.

includes the following minerals: sodium (0.020%), potassium (2.142%), magnesium (0.106%), calcium (0.029%), iron (0.001%), and manganese (0.002%) (Ralla et al., 2018).

4.2.1.2 Endosperm

The endosperm is the main part of the cereal grain. Structurally, all cereal grains have almost the same structure of endosperm but have different chemical compositions. The endosperm of mature grain consists of three different cell types, called sub-aleurone cells, central cells, and prismatic cells. All the cell types vary in their nutritional components: gluten protein, polysaccharides, starch, and lipids (Shewry et al., 2020). Mainly, the starchy endosperm consists of two regions: The first one is the sub-aleurone layer; it consists of the outermost cells that are located just beneath the aleurone cells. The second region is called the central region and makes up the rest of the endosperm (starch). The starchy endosperm comprises parenchyma cells that are elongated on the cross-sectional side and filled with starch granules (Juliano & Tuaño, 2019). The endosperm is mainly composed of proteins, starch, and small quantities of minerals and vitamins. For example, the endosperm of wheat has two different starch granule types. Type-A starch granules are >10 μm in diameter, lenticular in shape, constitute up to 10% of total starch, and hold 70% of the volume in the endosperm, while type-B starch granules are spherical and <10 μm in diameter (Orenday-Ortiz, 2018). Similarly, the endosperm of wheat comprises 8–15% proteins, of which 80% are composed of gluten. Gluten mainly consists of insoluble glutenins and soluble gliadins. Gliadin is composed of monomeric heterogeneous proteins, while glutenin is composed of high and low glutenin subunits that are connected by disulfide bonds (Landriscina et al., 2017). Rice mainly consists of an aleurone layer (outer part) and a starchy endosperm (inner part). It has been observed that endosperm and aleurone are distinct from each other in five di-ammonium phosphate (DAP). In mature seeds, all the cells that are present in the endosperm are dead, while cells present in the aleurone remain alive. The aleurone along the embryo makes up 7% weight of the grain (caryopsis), but these tissues contain approximately 80% of lipids, 80% of proteins, 60% of iron, and 90% of vitamin (B1), while the starchy endosperm, as the main energy source of rice, mainly comprises starch along with small quantities of proteins (Wu et al., 2016).

4.2.1.3 Germ

Germ is the main part of any cereal grain because it leads to the development of new plants under desired environmental conditions like the availability of light, water, and air. It can be separated during milling as a by-product. It is a miraculous by-product and has different applications like biological and pharmaceutical purposes and also as food. It is also used in snack foods, enriched germ bread, and in breakfast cereals for enrichment purposes (Ghafoor et al., 2017; Mahmoud et al., 2015). Naturally, there is a distinct line between the germ and endosperm that makes the separation of germ from grain easy. In the germ structure, two distinct parts are vital: the scutellum (storage organ) and the embryonic axis (Ghafoor et al., 2017). In addition to this, the embryonic axis is further divided into radicle (embryonic root) and plumule (embryonic leaves) that are joined to each other by a mesocotyl (short stem) (Juliano & Tuaño, 2019).

Wheat germ can be characterized by presence of fatty acids; phytosterols; minerals; and amino acids such as threonine, lysine, methionine, and tocopherols. On a percentage basis, this part provides 15 times more sugar, 3 times more proteins, and 6 times more mineral content (Koç & Özçıra, 2019) as compared to normal wheat flour. It approximately contains proteins (26–36%), lipids (10–15%), fiber (1.5–4.5%), and minerals (4%), as well as a fair amount of bioactive substances, like 4–18 mg/kg carotenoids, 300–740 mg/kg DM tocopherols, 10 mg/kg policosanols, 15–23 mg/kg thiamin, 24–50 mg/kg phytosterols and 6–10 mg/kg riboflavin (Youssef, 2015). Similarly, the germ plasm of sorghum contains protein (10.6–18.5%), starch (55.6–75.2%), crude fiber (1.0%–3.4%), sugar (0.8%–4.2%), ether extract (2.1–7.1%), ash (1.6–3.3%), and tannin (0.1–6.4%) (Fano, 2017).

4.2.2 Nutritional Composition

The main nutrients that are present in cereal grains are carbohydrates, proteins, lipids, minerals, vitamins, and dietary fiber. Carbohydrates are a great source of energy for humans and are made up of organic compounds (oxygen, carbon, hydrogen, and oxygen). Cereal carbohydrates are usually categorized in two broad groups: digestible carbohydrates (monosaccharides, starch, and polysaccharides) and non-digestible carbohydrates (complex oligosaccharides, non-starch polysaccharides, and resistant starches that are collectively called dietary fiber) (Navarro et al., 2019). In addition, proteins also play a significant role in energy provision, development, and growth in human beings. Other diversified functions of cereal protein are reported in the scientific literature such as transport of different nutrients in the human body, enzymatic activity, transport of other biochemical substances across membranes, and maintenance of buffering fluids. To achieve all of these functions, a fair amount of cereal protein consumption is required on daily basis. An inadequacy may cause depletion of the essential amino acids pool that may result in loss of muscular mass, increase the turnover of muscle proteins, and also stunt the growth process (Mæhre et al., 2018). Each amino acid in cereal protein serves as backbone segment with a carboxylic and amino group and a residue with functional and chemical groups (Grimaldo et al., 2019). Another prominent nutrient that is present in cereal grains is lipid. Lipids are hydrophobic small molecules that easily dissolve in organic solvents such as alcohols, chloroform, and ethers. They play a major role in the storage of energy. Moreover, lipids are divided into eight different classes, glycerolipids, sterols, fatty acyls, prenols, glycerophospholipids, polyketides, sphingolipids, and saccharolipids (Nakamura & Li-Beisson, 2016; Xicoy et al., 2019). A brief classification of cereal grain lipids is shown in Figure 4.5.

Furthermore, minerals and vitamins in cereal grains exist in minor amounts but play a significant role in the human body. Both play a role as a co-factor for different enzymes and a major role in maintaining osmotic pressure. The commonly found minerals in cereal grains are zinc, calcium, magnesium, and iron, while vitamins, commonly B and E, are present (Fernandesa et al., 2021). Dietary fiber is another important nutrient that exists in cereal grains; it has gained much attention in recent years due to its miraculous health effects on the human body. Dietary fibers refer to edible polymers of carbohydrates with at least three monomeric units that remain undigested in the small

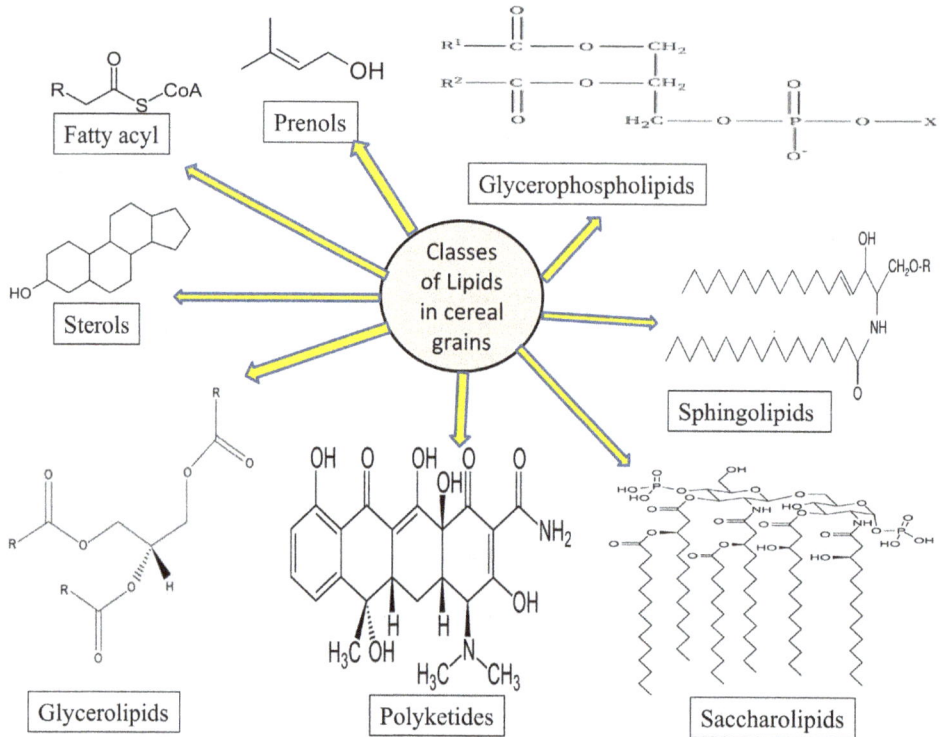

Figure 4.5 Classification of lipids present in cereal grains in different classes.

intestine because they are resistant to digestive enzymes. These are classified into polysaccharides (resistant starch, non-starch polysaccharides, and resistant oligosaccharides) or insoluble (cellulose and hemicellulose) and soluble (β-glucans) forms (Makki et al., 2018). A study showed that cereal products contributed 30.4% to whole dietary energy. Results showed that cereal and cereal products provided a specific percentage of six nutrients to the average Polish diet, carbohydrates (51%), iron (34.1%), manganese (64.1%), copper (31.3%), folate (33.6%), and dietary fiber (48.5%). A supply at the level of 10–20% was observed for calcium, poly-unsaturated fatty acids, vitamin B6, potassium, riboflavin, sodium, and niacin, and 20–30% for thiamin, zinc, phosphorous, and protein (Laskowski et al., 2019). The following sections elaborate on detailed research on all the major cereal nutrients that may be present in cereal grains.

4.2.2.1 Carbohydrates

Carbohydrates are a complex and diverse group of substances with a great range of physiological, chemical, and physical properties. Carbohydrates may exist as monosaccharides, oligosaccharides and polysaccharides in cereal grains. Molecular size (linkage type and degree of polymerization [DP]) are the basis for this classification (Knudsen et al., 2016). Carbohydrates with high molecular weight include starch, pentosans (arabinoxylans), non-starch polysaccharides (cellulose), β-glucans, and small quantities of oligosaccharides. Quantitatively, starch dominates other carbohydrates and mainly resides in the endosperm portion. Another method of carbohydrate classification demonstrated in research is on the basis of glycemic index (increase of the blood glucose level after digestion). The cell wall materials of cereal grains such as non-starch polysaccharides are non-glycemic in nature, while dextrin, oligosaccharides, and simple sugars are glycemic carbohydrates. Starch in cereal grains is a predominately glycemic carbohydrate. All the cereal grains (wheat, rice, corn, sorghum, oat, millet, rye, and barley) have different amounts and composition of carbohydrates and thus vary in their glycemic index value (Taylor et al., 2015).

On average, wheat contains starch (63–72%), amylose (23.4–27.6%), pentosan (6.6%), beta-glucan (1.4%), and total dietary fiber (14.6%) (Zhang & Hamaker, 2017). A study of different wheat types showed an average amount of 2.6% fiber in hard whole wheat flour, 0.42% fiber in hard white wheat, 2.5% fiber in soft whole wheat, and 0.36% fiber in soft white wheat. In wheat bran, 9.97 g/100 g fiber was present (Heshe et al., 2016). In another study, different wheat varieties showed different chemical compositions for carbohydrates, such as in the Benazir variety (79.7 g/100 g of carbohydrate), TH-83 (78.4 g/100 g), and Imdad (78.4 g/100 g), present based on particle size (<1.18–>0.43 mm) (Memon et al., 2020). For other cereals, brown rice contains carbohydrate (68 g/100 g), dietary fiber (6 g/100 g), and amylose (19.3 g/100 g) (RamyaBai et al., 2019). In another study, the number of carbohydrates in different rice samples was studied. Results showed carbohydrates in un-parboiled rice (43 g/100 g), parboiled rice (28.5 g/100 g), milled un-parboiled rice (46.5 g/100 g), and milled parboiled rice (39.4 g/100 g) (Kalita et al., 2021).

Moreover, the main components of oat starch grain are amylose and amylopectin (carbohydrates) that represent almost 98–99% (Ahmad & Ahmed, 2011; Punia et al., 2020). A study was carried out to check the chemical composition of different oat hull batches. One batch (Swe16) showed different percentages for carbohydrates such as lignin (25%), from which acid-insoluble (23.1%) and acid-soluble (2.3%) fiber, cellulose (17.2%), and hemicellulose (33.1%) can be extracted (Schmitz et al., 2020). Another study revealed the carbohydrate composition of oat, comprising starch (43–61%), amylose (16–27%), pentosan (7.7%), beta-glucan (3.9–6.8%), and total dietary fiber (9.6%) (Zhang & Hamaker, 2017). A study on three different genotypes of sweet corn indicated the average sugar concentration in sweet corn was 5%, while the average starch concentration of sugary corn was 27% lower than that of sugar-enhanced corn and 66% lower than that of shrunken corn (Szymanek, Tanaś & Kassar, 2015). Corn also contains starch (64–78%), pentosan (5.8–6.6%), amylose (24%), and total dietary fiber (13.4%) (Zhang & Hamaker, 2017).

4.2.2.2 Dietary Fiber

Dietary fibers are the carbohydrates that remain undigested in the small intestine and partially fermented in the large intestine. Dietary fibers are mainly composed of oligosaccharides and polysaccharides, like hemicellulose, cellulose, pectin substances, and resistant starches. They play a significant role in different physiological processes in the body and prevent many diseases (Singh et al., 2018; Tejada-Ortigoza et al., 2016). Studies have shown that dietary fiber plays a role in reducing the risk of cardiovascular disease and atherosclerosis in the human body (Ahmad et al., 2012a; Soliman, 2019). It has been studied that people who consume more dietary fiber in their daily routine are less prone to disease. Studies also showed that consumption of dietary fiber can reduce the risk of different diseases like type 2 diabetes, coronary heart disease, and some cancers and also reduce body weight (Ahmad et al., 2012b; Dahl & Stewart, 2015). All cereal grains contain a variable amount of dietary fiber and can be used to develop valuable fiber-based food products like granola bars with nutraceutical characteristics (Ahmad et al., 2017; Asif & Zaheer, 2016). Some other products from these fibrous materials can be used as prebiotics and offer health benefits similar to kefir (Ahmad & Khalid, 2018; Ahmed et al., 2013a; Ahmed et al., 2013b); often these prebiotics support the functioning of lactic acid bacteria in the gut in a synergistic manner to provide health benefits (Ahmed & Ahmad, 2017).

The amounts of crude and dietary fiber may vary in cereal grains, thus the different nature of these components. Rice contains crude fiber (0.2 g/100 g), total dietary fiber (4.11 g/100 g), and soluble fiber (0.92 g/100 g). Similarly, wheat contains crude fiber (0.3 g/100 g), total dietary fiber (12.48 g/100 g), and soluble fiber (2.84 g/100 g), while maize contains crude fiber (2.7 g/100 g), total dietary fiber (11.54 g/100 g), and soluble fiber (1.65 g/100 g) (Singh & Singh, 2015). In addition, a study was carried out to check the dietary fiber content in different millet varieties. Results showed that Ravi (finger millet variety) had total dietary fiber (13.01%), soluble dietary fiber (0.52%), and insoluble dietary fiber (12.49%), while Rawana had total dietary fiber (13.58%), soluble dietary fiber (0.57%), and insoluble dietary fiber (13.04%) (Jayawardana et al., 2019). Moreover, in another study, near- and mid-infrared spectroscopy techniques were used for the determination of dietary fiber in different wheat bran samples. Comparison between both techniques showed that near-infrared spectroscopy showed more accurate values as compared to mid-infrared (Hell et al., 2016). Furthermore, research was carried out to check the dietary fiber content in raw and extruded wheat bran. Raw wheat bran showed total fiber (43.20%) and soluble dietary fiber (9.82%), while extruded wheat bran products showed a higher percentage of TDF (44.16%) and 16.72% SDF (Yan et al., 2015).

4.2.2.3 Proteins

Proteins are necessary for the growth and development of living organisms and are found in almost all living cells. In the human body, dietary proteins break down into their constituent parts (amino acids) for digestion purposes. Proteins that perform the storage function in the body usually contain a higher amount of alkaline amino acids such as histidine, arginine, and lysine. Out of 20 amino acids, nine amino acids are considered essential and required by the body. These essential amino acids are tryptophan, histidine, phenylalanine, leucine, methionine, valine, threonine, isoleucine, and lysine. These amino acids can't be synthesized inside the human body, so they must be part of a balanced and healthy diet (Joye, 2019). A brief structure of essential amino acids is shown in Figure 4.6.

For example, the nutrient quality of wheat grain is determined by the presence of balanced amino acids and the amount of protein. In wheat flour, the most important protein is gluten, as it contains about 75% proteins based on dry weight. Isolated gluten consists of small quantities of different proteins that remain entrapped in the network of proteins. The total grain nitrogen that is contained by gluten proteins is up to 80%. Based on their sequential extraction in different solvents, gluten proteins are divided into two distinct groups: glutenins and gliadins (Rustgi et al., 2019). In wheat flour, both glutenins and

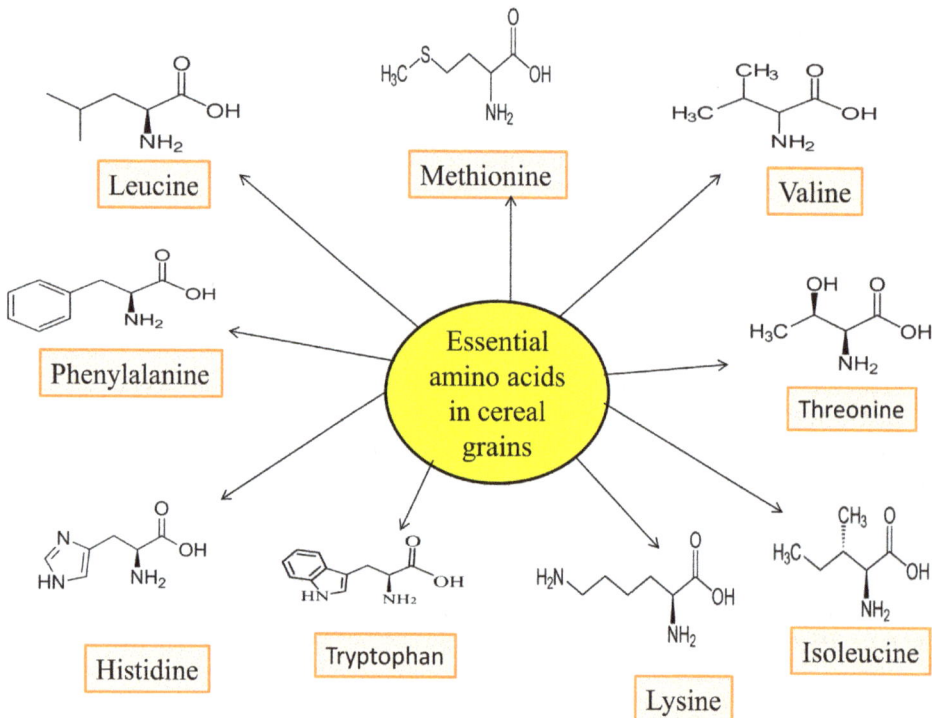

Figure 4.6 Structure of essential amino acids in cereal grains.

gliadins together account for up to 80–85% of the proteins, and they show unique properties in wheat dough like elasticity and extensibility. Glutenin is further divided into high and low molecular subunits, while gliadins are divided into four distinct groups (alpha, beta, gamma, and omega gliadins). Alpha (α) gliadins have the fastest mobility, while omega (ω) gliadins have the slowest (Barak et al., 2015). A study on different wheat varieties to study the amount of protein at different particle sizes was carried out. The results showed that proteins in different wheat varieties at particle size <1.18–>0.43 were as follows: Benazir (10.9%), TJ-83 (11.8%), and Imdad (11.4%) (Memon et al., 2020).

Moreover, the main proteins found in rice grains are glutelin, albumin, globulin, and prolamin. In milled and brown rice, the most prominent proteins are glutelin, albumin, and globulin, while prolamin is present in all rice fractions in small quantities. The amount of protein content in brown rice is 7.1–8.3%, in milled rice 6.3–7.1%, and bran 11–15%, which is comparable to red kidney beans (Hayat et al., 2014) and Ajwa dates (Khalid et al., 2016; Khalid et al., 2017). The distribution (%) of different proteins in rice is as follows: albumin (brown rice [5–10%], milled rice [4–6%], and rice bran [24–43%]), globulin (brown rice [7–17%], milled rice [6–13%], and rice bran [13–36%]), glutelin (brown rice [75–81%], milled rice [79–83%], and rice bran [22–45%]), and prolamin (brown rice [3–6%], milled rice [2–7%], and rice bran [1–5%]) (Amagliani et al., 2017). In another study, an analysis of rice was performed to check its nutritional profile. Results showed the number of proteins as follows: rice flour (7.32%), rice bran (14.7%), rice protein concentrate 1 (75%), rice protein concentrate 2 (75.5%), rice protein concentrate 3 (78.2%), rice endosperm protein hydrolysate 1 (75.6%), rice endosperm protein hydrolysate 2 (70.5%), rice bran protein hydrolysate 1 (34.6%), and rice bran protein hydrolysate 2 (32%) (Amagliani et al., 2017). Furthermore, oat groat (the kernel after the removal of the husk) contains almost 15–20% proteins. Protein content in oat increases from the inside of grain to the periphery, so that the starchy endosperm of oat contains 12% protein content, while the oat bran layer contains 18–26% protein content, and last, the germ contains a higher amount of protein, 29–38%. The main storage proteins that are present in oat are globulins (salt-soluble). The thermal denaturation temperature for globulin is 110°C. In oat protein, the most abundant amino acid is glutamine-glutamic acid. It contains almost 25% complete amino acid residue, followed by leucine (7.4%) and aspartic acid (8.9%) (Ibrahim et al., 2020; Mäkinen et al., 2017).

4.2.2.4 Lipids

Lipids are the main nutritional component of cereal grains. They are a great source of energy and play an important role in the signaling of hormones and different molecules in the body. The major groups that are present in plants are waxes, glycerols, sterols, terpenes, tools, phospholipids, free fatty acids, carotenes, phenolics, chlorophyll pigments, and hydrocarbons. The basic unit of lipids is fatty acids, and these differ in cereal plant lipids (Mumtaz et al., 2020). In plants, the main unsaturated fatty acids are linoleic acid, oleic acid, and α-linolenic acid. In the higher plants, these fatty acids represent 70% of the total fatty acids. The fatty acid profile of several cereal grains matches well with chia seed, flax, and coconut; this could be used to develop new food products by using a suitable combination of treatments (Hafeez et al., 2019) α-linolenic acid and linoleic acid consist of 50% of the total fatty acids, while oleic acid represents 10% of total fatty acids (He et al., 2020; Meï et al., 2015). Plant seeds are the main source of lipids, and fresh pericarp also contains lipids. In oil seeds, almost 95–98% triacylglycerols are present in the form of esters with glycerol. In addition, polar lipids like glycolipids, phosphatides, and glycolipids are also present (Cassim et al., 2019).

For example, wheat germ is the main source of wheat germ oil. Wheat germ contains total fat (9.4–9.7 g/100 g), fatty acids (saturated) (1.4 g/100 g), fatty acids (mono-unsaturated) (1.1 g/100 g), and fatty acids (poly-unsaturated) (4.3 g/100 g) (Boukid et al., 2018). It contains almost 8–14% oil and is used in different industries like food, cosmetics, and medical as an oil source. The fatty acid composition of wheat germ in another study was found to be palmitic acid (17.42%), myristic acid (0.13%), pentadecanoic acid (0.16%), palmitoleic acid (0.23%), and oleic acid (0.29%), with total saturated fatty acids (19.07%), total un-saturated fatty acids (80.93%), total mono-unsaturated fatty

Table 4.1   Composition of Fatty Acids in Different Cereal Grains

Cereals	Palmitic Acid (%)	Oleic Acid (%)	Linoleic Acid (%)	Alpha-Linolenic Acid (%)	Stearic Acid (%)	References
Wheat						
Wheat (bran)	18.9	16.4	58.6	5.5	0.4	(Kumar & Krishna, 2015)
Wheat (germ)	17.42	0.29	–	–	–	(Mahmoud et al., 2015)
Oat						
Oat (n-hexane extraction)	22.45	38.21	35.10	1.39	1.72	(Li et al., 2021)
Oat (Sub-critical extraction)	21.94	40.36	32.60	1.40	1.90	(Li et al., 2021)
Red sorghum oil	15.62	36.67	43.75	–	2.94	(Zhang et al., 2019)
Millet						
Pearl millet oil	16.79	27.07	47.50	2.15	5.02	(Slama et al., 2020)
Finger millet	18.17	22.19	48.05	–	1.20	(Nassarawa & Ahmad, 2019)
Fonio millet	18.20	22.22	46.33	–	1.13	(Nassarawa & Ahmad, 2019)
Pearl millet	18.34	22.26	45.57	–	1.27	(Nassarawa & Ahmad, 2019)
Rice bran oil	15.39	52.56	29.90	–	1.43	(Pinto et al., 2021)

acids (17.22%), and total poly-unsaturated fatty acids (63.71%) (Mahmoud et al., 2015). Similarly, in another study, the lipid composition of wheat bran oil was calculated. Results showed that wheat bran was composed of fat (3.35%) on a wet basis, while wheat bran oil was composed of free fatty acids (7.90%), palmitic acid (18.9%), stearic acid (0.4%), oleic acid (16.4%), linoleic fatty acid (58.6%), and alpha-linoleic acid (5.5%) (Kumar & Krishna, 2015).

Moreover, oat grains are also an incredible source of lipids. They also contain a high number of lipids as compared to other cereal grains. These lipids are a good source of energy and mostly made up of unsaturated fatty acids. The endosperm is a rich source of lipids in oat. In oat, the fat content ranges from 5–9% of total lipids (Rasane et al., 2015). In another study, the lipids percentage of oat cereal was calculated. Results showed that 4.6% lipid was present in oat cereal (Anthero et al., 2019). Furthermore, in another study, two different extraction methods were used for oat lipids. The fatty acid composition from both methods was calculated. Results showed that in conventional extraction (n-hexane), the fatty acid composition was as follows: palmitic acid (22.45%), stearic acid (1.72%), oleic acid (38.21%), linoleic acid (35.10%) and alpha-linolenic acid (1.39%), while in sub-critical extraction (dimethyl ether) the fatty acid composition was as follows: palmitic acid (21.94%), stearic acid (1.90%), oleic acid (40.36%), linoleic acid (32.60%) and alpha-linolenic acid (1.40%) (Zareef et al., 2021). In addition, in another study, the composition of fatty acid in pearl millet oil was studied. Results showed the fatty acid composition of millet oil was as follows: palmitic acid (16.79%), stearic acid (5.02%), oleic acid (27.07%), linoleic acid (47.50%), and linolenic acid (2.15%) (Slama et al., 2020). Moreover, a study was carried out to check the fatty acid composition of red sorghum oil. Results showed the fatty acid composition of sorghum oil was as follows: palmitic acid (15.62%), stearic acid (2.94%), oleic acid (36.67%), and linoleic acid (43.75%) (Zhang et al., 2019). Furthermore, in another study, the fatty acid composition of bran oil from different rice varieties was studied. Results showed the fatty acid composition in *J. ariete* (a rice variety) was: myristic acid (0.06%), palmitic acid (15.39%), stearic acid (1.43%), oleic acid (52.56%), linoleic acid (29.90%), arachidic acid (0.24%), gadoleic acid (0.11%), behenic acid (0.06%), and lignoceric acid (0.11%) (Pinto et al., 2021). The fatty acid composition of different cereal grains is represented in Table 4.1.

4.2.2.5 Minerals

Cereal grains are a major source of minerals in the daily human diet. Although they are present in very small quantities, they have a significant role in the body. Their deficiency can lead to severe diseases. Even in Europe, 19% of non-pregnant women and 21.7% of pre-school children are anemic due to iron deficiency, while older (5.6%) and middle-aged (4.8%) people are zinc deficient. So, daily mineral uptake is necessary, and the use of cereal grains in the daily diet is necessary (Henry et al., 2016; Kan, 2015). The main minerals that are present in cereal grains are iron, calcium, zinc, and magnesium, as well as manganese, copper, cadmium, nickel, chromium, and lead (Akinyele & Shokunbi, 2015; Ogunremi et al., 2020).

For example, a study was conducted to check the mineral contents in wheat bran, milled sorghum, and rice bran. Results showed the percentages of minerals as follows: wheat bran (iron [0.159 g/100 g] and zinc [0.0775 g/100 g]), rice bran (iron [0.169 g/100 g] and zinc [0.0477 g/100 g]), and milled sorghum (iron [0.0403 g/100 g] and zinc [0.0235 g/100 g]) (Nielsen & Meyer, 2016). In another study, mineral concentrations of different cereal grains were determined. Results showed the quantities of minerals as follows: rice (Mn [1.67 mgkg⁻¹], Fe [7.67 mgkg⁻¹], Cu [1.59 mgkg⁻¹], Zn [8.20 mgkg⁻¹], Cr [0.18 mgkg⁻¹], Cd [<0.01 mgkg⁻¹], Pb [<0.08 mgkg⁻¹], and Ni [0.12 mgkg⁻¹]), yellow maize [Mn [5.42 mgkg⁻¹], Fe [14.83 mgkg⁻¹], Cu [2.39 mgkg⁻¹], Zn [24.51 mgkg⁻¹], Cr [0.02 mgkg⁻¹], Cd [<0.01 mgkg⁻¹], Pb [<0.08 mgkg⁻¹], and Ni [0.08 mgkg⁻¹]), and wheat [Mn [32.00 mgkg⁻¹], Fe [30.75 mgkg⁻¹], Cu [3.64 mgkg⁻¹], Zn [22.74 mgkg⁻¹], Cr [0.04 mgkg⁻¹], Cd [<0.01 mgkg⁻¹], Pb [<0.08 mgkg⁻¹], and Ni [0.06 mgkg⁻¹]) (Akinyele & Shokunbi, 2015).

Moreover, in another study, the mineral content of different cereals like barley, corn, and oat was calculated. The quantities of calcium, iron, and zinc were calculated in cereal grains. The results showed the quantities of minerals as follows: barley (iron [8.13 mg/100 g], zinc [1.16 mg/100 g], and calcium [23.2 mg/100 g]), yellow corn (iron [3.18 mg/100 g], zinc [3.72 mg/100 g], and calcium [5.00 mg/100 g]), black cornmeal (iron [9.01 mg/100 g], zinc [5.19 mg/100 g], and calcium [54.7 mg/100 g]), and oat (iron [4.55 mg/100 g], zinc [3.17 mg/100 g], and calcium [40.6 mg/100 g]) (Castro-Alba et al., 2019). Furthermore, a study was carried out to check the minerals in different oat genotypes. Results showed the minerals in genotype TL6 as follows: phosphorus (3391 mg/kg), calcium (1232 mgkg⁻¹), iron (40.2 mgkg⁻¹), zinc (24.31 mgkg⁻¹), manganese (38.62 mgkg⁻¹), and copper (1.77 mgkg⁻¹). Genotype TL7 showed mineral contents as follows: phosphorus (3127 mg/kg), calcium (1098 mg/kg), iron (29.98 mg/kg), zinc (21.75 mg/kg), manganese (34.94 mg/kg), and copper (2.20 mg/kg) (Özcan et al., 2017). The mineral content in different cereal grains is also represented in Table 4.2.

4.2.2.6 *Vitamins*

Vitamins are essential for growth and development. They are required in small quantities but have a major role in body function. Their deficiency can cause a problem in biological pathways in the body (Maqbool et al., 2018). Cereal grains are a staple food and an amazing source of vitamins (Garg et al., 2021). Whole cereal grains are a great source of vitamins A, thiamine, riboflavin, niacin, pantothenic acid, pyridoxine, folic acid, vitamin E, and vitamin K, but they are not a good source of vitamin B_{12} or vitamins C and D (Figure 4.7).

For example, a study was carried out on cereal grains to check the quantities of vitamins. Results showed the quantities of vitamins as follows: vitamin B_1 (0.97 mg/100 g), vitamin B_2 (1.2 mg/100 g), vitamin B_3 (13 mg/100 g), vitamin B_6 (1.2 mg/100 g), vitamin B_9 (334 µg/100 g), and vitamin B_{12} (2.1 µg/100 g) (Radu et al., 2016). In another study, the quantities of riboflavin and thiamine in different cereal grains were calculated. Results showed the quantities of different cereals as follows: winter wheat (thiamine [2.31 mg/kg] and riboflavin [0.74 mg/kg]), spring wheat (thiamine [2.22 mg/kg] and riboflavin [0.85 mg/kg]), winter rye (thiamine [1.76 mg/kg] and riboflavin [1.06 mg/kg]), winter triticale (thiamine [1.83 mg/kg] and riboflavin [0.91 mg/kg]), winter barley (thiamine [2.21 mg/kg] and riboflavin [0.80 mg/kg]), and oats (thiamine [2.71 mg/kg] and riboflavin [1.00 mg/kg]) (Witten & Aulrich, 2018). Moreover, in another study, vitamin quantities in sweet white corn were studied as follows: total ascorbic acid (6.8 mg/100 g), thiamin (0.2 mg/100 g), riboflavin (0.06 mg/100 g), niacin (1.7 mg/100 g), vitamin B6 (0.055 mg/100 g), folate (46 mg/100 g), vitamin E (0.07 mg/100 g), and vitamin K (0.3 mg/100 g) (Siyuan et al., 2018).

Table 4.2 Mineral Contents in Different Cereal Grains

Cereal Grains	Macronutrients					Micronutrients			References
	Sodium (Na) (µg/g)	Calcium (Ca) (µg/g)	Magnesium (Mg) (µg/g)	Potassium (K) (µg/g)	Iron (Fe) (µg/g)	Zinc (Zn) (µg/g)	Copper (Cu) (µg/g)		
Wheat genotypes	1.22–7.14	0.65–2.34	4.78–8.36	16.47–44.58	0.70–2.45	0.78–1.01	3.35–15.79		(Sharanappa et al., 2016)
Rice cultivars	–	36.0–37.5	227.9–236.0	–	22.2–23.3	54.2–54.9	17.1–17.2		(Chaturvedi et al., 2017)
Millet	729.2	2103.3	17.43	22922.82	38.32	3.34	10.26		(Krishnan & Meera, 2018)
Corn	–	1611–8557	1404–1762	–	1.98–87.23	204.14–266.76	6.22–8.031		(Ahmadi & Ziarati, 2015)

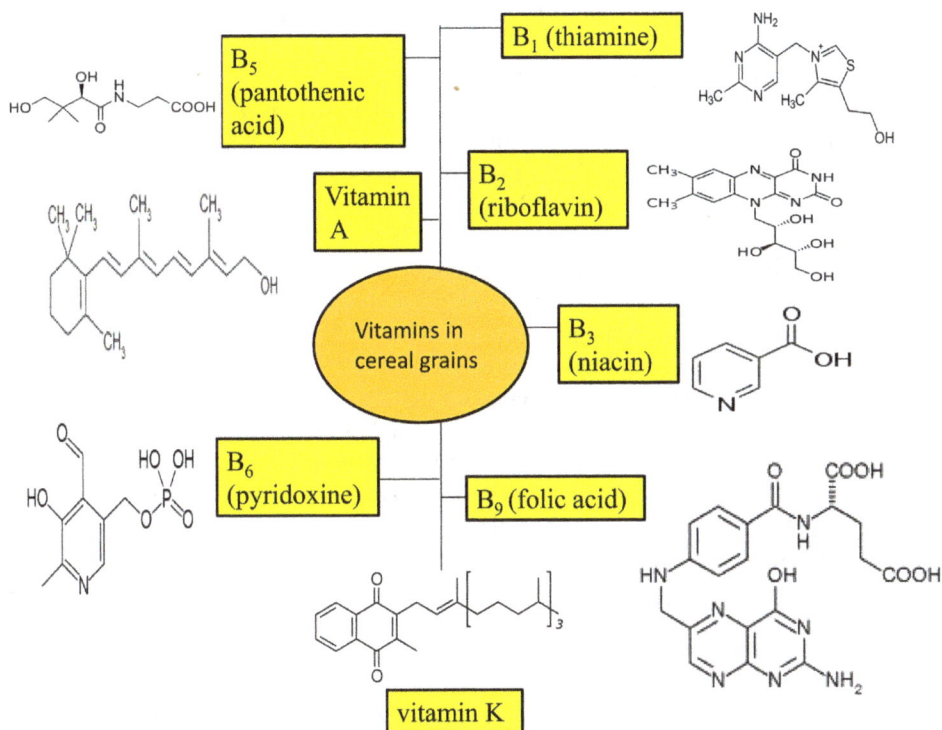

Figure 4.7 Vitamins in different cereal grains.

4.2.3 Bioactive Substances

Secondary metabolites are chemical substances that are produced within different plant parts and do not take part directly in the normal reproduction, growth, and development of plants. Bioactive substances are secondary metabolites that show different health benefits in human beings as well as animals (Zhao et al., 2015). Secondary metabolites like alkaloids, phenolic compounds, mycotoxins, plant growth factors, antibiotics, and different pigments are the most common bioactive substances that are produced inside the plant body (Espitia-Hernández et al., 2020). In addition, cereal grains are a great source of minerals; fiber; and significant quantities of bioactive compounds like inulin, vitamin E, phenolics, phytates, b-glucan, carotenoids, lignans, and sterols (Idehen et al., 2017). Some of these bioactive substances act as radical scavengers, while others work with enzymes as cofactors. They also play a significant role in controlling or reducing the risks of different diseases in the human body due to their synergistic effects (Singh & Sharma, 2017). The following section looks at bioactive substances that are present in cereal grains.

4.2.3.1 Polyphenols

Polyphenols are natural bioactive substances; chemically they resemble phenolic compounds and have great antioxidant properties. These substances are mainly present in whole cereal grains, green tea, fruits, and vegetables (Singla et al., 2019). According to the most recent definition, polyphenols are defined as compounds exclusively derived from the phenylpropanoid pathway, containing more than one phenolic unit and deprived of nitrogen (N)-based functions (Mirza-Aghazadeh-Attari et al., 2020). Moreover, they are categorized as substances with phenolic structures. There are more than 8000 different types of polyphenols that have been studied, mainly divided into four groups:

flavonoids, phenolic acids, lignans, and stilbenes (Gao et al., 2021). The following looks at polyphenols that are present in cereal grains.

A study was carried out to check polyphenols in wheat seed. The results showed that the husk of wheat had the maximum total phenolic content (TPC), followed by wheat bran. Total proanthocyanidin contents in the wheat husk were similar to those present in wheat bran. The contents of polyphenols in the hull were different from those of endosperm and bran. The main phenolics in the husk were present in ranges as follows: hyperin (53–274 mg/100 g), rutin (62–173 mg/100 g), and vitexin (101.65–188.78 mg/100 g) (Zhang et al., 2017). Similarly, in rice grains, polyphenols are mainly categorized into three subgroups: anthocyanins, which are only present in dark purple and black rice grains; proanthocyanidins, which mainly contain epicatechin and catechin units in red rice grains and have strong antioxidant properties; and phenolic acids, which are the most common bioactive substances in whole cereal grains (Shao & Bao, 2015). In addition, in another study, the total phenol content of different millet varieties was studied. Results showed the total phenol in millet varieties as follows: the concentration of total phenol in finger millet was 2.61 mg/g, in pearl millet 4.79 mg/g, and in fonio millet 1.96 mg/g (Nassarawa & Ahmad, 2019). Moreover, a study was carried out to check the phenolic acid content and avenanthramides in 20+ oat products. Results showed that oat products provided total phenolic acids between 15.79 and 25.05 mg and avenanthramides between 1.1 and 2 mg of a total 40-gram portion, while an 11-gram portion of oat provided 1.2 mg of avenanthramide content and 16 mg of total phenolic acid content (Soycan et al., 2019).

4.2.3.2 Terpenes

Plant terpenoids have biological functions in facilitating plant interactions and allowing them to tolerate biotic and abiotic stress (Sun et al., 2017). Terpenes are made up of isoprenoid units, while these units are made up of five-carbon compounds. In terpenes, these isoprenoid units arrange themselves in head-to-tail manner. Some vitamins (A, E, and K) attached with side chains are also part of terpenes (Kandi et al., 2015). These are mainly present in 25,000 compounds that are isolated from different living organisms. Isoprene structures in terpenes define their nomenclature. In addition, these substances are characterized as sesquiterpenes, monoterpenes, diterpenes, triterpenes, and polyterpenes. Sesquiterpenes, monoterpenes, and diterpenes are considered secondary metabolites because they are not necessary for viability (Baccouri & Rajhi, 2021). The following are some examples of terpene compounds present in different cereal grains.

For example, a study was conducted to identify the terpene synthase gene family of foxtail millet crops. The genome of *Setaria italica* (foxtail millet) contains 32 terpene synthase genes. From these 32 genes, 17 were characterized biochemically. The findings of this research increase the known chemical space of terpene metabolism to enable further studies of terpenoid-mediated stress resilience in crops (Karunanithi et al., 2020). Similarly, in another study, phytochemical analysis and antioxidant properties of different millet varieties were studied. Results showed the concentration of terpenoid in different millet varieties as follows: The concentration of terpenoid in finger millet was 11.92 mg/g, in pearl millet 8.49 mg/g, and in fonio millet 5.71 mg/g (Nassarawa & Ahmad, 2019). Moreover, in another study, phytochemical analysis of sorghum was carried out. Different extraction methods were used to check the presence of phytochemicals in sorghum. In petroleum ether extraction and ethanol extraction of sorghum, the presence of terpenoids was checked. Results showed that terpenoids in sorghum extracts of both extraction techniques were present (Felicia & Deborah, 2021).

4.2.3.3 Alkaloids

Plant alkaloids are a unique group of chemical components and are also considered one of the major groups of natural products. Historically, they have been used by human beings for at least 3000 years as teas, as liquids for remedies, and for medicinal purposes, but the substances responsible for these activities were not classified and identified until the 19th century (Bribi, 2018). Oxygen is the main component in the molecular structure of most alkaloids; they usually show colorless crystal properties in ambient environments. Oxygen-free alkaloids, like coniine/nicotine, are usually oily, volatile, and colorless, while sanguinarine and berberine are some colored alkaloids. Most alkaloids act as weak bases, but some are amphoteric, such as theophylline and bromine (Babbar, 2015). Alkaloids are the most important class of bioactive substances and possess significant biological characteristics like antioxidants, analgesics, and muscle relaxants (Roy, 2017). Following is a discussion of alkaloids present in cereal grains.

For example, a study was carried out to check the total alkaloid content in polished and brown rice. Results showed the ranges of alkaloids for both brown and polished rice as follows: brown rice 15.22–49.75 mg/100 g and polished rice 10.07–35.47 mg/100 g (Zeng et al., 2016). Similarly, in another study, two new alkaloids, oryzadiamine A and oryzadiamine B, were isolated from rice (*Oryza sativa*) mutated with yellow grain. Both A and B oryzadiamines are new natural substances that possess unique nitrogen (N)-containing heterocyclic rings. It was found that oryzadiamine A may impart a yellow color to the grain (Nakano et al., 2020). Moreover, in another study, the liquid chromatography-mass spectrometry (LC-MS) method was used to calculate the tropane alkaloids in maize (*Zea mays*) crop. The results showed that extraction-recovered scopolamine and atropine were 85.5% and 65.7%, respectively (Vuković et al., 2018). In addition, in another study, the alkaloid content of different millet varieties was studied. Results showed the alkaloid in millet varieties as follows: the concentration of alkaloid in finger millet was 72.83%, in pearl millet 25.54%, and in fonio millet 57.64% (Nassarawa & Ahmad, 2019). Furthermore, a study was carried out to check ergot alkaloids in different grain products. The analytical method was applied to measure ergot alkaloids in different rye-based food products. Mostly, alkaloids are present in rye flour. The profile of alkaloids was dominated by ergocristine at 44.7 mgkg[-1], while the least abundant alkaloid was ergocorninie at 0.2 mg/kg (Bryła, Szymczyk, Jędrzejczak & Roszko, 2015).

4.2.3.4 Saponins

The name of saponins is derived from their soap-like characteristics, and they are mainly glycoside compounds. Structurally, they are made up of hydrophilic saccharide chains that are attached to hydrophobic steroid structures. Saponins are mainly classified into three main groups: steroidal-glycoalkaloid, tri-terpenoid, and steroidal saponins (Singh & Kaur, 2018).

The antioxidant properties and phytochemical analysis of different millet varieties were studied. Results showed the concentration of saponins in different millet varieties as follows: the concentration of saponins in finger millet was 28.00 mg/g, in pearl millet 38.64 mg/g, and in fonio millet 23.82 mg/g (Nassarawa & Ahmad, 2019). Similarly, in the study, sorghum extracts were prepared by using different organic solvents, and the presence of saponins in organic solvent extracts was studied. Results showed that saponins were present in petroleum ether and ethanol extracts of sorghum (Felicia & Deborah, 2021).

4.3 FACTORS THAT AFFECT THE CHEMICAL COMPOSITION OF CEREAL GRAIN PRODUCTS

4.3.1 Effect of Milling

Milling is the process in which cereal grains are converted into flour. In the post-production of cereal grains, milling is an intermediate and very important step. The basic purpose of milling is to remove the hull or husk and sometimes also the bran layer from cereal grains. This process yields an edible portion that has no impurities in it and converts the cereal grains into powder form with different particle sizes. The degree of milling affects the nutritional composition of cereal grains

(Oghbaei & Prakash, 2016). There are two types of milling: wet milling and dry milling. In industries, the wet milling process is used for the separation of a desired or main component from the cereal grains by using different mechanical, physical, chemical, and biological methods. Currently, the wet milling method is mainly used for wheat and corn, but it can also be applied to other cereal grains like oats, sorghum, barley, and rice. The main fractions that are obtained from the wet milling process have broad uses in food and non-food products (Wronkowska, 2016). Dry milling is mostly used in industries for milling of cereal grains, and it is the oldest method (Papageorgiou & Skendi, 2018).

Milling affects the chemical composition of cereal grains. It has a great effect on the yield, nutrients, bioactive compounds, and so on. For example, different milling processes were used for different wheat varieties, and their nutritional composition was studied. It showed that in roller mill, wheat variety Zhengmai 366 showed moisture 12.95 g/100 g, ash 0.65 g/100 g, protein 11.3 g/100 g, and wet gluten 26.9 g/100 g, and wheat variety Yumai 57 showed moisture 13.43 g/100 g, ash 0.61 g/100 g, protein 10.3 g/100 g, and wet gluten 30.8 g/100 g. In a stone mill, wheat variety Zhengmai 366 showed moisture 10.75 g/100 g, ash 1.18 g/100 g, protein 14.8 g/100 g, and wet gluten 29.6 g/100 g, and Yumai 57 showed moisture 10.40 g/100 g, ash 1.13 g/100 g, protein 13.4 g/100 g, and wet gluten 22.5 g/100 g (Liu et al., 2015). In another study, differences in the composition of the wheat (*Triticum*) flour and grain were studied. Results showed the chemical composition of wheat grain as follows: protein (13.3%), starch (64.1%), ash (1.9%), and lipids (2.2%), while wheat flour showed protein (11.5%), starch (74.1%), ash content (0.8%), and lipids (1.4%) (Fraś et al., 2016).

Moreover, a study was carried out on different rice varieties to check their chemical composition after the wet and dry milling process. The results showed that these milling methods had different effects on the chemical composition. In the wet milling method, the rice flour showed a great amount of carbohydrate and a low quantity of protein and ash in it. In addition, in this method, rice flour showed high amylose and low lipid quantity, while dry milling produced flour with low crystallinity as compared to the wet milling method (Leewatchararongjaroen & Anuntagool, 2016). Similarly, the degree of milling affects the chemical composition of rice greatly. The proximate composition of rice under different degrees of milling (DoM) was as follows: at 4% DoM, moisture (10.76%), protein (8.76%), fat (1.12%), ash (0.69%), crude fiber (0.50%), and carbohydrate (78.77%); at 6% DoM, moisture (10.91%), protein (8.58%), fat (0.67%), ash (0.45%), crude fiber (0.48%), and carbohydrate (79.97%); and at 8% DoM, moisture (10.25%), protein (8.25%), fat (0.39%), ash (0.40%), crude fiber (0.40%), and carbohydrate (81.31%) (Nambi et al., 2017). Moreover, milling also causes the loss of vitamins of cereal grains. In another study, the comparison between whole and milled rice was as follows: whole rice: vitamin B_1 (0.34%), B_2 (0.09%), B_3 (4.62%), B_5 (0.92%), B_6 (1.3%), and B_9 (0.03%) and milled rice: vitamin B_1 (0.07%), B_2 (0.03%), B_3 (1.60%), B_5 (0.45%), B_6 (0.75%), and B_9 (0.01%) (Garg et al., 2021).

4.3.2 Environmental Factors

Environmental factors also affect the chemical composition of cereal grains. The main factors that influence the composition of cereal grains are soil nutrients, the temperature of air, light, water, and carbon dioxide. All these environmental factors play a great role in the growth and development of cereal grains (Patindol et al., 2015). For example, a study was conducted to check the chemical composition and grain yield of oat varieties in different environments of Turkey. Six different environmental conditions were measured for 25 oat genotypes. The oat varieties showed significant variation in grain weight (21.8–34.2 g), grain yield (2.15–5.81 t ha^{-1}), percentage of groat (70.1–73.6%), test weight (40.8–46.7 kg), protein content (12.0–13.3%), starch content (42.7–49.6%), acid detergent fiber (13.6–16.4%), ash content (2.34–2.77%), β-glucan (2.93–3.56%), fat content (5.69–6.80%), and neutral dietary fiber (31.5–34.4%) (Mut et al., 2018). In another study, the effect of growing location and varieties on the content of β-glucan in oat was studied. For this purpose, eight different oat varieties were grown in 20 different locations for 3 consecutive years. It was found that oat cultivars greatly influence the amount of beta glucan in oat. Among the eight cultivars, Derby had the lowest β-glucan content (4.37% w/w), and the highest β-glucan content was recorded in HiFi (5.82% w/w) (Herrera et al., 2016). Also, in another study, the effect of weather conditions on the chemical composition of wheat grains was studied. Results showed that when oat was grown in normal weather conditions, the hulls showed chemical composition as follows: lignocellulose (84%), lignin (25%), hemicellulose (35%), and cellulose (23%). When the oat was grown in a drier and warmer season, the amount of lignocellulose was reduced to 25%. Additionally, phenolic content was reduced by 60% in oat (Schmitz et al., 2020).

4.3.3 Cereal Crop Varieties

The chemical composition of cereal grains may vary according to the variety. For example, in China, three waxy wheat varieties were grown, and their chemical composition was studied. Results showed the composition of wheat varieties as follows: variety 1 (NW1) showed total starch (54.7%), protein (15.2%), lipid (0.15%), and moisture (9.3%), while variety 2 (TW1) showed total starch (54.1%), protein (14.1%), lipid (0.30%), and moisture (10.2%), and variety 3 (YW1) showed total starch (55.0%), protein (13.7%), lipid (0.24%), and moisture (8.9%) (Wang et al., 2015). Moreover, in another study, the chemical composition of aromatic and non-aromatic rice varieties was studied. Results showed the chemical composition of aromatic rice variety as follows: Gopal Bhog (aromatic variety) showed moisture (10.11%), fat (0.72%), protein (8.78%), ash (0.52%), fiber (0.64%), and carbohydrate (79.87%), while the chemical composition of the non-aromatic variety was as follows: Sarbati (non-aromatic variety) showed moisture (11.25%), fat (0.06%), protein (6.87%), ash (0.35%), fiber (0.64%), and carbohydrate (81.47%) (Verma & Srivastav, 2017). Furthermore, research was carried out to check the chemical composition of different oat cultivars. Results showed the chemical composition of oat cultivars as follows: OS-7 (oat cultivar) showed moisture (8.1%), ash (3.5%), fat (4.9%), protein (12.9%), crude fiber (10.9%), and carbohydrate (59.7%), while Kent showed moisture (7.9%), ash (2.9%), fat (5.1%), protein (13.9%), crude fiber (13.1%), and carbohydrate (57.4%) (Sandhu et al., 2017). Similarly, another study was carried on different corn varieties, and their chemical composition was studied. Results showed the chemical composition of corn cultivars as follows: white corn flour (WCF) showed ash (1.04%), moisture (9.74%), fat (12.90%), crude fiber (1.50%), protein (13.18%), and carbohydrate (16.28%), while yellow corn flour (YCF) showed ash (1.00%), moisture (9.75%) fat (13.50%), crude fiber (1.05%), protein (12.32%), and carbohydrate (62.38%) (Oladapo et al., 2017). Additionally, the chemical composition of flour from different millet varieties was studied. Results showed the composition of millet varieties as follows: HC-10 variety showed moisture (6.5%), ash (1.65%), fat (6.5%), protein (9.7%), fiber (3.1%), and carbohydrate (72.5%); HHB-67 variety showed moisture (6.9%), ash (1.65%), fat (6.6%), protein (9.9%), fiber (3.4%), and carbohydrate (71.5%); and W-445 variety showed moisture (7.6%), ash (1.90%), fat (5.1%), protein (11.3%), fiber (3.2%), and carbohydrate (70.9%) (Siroha et al., 2016). Moreover, in a similar study, the proximate composition of different millet varieties was also studied. Foxtail showed free lipid (46.8 mg/g), protein (10.5%), ash (3.5%), insoluble dietary fiber (23.0%), and soluble dietary fiber (2.7%), and finger millet showed free lipid (9.3 mg/g), protein (6.2%), ash (8.0%), insoluble dietary fiber (20.4%), and soluble dietary fiber (3.7%) (Bora et al., 2019). The proximate composition of different cereal varieties is also shown in Table 4.3.

4.3.4 Post-Harvest Storage

The time after the harvest to the proper utilization or consumption of cereal grains is called the post-harvest storage period. It plays a significant role in the nutritional and chemical composition of cereal grains (Lufu et al., 2020). Studies showed that post-harvest storage also affects the

Table 4.3 Proximate Composition Of Different Cereal Crop Varieties

Cereal Crops	Moisture (%)	Carbohydrates (%)	Proteins (%)	Lipids (%)	Ash (%)	References
Wheat	8.9–10.2	54.1–55.0	13.7–15.2	0.15–0.30	–	(Wang et al., 2015)
Rice	8.90–13.57	75.87–82.70	6.87–9.51	0.06–0.92	0.35–0.73	(Verma & Srivastav, 2017)
Oat	6.7–8.2	55.7–59.9	12.9–14.4	4.2–5.3	2.6–3.9	(Sandhu et al., 2017)
Corn	9.74–10.94	16.28–62.38	12.32–13.50	12.90–14.20	0.90–1.04	(Oladapo et al., 2017)
Millet	6.5–7.7	69.6–72.5	9.7–11.3	5.1–7.2	1.65–1.90	(Siroha et al., 2016)
Sorghum	5.94	74.27	5.54	–	3.05	(Felicia & Deborah, 2021)
Barley	13.9	54.1	12.0	–	2.84	(Panizo-Casado et al., 2020)

chemical composition of cereal grains. For example, red popcorn grains were subjected to different temperatures after harvesting, and their chemical composition was studied at different temperatures. Results showed the chemical composition of red popcorn grain at different temperatures as follows: At 30°C, amylose (22.25%), protein (0.89%), and fat (0.58%); at 40°C, amylose (23.03%), protein (0.75%), and fat (0.57%); at 70°C, amylose (21.48%), protein (0.77%), and fat (0.58%); and at 100°C, amylose (21.32%), protein (0.80%), and fat (0.59%)] (Ziegler et al., 2020). Moreover, in another study, the chemical composition of Ozgon rice endosperm at intervals of different days was studied. Results showed the chemical composition of rice endosperm at different days as follows: on the 4th day, moisture (11.68 g/100 g), carbohydrate (78.40 g/100 g), starch (67.53 g/100 g), protein (9.12 g/100 g), fat (0.23 g/100 g), and ash (0.57 g/100 g); on the 6th day, moisture (9.88 g/100 g), carbohydrate (77.75 g/100 g), starch (69.56 g/100 g), protein (10.85 g/100 g), fat (0.49 g/100 g), and ash (1.03 g/100 g); and on the 12th day, moisture (9.23 g/100 g), carbohydrate (79.86 g/100 g), starch (57.03 g/100 g), protein (8.53 g/100 g), fat (1.28 g/100 g) and ash (1.10 g/100 g) (Smanalieva et al., 2015). Furthermore, a study was carried out to check the proximate composition of whole and de-hulled millet treated with heat (150°C to 170°C/90s) at intervals of different days. Results showed the chemical composition of de-hulled and whole millet at 90 days as follows: de-hulled millet showed fat (2.54%), protein (6.62%), moisture (11.77%), ash (1.40%), fiber (2.19%), and carbohydrates (75.34%), while whole millet showed fat (2.82%), protein (6.81%), moisture (11.07%), ash (1.67%), fiber (2.71%), and carbohydrates (70.33%) (Huang et al., 2021).

4.4 RECENT APPROACHES TO ACHIEVE BETTER COMPOSITION

Cereal grains are a main source of food and play a major role in fulfilling the dietary needs of human beings. They provide a great amount of nutrition that is required by the body for growth and development. In recent years, it has been observed that cereal grains have not been fulfilling dietary requirements due to different environmental factors (Daryanto et al., 2016). To overcome the effect of environmental factors on cereal crops, a new technique called "envirotyping" has been introduced. This technique plays a part in the modeling of crops and also the prediction of phenotypes. The driving force for envirotyping is support systems and information. This technique has multiple applications such as prediction of phenotype, environmental characterization, agronomic genomics, and in the development of four-dimensional crop profiles involving envirotype (E), genotype (G), time (T), and phenotype (P). In the future, this technique will need to zoom in to individual plants and experimental plots to assimilate phenotypic, envirotypic, and genotypic information for developing a sustainable crop system (Xu, 2016). Moreover, another technique that has been used in recent years to improve the chemical composition of cereal grains is planted genome engineering. This technique has made it possible to make changes in the genome of plant cereals to acquire desired traits and nutritional composition (Bilichak et al., 2020; Sedeek et al., 2019). Furthermore, to minimize nutritional losses in cereal grains, a recent advancement in milling techniques has also been observed. For example, cryogenic milling has gained the attention of milling industries. This milling process gives a high production rate and fine flour particle size along with minimal nutritional losses (De Bondt et al., 2020). In addition, the use of nanotechnology to produce nano-fertilizers has also gained much attention. Conventional fertilizers have been used for cereal plants, but due to their low uptake potential, they are not fulfilling the nutrient requirements of cereal plants, ultimately leading to the loss of nutritional composition of grains. To overcome this issue, nano-fertilizers have been prepared that have great uptake potential and can easily reach the target and ultimately enhance the chemical composition of cereal grains (Elemike et al., 2019; Iqbal, 2019).

4.5 CONCLUSION

The main cereal grains that are consumed on a daily basis all over the world are wheat (*Triticum*), corn (*Zea mays*), oat (*Avena sativa*), rice (*Oryza sativa*), sorghum (*Sorghum bicolor*), and millet (*Pennisetum glaucum*). They fulfill the daily dietary requirements due to their high nutritional profile. This chapter has shown that structurally, all the cereal grains have three main distinct parts: bran, endosperm, and germ. Bran is the outer layer and mainly contains bioactive substances, minerals, vitamins, and dietary fiber. The endosperm is the major food reserve tissue in cereal grains, while the germ is the part that grows into a new plant in suitable and desired conditions. In addition, the nutritional composition of all cereal grains varies from one to the other. They all contain mainly carbohydrates, proteins, lipids, minerals, vitamins, and bioactive substances, but the percentages of these nutrients vary in all cereal grains. Among all nutritional components, carbohydrates are abundantly found in all cereal grains, after which are proteins, lipids, vitamins, and minerals. Furthermore, the chapter showed that bioactive substances (secondary metabolites) like polyphenols, terpenes, alkaloids, and saponins are also present in cereal grains in low quantities, but they have great importance in the daily diet due to multiple benefits. This chapter also showed that the chemical composition of cereal grains depends upon many factors, like the degree of milling, environmental factors, cereal grain varieties, and post-harvest storage techniques. Moreover, in recent years, different technologies like envirotyping, plant genome engineering, cryogenic milling, and nano-fertilizers have been used to improve the chemical composition of cereal grains.

REFERENCES

Ahmad, A., and Ahmed, Z., 2011. Tailoring oat for future foods. In: Murphy, D. L. (ed.), *Oats: Cultivation, uses and health effects*. London: Nova Science Publishers.

Ahmad, A., and Anjum, F. M., 2010. *Perspective of β-glucan: Extraction and utilization*. Berlin: VDM Publishers.

Ahmad, A., and Kaleem, M., 2018. β-Glucan as a food ingredient. In: Grumezesu, A. M., and Holban, A. M. (Eds.), *Biopolymers for food design* (pp. 351–381). New York: Elsevier.

Ahmad, A., Anjum, F. M., Zahoor, T., Nawaz, H., and Dilshad, S. M. R., 2012b. Beta glucan: A valuable functional ingredient in foods. *Critical Reviews in Food Science and Nutrition*, 52(3), 201–212.

Ahmad, A., Anjum, F., M, Zahoor, T., and Nawaz, H., 2009. Extraction of β-glucan from oat and its interaction with glucose and lipoprotein profile. *Pakistan Journal of Nutrition*, 8(9), 1486–1492.

Ahmad, A., Bushra, M., Muhammad, A., Shaukat, B., Muhammad, A., and Tahira, T., 2012a. Perspective of β-glucan as functional ingredient for food industry. *Journal of Nutrition and Food Science*, 2(2), 133–139.

Ahmad, A., Irfan, U., Amir, R., and Abbasi, K., 2017. Development of high energy cereal and nut granola bar. *International Journal of Agriculture and Biological Sciences*, 1(3), 13–20.

Ahmadi, A., and Ziarati, P., 2015. Chemical composition profile of canned and frozen sweet corn (*Zea mays L.*) in Iran. *Oriental Journal of Chemistry*, 31(2), 1065–1070.

Ahmed, Z., and Ahmad, A., 2017. Chapter 8—Biopolymer produced by the lactic acid bacteria: Production and practical application. In: Holban, A. M., and Grumezescu, A. M. (Eds.), *Microbial production of food Ingredients and additives* (pp. 217–257). London: Academic Press.

Ahmed, Z., Wang, Y., Ahmad, A., Khan, S. T., Nisa, M., Ahmad, H., and Afreen, A., 2013a. Kefir and health: A contemporary perspective. *Critical Reviews in Food Science and Nutrition*, 53(5), 422–434.

Ahmed, Z., Wang, Y., Anjum, N., Ahmad, A., and Khan, S. T., 2013b. Characterization of exopolysaccharide produced by *Lactobacillus kefiranofaciens* ZW3 isolated from Tibet kefir—Part II. *Food Hydrocolloids*, 30(1), 343–350.

Akinyele, I., and Shokunbi, O., 2015. Concentrations of Mn, Fe, Cu, Zn, Cr, Cd, Pb, Ni in selected Nigerian tubers, legumes and cereals and estimates of the adult daily intakes. *Food Chemistry*, 173, 702–708.

Amagliani, L., O'Regan, J., Kelly, A. L., and O'Mahony, J. A., 2017. The composition, extraction, functionality and applications of rice proteins: A review. *Trends in Food Science and Technology*, 64, 1–12.

Anthero, A. G., Lima, J. M., Cleto, P. B., Jorge, L. M., and Jorge, R. M., 2019. Modeling of maceration step of the oat (*Avena sativa*) malting process. *Journal of Food Process Engineering*, 42(7), e13266.

Asif, A., and Zaheer, A., 2016. Nutraceuticals aspects of β-glucan with application in food products. In: Grumezescu, A. (Ed.), *Nutraceutical*. New York: Academic Press.

Babbar, N., 2015. An introduction to alkaloids and their applications in pharmaceutical chemistry. *The Pharma Innovation Journal*, 4(10), 74–75.

Baccouri, B., and Rajhi, I., 2021. *Potential antioxidant activity of terpenes*. In: Shagufta and Areej (Eds.), *InTechOpen*. Croatia: InTechOpen

Barak, S., Mudgil, D., and Khatkar, B., 2015. Biochemical and functional properties of wheat gliadins: A review. *Critical Reviews in Food Science and Nutrition*, 55(3), 357–368.

Bhosale, S., and Vijayalakshmi, D., 2015. Processing and nutritional composition of rice bran. *Current Research in Nutrition and Food Science Journal*, 3(1), 74–80.

Bilichak, A., Gaudet, D., and Laurie, J., 2020. Emerging genome engineering tools in crop research and breeding. In: Vaschetto, L. (Ed.), *Cereal genomics* (pp. 165–181). New York: Springer.

Bora, P., Ragaee, S., and Marcone, M., 2019. Characterisation of several types of millets as functional food ingredients. *International Journal of Food Sciences and Nutrition*, 70(6), 714–724.

Boukid, F., Folloni, S., Ranieri, R., and Vittadini, E., 2018. A compendium of wheat germ: Separation, stabilization and food applications. *Trends in Food Science and Technology*, 78, 120–133.

Bribi, N., 2018. Pharmacological activity of alkaloids: A review. *Asian Journal of Botany*, 1(1), 6.

Bryła, M., Szymczyk, K., Jędrzejczak, R., and Roszko, M., 2015. Application of liquid chromatography/ion trap mass spectrometry technique to determine ergot alkaloids in grain products. *Food Technology and Biotechnology*, 53(1), 18–28.

Cassim, A. M., Gouguet, P., Gronnier, J., Laurent, N., Germain, V., Grison, M., Boutté, Y., Gerbeau-Pissot, P., Simon-Plas, F., and Mongrand, S., 2019. Plant lipids: Key players of plasma membrane organization and function. *Progress in Lipid Research*, 73, 1–27.

Castro-Alba, V., Lazarte, C. E., Bergenståhl, B., and Granfeldt, Y., 2019. Phytate, iron, zinc, and calcium content of common Bolivian foods and their estimated mineral bioavailability. *Food Science and Nutrition*, 7(9), 2854–2865.

Chaturvedi, A. K., Bahuguna, R. N., Shah, D., Pal, M., and Jagadish, S. V., 2017. High temperature stress during flowering and grain filling offsets beneficial impact of elevated CO2 on assimilate partitioning and sink-strength in rice. *Scientific Reports*, 7(1), 1–13.

Chaudhari, P. R., Tamrakar, N., Singh, L., Tandon, A., and Sharma, D., 2018. Rice nutritional and medicinal properties: A. *Journal of Pharmacognosy and Phytochemistry*, 7(2), 150–156.

Coda, R., Katina, K., and Rizzello, C. G., 2015. Bran bioprocessing for enhanced functional properties. *Current Opinion in Food Science*, 1, 50–55.

Dahl, W. J., and Stewart, M. L., 2015. Position of the academy of nutrition and dietetics: Health implications of dietary fiber. *Journal of the Academy of Nutrition and Dietetics*, 115(11), 1861–1870.

Daryanto, S., Wang, L., and Jacinthe, P.-A., 2016. Global synthesis of drought effects on maize and wheat production. *PloS One*, 11(5), e0156362.

De Bondt, Y., Liberloo, I., Roye, C., Windhab, E. J., Lamothe, L., King, R., and Courtin, C. M., 2020. The effect of wet milling and cryogenic milling on the structure and physicochemical properties of wheat bran. *Foods*, 9(12), 1755.

De Carvalho, C. C., and Caramujo, M. J., 2018. The various roles of fatty acids. *Molecules*, 23(10), 2583.

de Sousa, M. F., Guimarães, R. M., de Oliveira Araújo, M., Barcelos, K. R., Carneiro, N. S., Lima, D. S., Dos Santos, D. C., de Aleluia Batista, K., Fernandes, K. F., and Lima, M. C. P. M., 2019. Characterization of corn (*Zea mays* L.) bran as a new food ingredient for snack bars. *LWT-Food Science and Technology*, 101, 812–818.

Deroover, L., Tie, Y., Verspreet, J., Courtin, C. M., and Verbeke, K., 2020. Modifying wheat bran to improve its health benefits. *Critical Reviews in Food Science and Nutrition*, 60(7), 1104–1122.

Elemike, E. E., Uzoh, I. M., Onwudiwe, D. C., and Babalola, O. O., 2019. The role of nanotechnology in the fortification of plant nutrients and improvement of crop production. *Applied Sciences*, 9(3), 499.

Espitia-Hernández, P., Chávez González, M. L., Ascacio-Valdés, J. A., Dávila-Medina, D., Flores-Naveda, A., Silva, T., Ruelas Chacón, X., and Sepúlveda, L., 2020. Sorghum (*Sorghum bicolor* L.) as a potential source of bioactive substances and their biological properties. *Critical Reviews in Food Science and Nutrition*, 1–12.

Fano, D., 2017. Breeding of sorghum for high lysine in the seed. *International Journal of Agriculture Sciences*, 9(43), 4702–4707.

Felicia, O. T., and Deborah, A. K., 2021. Sorghum extract: Phytochemical, proximate, and GC-MS analyses. *Foods and Raw Materials*, 9(2), 371–378.

Fernandesa, C. G., Sonawaneb, S. K., and Arya. 2021. Cereal based functional beverages: A review. *Journal of Microbiology, Biotechnology and Food Sciences*, 2021, 914–919.

Fraś, A., Gołębiewska, K., Gołębiewski, D., Mańkowski, D. R., Boros, D., and Szecówka, P., 2016. Variability in the chemical composition of triticale grain, flour and bread. *Journal of Cereal Science*, 71, 66–72.

Gao, X., Xu, Z., Liu, G., and Wu, J., 2021. Polyphenols as a versatile component in tissue engineering. *Acta Biomaterialia*, 119, 57–74.

Garg, M., Sharma, A., Vats, S., Tiwari, V., Kumari, A., Mishra, V., and Krishania, M., 2021. Vitamins in cereals: A critical review of content, health effects, processing losses, bioaccessibility, fortification, and biofortification strategies for their improvement. *Frontiers in Nutrition*, 8, 254.

Ghafoor, K., Özcan, M. M., AL-Juhaımı, F., Babıker, E. E., Sarker, Z. I., Ahmed, I. A. M., and Ahmed, M. A., 2017. Nutritional composition, extraction, and utilization of wheat germ oil: A review. *European Journal of Lipid Science and Technology*, 119(7), 1600160.

Gharibzahedi, S. M. T., and Jafari, S. M., 2017. The importance of minerals in human nutrition: Bioavailability, food fortification, processing effects and nanoencapsulation. *Trends in Food Science and Technology*, 62, 119–132.

Grimaldo, M., Roosen-Runge, F., Zhang, F., Schreiber, F., and Seydel, T., 2019. Dynamics of proteins in solution. *Quarterly Reviews of Biophysics*, 52(e7), 1–63.

Hafeez, A., Ahmad, A., Amir, R. M., and Kaleem, M., 2019. Quality evaluation of coconut–flaxseed balls enriched with chiaseeds. *Journal of Food Processing and Preservation*, 43(11), e14184.

Hayat, I., Ahmad, A., Ahmed, A., Khalil, S., and Gulfraz, M., 2014. Exploring the potential of red kidney beans (*Phaseolus vulgaris* L.) to develop protein-based product for food applications. *Journal of Animal and Plant Sciences*, 24(3), 860–868.

He, M., Qin, C.-X., Wang, X., and Ding, N.-Z., 2020. Plant unsaturated fatty acids: Biosynthesis and regulation. *Frontiers in Plant Science*, 11, 390.

Hell, J., Prückler, M., Danner, L., Henniges, U., Apprich, S., Rosenau, T., Kneifel, W., and Böhmdorfer, S., 2016. A comparison between near-infrared (NIR) and mid-infrared (ATR-FTIR) spectroscopy for the multivariate determination of compositional properties in wheat bran samples. *Food Control*, 60, 365–369.

Henry, R. J., Rangan, P., and Furtado, A., 2016. Functional cereals for production in new and variable climates. *Current Opinion in Plant Biology*, 30, 11–18.

Herrera, M. P., Gao, J., Vasanthan, T., Temelli, F., and Henderson, K., 2016. β-glucan content, viscosity, and solubility of Canadian grown oat as influenced by cultivar and growing location. *Canadian Journal of Plant Science*, 96(2), 183–196.

Heshe, G. G., Haki, G. D., Woldegiorgis, A. Z., and Gemede, H. F., 2016. Effect of conventional milling on the nutritional value and antioxidant capacity of wheat types common in Ethiopia and a recovery attempt with bran supplementation in bread. *Food Science and Nutrition*, 4(4), 534–543.

Hou, Y., Yin, Y., and Wu, G., 2015. Dietary essentiality of "nutritionally non-essential amino acids" for animals and humans. *Experimental Biology and Medicine*, 240(8), 997–1007.

Huang, H. H., Dikkala, P. K., Sridhar, K., Yang, H. T., Lee, J. T., and Tsai, F. J., 2021. Effect of heat and γ-irradiation on fungal load, pasting, and rheological characteristics of three whole and dehulled millets during storage. *Journal of Food Processing and Preservation*, 45(4): e15355.

Ibrahim, M. S., Ahmad, A., Sohail, A., and Asad, M. J., 2020. Nutritional and functional characterization of different oat (*Avena sativa* L.) cultivars. *International Journal of Food Properties*, 23(1), 1373–1385.

Idehen, E., Tang, Y., and Sang, S., 2017. Bioactive phytochemicals in barley. *Journal of Food and Drug Analysis*, 25(1), 148–161.

Ignjatovic-Micic, D., Vancetovic, J., Trbovic, D., Dumanovic, Z., Kostadinovic, M., and Bozinovic, S., 2015. Grain nutrient composition of maize (*Zea mays* L.) drought-tolerant populations. *Journal of Agricultural and Food Chemistry*, 63(4), 1251–1260.

Iqbal, M. A., 2019. *Nano-fertilizers for sustainable crop production under changing climate: A global perspective*. Croatia: InTechOpen.

Jayawardana, S. A. S., Samarasekera, J. K. R. R., Hettiarachchi, G. H. C. M., Gooneratne, J., Mazumdar, S. D., and Banerjee, R., 2019. Dietary fibers, starch fractions and nutritional composition of finger millet varieties cultivated in Sri Lanka. *Journal of Food Composition and Analysis*, 82, 103249.

Joye, I., 2019. Protein digestibility of cereal products. *Foods*, 8(6), 199.

Juliano, B. O., and Tuaño, A. P. P., 2019. *Gross structure and composition of the rice grain* (4th ed.). AACC International, USA.

Kalita, T., Gohain, U. P., and Hazarika, J., 2021. Effect of different processing methods on the nutritional value of rice. *Current Research in Nutrition and Food Science Journal*, 9(2), 683–691.

Kan, A., 2015. Characterization of the fatty acid and mineral compositions of selected cereal cultivars from Turkey. *Records of Natural Products*, 9(1), 124–134.

Kandi, S., Godishala, V., Rao, P., and Ramana, K., 2015. Biomedical significance of terpenes: An insight. *Biomedicine*, 3(1), 8–10.

Karunanithi, P. S., Berrios, D. I., Wang, S., Davis, J., Shen, T., Fiehn, O., Maloof, J. N., and Zerbe, P., 2020. The foxtail millet (*Setaria italica*) terpene synthase gene family. *The Plant Journal*, 103(2), 781–800.

Khalid, S., Ahmad, A., Masud, T., Asad, M., and Sandhu, M., 2016. Nutritional assessment of Ajwa date flesh and pits in comparison to local varieties. *Journal of Plant and Animal Sciences*, 26(4), 1072–1080.

Khalid, S., Khalid, N., Khan, R. S., Ahmed, H., and Ahmad, A., 2017. A review on chemistry and pharmacology of Ajwa date fruit and pit. *Trends in Food Science and Technology*, 63, 60–69.

Klerks, M., Bernal, M. J., Roman, S., Bodenstab, S., Gil, A., and Sanchez-Siles, L. M. J. N., 2019. Infant cereals: Current status, challenges, and future opportunities for whole grains. *Nutrients*, 11(2), 473.

Knudsen, K. B., Lærke, H., Ingerslev, A., Hedemann, M., Nielsen, T., and Theil, P., 2016. Carbohydrates in pig nutrition—Recent advances. *Journal of animal science*, 94(suppl_3), 1–11.

Koç, G. Ç., and Özçıra, N., 2019. Chemical composition, functional, powder, and sensory properties of tarhana enriched with wheat germ. *Journal of Food Science and Technology*, 56(12), 5204–5213.

Krishnan, R., and Meera, M. S., 2018. Pearl millet minerals: Effect of processing on bioaccessibility. *Journal of Food Science and Technology*, 55(9), 3362–3372.

Kumar, G. S., and Krishna, A. G., 2015. Studies on the nutraceuticals composition of wheat derived oils wheat bran oil and wheat germ oil. *Journal of Food Science and Technology*, 52(2), 1145–1151.

Landriscina, L., D'Agnello, P., Bevilacqua, A., Corbo, M. R., Sinigaglia, M., and Lamacchia, C., 2017. Impact of gluten-friendly™ technology on wheat kernel endosperm and gluten protein structure in seeds by light and electron microscopy. *Food Chemistry*, 221, 1258–1268.

Laskowski, W., Górska-Warsewicz, H., Rejman, K., Czeczotko, M., and Zwolińska, J., 2019. How important are cereals and cereal products in the average Polish diet. *Nutrients*, 11(3), 679.

Leewatchararongjaroen, J., and Anuntagool, J., 2016. Effects of dry-milling and wet-milling on chemical, physical and gelatinization properties of rice flour. *Rice Science*, 23(5), 274–281.

Li, H., Yan, S., Yang, L., Xu, M., Ji, J., Mao, H., Song, Y., Wang, J., and Sun, B., 2021. Starch gelatinization in the surface layer of rice grains is crucial in reducing the stickiness of parboiled rice. *Food Chemistry*, 341, 128202.

Liu, C., Liu, L., Li, L., Hao, C., Zheng, X., Bian, K., Zhang, J., and Wang, X., 2015. Effects of different milling processes on whole wheat flour quality and performance in steamed bread making. *LWT-Food Science and Technology*, 62(1), 310–318.

Los, A., Ziuzina, D., and Bourke, P., 2018. Current and future technologies for microbiological decontamination of cereal grains. *Journal of Food Science*, 83(6), 1484–1493.

Lufu, R., Ambaw, A., and Opara, U. L., 2020. Water loss of fresh fruit: Influencing pre-harvest, harvest and postharvest factors. *Scientia Horticulturae*, 272, 109519.

Mæhre, H. K., Dalheim, L., Edvinsen, G. K., Elvevoll, E. O., and Jensen, I.-J., 2018. Protein determination—Method matters. *Foods*, 7(1), 5.

Mahmoud, A. A., Mohdaly, A. A., and Elneairy, N. A., 2015. Wheat germ: An overview on nutritional value, antioxidant potential and antibacterial characteristics. *Food and Nutrition Sciences*, 6(02), 265.

Mäkinen, O. E., Sozer, N., Ercili-Cura, D., and Poutanen, K., 2017. Protein from oat: Structure, processes, functionality, and nutrition. In: Nadathur, S., Wanasundara, J. P. D., and Scanlin, L. (Eds.), *Sustainable Protein Sources*. (pp. 105–119), Elsevier, USA.

Makki, K., Deehan, E. C., Walter, J., and Bäckhed, F., 2018. The impact of dietary fiber on gut microbiota in host health and disease. *Cell Host and Microbe*, 23(6), 705–715.

Maqbool, M. A., Aslam, M., Akbar, W., and Iqbal, Z., 2018. Biological importance of vitamins for human health: A review. *Journal of Agriculture and Basic Sciences*, 2(3), 50–58.

Meï, C., Michaud, M., Cussac, M., Albrieux, C., Gros, V., Maréchal, E., Block, M. A., Jouhet, J., and Rébeillé, F., 2015. Levels of polyunsaturated fatty acids correlate with growth rate in plant cell cultures. *Scientific Reports*, 5(1), 1–9.

Memon, A. A., Mahar, I., Memon, R., Soomro, S., Harnly, J., Memon, N., Bhangar, M. I., and Luthria, D. L., 2020. Impact of flour particle size on nutrient and phenolic acid composition of commercial wheat varieties. *Journal of Food Composition and Analysis*, 86, 103358.

Merali, Z., Collins, S. R., Elliston, A., Wilson, D. R., Käsper, A., and Waldron, K. W., 2015. Characterization of cell wall components of wheat bran following hydrothermal pretreatment and fractionation. *Biotechnology for Biofuels*, 8(1), 1–13.

Mirza-Aghazadeh-Attari, M., Ekrami, E. M., Aghdas, S. A. M., Mihanfar, A., Hallaj, S., Yousefi, B., Safa, A., and Majidinia, M., 2020. Targeting PI3K/Akt/mTOR signaling pathway by polyphenols: Implication for cancer therapy. *Life Sciences*, 255, 117481.

Mumtaz, F., Zubair, M., Khan, F., and Niaz, K., 2020. Analysis of plants lipids. In: Silva, A. S., Nabavi, S. F., Saeedi, M., and Nabavi, S., M. (Eds.), *Recent advances in natural products analysis*. (pp. 677–705), Elsevier, USA.

Mut, Z., Akay, H., and Erbaş Köse, Ö. D., 2018. Grain yield, quality traits and grain yield stability of local oat cultivars. *Journal of Soil Science and Plant Nutrition*, 18(1), 269–281.

Nakamura, Y., and Li-Beisson, Y., 2016. *Lipids in plant and algae development* (Vol. 86). Springer, USA.

Nakano, H., Ono, H., Kaji, R., Sakai, M., Doi, S., and Kosemura, S., 2020. Oryzadiamines A and B, alkaloids from *Oryza sativa* with yellow grain. *Tetrahedron Letters*, 61(8), 151519.

Nambi, V. E., Manickavasagan, A., and Shahir, S., 2017. Rice milling technology to produce brown rice. In: Manickavasagan, A., Santhakumar, C., and Venkatachalapathy, N. (Eds.), *Brown Rice*. (pp. 3–21), Springer, USA.

Nassarawa, S. S., and Ahmad, G., 2019. Comparative of phytochemical and antioxidant properties of selected millet varieties in Katsina metropolis, Nigeria. *Annals Food Science and Technology*, 20(4), 820–827.

Navarro, D. M., Abelilla, J. J., and Stein, H. H., 2019. Structures and characteristics of carbohydrates in diets fed to pigs: A review. *Journal of Animal Science and Biotechnology*, 10(1), 1–17.

Nielsen, A. V., and Meyer, A. S., 2016. Phytase-mediated mineral solubilization from cereals under in vitro gastric conditions. *Journal of the Science of Food and Agriculture*, 96(11), 3755–3761.

Ocheme, O. B., Adedeji, O. E., Chinma, C. E., Yakubu, C. M., and Ajibo, U. H., 2018. Proximate composition, functional, and pasting properties of wheat and groundnut protein concentrate flour blends. *Food Science and Nutrition*, 6(5), 1173–1178.

Oghbaei, M., and Prakash, J., 2016. Effect of primary processing of cereals and legumes on its nutritional quality: A comprehensive review. *Cogent Food and Agriculture*, 2(1), 1136015.

Ogunremi, O. R., Agrawal, R., and Sanni, A., 2020. Production and characterization of volatile compounds and phytase from potentially probiotic yeasts isolated from traditional fermented cereal foods in Nigeria. *Journal of Genetic Engineering and Biotechnology*, 18, 1–8.

Oladapo, A., Adepeju, A., Akinyele, A., and Adepeju, D., 2017. The proximate, functional and anti-nutritional properties of three selected varieties of maize (yellow, white and pop corn) flour. *International Journal of Scientific Engineering and Science*, 1(2), 23–26.

Olsen, O. A., 2020. The modular control of cereal endosperm development. *Trends in Plant Science*, 25(3), 279–290.

Orenday-Ortiz, J. M., 2018. *Morphological and structural characterization of super soft wheat as observed by scanning electron microscopy.* (Master of Science in Food Science Masters Thesis), Washington State University,

Özcan, M. M., Bağcı, A., Dursun, N., Gezgin, S., Hamurcu, M., Dumlupınar, Z., and Uslu, N., 2017. Macro and micro element contents of several oat (*Avena sativa* L.) genotype and variety grains. *Iranian Journal of Chemistry and Chemical Engineering*, 36(3), 73–79.

Panizo-Casado, M., Déniz-Expósito, P., Rodríguez-Galdón, B., Afonso-Morales, D., Ríos-Mesa, D., Díaz-Romero, C., and Rodríguez-Rodríguez, E. M., 2020. The chemical composition of barley grain (Hordeum vulgare L.) landraces from the Canary Islands. *Journal of Food Science*, 85(6), 1725–1734.

Papageorgiou, M., and Skendi, A., 2018. Introduction to cereal processing and by-products. In: Galanakis, C. M. (Ed.), *Sustainable recovery and reutilization of cereal processing by-products.* (pp. 1–25), Elsevier, USA.

Patindol, J. A., Siebenmorgen, T. J., and Wang, Y. J., 2015. Impact of environmental factors on rice starch structure: A review. *Starch-Stärke*, 67(1–2), 42–54.

Pinto, T. I., Coelho, J. A., Pires, B. I., Neng, N. R., Nogueira, J. M., Bordado, J. C., and Sardinha, J. P., 2021. Supercritical carbon dioxide extraction, antioxidant activity, and fatty acid composition of bran oil from rice varieties cultivated in Portugal. *Separations*, 8(8), 115.

Punia, S., Sandhu, K. S., Dhull, S. B., Siroha, A. K., Purewal, S. S., Kaur, M., and Kidwai, M. K., 2020. Oat starch: Physico-chemical, morphological, rheological characteristics and its applications-A review. *International Journal of Biological Macromolecules*, 154, 493–498.

Quentin, A. G., Pinkard, E. A., Ryan, M. G., Tissue, D. T., Baggett, L. S., Adams, H. D., Maillard, P., Marchand, J., Landhäusser, S. M., and Lacointe, A., 2015. Non-structural carbohydrates in woody plants compared among laboratories. *Tree Physiology*, 35(11), 1146–1165.

Radu, A. I., Kuellmer, M., Giese, B., Huebner, U., Weber, K., Cialla-May, D., and Popp, J., 2016. Surface-enhanced Raman spectroscopy (SERS) in food analytics: Detection of vitamins B2 and B12 in cereals. *Talanta*, 160, 289–297.

Ralla, T., Salminen, H., Edelmann, M., Dawid, C., Hofmann, T., and Weiss, J., 2018. Oat bran extract (*Avena sativa* L.) from food by-product streams as new natural emulsifier. *Food Hydrocolloids*, 81, 253–262.

RamyaBai, M., Wedick, N. M., Shanmugam, S., Arumugam, K., Nagarajan, L., Vasudevan, K., Gunasekaran, G., Rajagopal, G., Spiegelman, D., and Malik, V., 2019. Glycemic index and microstructure evaluation of four cereal grain foods. *Journal of Food Science*, 84(12), 3373–3382.

Rasane, P., Jha, A., Sabikhi, L., Kumar, A., and Unnikrishnan, V., 2015. Nutritional advantages of oats and opportunities for its processing as value added foods-A review. *Journal of Food Science and Technology*, 52(2), 662–675.

Roy, A., 2017. A review on the alkaloids an important therapeutic compound from plants. *International Journal of Plant Biotechnology*, 3(2), 1–9.

Rustgi, S., Shewry, P., Brouns, F., Deleu, L. J., and Delcour, J. A., 2019. Wheat seed proteins: Factors influencing their content, composition, and technological properties, and strategies to reduce adverse reactions. *Comprehensive Reviews in Food Science and Food Safety*, 18(6), 1751–1769.

Sandhu, K. S., Godara, P., Kaur, M., and Punia, S., 2017. Effect of toasting on physical, functional and antioxidant properties of flour from oat (*Avena sativa* L.) cultivars. *Journal of the Saudi Society of Agricultural Sciences*, 16(2), 197–203.

Schmitz, E., Nordberg Karlsson, E., and Adlercreutz, P., 2020. Warming weather changes the chemical composition of oat hulls. *Plant Biology*, 22(6), 1086–1091.

Sedeek, K. E., Mahas, A., and Mahfouz, M., 2019. Plant genome engineering for targeted improvement of crop traits. *Frontiers in Plant Science*, 10, 114.

Shao, Y., and Bao, J., 2015. Polyphenols in whole rice grain: Genetic diversity and health benefits. *Food Chemistry*, 180, 86–97.

Sharanappa, T., Chetana, R., & Suresh Kumar, G., 2016. Evaluation of genotypic wheat bran varieties for nutraceutical compounds. *Journal of Food Science and Technology*, 53(12), 4316–4324.

Shewry, P. R., Wan, Y., Hawkesford, M. J., and Tosi, P., 2020. Spatial distribution of functional components in the starchy endosperm of wheat grains. *Journal of Cereal Science*, 91, 102869.

Singh, A., and Sharma, S., 2017. Bioactive components and functional properties of biologically activated cereal grains: A bibliographic review. *Critical Reviews in Food Science and Nutrition*, 57(14), 3051–3071.

Singh, A., and Singh, S., 2015. Dietary fiber content of Indian diets. *Asian Journal of Pharmaceutical and Clinical Research*, 8(3), 58–61.

Singh, A., Kaur, V., and Kaler, R., 2018. A review on dietary fiber in cereals and its characterization. *Journal of Applied and Natural Science*, 10(4), 1216–1225.

Singla, R. K., Dubey, A. K., Garg, A., Sharma, R. K., Fiorino, M., Ameen, S. M., Haddad, M. A., and Al-Hiary, M., 2019. Natural polyphenols: Chemical classification, definition of classes, subcategories, and structures. *Journal of AOAC International*, 102(5), 1397–1400.

Siroha, A. K., Sandhu, K. S., and Kaur, M., 2016. Physicochemical, functional and antioxidant properties of flour from pearl millet varieties grown in India. *Journal of Food Measurement and Characterization*, 10(2), 311–318.

Siyuan, S., Tong, L., and Liu, R., 2018. Corn phytochemicals and their health benefits. *Food Science and Human Wellness*, 7(3), 185–195.

Slama, A., Cherif, A., Sakouhi, F., Boukhchina, S., and Radhouane, L., 2020. Fatty acids, phytochemical composition and antioxidant potential of pearl millet oil. *Journal of Consumer Protection and Food Safety*, 15(2), 145–151.

Smanalieva, J., Salieva, K., Borkoev, B., Windhab, E. J., and Fischer, P., 2015. Investigation of changes in chemical composition and rheological properties of Kyrgyz rice cultivars (Ozgon rice) depending on long-term stack-storage after harvesting. *LWT-Food Science and Technology*, 63(1), 626–632.

Soliman, G. A., 2019. Dietary fiber, atherosclerosis, and cardiovascular disease. *Nutrients*, 11(5), 1155.

Soycan, G., Schär, M. Y., Kristek, A., Boberska, J., Alsharif, S. N., Corona, G., Shewry, P. R., and Spencer, J. P., 2019. Composition and content of phenolic acids and avenanthramides in commercial oat products: Are oats an important polyphenol source for consumers? *Food Chemistry: X*, 3, 100047.

Sun, Y., Huang, X., Ning, Y., Jing, W., Bruce, T. J., Qi, F., Xu, Q., Wu, K., Zhang, Y., and Guo, Y., 2017. TPS46, a rice terpene synthase conferring natural resistance to bird cherry-oat aphid, *Rhopalosiphum padi* (Linnaeus). *Frontiers in Plant Science*, 8, 110.

Szymanek, M., Tanaś, W., and Kassar, F. H., 2015. Kernel carbohydrates concentration in sugary-1, sugary enhanced and shrunken sweet corn kernels. *Agriculture and Agricultural Science Procedia*, 7, 260–264.

Tacer-Caba, Z., Nilufer-Erdil, D., and Ai, Y., 2015. *Chemical composition of cereals and their products*. In: Bhavbhuti M. M., and Peter, C. K. C. (Eds.). Springer, Germany.

Taylor, J. R., Emmambux, M. N., and Kruger, J., 2015. Developments in modulating glycaemic response in starchy cereal foods. *Starch-Stärke*, 67(1–2), 79–89.

Tejada-Ortigoza, V., Garcia-Amezquita, L. E., Serna-Saldívar, S. O., and Welti-Chanes, J., 2016. Advances in the functional characterization and extraction processes of dietary fiber. *Food Engineering Reviews*, 8(3), 251–271.

Tosi, P., He, J., Lovegrove, A., Gonzáles-Thuillier, I., Penson, S., and Shewry, P. R., 2018. Gradients in compositions in the starchy endosperm of wheat have implications for milling and processing. *Trends in Food Science and Technology*, 82, 1–7.

Verma, D. K., and Srivastav, P. P., 2017. Proximate composition, mineral content and fatty acids analyses of aromatic and non-aromatic Indian rice. *Rice Science*, 24(1), 21–31.

Vuković, G., Bursić, V., Stojanović, T., Petrović, A., Gvozdenac, S., Starović, M., Kuzmanović, S., and Aleksić, G., 2018. LC-MS/MS determination of tropane alkaloids in maize crop. *Contemporary Agriculture*, 67(3–4), 221–226.

Wang, J., Tang, J., Ruan, S., Lv, R., Zhou, J., Tian, J., Cheng, H., Xu, E., and Liu, D., 2021. A comprehensive review of cereal germ and its lipids: Chemical composition, multi-objective process and functional application. *Food Chemistry*, 130066.

Wang, J., Vanga, S. K., Saxena, R., Orsat, V., and Raghavan, V., 2018. Effect of climate change on the yield of cereal crops: A review. *Climate*, 6(2), 41.

Wang, S., Wang, J., Zhang, W., Li, C., Yu, J., and Wang, S., 2015. Molecular order and functional properties of starches from three waxy wheat varieties grown in China. *Food Chemistry*, 181, 43–50.

Witten, S., and Aulrich, K., 2018. Effect of variety and environment on the amount of thiamine and riboflavin in cereals and grain legumes. *Animal Feed Science and Technology*, 238, 39–46.

Wronkowska, M., 2016. Wet-milling of cereals. *Journal of Food Processing and Preservation*, 40(3), 572–580.

Wu, X., Liu, J., Li, D., and Liu, C. M., 2016. Rice caryopsis development II: Dynamic changes in the endo-sperm. *Journal of Integrative Plant Biology*, 58(9), 786–798.

Xicoy, H., Wieringa, B., and Martens, G. J., 2019. The role of lipids in Parkinson's disease. *Cells*, 8(1), 27.

Xu, Y., 2016. Envirotyping for deciphering environmental impacts on crop plants. *Theoretical and Applied Genetics*, 129(4), 653–673.

Yan, X., Ye, R., and Chen, Y., 2015. Blasting extrusion processing: The increase of soluble dietary fiber con-tent and extraction of soluble-fiber polysaccharides from wheat bran. *Food Chemistry*, 180, 106–115.

Youssef, H. M., 2015. Assessment of gross chemical composition, mineral composition, vitamin composition and amino acids composition of wheat biscuits and wheat germ fortified biscuits. *Food and Nutrition Sciences*, 6(10), 845.

Yu, S., and Tian, L., 2018. Breeding major cereal grains through the lens of nutrition sensitivity. *Molecular Plant*, 11(1), 23–30.

Zareef, M., Arslan, M., Hassan, M. M., Waqas, M., Ali, S., Li, H., Ouyang, Q., Wu, X., Hashim, M. M., and Chen, Q., 2021. Recent advances in assessing qualitative and quantitative aspects of cereals using non-destructive techniques: A review. *Trends in Food Science and Technology*, 116, 815–828.

Zeng, Y., Sun, D., Du, J., Pu, X., Yang, S., Yang, X., Yang, T., and Yang, J., 2016. Identification of QTLs for resistant starch and total alkaloid content in brown and polished rice. *Genetics and Molecular Research*, 15(3), 10.4238.

Zhang, G., and Hamaker, B. R., 2017. The nutritional property of endosperm starch and its contribution to the health benefits of whole grain foods. *Critical Reviews in Food Science and Nutrition*, 57(18), 3807–3817.

Zhang, W., Zhu, Y., Liu, Q., Bao, J., and Liu, Q., 2017. Identification and quantification of polyphenols in hull, bran and endosperm of common buckwheat (*Fagopyrum esculentum*) seeds. *Journal of Functional Foods*, 38, 363–369.

Zhang, Y., Li, M., Gao, H., Wang, B., Tongcheng, X., Gao, B., and Yu, L., 2019. Triacylglycerol, fatty acid, and phytochemical profiles in a new red sorghum variety (Ji Liang No. 1) and its antioxidant and anti-inflammatory properties. *Food Science and Nutrition*, 7(3), 949–958.

Zhao, L., Pan, T., Cai, C., Wang, J., and Wei, C., 2016. Application of whole sections of mature cereal seeds to visualize the morphology of endosperm cell and starch and the distribution of storage protein. *Journal of Cereal Science*, 71, 19–27.

Zhao, Y., Wu, Y., and Wang, M., 2015. Bioactive substances of plant origin. In: Cheung, P. C. K and Mehta, B. M., Handbook of Food Chemistry. Springer, USA.

Ziegler, V., da Silva Timm, N., Ferreira, C. D., Goebel, J. T., Pohndorf, R. S., and de Oliveira, M., 2020. Effects of drying temperature of red popcorn grains on the morphology, technological, and digestibility proper-ties of starch. *International Journal of Biological Macromolecules*, 145, 568–574.

CHAPTER 5

Health-Endorsing Properties of Cereal Grains

Sunanda Biswas and Mohammad Javed Ansari

CONTENTS

DOI: 10.1201/9781003252023-5

5.1 INTRODUCTION

Cereal grains are essential food commodities for consumption and income generation. They are edible seeds consumed since early times, and they have been an integral part of the human diet. Wheat, rice, oats, millets, maize, sorghum, rye, and barley are consumed by people, and among these, wheat and rice are the most valuable cereal crops, about half of the total cereal production of world. The edible parts of cereal grains includes bran, germ, and endosperm. Generally, cereals are taken as staple foods and essential nutrients. Cereals and their products are indispensable sources of carbohydrate, energy, protein, and fiber and contain several micronutrients. People consume cereals in gruels, chapattis, nan, parathas, bread, biscuits, cakes, idlis, dosas, burgers, pasta, pizzas, steamed rice, biriyanis, ready-to-eat breakfast cereals, porridges, and many others. Generally, cereals are cheap to produce, low in price, and easy to store and transport compared to other foods. They are mainly consumed as affordable sources of energy by the population.

The consumption of cereal grains worldwide has been significant to meet the food demand. People mainly eat cereal grains, rice, wheat, corn, and oats. Other cereals consumed in low quantities include triticale, barley, sorghum, and millets. Wheat ranks in the top position, with leading consumption among cereal grains globally. Evidence suggests that daily intake of cereals, especially whole-grains, has a role in preventing chronic diseases such as coronary heart diseases, diabetes mellitus, colorectal cancer, and obesity. Several factors may be involved with these functions, such as their micronutrient, fiber, and glycemic index. Consumption of whole grain cereals is interrelated with several positive health effects. To increase the consumption of whole-grain foods, it may be helpful to have a quantitative recommendation. Additionally, a more comprehensive range of fast and easy-to-prepare foods may also help to increase intake of whole grains.

Cereals contain a significant amount of energy, dietary fiber, proteins, vitamins, minerals, and antioxidants essential for human health and are accepted as nutraceuticals and functional foods. Cereals like rice, wheat, maize, oats, and so on are used to prepare foods as conventional foods and used in a regular diet, important for aiding physiological functions and providing nutrition (Saikia et al., 2011). It is also important to remember that proper consumption habits according to nutrition knowledge can drastically reduce healthcare expenditures. This chapter explores the health-promoting properties of cereal grains and the potential power of cereals for lowering risks of various diseases.

5.2 NUTRITIONAL COMPONENTS PRESENT IN CEREALS

Cereals are scientifically defined as grains or edible seeds that belong to the *Gramineae* family. Pseudocereals are not from grass family members but can substitute for cereals in allergic persons (Blaise & Alexander, 2010). Cereals are sources of good carbohydrate content and energy (Pol et al., 2020). The healthy eating pyramid also recommends high-quality cereals and baked products (Saura et al., 2020). There are several benefits of cereal grains, and they contribute as an important part of daily human diet. The major cereals consumed are wheat, rice, oats, corn, and barley, with wheat heading the list. Besides cereal grains commonly consumed globally, buckwheat and wild rice are two other non-cereals (pseudograins). The cereals are classified as follows:

- *True Cereals*: Barley, maize (corn), brown rice, millets, oat, rye, sorghum, teff, and wheat.
- *Pseudocereals*: Amaranth, buckwheat, and quinoa.

Whole grains and pseudocereals have adequate nutritional importance in the human diet and healthcare. For this reason, cereal whole grains and pseudocereals have significant attraction for consumers and commercial foods, with the development of new foods with increasing interest from food scientists, nutritionists, and technologists (Patil & Khan, 2011).

5.2.1 Nutritional Components Present in Whole Grains

Food researchers are looking forward to developing high-yielding varieties. Food nutritionists are focusing on the dietary recommendations of cereals with health benefits. The high fiber content in whole-grain foods has significant benefits to consumer health. A whole-grain ingredient is known as "the caryopsis, with starchy endosperm, bran, and germ present in significantly similar proportions as in the intact caryopsis." Whole grains have three major edible parts: bran, germ, and endosperm. The essential biologically active elements come from the bran and germ proportions of the grain. My Pyramid (also called the U.S. food guidance system) recommends a daily whole grain intake of three servings (Slavin, 2004). Cereal grain contains a good amount of B-vitamins (including riboflavin and thiamine), minerals (including Ca and Mg), and essential amino acids (including lysine and arginine) (Table 5.1) (Zettel & Hitzmann, 2018). The whole grain possesses good antioxidant quantity due to folates, phenolic compounds, avenanthramides, avenalumic acid (McMackin et al., 2013), dietary fiber, antinutrients (phytic acid and tannins), lignans, and tocotrienols. Among cereal grains, wheat is a significant source of dietary fiber when compared to other commonly consumed whole grain cereals (rice, maize, oats). Research has shown that lower dietary fiber intake may lead to constipation, colon cancer, Crohn's disease, ulcers, gallstones, obesity, coronary heart disease, and many others (Fernandes & de las Mercedes, 2017). Cereal grains are also good sources of oligosaccharides, which aid in the growth of bifidobacteria in the human large intestine (Vuksan et al., 2007), which helps the body perform essential functions such as digestion and warding off harmful bacteria. Other nutrient components are phenolic compounds, carotenoids, phytic acid, phytosterols, and tocols (Gosine & McSweeney, 2019).

5.2.2 Nutritional Components Present in Pseudocereals

Pseudocereals are nongrasses used in much the same way as cereals (true cereals are grasses). Botanically, pseudocereals are dicots; however, they are not categorized as true cereals. They are called psuedocereals because their seeds or fruits have parity with true cereals' composition and functional properties. Therefore, they can be easily used like cereals in ground flour or whole cereals. Amaranthus, buckwheat, quinoa, and chia are examples of such pseudocereals. Pseudocereals can be easily used in ground flour or entire cereals, like cereals. The nutrient composition of four pseudocereal grains (amaranthus, buckwheat, quinoa, and chia) is given in Table 5.2. Pseudocereals are rich in nutrients, containing many nutraceutical components. They are a rich source of proteins, starch, bioactive compounds, lipids, and minerals (USDA, 2019). The protein content and biological value of proteins are better than or comparable to other protein sources and can contribute to many functions of the human body.

5.3 BIOACTIVE COMPOUNDS PRESENT IN CEREALS

Consumption of whole-grain products has gained preference owing to their richness in functional components, such as phytochemicals and dietary fiber. These active components are referred to as bioactive compounds. The bran layer of grains contains a higher amount of bioactive compounds than the endosperm parts. Several factors, such as food source, interactions among phytochemicals and with food, and the gastrointestinal environment interfere with the bioavailability of nutrients and bioactive components (Bohn, 2014), and it does not necessarily rely on the number of active metabolites in target tissues and concentration of the bioactive compounds of food (Acosta, 2009). Various processing techniques (such as germination, fermentation, extrusion) increase the biological activity of grains, thus increasing the bioavailability of nutritional health-promoting compounds. Processing possesses both positive and negative impacts on bioactive compounds; for example, during thermal treatments due to high temperature, thermolabile compounds get decomposed, or polymerization

Table 5.1 Nutritional Composition of Selected Whole Grain Cereals (per 100 g) (USDA National Nutrient Database)

Grain Cereal	Energy (KJ)	Total Carbohydrate (g)	Protein (g)	Total Fat (g)	Fiber, Total Dietary (g)	Calcium (mg)	Iron (mg)	Magnesium (mg)	Thiamin (mg)	Riboflavin (mg)	Folate, Total (µg)	Lysine (g)	Arginine (g)
Barley (hulled)	1480	73.5	12.5	2.3	17.3	33	3.6	133	0.646	0.285	19	0.465	0.625
Corn (yellow)	1530	74.3	9.4	4.7	7.3	7	2.71	127	0.385	0.201	19	0.265	0.47
Millet, raw	1580	72.8	11	4.2	8.5	8	3.01	114	0.421	0.29	85	0.212	0.382
Oats	1630	66.3	16.9	6.9	10.6	54	4.72	117	0.763	0.139	56	0.701	1.19
Rice (brown, long grain)	1530	76.2	7.54	3.2	3.6	9	1.29	116	0.541	0.095	23	0.303	0.602
Rye	1410	75.9	10.3	1.63	15.1	24	2.63	110	0.316	0.251	38	0.286	0.454
Sorghum	1380	72.1	10.6	3.46	6.7	13	3.36	165	0.332	0.096	20	0.229	0.355
Teff	1540	73.1	13.3	2.38	8	180	7.63	184	0.39	0.27	NA	0.376	0.517
Triticale	1410	72.1	13	2.09	NA	37	2.57	130	0.416	0.134	73	0.365	0.671
Wheat (soft white)	1420	75.4	10.7	1.99	12.7	34	5.37	90	0.41	0.107	41	NA	NA
Spelt	1410	70.2	14.6	2.43	10.7	27	4.44	136	0.364	0.113	45	0.409	0.687
Bulgur	1430	75.9	12.3	1.33	12.5	35	2.46	164	0.232	0.115	27	0.339	0.575

Table 5.2 Nutritional Composition of Selected Pseudocereals (per 100 g) (USDA National Nutrient Database)

Grain Cereal	Energy (KJ)	Total Carbohydrate (g)	Protein (g)	Total Fat (g)	Fiber, Total Dietary (g)	Calcium (mg)	Iron (mg)	Magnesium (mg)	Thiamin (mg)	Riboflavin (mg)	Folate, Total (µg)	Lysine (g)	Arginine (g)
Buckwheat	1440	71.5	13.2	3.4	10	18	2.2	231	0.101	0.425	30	0.672	0.982
Quinoa	1540	64.2	14.1	6.07	7	47	4.57	197	0.36	0.318	184	0.766	1.09
Chia	2030	42.1	16.5	30.7	34.4	631	7.72	335	0.62	0.17	49	0.97	2.14
Amaranthus	1550	65.2	13.6	7.02	6.7	159	7.61	248	0.116	0.2	82	0.747	1.06

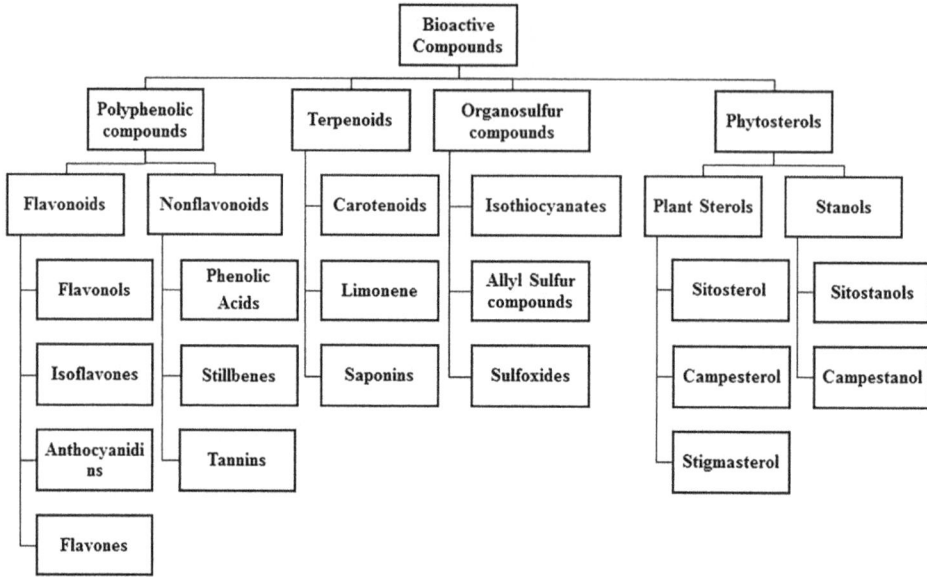

Figure 5.1 Classification of bioactive compounds.

(*Source:* Hu et al., 2016)

of phenolic compounds occurs due to the interaction of phenols with Maillard reaction by-products, resulting in a loss in antioxidant activity of phenolic compounds (Atlan et al., 2009). In contrast, milling, extrusion, fermentation, and the action of hydrolytic enzymes during germination facilitate the release of bound phenolic compounds by breaking the cell wall and releasing cellular constituents (Awika et al., 2003).

Bioactive compounds are classified into polyphenolic compounds, organosulfur compounds, terpenoids, and phytosterols, which vary in chemical structure, concentration, and nature of distribution in foods (Figure 5.1). These phytochemicals possess antimutagenic, antioxidant, and other biological activities (Wang et al., 2014). The concentration of these compounds in cereals depends on the genus, species, and localization site, as most bioactive components are present in the bran layer of grains.

Polyphenolic compounds are characterized as flavonoids and nonflavonoids. Primary polyphenols present in cereals are phenolic acids, alkylresorcinols, and avenanthramides. Phenolic acids are aromatic compounds widely present in cereals, oilseeds, legumes, fruits, and vegetables, and these occur in free, conjugated, and ester-bound forms. Bound phenolic acids are predominant over free phenolic acids, and they form bridges between the polymer chains in the kernel. Two groups of phenolic acids in the bound form are:

- Derivatives from dihydroxybenzoic acids include vanillic, protocatechuic, and syringic acids.
- Derivatives from hydroxycinnamic acids include ferulic, caffeic, ρ-coumaric, and sinapic acids.

5.3.1 Bioactive Compounds Present in Whole Grains

The exterior layer of cereal grain contains a good amount of bioactive compounds, like phenolic compounds, phytosterols, tools, and carotenoids (Mattila, 2005). Bioactive compounds present in whole grain cereals are listed in Table 5.3. Whole grain cereals are richer in phenolic compounds. All grain phenolic compounds, including alkylresorcinols, lignans, and phenolic acids, are absorbed and metabolized in humans and show their protective physiological effect (Ross et al., 2003a).

Table 5.3 Bioactive Compounds Present in Selected Whole Grain Cereals

Name of Whole Grain Cereal	Bioactive Compounds Present	References
Wheat	Caffeic acid, ferulic acid, salicylic acid, syringic acid, vanillic acid, alkylresorcinols, β-tocotrienol, arabinoxylans, lutein, zeaxanthin, phytosterols	(Dykes & Roony, 2007), (Liu, 2007)
Barley	Ferulic acid, salicylic acid, ρ-hydroxybenzoic acid, caffeic acid, vanillic acid, α-tocotrienols, β-glucans, arabinoxylans, 3-deoxyanthocyanidins, γ-aminobutyric acid	(Bartlomiej et al., 2012), (Liu, 2007), (Wang et al., 2014)
Rice	ρ-coumaric acid, sinapic acid, caffeic acid, ρ-hydroxybenzoic acid, gallic acid, protocatechuic acid, γ-oryzanol, γ-aminobutyric acid, phytosterols	(Liu, 2007), (Dykes & Roony, 2007), (Wang et al., 2014)
Rye	Ferulic acid, ρ-coumaric acid, sinapic acid, caffeic acid, alkylresorcinols, α-tocopherols, phytosterols, arabinoxylans	(Liu, 2007), (Dykes & Roony, 2007), (Wang et al., 2014)
Maize	ρ-coumaric acid, syringle acid, lignan, carotenoid, γ-aminobutyric acid	(Liu, 2007), (Dykes & Roony, 2007)
Millet	Cinnamic acid, ρ-hydroxybenzoic ρ-coumaric acid, acid, vanillic acid	(Liu, 2007), (Dykes & Roony, 2007), (Wang et al., 2014)
Oat	Caffeic acid, vanillic acid, ferulic acid, ρ-coumaric acid, ρ-hydroxybenzoic acid, aventhramides, β-glucans, lignans, phytosterols	(Liu, 2007), (Wang et al., 2014)
Sorghum	ρ-coumaric acid, caffeic acid, salicylic acid, syringic acid, protocatechuic acid	(Liu, 2007), (Dykes & Roony, 2007), (Wang et al., 2014)
Corn	p-hydroxybenzoic acid, vannilic acid, protocatechuic acid, caffeic acid, syringic acid, quinic acid, p-coumaric acid, ferulic acid	(Sheng et al., 2018)
Bulgur	Gallic acid, 3,4 hydroxybenzoic acid, caffeic acid, ferulic acid and epicatechin.	(Caba et al., 2012)

Bound phenolics are not digested in the small intestine but can enter portal circulation (Slavin, 2004), absorbed slowly by the enzymes (β-glucosidases, esterases) present in the colon, and show preventive action against colon cancer. Consequently, enhancing bound phenolics during processing exerts profound favorable health effects after absorption (Ti, 2015).

However, the bioavailability and the impact of processing on phenolic acid content depend on the type of phenolic acids that can chelate ions of transition metals, thus reducing the generation of free radicals (Abdel-Aal et al., 2012), preventing the oxidation of methyl linoleate (Kahkonen et al., 1999) and decreasing low-density lipoproteins (LDLs) in human blood (Abdel-Aal & Gamel., 2008). However, the stability and bioavailability of individual phenolic compounds are greatly influenced by processing methods and the chemical nature of phenolic acids. The predominant phenolic acid in the whole grain is ferulic acid at all stages of grain development (McKeehen et al., 1999). It acts as a potent antioxidant and anti-inflammatory and plays a role as a chemo-protectant by preventing peroxynitrite-mediated nitration of tyrosine residues in collagen and suppresses lipid peroxidation in the microsomal membrane (Verma et al., 2009). In grains, other phenolic acids are vanillic acid, caffeic acid, ρ-coumaric, sinapic, and syringic acid.

Avenanthramides are nitrogen-containing polyphenols in oats that have anti-inflammatory, antipruritic, antioxidation, antiproliferation, and antiatherogenic activities (Mattila et al., 2005). Avenanthramides are classified into two groups: amides of different cinnamic acids and cinnamic acids substituted with anthranilic acid. Out of 40 known structures of avenanthramides, the most common forms are avenanthramide A (2f), B (2p), and C (2c) of group 2. Avenanthramide C is thermostable, shows higher antioxidant activities, and is more abundant in oats than the A and B forms, whereas avenanthramide A possesses fourfold higher bioavailability than the others. However, their levels in oats depend on the variety, environmental stress, and geographical parameters. Research has shown that avenanthramides are present mainly in the outer layer than the whole oats, whereas a few researchers found higher concentrations in oat flakes of whole grain oats (27 mg/kg) than oat bran (13 mg/kg). The concentration of A, B, and C avenanthramides reported in oat flakes of whole oats was 9.0, 8.6, and 9.0 mg/kg, respectively; oat bran contains 4.4, 4.1, and 4.3 mg/kg of fresh weight, respectively. Several health benefits of avenanthramides are enhancing cognitive abilities, reducing inflammation, relaxing blood vessels, interfering with cancer proliferation, and regulating blood sugar (Tripathi et al., 2018).

Alkylresorcinols are the phenolic lipids composed of a resorcinol-type phenolic ring to which a long odd-numbered aliphatic chain is attached at the fifth position. These phenolic compounds are abundant in whole grains such as: rye (0.36–3.2 mg/g), triticale (0.58–1.63 mg/g), wheat (0.32–1.01 mg/g), and barley (0.04–0.5 mg/g). Alkylresorcinol metabolites 3,5-dihydroxy phenyl propionic acid and 3,5-dihydroxybenzoic acid (DHBA) act as nutritional biomarkers to identify the intake of whole wheat and rye. There is a higher concentration of alkylresorcinols in plasma on ingestion of whole grains than refined grains; therefore, they could be used as a marker to detect bran in flour and other cereal products (Ross et al., 2003b). These compounds interfere with cancer progression, reduce mutagenic activity, inhibit metabolic enzyme in vitro, and have antimicrobial and antioxidant activities. Some researchers reported the loss of alkylresorcinols on processing (such as extrusion, fermentation, and baking). Besides polyphenolic compounds, other bioactive components in cereals are γ-aminobutyric acid, phytosterols, carotenoids, arabinoxylans, and β-glucans.

γ-aminobutyric acid (GABA) is a naturally occurring nonprotein amino acid in whole grains (brown rice, maize, barley), lentils, soy, nuts, beans, fish, fruits, and vegetables. GABA is synthesized by α-decarboxylation of glutamic acid catalyzed by glutamate decarboxylase (GAD) (Mayer et al., 1990). It acts as an inhibitory neurotransmitter, induces diuretic and hypotensive effects, and inhibits the proliferation of cancer cells. The concentration of GABA in grains and legumes could be increased by the biological activation of grains and legumes. Several health benefits exhibited by GABA are preventing headaches, regulating blood sugar levels, relieving constipation, reducing anxiety, and lowering risk of developing colon cancer and Alzheimer's disease (Hagiwara et al., 2004).

Phytosterols are mainly present in whole cereal grains (rice, wheat, oat), vegetable oil, legumes, nuts, and vegetables. In general, out of all plant sterols, β-sitosterols are higher in concentration (62%), followed by campesterol (21%) and stigmasterol (4%). In contrast, stanols such as β-sitostanol (4%) and campestanol (2%) are present in minimal amounts (Jiang & Wang, 2005). β-sitosterols show antipyretic and anti-inflammatory properties, control blood sugar, and exhibit immune-modulating effects.

Arabinoxylans are found in all cereal grains but are abundant in rye, followed by wheat, barley, oat, and rice. Arabinoxylans are localized in the husk, bran layer, and cell wall of endosperm in different concentrations. However, the attention and structure of these compounds depend on the cereal variety and type, geographical variation, and location. These polysaccharides are diverse, and arabinoxylans of endosperm are soluble in water. In contrast, those present in the bran layer from covalent linkages with ferulic acid result in water-insoluble arabinoxylans, which exhibit high antioxidant and antimicrobial activities. However, consumption of cereal arabinoxylans showed a significant impact on health by lowering serum cholesterol and serum insulin by the production of short-chain fatty acids, reducing postprandial and fasting glucose, increasing absorption of minerals (magnesium and calcium), and enhancing immunity, thus preventing chronic diseases (Singh & Sharma, 2017).

β-glucan is a naturally present polysaccharide comprising β-D-glucose units. It is found in cereals, bacteria, yeast, and fungi cell walls. It is predominantly present in the bran layer of oats and barley.

β-glucans present in cereals consist of β-1,3 and β-1,4 linkages of β-D-glucose units and are water soluble. In contrast, those in cell walls of bacteria, yeast, and fungi constitute β, 1–3 linkages of β-D-glucose units and are water insoluble. The intestine enzymes do not digest these polysaccharides, but they are fermented in the colon by the gut bacteria, thus acting as a substrate for the gut microflora. On anaerobic fermentation, short-chain fatty acids are produced that exhibit several health advantages, like reduction of heart disease, diabetes mellitus, and obesity risk (El-Khoury et al., 2012).

5.3.2 Bioactive Compounds Present in Pseudocereals

The major bioactive components in pseudocereals are phenolic groups present in the outermost layers of amaranth and quinoa seeds (Table 5.4). These phenolic compounds can prevent oxidative damage to tissue cells (Rodas & Bressani, 2009). However, the antioxidants in pseudocereals are helpful in the prevention of diseases (Majewska et al., 2011). Polyphenol compounds play a role in preventing cancer and cardiovascular disease due to the higher concentration of functional components (such as phytosterol) in pseudocereals. The flavonoids, tocopherols, and antioxidants in amaranth provide antioxidant properties. Although amaranth grains also contain phytic acid and tannins, they have a defense mechanism system due to protecting tissues from free radical mechanisms. Due to their unique antioxidant properties, amaranth grains can improve human health.

Buckwheat is considered a super grain because it has better polyphenolic content than amaranth (Alvarez-Jubete et al., 2010). The primary antioxidant compounds in buckwheat are protocatechuic acid, vanillic acid, and rutin. Besides these antioxidants, it has other bioactive constituents, such as quercetin, caffeic, vanillic acid, and luteolin (Klvarez-Jubete et al., 2010).

Quinoa grains include phenolics, fagopyritols, carotenoids, phytosterols, and other polyphenolic components (Renzetti & Rosell, 2016). The flavonoid content of these grains is about 144 mg/100 g. The phenolics and polyphenols (gallic acid, vanillic acid, quercetin) have ultimate antioxidant properties. The quercetin content present in quinoa is even higher than in cranberries. Quinoa seeds are bitter due to an increased number of antioxidants.

Chia is a good source of phenolic compounds with antioxidant activities. The most common polyphenols in chia are derivatives of cinnamic acid and flavonoids. The phenolic compounds content in chia contain 0.88–1.6 mg of gallic acid equivalent per g. The potential antioxidants in chia seeds are chlorogenic acid, caffeic acid, kaempferol, ferulic acid, quercetin, myricetin, and rosmarinic acids that exhibit antihypertensive, anticarcinogenic, neuroprotective, hepatoprotective, and anti-ageing activities (Shahidi & Naczk, 1995). Some isoflavones are also present in chia seeds, such as genistin, glycitin, daidzin, glycitein, and genistein (Orona-Tamayo et al., 2015).

Table 5.4 Bioactive Compounds Present in Pseudocereals

Name of Pseudocereals	Bioactive Compounds Present	References
Buckwheat	Rutin, quercetin, γ-aminobutyric acid, 2"-hydroxynicotianamine, epicatechin	(Aoyagi, 2006), (Zielińska & Zieliński, 2009)
Quinoa	Ferulic acid, vanillic acid, p-coumaric acid, protocatechuic acid, quercetin-glucuronide, caffeic acid, gallic acid, cinnamic, myricetin, kaempferol, rutin, hesperidin, neohesperidin	(Gomez-Caravaca et al., 2014), (Thakur & Kumar, 2019)
Chia	Gallic acid, caffeic acid, p-coumaric acid, ferulic acids, chlorogenic, kaempferol, cinnamic, quercetin, epicatechin, rutin, apigenin	(Kulczyński et al., 2019)
Amaranthus	Rutin, isoquercetin, quercetin, nicotiflorin, anthocyanins	(Thakur & Kumar, 2019)

5.4 HEALTH BENEFITS OF CEREALS

Cereals (rice, wheat, sorghum, oats, maize, rye, millets, teff, and barley) and pseudocereals (amaranth, quinoa, chia, and buckwheat) both have excellent nutritional value, which is beneficial for human health. Nutrients such as carbohydrates, proteins, and fats in cereals can supply energy to consumers. At the same time, vitamin E, vitamin B, calcium, magnesium, zinc, and selenium play a vital role in combatting micronutrient deficiencies.

All cereal grains consist of (1) the multi-layered outer portion known as bran, which is rich in dietary fibers; (2) the germ portion, which is rich in micronutrients and lipids; and (3) the endosperm, which is rich in starch and protein. They are healthy choices if consumed as a whole grain due to the presence of bran, which is rich in antioxidants, phytochemicals, minerals, vitamins, and dietary fiber.

Bran's high level of bioactive compounds is present in the aleurone layer. Bioactive compounds like vitamins, minerals (like zinc, selenium, and magnesium), phytochemicals, and complex carbohydrates (lignans, β-glucans, inulin, resistant starch) have been related to various health functions and potential disease-protective mechanisms (Liu, 2007). Lignans have been linked to lower risk of cancer and heart diseases. Phytic acid is found to reduce the glycemic index (GI) and may help to control diabetes. It also assists in safeguarding the colon against the instigation of cancer cells. Tocotrienols, saponins, and oryzanol can lower blood cholesterol levels.

In contrast, different research showed phenolic compounds have antioxidant effects (Gry et al., 2007) due to the power of free radical removing and metal-chelation activities. The γ-oryzanol in rice; alkyl resorcinol in rye; β-glucans in oats and barley; and avenanthramide, avenacosides, and saponins in oats (Fardet, 2010) have beneficial health roles for humans. Recently, anthocyanins in colored cereal grains (black, purple, blue, pink, red, and brown) have been claimed to provide oxidation, anticancer, glycaemic and body-weight regulation, neuroprotection, renal protection, hypolipidemia, hepatoprotection, and antiageing activities (Zhu, 2018).

Refined cereals consist mainly of the endosperm. A high proportion of dietary value, vitamins, minerals, and phytochemicals located in the outermost layers and skin are lost during processing, milling, and refining (Fardet, 2010); for example, a loss in phenolic compounds has been observed after the milling of maize (Butts-Wilmsmeyer et al., 2018). Various phenolic compounds, flavonoids, carotenoids, tocopherols, sterols, minerals, and antinutrients are lost during the processing of different cereals (Oghbaei and Prakash, 2013). The products prepared from such flours are suitable for people suffering from micronutrient deficiencies.

Also, refined grains containing low insoluble fibers are recommended for people with inflammatory bowel disease (IBD), because there is an increased requirement for energy and protein in some patients (Forbes, 2017). These disorders cause chronic inflammation in the gastrointestinal tract. In the IBD, the immune system works inappropriately, causing inflammation and symptoms like diarrhea, abdominal pain, cramping, rectal bleeding, loss of body weight, and general fatigue. Typically, intake of dietary fiber is avoided in patients with intestinal inflammation, but dietary fiber like germinated barley, *Plantago ovata* seeds, and husk (also known as psyllium) may be helpful in the maintenance of remission in some patients with ulcerative colitis (Forbes, 2017). However, dietary fiber is ineffective in maintaining gastrointestinal microflora in Crohn's disease.

5.4.1 Health Benefits of Whole Grains

Extensive research work has been conducted to establish the health benefits of whole grain. Research has confirmed that intake of whole grains is related to a reduced risk of cardiovascular diseases and the occurrence of different kinds of cancers and mortality from infectious diseases, respiratory diseases, and diabetes (Aune et al., 2016). It also supports cognitive control (Edwards et al., 2017). This section discusses the health benefits of whole-grain. The nutritional aspects and health benefits of whole-grain cereals are summarized in Table 5.5.

Table 5.5 Nutritional Aspects and Health Benefits of Whole-Grain Cereals

Name of Whole Grain Cereal	Nutritional Aspects	Health Benefits	References
Barley	Barley is consumed as food, feed, and malt. As food, it is consumed in the form of flour or porridge. Pearl barley is used in soups and stews.	B-glucans are a primary soluble fiber that cures hypercholesterolemia, chemically induced colon cancer, and hypoglycemia.	(Das et al., 2012)
Corn	Sweet corn is rich in carbohydrates and sugars, so it is consumed as a cheap form of starch and major energy source.	Sweet corn can protect the human body against cancer, diabetes mellitus, heart diseases, and neurodegenerative diseases (Alzheimer's).	(Swapna et al., 2020)
Millet	Millets are a significant food source in dry and semidry regions of the world. They provide all nutrients and are rich in essential amino acids, vitamins, and minerals.	The high fiber content of millets improves constipation. Millets can improve hypoglycemia and hypocholesterolemia and act as an antiulcerative.	(Kulkarni et al., 2018)
Oats	Oats contain high content of β-glucan and compounds rich in antioxidant activity.	Soluble fibers in oats act as prebiotics to maintain a healthy colon wall.	(Das et al., 2012)
Rice (brown, long-grain)	Brown rice is a whole grain that contains three parts of the grain kernel: bran, germ, and endosperm. Rice is mainly consumed as human food, including breakfast cereals.	Rice is rich in fiber and antioxidants and gluten-free and has a low glycemic index that lowers blood glucose levels and cholesterol in the body.	(Upadhyay & Karn, 2018)
Rye	Rye is rich in fiber content and contains bioactive compounds like lignans, arabinoxylan, phytosterols, folates, phenolic compounds, tocopherols, and tocotrienols.	Rye bread is rich in antioxidants, dietary fibers, and arabinoxylans to prevent colon, colorectal, stomach, and pancreatic cancers. Rye stimulates weight loss, lowers the risk of type 2 diabetes, prevents gallstones, improves the cardiovascular condition, and promotes gastrointestinal health.	(Jonsson et al., 2018)
Sorghum	An important staple crop providing carbohydrates, proteins, fiber, vitamins, minerals, and phytochemicals (such as phenolic acids, tannins, phytosterols, and anthocyanins).	Sorghum stimulates weight loss, lowers the risk of type 2 diabetes, reduces the risk of specific types of cancers, and prevents diverticular disease and erosions in the tract.	(Rao et al., 2015)
Teff	Teff is gluten free and also considered a rich source of essential fatty acids, fiber, minerals, and phytochemicals.	Teff helps to reduce diabetes and prevent anemia and celiac diseases.	(Gebru et al., 2020)
Wheat	Wheat bran and whole wheat–based cereal products are rich sources of dietary antioxidants and phenolic acids. Wheat is used for human consumption for its unique properties, and a large range of ingredients and foods can be prepared from it.	Whole wheat helps in protection against constipation, heart diseases, obesity, diabetes mellitus, and colon diseases.	(Kumar et al., 2011)
Spelt	Spelt is a rich source of minerals such as copper, iron, zinc, magnesium, and phosphorus.	Spelt plays an important role in glycemic control. The dietary fiber presents in spelt can modulate postprandial glycemia.	(Biskup et al., 2017)

5.4.1.1 Relief in Constipation and Related Disorders

Dietary fibers are present in whole-grain cereals. These are carbohydrate polymers with a degree of polymerization of not less than three so that they do not get hydrolyzed by endoenzymes in the small intestine of humans (Spiller, 2001). They include undigested polysaccharides, lignins, fructooligosaccharides, and associated plant substances. A lack of dietary fiber (such as β-glucan and inulin) in the daily diet is associated with constipation, hemorrhoids, heart disease, gall bladder disease, appendicitis, diverticular disease, hiatus hernia, gastro-esophageal reflux, abnormally dilated veins, and deep vein thrombosis. The functional level of dietary fibers also depends on the level of fermentation by microorganisms in the gut.

Dietary fibers (both soluble and insoluble) have β-1–4 covalent bonds, which cannot be broken by digestive enzymes; however, some fibers are fermented by microorganisms in the colon. Many soluble dietary fibers can hold water and are fermentable in the colon. Therefore, a substantial quantity of bacterial mass accumulates to form soft and bulky feces because of soluble fiber intake, affecting feces' transit time, relieving constipation, and easing or preventing hemorrhoids (Hillemeier, 1995). Also, undegraded or unfermented fibers add to the volume of feces. The minor fermentable fibers (wheat products, usually bran, methylcellulose and carboxymethylcellulose, corn bran, psyllium husk/ispaghula, or some hydrocolloids like gum acacia) show high water-holding capacity and thereby work as efficient laxatives and bring relief from constipation (Spiller, 2001). In one such study, it was found that the consumption of whole grains relieved constipation and showed improved gut microbiota in comparison to intake of refined grains (Vanegas et al., 2017). Due to ease in the defecation process, the feces get removed from the digestive system faster and more efficiently, reducing the exposure period of gut walls to harmful and toxic/carcinogenic compounds.

5.4.1.2 Prebiotic Effect

Whole grains and dietary fiber help in the growth of probiotic microorganisms (the beneficial bacteria, e.g., lactobacilli and bifidobacteria) in the gut (Quigley, 2011). Oligosaccharides, β-glucan, resistant starch, and arabinoxylan can selectively stimulate the growth of probiotic microorganisms in the colon and are known as prebiotics. Intestinal microflora ferment soluble dietary fibers in the large intestine, whereas insoluble dietary fiber shows very little fermentation. The primary effect of some soluble dietary fibers like inulin (an oligofructose) on gut ecology is to stimulate bifidobacterial growth (Roberfroid, 2007).

It has been reported that *bifidobacteria* may increase by 1 log compared to the baseline (Liu, 2007) by adding 15 g inulin/day to the diet for 15 days. Oats also have similar effects on gut microbiology due to the presence of β-glucan. A 6-week research study on the intake of whole-grain rye and wheat in overweight adults also showed improvement in the markers representing gut health (Vuholm et al., 2017). It was found in a study that whole-grain wheat consumption for 4–8 weeks resulted in a quadruple increase in serum dihydro ferulic and a double increase in fecal ferulic acid in comparison to refined wheat consumption in a randomized controlled trial on adults with unhealthy lifestyles. After whole-grain wheat consumption, increased fecal ferulic acid was correlated with increased *Firmicutes* and *Bacteroidetes*. It reduced the role of dietary fibers and other bioactive compounds of whole grain, which might affect the gut microbiology and health of the consumer.

5.4.1.3 Diabetes Management

Diabetes mellitus is characterized by either deficient production of insulin (type 1 diabetes) or combined resistance to insulin-secretory response or insulin action (type 2 diabetes), which changes carbohydrate, protein, and lipid metabolism in the body, resulting in hyperglycemia. It is regarded as the most recurrent endocrine disorder (Saleh et al., 2013). The age-standardized disability-adjusted life-years rate for the corresponding period for diabetes has increased by 39.6% (Tandon et al.,

2018). Increased intake of whole grains can help manage healthy blood glucose-insulin levels and lower the risk of diabetes (Kyrø et al., 2018). A viscosity created by soluble fibers slows the transit time of semi-digested food (chime) in the digestive tract, which causes a reduction in nutrients along with a delay in stomach emptying, hence spreading out the absorption of sugar over a more extended period. Soluble dietary fiber also attaches bile and cholesterol in the digestive region, preventing their recirculation and reabsorption by the human body (Jenkins et al., 2002).

Based on in vivo digestibility, different carbohydrate-rich foods may be classified concerning their effect on postmeal glycemia (Foster-Powell et al., 2002): (1) low glycemic index foods (<55), (2) medium glycemic index foods (55–69), and high glycemic index foods (>70). Several perspective observational studies have shown that consuming a low-GI diet can safeguard against the generation of colon cancer, obesity, and breast cancer. In contrast, the consumption of a diet over many years with a high glycemic load (= GI × dietary carbohydrate content) is independently linked with a high risk of developing heart disease, type 2 diabetes, and certain cancers. Low-GI foods are beneficial for enhancing glycemic control in diabetic subjects as well as in controlling metabolic disorders (Brand-Miller et al., 2009). In a study, it has been observed that after consumption of low-GI products, serum glucose levels and insulin production decreased (Good et al., 2008).

Among new product development scientists, there is an appreciable interest in reducing the GI of highly desirable starchy products by evaluating and promoting the use and consumption of various millets. The constant intake of low-GI millets promotes only a slight rise in blood glucose levels (Giuberti & Gallo, 2018). Sufficient levels of magnesium content enhance the efficiency of glucose receptors and insulin in the body in millets, which helps control diabetes (Rao et al., 2017). The presence of soluble dietary fiber in millets increases the viscosity of the bowel contents, affecting insulin resistance and postprandial (after-the-meal) blood sugar.

Pearl millet grown in several African and Asian countries is known to have a low GI; hence it can be used as replacement food for control of weight and to lower the risk of diabetes mellitus (Martins et al., 2018).

Based on in vitro starch digestibility, starch has been categorized into resistant starch (RS), rapidly digestible starch (RDS), and slowly digestible starch (SDS) (Dona et al., 2010). SDS and RDS are likely to be completely digested in the human small intestine but at variable rates. Because SDS results in a comfortable blood glucose level with time and controls hunger (Lehmann & Robin, 2007), it is usually linked with managing diabetes. RS is that portion of starch that remains indigestible in the small intestine of healthy human beings. Still, it may be partially or fully fermented in the large intestine, generating short-chain fatty acids, like butyric acid, to fight against cancer (Fuents-Zaragoza et al., 2011). As stated by the physicochemical parameters and the source of enzyme resistance, RS has been classified as (Dupuis et al., 2014):

1. RS1 is naturally occurring starch physically entrapped in a matrix inaccessible to enzymes, found in partly milled grains, seeds, and legumes;
2. RS2 is a native granular indigestible starch found in raw banana and raw potato;
3. RS3 is starch formed by the process of retrogradation that takes place naturally during the process of cooking and cooling of starch, for example, in cooked and cooled chapati and rice;
4. RS4 is chemically modified starch that is commercially manufactured; and
5. RS5 is formed due to the complex formation of amylose with lipid.

A research study (Dona et al., 2010) suggested that a change in the physical nature of the suspended starch samples (substrate) determines the kinetics of digestion. Multiple nutritional reports indicated that the consumption of RS can prevent colorectal cancer, promote hypoglycemic effects, lower plasma cholesterol and triglyceride concentrations, inhibit fat accumulation, and enhance vitamin and mineral absorption (Raigond et al., 2015). The latest dietary guidelines recommend the intake of slowly digestible carbohydrates such that RS is at least 14% of the total starch consumed. Researchers are putting in efforts to increase the content of RS to realize its potential in healthcare.

5.4.1.4 Weight Management and Preventing Obesity

Obesity is significantly correlated with the high consumption of carbohydrates. It contributes to a risk factor for diabetes (Tandon et al., 2018) and different organ cancers (e.g., esophagus, colorectum, breast, endometrium, and kidney) (Key et al., 2004). Persons on a whole-grain diet rather than refined flour products have healthier body weights and gain less weight over time. Higher consumption of dietary fibers has been correlated with reduced body weight and fat in women (Tucker & Thomas, 2009). Women consuming a minimum of one serving of whole grain per day had a notably lower waist circumference and body mass index (BMI) than the women with no whole-grain consumption (Good et al., 2008).

Whole grains are a rich source of dietary fiber whose intake decreases obesity (Alfieri et al., 1995). Soluble dietary fiber reduces the calories provided through a diet because of the feeling of fulness, thereby promoting satiety for a longer duration than intake of starch and simple sugars (Pereira & Ludwig, 2001). High-fiber foods have less energy than high-fat and high-protein food. Hence, foods with high fiber can be used to reduce body weight. In some research, it has been shown that β-glucan increases satiety, influences the absorption efficiency in the small intestine, lowers cholesterol level, and reduces body weight (Rondanelli et al., 2011). Increased whole-grain intake was also found to reduce waist circumference and body mass index from data of 13,000 adults (19–50 years old) (O'Neil et al., 2010).

Refined grain flours are a concentrated source of calories with high GI values. Another study showed that subcutaneous and visceral adipose tissues were directly correlated with refined grain consumption while negatively related to whole-grain consumption (McKeown et al., 2010). In randomized cross-over trials, it was reported that without inducing significant changes in the gut microbiome, regular consumption of whole grain may lower body weight and low-grade systemic inflammation (Roager et al., 2017). In a randomized controlled trial related to polyphenols (present in the bound form to the cereal dietary fiber) in reducing inflammation in subjects, it was reported that issues leading to unhealthy dietary and lifestyle were suffering from overweight or obesity (Vitaglione et al., 2015).

5.4.1.5 Protection against Cardiovascular Diseases

Various studies have shown that whole grains can reduce the probability of occurrence of heart disease. Soluble fibers from barley bran, oat bran, and psyllium husk have been reported to tend to decrease blood lipid levels (Martos et al., 2010). The Federal Food and Drug Administration (1997) approved a health claim stating there is a reduction in the level of blood cholesterol with the daily intake of 3 g of β-glucan soluble fiber. Also, due to the presence of magnesium in cereals (particularly millets), there are reports of lowering blood pressure and heart attack, particularly in patients who have atherosclerosis. Furthermore, vasodilation potassium in cereals and millets assists in reducing blood pressure, thereby reducing heart-related problems, and dietary fiber also plays a significant role in attenuating level of cholesterol.

Interestingly, microflora present in the human digestive system can convert the plant lignans present in millets into animal lignans, which may protect against certain cancers and heart disease (Rao et al., 2017). In a research study (Heidemann et al., 2008), it was revealed that the planned intake of a balanced diet (based on whole grains, fruits, vegetables, poultry, and fish) reduced the incidence of cardiovascular disease and reduced related deaths significantly. After analyzing the diet records of 27,000 men (aged between 40 and 75 years) over 14 years, it was concluded that the regular consumption of whole grains (40 g/day) could lower the chances of heart diseases by 20%. Phenolic compounds present in whole grains can provide protective effect against heart diseases by preventing oxidation of low-density lipoprotein-cholesterol (Jensen et al., 2004). The ability of free radical scavenging by simple phenols and their derivatives can prevent and treat many free radical–mediated degenerative diseases, such as atherosclerosis and ischemia, cancer, inflammation, diabetes, infection, radioactive damage, and even Parkinsonism. Also, some properties of phenolics like antiplatelet aggregation and vasodilation might exert a protective effect on the heart.

5.4.1.6 Protective Effect for Certain Cancers

About 20% of the cases of cancer are caused by being obese, along with the high risk of malignant tumors that develop into cancers, which is influenced by the level of physical activity, diet pattern, change in weight, and body fat distribution (Pergola & Silvestris, 2013). In obese people, insulin and insulin-like growth factor-I, steroids, and adipokines are the four main potential factors that can cause cancer. A recent study observed that higher-BMI individuals with the most remarkable dietary potential for hyperinsulinemia or chronic inflammation (characterized by a greater level of insulin in the blood than usual) had the highest rise in multiple myeloma (Lee et al., 2019). Also, obesity is linked with the secretion of rich levels of proinflammatory cytokines by immune cells and endogenous growth factors. Reports from the International Agency for Research into Cancer (2002) and the World Cancer Research Fund (2007) have shown a strong link between obesity and the occurrence of different types of cancers like postmenopausal breast, endometrial, esophageal colorectal, adenocarcinoma, prostate, and renal (Wiseman, 2008)

Dietary habits have an dominant role in preventing and producing cancer (WCRF, 2018). Several types of research have stated that intake of whole grains is persistently related to a lower risk of cancer mortality. Consumption of grains and dietary patterns are related to different cancers in the human body. Cancers such as colorectal, colon, rectal, pancreatic, gastric, breast, prostate, and esophageal are related to the consumption of whole grains (Gaesser, 2020). Whole grain contains cereal fiber, associated with reduced cancer risk (McKeown et al., 2009). Cereal fiber elevates fecal bulk and reduces gastrointestinal transit time (De Vries et al., 2015), which could reduce absorption of carcinogens. Whole grain consumption like wheat increases the production of butyrate, a significant energy source of normal human colon cells (Topping & Cliffton, 2001). Butyrate has also shown effectivity against the growth of cancerous cells, mainly by inducing apoptosis (Young et al., 2005). It is protective against colorectal cancer (Johnson et al., 1994). Whole grain consumption is related lower body mass index and central adiposity (Harland & Garton, 2008), which could reduce adiposity-related cancers. So consumption of whole grains helps reduce the risk of stomach and colon cancers (Haas, 2009). Consumption of millet grains may also be practiced in the regular diet, as millets are known to be rich in compounds like tannins, phenolic acids, and phytates (Thompson, 1993), which have been found to reduce the risk of colon and breast cancer in animal studies (Graf Eaton, 1990). Recent research has revealed that increased consumption of dietary fiber (DF) is the best and most practical mode to control the onset of breast cancer. It is believed that by eating more than 30 g of DF/day, the possibility of breast cancer may be reduced by more than 50%. DF is also present in millets (e.g., sorghum) and the phenolic compounds that can also help lower the incidence of esophageal cancer (Van Rensburg, 1981).

5.4.1.7 Protection against Gluten Intolerance

Gluten-intolerant individuals may encounter three types of medical conditions associated with gluten: celiac disease, allergy to wheat, and nonceliac gluten sensitivity. Celiac disease, wheat allergy, and gluten sensitivity show gastrointestinal symptoms, which range from chronic diarrhea to intestinal pain (Sapone et al., 2012). The major gluten sources are wheat, rye, barley, and oats. Management through a gluten-free diet using alternative sources (such as rice, millets, pseudocereals, legumes, and any other gluten-free developed products) is the only solution for gluten intolerance.

5.4.2 Health Benefits of Pseudocereals

Pseudocereals can replace other cereal crops for improving nutrition, health, and economic status of people. The regular intake is beneficial for preventing malnutrition in children and immunological disorders in adults due to their unique health benefits. The nutritional aspects and health benefits of pseudocereals are listed in Table 5.6.

5.4.2.1 Glycemic Properties

Pseudocereals have a low glycemic index, reducing cardiovascular diseases, diabetes, many types of cancers, and insulin resistance. Foods with high GI have a score >70, a moderate glycemic range is 56–69, and a low GI score is <55. The GI of buckwheat is 45, which helps design foods with low GI. The GIs of amaranth and quinoa grains are 40 and 53, respectively. Studies on diabetic rats indicate that amaranth grains can effectively reduce serum glucose levels. Amaranth grains help prevent hyperglycemia and other diabetic complications (Shukla et al., 2003).

5.4.2.2 Antioxidant Agents

The peptides in pseudocereals help enhance their antioxidant characteristics. Amaranth grains have antioxidant and antithrombotic properties. Among pseudocereals, amaranth grains contain a good quantum of lysine and some other peptides, along with potassium and iron. These bioactive components in amaranth help prevent inflammation and diseases, such as arthritis and gout. The peptides

Table 5.6 Nutritional Aspects and Health Benefits of Pseudocereals

Name of Pseudocereal	Nutritional Aspects	Health Benefits	References
Buckwheat	Buckwheat contains high amounts of amino acids, fibers, and minerals. It is also rich in rutin, polyphenols, and catechins that act as potential antioxidants.	Buckwheat lowers the plasma cholesterol level; acts as a neuroprotectant, anti-inflammatory, anticancer, antidiabetic; and improves hypertension conditions. It also helps in prevention of obesity and diabetes.	(Das etal., 2012)
Quinoa	A high concentration of protein, including all essential amino acids, unsaturated fatty acids, vitamins, and minerals, is present in quinoa. It also has a low glycemic index.	Quinoa supplementation showed its significant beneficial effects on metabolic, cardiovascular, and gastrointestinal health of humans.	(Graf et al., 2015)
Chia	Chia contains good amounts of proteins, phytochemicals, and dietary fiber. It has also a high amount of polyunsaturated fatty acids, including omega-3 and omega-6 fatty acids.	Chia helps to prevent many diseases like cardiovascular diseases, diabetes, obesity, and cancer, and it also strengthens the immune system. Proteins, dietary fibers, and antioxidants of chia seed are used to keep a person healthy.	(Pal & Raj, 2020)
Amaranthus	Amaranthus grain has an appreciable amount of protein with high lysine content.	Amaranth grain shows its health function in decreasing plasma cholesterol levels, reducing blood glucose levels and anemia, and exerting antitumor activity.	(Maurya & Arya, 2018)

of amaranth can inhibit free radicals and mutations of healthy cells (Alvarez-Jubete et al., 2010). Similarly, buckwheat has antioxidant components (such as quercetin and rutin), which help prevent cancer and heart-related diseases. Moreover, these antioxidants are essential for maintaining cardio-vascular functioning in the human body (Larson et al., 2012).

Quinoa seeds have significant antioxidant properties due to the presence of lunasin, which can easily get attached to cancer cells and then split the cells away from the healthy cells (Hermandez-Ledesma, 2013). Laboratory studies on animals by feeding quinoa seeds indicate that the discrete component lunasin separates cancerous cells without harming healthy cells (Ranilla, 2009). Studies on quercetin in quinoa show that it can scavenge free radicals, and therefore it helps prevent many types of cancer, especially lung cancer (Murakami et al., 2008).

Chia seed contains bioactive compounds like caffeic acid, chlorogenic acid, quercetin, myric-etin, and kaempferol, which act as antioxidants, and it has cardiac and hepatic protective effects, as well as anti-ageing and anti-carcinogenic characteristics. So, chia is used for therapeutic purposes in the control of diabetes, dyslipidemia, hypertension, and cancer (Ullah et al., 2016).

5.4.2.3 Improves Bone Quality

Due to calcium in amaranth, it is a fantastic food in preventing osteoporosis and maintaining bone strength. Calcium also plays a significant role in the proper functioning of the heart, muscles, and nerves, particularly in old age. Buckwheat with trace amount minerals (i.e., zinc and selenium) is essential for the human body for strong bones, nails, and teeth. Quinoa seeds are rich in phospho-rous, magnesium, and manganese, which help maintain and repair cells and tissues; filter waste from kidneys; build strong bones and teeth; and produce DNA and RNA, the body's genetic build-ing materials (Ikeda & Yamashita, 1994).

5.4.2.4 Prevents Atherosclerosis

The pseudocereal amaranth has high potassium content, which helps regulate body fluids, lowers blood pressure, decreases stress on the heart, and preserves bone mineral density. It also helps in preventing the formation of kidney stones and prevents atherosclerosis. Amaranth seeds are rich in phytosterols, a family of molecules related to cholesterol. Therefore, low-density lipoproteins can be reduced by the consumption of amaranth grains. Quinoa seeds are rich in fatty acid butyrate, making a healthy gut (Aguilar et al., 2014).

5.4.2.5 Improves Vision

Carotenoids in pseudocereals protect the healthy cells in the eye, reduce muscle degeneration, and prevent cataract formation. With regular pseudocereals, oxygen stress is concentrated in the eye portion. This has makes vision healthy and strong for many years (Schulze et al., 2007).

5.4.2.6 Eliminates Varicose Veins

Flavonoids like rutin in amaranth strengthen the walls of blood capillaries, which helps eradicate vari-cose veins. Vitamin C plays a vital role in collagen production and thus improves the functioning of rutin; also, it assists in repairing and strengthening the walls of blood vessels (Kalinova & Dadakova, 2009).

5.4.2.7 Prevents Asthma Attacks

Buckwheat has a high amount of vitamin E and Mg that prevent asthmatic disorders in children (Colin et al., 2004). Studies have indicated that children consuming fewer cereal grains develop asthmatic disease. Seeds like buckwheat and quinoa have anti-inflammatory properties.

5.4.2.8 Improves Digestion

The dietary fiber content in pseudocereals ranges from 1.1 to 17.3%, depending on the variety. Seventy-eight percent of the fiber in pseudocereals is insoluble and made up of pectic polysaccharides, which help improve digestion. The dietary fiber of amaranth has many gastrointestinal benefits (i.e., efficient uptake of minerals) (Alvarez-Jubete et al., 2010). The dietary fiber of buckwheat is helpful in bowel movement in the digestive tract and improves peristaltic motion.

5.4.2.9 Diabetes Management

The dietary fibers in pseudocereals are beneficial in maintaining blood sugar levels. Buckwheat grains consist of soluble dietary fibers, which prevent diabetes. It contains chiroinositol, which is similar in function to insulin. It helps control type I diabetes. Buckwheat also contains Mg, a component of 300 enzymes present in the human body, and helps regulate type 2 diabetes (Schoenlechner et al., 2008).

5.4.2.10 Gluten-Free

A gluten-free diet does not contain proteins, such as gliadin and glutenin, that form gluten. These proteins are available in wheat, rye, and barley, and the combination of these proteins from gluten results in many celiac conditions. Gliadin and glutenin are not present in pseudocereals, and these are proving a suitable alternative grain for people suffering from gluten intolerance. Quinoa is very helpful due to iron, protein, fiber, and Ca to replace gluten in intolerant persons (Lee et al., 2009).

5.4.2.11 Lower Cholesterol Level

Amaranth seed oil contains a bioactive component called squalene ranging from 1.9 to 11.19%. Squalene is also found in quinoa seeds in the range of 3.39–5.84%. Squalene is a triterpene, which is highly unsaturated and helps lower cholesterol levels in the body by removing sterols through feces. Amaranth leaves have an essential bioactive component called phytosterol, similar to cholesterol. It prevents the absorption of cholesterol in the intestine, thereby lowering the levels of low-density lipoproteins (Alvarez-Jubete et al., 2010). Amaranth is also rich in phylloquinone, which aids in boosting a healthy heart. Moreover, the dietary fibers in amaranth are beneficial in reducing low-density lipoproteins. Vitamin E in amaranth has cholesterol-lowering activity (Narwade et al., 2018).

Buckwheat is a superfood because of is a rich source of phytonutrients. Among the phytonutrients in buckwheat, flavonoids are present in significant amounts. The flavonoid rutin is beneficial in increasing high-density lipoproteins (good cholesterol) and reducing coronary heart diseases. The other role of flavonoids is to prevent platelets from clotting, which can cause cardiovascular and coronary heart diseases (Alvarez-Jubete et al., 2010). Quinoa seeds contain 8% alpha-linoleic acid and oleic acid, healthy fatty acids. Diets rich in alpha-linoleic acid reduce heart diseases by lowering blood cholesterol levels.

5.4.2.12 Reduces Risk of Cancer

Pseudocereals contain flavonoids and phytosterols, nutraceuticals, amino acids, and peptides (proteins), which can provide many biological functions to prevent cancer and other diseases (Kim et al., 2007). The bioactive peptides in pseudocereals are essential in preventing many gastric, breast, and colon cancers (De Lumen, 2008). Besides peptides, excellent dietary fiber intake in the diet lowers the risk of breast cancer, especially in postmenopausal women. Research workers have found that the chances of breast cancer in Swedish women were reduced by half with regular consumption of dietary fiber (Li et al., 2017). Plant lignans are essential in the defence mechanism against hormone-based cancers (Durazzo et al., 2013). Buckwheat can play a vital role in cancer prevention, both colon and gastric (Steadman et al., 2001).

5.5 CONCLUSION

Cereal products are staple diet throughout the world. Cereals contribute a considerable amount of carbohydrates, proteins, fiber, B vitamins, sodium, magnesium, zinc, and selenium. Whole-grain cereals consumed in combination with each other or with legumes cater to the health requirements of mass populations. Consuming whole grains as a part of healthy diet may help in preventing type 2 diabetes, (colon/colorectal) cancer, and cardiovascular diseases and helps in reducing weight gain, obesity, respiratory diseases, and infectious diseases. The intake of whole cereals exerts protective effects by reducing inflammation, increasing the viscosity of chime, enhancing the response of insulin, improving blood lipid profiles, and maintaining gut health. Pseudocereals, as a rich source of nutrients and bioactive compounds (antioxidants), play a major role in prevention of cancers, cardiovascular diseases, celiac disease, diabetes, and so on. However, the role of food products made from refined flour in overcoming several micronutrient and energy deficiencies is also important. Processing of cereals breaks down complex carbohydrates and proteins, reduces the antinutritional factors that interfere with mineral bioavailability, and breaks cellular matrices to release bound phenolic compounds. Thus, the consumer should have a balance in the intake of whole-grain cereals, refined flours, pseudocereals, and processed products in line with the individual's activity level and state of health.

REFERENCES

Abdel-Aal, E.S.M., Choo, T.M., Dhillon, S., Rabalski, I. 2012. Free and bound phenolic acids and total phenolic in black, blue, and yellow barley and their contribution to free radical scavenging capacity. *Cereal Chemistry*, 89(4), 198–204.

Abdel-Aal, E.S.M., Gamel, T.H. 2008. Effects of selected barley cultivars and their pearling fractions on the inhibition of human LDL oxidation in vitro using a modified conjugated dienes method. *Cereal Chemistry*, 285(6), 730–737.

Acosta, E. 2009. Bioavailability of nanoparticles in nutrient and nutraceutical delivery. *Current Opinion in Colloid and Interface Science,* 14(1), 3–15.

Aguilar, E.C., Leonel, A.J., Teixeira, L.G., Silva, A.R., Silva, J.F., Pelaez, J.M.N., Capettini, L.S.A., Lemos, V.S., Santos, R.A.S., Alvarez-Leite, J.I. 2014. Butyrate impairs atherogenesis by reducing plaque inflammation and vulnerability and decreasing NF-κB activation. *Nutrition Metabolism and Cardiovascular Diseases*, 24, 606–613.

Alfieri, M.A. H., Pomerleau, J., Grace, D.M., Anderson, L. 1995. Fiber intake of normal weight, moderately obese and severely obese subjects. *Obesity Research*, 3(6), 541–547.

Altan, A., McCarthy, K. L., Maskan, M. 2009. Effect of extrusion process on antioxidant activity, total phenolics and β-glucan content of extrudates developed from barley-fruit and vegetable by-products. *International Journal of Food Science and Technology*, 44(6), 1263–1271.

Alvarez-Jubete, L., Wijngaarda, H., Arendt, E.K., Gallagher, E. 2010. Polyphenol composition and *in vitro* antioxidant activity of amaranth, quinoa buckwheat and wheat as affected by sprouting and baking. *Food Chemistry*, 119(2), 770–778.

Aoyagi, Y. 2006. An angiotensin-I converting enzyme inhibitor from buckwheat (*Fagopyrum esculentum* Moench) flour. *Phytochemistry*, 67(6), 618–621.

Aune, D., Keum, N., Giovannucci, E., Fadnes, L.T., Boffetta, P., Greenwood, D.C., Tonstad, S., Vatten, L.J., Riboli, E., Norat, T. 2016. Whole grain consumption and risk of cardiovascular disease, cancer, and all cause and cause-specific mortality: Systematic review and dose-response meta-analysis of prospective studies. *British Medical Journal*, 353, i2716.

Awika, J.M., Dykes L., Gu, L.W., Rooney, L.W., Prior, R.L. 2003. Processing of sorghum (*Sorghum bicolor*) and sorghum products alters procyanidin oligomer and polymer distribution and content. *Journal of Agricultural and Food Chemistry*, 51(18), 5516–5521.

Bartlomiej, S., Justyna, R., Ewa, N. 2012. Bioactive compounds in cereal grains occurrence, structure, technological significance and nutritional Benefits. A review. *Food Science and Technology International*, 18(6), 559–568.

Biskup, I., Gajcy, M., Fecka, I. 2017. The potential role of selected bioactive compounds from spelt and common wheat in glycemic control. *Advances in Clinical and Experimental Medicine,* 26(6), 1013–1019.

Blaise, P.N.P., Alexander, M., Schehl, B.D., Zarnkow, M., Gastl, M., Herrmann, M., Zannini, E., Arendt, E.K. 2010. Processing of top fermented beer brewed from 100% buckwheat malt with sensory and analytical characterization. *Journal of Institute of Brewing*, 116(3), 265–274.

Bohn, T. 2014. Dietary factors affecting polyphenol bioavailability. *Nutrition Reviews*, 72(7), 429–452.

Brand-Miller, J.C., Stockmann, K., Atkinson, F., Petocz, P., Denyer, G. 2009. Glycaemic index, postprandial glycemia, and the shape of the curve in healthy subjects: Analysis of a database of more than 1,000 foods. *American Journal of Clinical Nutrition*, 89(1), 97–105.

Butts-Wilmsmeyer, C.J., Mumm, R.H., Rausch, K.D. Kandhola, G., Yana, N.A., Happ, M.M., Ostezan, A., Wasmund, M., Bohn, M.O. 2018. Changes in phenolic acid content in maize during food product processing. *Journal of Agricultural and Food Chemistry*, 66(13), 3378–3385.

Caba, Z.T., Boyacioglu, M.H., Boyacioglu, D. 2012. Bioactive healthy components of bulgur. *International Journal of Food Sciences and Nutrition*, 63(2), 250–256.

Colin, W., Harold, C., Charles, E.W. 2004. *Encyclopedia of grain science*;1700. Amsterdam: Elsevier Academic Press.

Das, A., Raychaudhuri, U., Chakraborty, R. 2012. Cereal based functional food of Indian subcontinent: A review. *Journal of Food Science and Technology*, 49(6), 665–672.

De Lumen, B.O. 2008. Lunasin: A novel cancer preventive seed peptide that modifies chromatin. *Journal AOAC International*, 91(4), 932–935.

De Vries, J., Miller, P.E., Verbeke, K. 2015. Effects of cereal fiber on bowel function: A systematic review of intervention trials. *World Journal of Gastroenterology*, 21, 8952–8963.

Dona, A.C., Pages, G., Gillber, R.G., Kuchel, P.W. 2010. Digestion of starch: *In vivo* and *in vitro* kinetic models used to characterize oligosachharide or glucose release. *Carbohydrate Polymers*, 80(3), 599–617.

Dupuis, J.H., Liu, Q., Yada, R.Y. 2014. Methodologies for increasing the resistant starch content of food starches: A review. *Comprehensive Reviews in Food Science and Food Safety,* 13(6), 1219–1234.

Durazzo, A., Zaccaria, M., Polito, A., Maiani, G., Carcea, M. 2013. Lignan content in cereals, buckwheat and derived foods. *Foods*, 2(1), 53–63.

Dykes, L., Rooney, L.W. 2007. Phenolic compounds in cereal grains and their health benefits. *Cereal Foods World*, 52(3), 105–111.

Edwards, C., Walk, A., Baumgartner, N., Chojnacki, M., Covello, A., Evensen, J., Thompson, S., Holscher, H., Khan, N. 2017. Relationship between whole grain consumption and selective attention: A behavioral and neuroelectric approach. *Journal of the Academy of Nutrition and Dietetics*, 117(9), A93–A101.

El Khoury, D., Cuda, C., Luhovyy, BL., Anderson, G.H. 2012. Beta-glucan: Health benefits in obesity and metabolic syndrome. *Journal of Nutrition and Metabolism* 2012, Article ID 851362, 28 pages.

Fardet, A. 2010. New hypotheses for the health-protective mechanisms of whole-protective mechanisms of whole grain cereals: What is beyond fiber? *Nutrition Research Reviews*, 23(1), 65–134.

Fernandes, S.S., de las Mercedes Salas-Mellado, M. 2017. Addition of chia seed mucilage for reducing fat content in bread and cakes. *Food Chemistry*, 227, 237–244.

Food and Drug Administration. 1997. Food labelling: Health claims; oats and coronary heart disease. *Rules and Regulations, Federal Register*, 62, 3584–3601.

Forbes, A., Escher, J., Hebuterne, X., Kłek, S., Krznaric, Z., Schneider, S., Shamir, R., Stardelova, K., Wierdsma, N., Wiskin, A.E., Bischoff, S.C. 2017. ESPEN guideline: Clinical nutrition in inflammatory bowel disease. *Clinical Nutrition*, 36, 321–347.

Foster-Powell, K., Holt, H.A.S., Brand-Miller, J.C. 2002. International table of glycemic index and glycaemic load values: 2002. *American Journal of Nutrition*, 76, 5–56.

Fuentes-Zaragoza, E., Sánchez-Zapata, E., Sendra, E. Sayas, E., Navarro, C., Fernández-López, J., Pérez-Alvarez, J. 2011. Resistant starch as prebiotic: A review. *Starch*, 63(7), 406–415.

Gaesser, G. A. 2020. Whole grains, refined grains, and cancer risk: A systematic review of meta-analyses of observational studies. *Nutrients*, 12(12), 3756.

Gebru, Y. A., Sbhatu, D.B., Kim, K.P. 2020. Nutritional composition and health benefits of teff (*Eragrostis tef (Zucc.) Trotter*). *Hindawi Journal of Food Quality*, 2020, 9595086.

Giuberti, G., Gallo, A. 2018. Reducing the glycaemic index and increasing the slowly digestible starch content in gluten-free cereal-based foods: A review. *International Journal of Food Science and Technology*, 53(1), 50–60.

Gomez-Caravaca, A.M., Iafelice, G., Verardo, V., Marconi, E., Caboni, M.F. 2014. Influence of pearling process on phenolic and saponin content in quinoa (*Chenopodium quinoa* Willd). *Food Chemistry*, 157, 174–178.

Good C.K.; Holschuh, N.; Albertson, A.M.; Eldridgr, A.L. 2008. Whole grain consumption and body mass index in adult women: An analysis of NHANES 1999–2000 and the USDA pyramid servings database. *Journal of American College Nutrition*, 27(1), 80–87.

Gosine, L; McSweeney, M.B. 2019. Consumers' attitudes towards alternative grains: A conjoint analysis study. *International Journal of Food Science and Technology*, 54(5), 1588–1596.

Graf, B. L., Rojas-Silva, P., Rojo, L. E., Delatorre-Herrera, J., Baldeón, M. E., Raskin, I. 2015. Innovations in health value and functional food development of quinoa (*Chenopodium quinoa* Willd.). *Comprehensive Reviews in Food Science and Food Safety*, 14(4), 431–445.

Graf, E., Eaton, J.W. 1990. Antioxidant functions of phytic acid. *Free Radical Biology and Medicine*, 8(1), 61–69.

Gry, J., Black, L., Eriksen, F.D., Pilegaard, K., Plumb, J., Rhodes, M., Sheehan, D., Kiely, M., Kroon, P.A. 2007. EuroFIR-BASIS: A combined composition and biological activity database for bioactive compounds in plant-based foods. *Trends in Food Science and Technology*, 18, 434–444.

Haas, P. Machado MJ, Anton AA, Silva AS, de Francisco A. 2009. Effectiveness of whole grain consumption in the prevention of colorectal cancer: Meta-analysis of cohort studies. *International Journal of Food Science and Nutrition*, 60(6), 1–13.

Hagiwara, H., Seki, T., Ariga, T. 2004. The effect of pre-germinated brown rice intake on blood glucose and PAI-1 levels in streptozotocin-induced diabetic rats. *Bioscience, Biotechnology, and Biochemistry*, 68(2), 444–447.

Harland, J.I., Garton, L.E.2008. Whole-grain intake as a marker of healthy body weight and adiposity. *Public Health Nutrition*, 11, 554–563.

Heidemann, C., Schulze, M.B., Franco, O.H., van Dam, R.M., Mantzoro, C.S., Hu, F.B. 2008. Dietary patterns and risk of mortality from cardiovascular disease, cancer, and all-causes in a prospective cohort of women. *Circulation*, 118(3), 230–237.

Hermandez-Ledesma, B., Hsieh, C.C., de Lumen, B.O. 2013. Chemo preventive properties of peptide lunasin: A review. *Protein and Peptide Letter*, 20(4), 424–432.

Hillemeier, C. 1995. An overview of the effects of dietary fiber on gastrointestinal transit. *Pediatrics*, 96 (5 Pt 2), 997–999.

Hu, B., Liu, X., Zhang, C., Zeng, X. 2016. Food macro molecule-based nano-delivery systems for enhancing the bioavailability of polyphenols. *Journal of Food and Drug Analysis*, 25(1), 1–13.

Ikeda, S., Yamashita, Y. 1994. Buckwheat as dietary source of zinc, copper, and manganese. *Fagopyrum*, 14, 29–34.

Jenkins, D.J.A., Kendall, C.W.C., Augustin, L.S.A., Martini, M.C., Axelsen, M., Faulkner, D., Vidgen, E., Parker, T., Lau, H., Connelly, P.W., Teitel, J., Singer, W., Vandenbroucke, A.C., Leiter, L.A., Josse, R.G. 2002. Effect of wheat bran on glycemic control and risk factors for cardiovascular disease in type 2 diabetes. *Diabetes Care*, 25, 1522–1528.

Jensen, M.K., Koh-Banerjee, P.; Hu, F.B., Franz, M., Sampson, L., Grønbaek, M., Rimm, E.B. 2004. Intakes of whole grains, bran, and germ and the risk of coronary heart disease in men. *American Journal of Clinical Nutrition*, 80(6), 1492–1499.

Jiang, Y., Wang, T. 2005. Phytosterol in cereal by-product. *Journal of the American Oil Chemists Society*, 82, 439–444.

Johnson, I.T., Williamson, G., Musk, S.R. 1994. Anticarcinogenic factors in plant foods: A new class of nutrients? *Nutrition Research Reviews*, 7, 175–204.

Jonsson, K., Andersson, R., Bach Knudsen, K.E., Hallmanns, G., Hanhineva, K., Katina, K., Kolemainen, M., Kyrø, C., Langton, M., Nordlund, E., Laerke, H.N., Olsen, A., Poutanen, K., Tjønneland, A., Landberg, R. 2018. Rye and health: Where do we stand and where do we go? *Trends in Food Science and Technology*, 79, 78–87.

Kahkonen, M.P., Hopia, A.L., Vuorela, H.J., Rauha, J.P., Pihlaja, K., Kujala, T.S., Heinonen, M. 1999. Antioxidant activity of plant extracts containing phenolic compounds. *Journal of Agricultural and Food Chemistry*, 47, 3954–962.

Kalinova, J., Dadakova, E. 2009. Rutin and total quercetin content in Amaranth (Amaranthus spp.). *Plant Foods for Human Nutrition*, 64(1), 68–74.

Key, T.J., Sachtzkin, A., Willett, W.C., Allen, N.E. 2004. Diet, Nutrition and the Prevention of Cancer. *Public Health Nutrition*, 7(1A):187–200.

Kim, S.H., Cui, C.B., Kang, I.J., Kim, S.Y., Ham, S.S. 2007. Cytotoxic effect of buckwheat (*Fagopyrum esculentum Moench*) hull against cancer cells. *Journal of Medicinal Food*10(2), 232–238.

Klavarez-Jubete, L., Arendt, E. K., Gallagher, E. 2010. Nutritive value of pseudocereals and their increasing use as functional gluten-free ingredients. *Trends in Food Science and Technology*, 21(2), 106–113.

Kulczyński, B., Kobus-Cisowska, J., Taczanowski, M., Kmiecik, D., Gramza-Michałowska, A. 2019. The chemical composition and nutritional value of chia seeds-current state of knowledge. *Nutrients*, 11(6), 1242.

Kulkarni, D.B., Sakhale, B.K., Giri, N.A. 2018. A potential review on millet grain processing. *International Journal of Nutrition Sciences*, 3(1), 1018–1022.

Kumar, P., Yadava, R.K., Gollen, B., Kumar, S., Verma, R.K., Yadav, S. 2011. Nutritional contents and medicinal properties of wheat: a review. *Life Sciences and Medicine Research*, 2011, 22.

Kyrø, C., Tjønneland, A., Overvad, K., Olsen, A., Landberg, R. 2018. Higher whole-grain intake is associated with lower risk of type 2 diabetes among middle-aged men and women: the Danish diet, cancer, and health cohort. *The Journal of Nutrition*, 148 (9), 1434–1444.

Larson, A.J., Symons, J.D., Jalili, T. 2012. Therapeutic potential of quercetin to decrease blood pressure: Review of efficacy and mechanisms. *Advances in Nutrition*, 3(1), 39–46.

Lee, A.R., Ng, D.L., Dave, E., Ciaccio, E.J., Green, P.H. 2009. The effect of substituting alternative grains in the diet on the nutritional profile of the gluten-free diet. *Journal of Human Nutrition and Dietetics*, 22, 359–363.

Lee, D.H., Fung, T.T., Tabung, F.K., Colditz, G.A., Ghobrial, I.M., Rosner, B.A., Giovannucci, E.L., Birmann, B.M. 2019. Dietary pattern, and risk of multiple myeloma in two large prospective US cohort studies. *JNCI Cancer Spectrum*, 3(2): pkz025.

Lehmann, U. Robin, F.2007. Slowly digestible starch its structure and health implications: a review. *Trends in Food Science and Technology*, 18(7), 346–355.

Li, F., Zhang, X., Li, Y., Lu, K., Yin, R., Ming, J. 2017. Phenolics extracted from tartary (*Fagopyrum tartaricum L. Gaerth*) buckwheat bran exhibit antioxidant activity and an antiproliferative effect on human breast cancer MDA-MB-231 cells through the P38/MAP kinase pathway. *Food Function*, 8, 177–188.

Liu, R.H. 2007. Whole grain phytochemicals and health. *Journal of Cereal Sciences*, 46 (3), 207–219.

Majewska, M., Skrzycki, M., Podsiad, M., Czeczot, H. 2011. Evaluation of antioxidant potential of flavonoids: an in vitro study. *Acta Poloniae Pharmaceutica*, 68(4), 611–615.

Martins, A.M.D., Pessanha K.L.F., Pacheco, S., Rodrigues, J.A.S., Carvalho, C.W.P. 2018. Potential use of pearl millet *(Pennisetum glaucum* (L.) R.Br.) in Brazil: food security, processing, health benefits, and nutritional products. *Food Research International*, 109, 175–186.

Martos, M.V., Marcos, M.C.L., L'opez, J.F., Sendra, E., Lopez-Vargas, J.H., Perez-Alvarez, J.A. 2010. Role of fiber in cardiovascular disease: a review. *Comprehensive Reviews in Food Science and Food Safety*, 9(2), 240–258.

Mattila, P., Pihlava, J., Hellstrom, J. 2005. Contents of phenolic acids, alkyl-and alkenyl resorcinol, and avenanthramides in commercial grain products. *Journal of Agricultural and Food Chemistry*, 53(21), 8290–8295.

Maurya, N.K., Arya, P. 2018. Amaranthus grain nutritional benefits: A review. *Journal of Pharmacognosy and Phytochemistry*, 7(2), 2258–2262.

Mayer, R., Cherry, J., Rhodes, D. 1990. Effects of heat shock on amino acid metabolism of cowpea cells. *Plant Physiology*, 94(2), 796–810.

McKeehen, J.D., Busch, R.H., Fulcher, R.G. 1999. Evaluation of wheat (*Triticum aestivam* L.) phenolic acids during grain development and their contribution to fusarium resistance. *Journal of Agricultural and Food Chemistry*, 47(4), 1476–1482.

McKeown, N.M., Troy, L.M., Jacques, P.F., Hoffmann, U., O'Donnell, C.J., Fox, C.S. 2010. Whole grain, and refined grain intakes are differentially associated with abdominal visceral and subcutaneous adiposity in healthy adults: The Framingham heart study. *American Journal of Clinical Nutrition*, 92(5), 1165–1171.

McKeown, N.M., Yoshida, M., Shea, M.K., Jacques, P.F., Lichtenstein, A.H., Rogers, G., Booth, S.L., Saltzman, E. 2009. Whole-grain intake and cereal fiber are associated with lower abdominal adiposity in older adults. *Journal of Nutrition,* 139, 1950–1955.

McMackin, E, Dean, M., Woodside, J.V., McKinley, M.C. 2013. Whole grains and health: attitudes to whole grains against a prevailing background of increased marketing and promotion. *Public Health Nutrition*, 16(4), 743–751.

Murakami, A., Ashida, H., Terao, J. 2008. Multitargeted cancer prevention by quercetin. *Cancer Letters*, 269(2), 315–325.

Narwade, S., Pinto, S. 2018. Amaranth—a functional food. *Concept of Dairy and Veterinary Sciences*, 1(3), 72–77.

Oghbaei, M., Prakash, J. 2013. Effect of fractional milling of wheat on nutritional quality of milled fractions. *Trends in Carbohydrate Research*, 5, 53–58.

O'Neil, C.E., Zanovec, M., Cho, S.S., Nicklas, T.A. 2010. Whole grain and fiber consumption are associated with lower body weight measures in us adults: national health and nutrition examination survey 1999–2004. *Nutrition Research*, 30(12), 815–822.

Orona-Tamayo, D., Valverde, M.E., Rendon, B.N., Lopez, O.P. 2015. Inhibitory activity of Chia (*Salvia hispanica L.*) Protein fractions against angiotensin I-converting enzyme and antioxidant capacity. *LWT-Food Science and Technology*, 64(1), 236–242.

Pal, D., Raj, K. 2020. Chia seed in health and disease prevention: prevent usage and future perspectives. *IJPSR*, 11(9), 4123–4133.

Patil, S.B., Khan, M.K. 2011. Germinated brown rice as a value-added rice product: a review. *Journal of Food Science and Technology*, 48(6), 661–667.

Pereira, M.A., Ludwig, D.S. 2001. Dietary Fiber and body-weight regulation: observations and mechanisms. *Paediatric Clinics of North America*, 48(4), 969–980.

Pergola, G.D., Silvestris, F.2013. Obesity as a major risk factor for cancer. *Journal of Obesity*, 2013, 291546.

Pol, K., Graaf de K., Diepeveen-de Bruin M., Balvers, M., Mars, M. 2020. The effect of replacing sucrose with L-arabinose in drinks and cereal foods on blood glucose and plasma insulin responses in healthy adults. *Journal of Functional Foods*, 73, 1–9.

Quigley, E.M. 2011. Prebiotics and probiotics: their role in the management of gastrointestinal disorders in adults. *Nutrition in Clinical Practice*, 27(2), 195–200.

Raigond, P., Ezekeil, R., Raigond, B. 2015. Resistant starch in food: a review. *Journal of the Science of Food and Agriculture*, 95, 1968–1978.

Ranilla, L.G., Apostolidis, E., Genovese, M.I., Lajolo, F.M., Shetty, K. 2009. Evaluation of indigenous grains from the Peruvian Andean region for anti-diabetes and antihypertension potential using *in vitro* methods. *Journal of Medicinal Foods*, 12(4), 704–713.

Rao, B.D., Bhaskarachary, K., Christina, G.D.A. 2017. *Nutritional and Health Benefits of Millets* (112 ICAR). Rajendranagar: Indian Institute of Millets Research

Rao, B.D., Kalpana, K., Srinivas, K., Patil, J.V. 2015. Development and standardization of sorghum-rich multigrain flour and assessment of its storage stability with addition of TBHQ. *Journal of Food Processing and Preservation*, 39(5), 451–457.

Renzetti, S., Rosell, C. M. 2016. Role of enzymes in improving the functionality of proteins in non-wheat dough systems. *Journal of Cereal Sciences*, 67, 35–45.

Roager, H.M., Vogt, J.K., Kristensen, M., Hansen, L.B.S., Ibrügger, S., Mærkedahl, R.B., Bahl, M.I., Lind, M.V., Nielsen, R.L., Frøkiær, H., Gøbel, R.J., Landberg, R., Ross, A.B., Brix, S., Holck, J., Meyer, A.S., Sparholt, M.H., Christensen, A.F., Carvalho, V., Hartmann, B., Holst, J.J., Rumessen, J.J., Linneberg, A., Sicheritz-Pontén, T., Dalgaard, M.D., Blennow, A., Frandsen, H.L., Villas-Bôas, S., Kristiansen, K., Vestergaard, H., Hansen, T., Ekstrøm, C.T., Ritz, C., Nielsen, H.B., Pedersen, O.B., Gupta, R., Lauritzen, L., Licht, T.R. 2017. Whole grain-rich diet reduces body weight and systemic low-grade inflammation without inducing major changes of the gut microbiome: a randomized cross-over trial. *Gut*, 68(1), 83–93.

Roberfroid, M. 2007. Prebiotics: the concept revisited. *The Journal of Nutrition*, 137(3), 830S–837S.

Rodas, B., Bressani, R. 2009. The oil, fatty acid and squalene content of varieties of raw and processed amaranth grain. *Archivos Latinoamericanos de Nutrition*, 59(1), 82–87.

Rondanelli, M., Opizzi, A., Monteferrario, F., Klersy, C., Cazzola, R., Cestaro, B. 2011. Beta-glucan or rice bran-enriched foods: A comparative crossover clinical trial on lipidic pattern in mildly hypercholesterolemic men. *European Journal of Clinical Nutrition*, 65(7), 864–871.

Ross, A.B., Kamal-Eldin, A., Lundin, E.A., Zhang, J.X., Hallmans, G., Aman P. 2003a. Cereal alkylresorcinols are absorbed by humans. *Journal of Nutrition*, 133(7), 2222–2224.

Ross, A.B., Shephered, M. J., Schüpphaus, M., Sinclair, V., Alfaro, B., Kamal-Eldin, A., Aman, P. 2003b. Alkyl resorcinols in cereals and cereal products. *Journal of Agricultural and Food Chemistry*, 51(14), 4111–4118.

Saikia D., Deka S. C. 2011. Staple food to nutraceuticals. *International Journal of Food Research*, 18, 21–30

Saleh, A.S. M., Zhang, Q., Chen, J., Shen, Q. 2013. Millet grains: nutritional quality, processing, and potential health benefits. *Comprehensive Reviews in Food Science and Food Safety*, 12(3), 281–295.

Sapone, A., Bai, J.C., Ciacci, C. 2012. Spectrum of gluten-related disorders: consensus on new nomenclature and classification. *BMC Medicine*, 10, 13–20.

Saura, J. R., Reyes-Menendez, A., Thomas, S. B. 2020. Gaining a deeper understanding of nutrition using social networks and user-generated content. *Internet Interventions*, 20, 1–9.

Schoenlechner, R., Siebenhandl, S., Berghofer, E. 2008. Pseudocereals Chapter 7; In: *Gluten-free Cereal Products and Beverages*; 464. Arendt, E. and Dal Bello, F. (Eds); New York: Elsevier.

Schulze, M.B., Schulze, M., Heidemann, C., Schienkiewitz, A., Hoffmann, K., Boeing, H. 2007. Fiber and magnesium intake and incidence of type 2 diabetes: a prospective study and meta-analysis. *Archives of Internal Medicine*, 167(9), 956–965.

Shahidi, F., Nazca, M. 1995. Phenolic compounds in grains. In: *Food Phenolics, Source, Chemistry, Effects, Applications*; 36–45. Lancaster, PA: Technomic Publishing Company.

Sheng, S., Li, T., Liu, R.H. 2018. Corn phytochemicals and their health benefits. *Food Science and Human Wellness*, 7(3), 185–195.

Shukla, S., Pandey, V., Pachauri, G., Dixit, B.S., Banerji, R., Singh, S.P. 2003. Nutritional contents of different foliage cuttings of vegetable amaranth. *Plant Foods for Human Nutrition*, 8, 1–8.

Singh A., Sharma, S. 2017. Bioactive components and functional properties of biologically activated cereal grains: a bibliographic review. *Critical Reviews in Food Science and Nutrition*, 57(14), 3051–3071.

Slavin, J. 2004. Whole grains and human health. *Nutrition Research Reviews*, 17(1), 99–110.

Spiller, G.A. 2001. *Handbook of Dietary Fiber in Human Nutrition*. Boca Raton, FL: CRC Press.

Steadman, K.J., Burgoon, M.S., Lewis, B.A., Edwardson, S.E., Obendorf, R.L. 2001. Minerals, phytic acid. Tannin and rutin in buckwheat seed milling fractions. *Journal of the Science of Food and Agriculture*, 81(11), 1094–1100.

Swapna, G., Jadesha, G., Mahadevu, P. 2020. Sweet corn—a future healthy human nutrition food. *International Journal of Current Microbiology and Applied Sciences*, 9(7), 3859–3865.

Tandon, N., Anjana, R.M., Mohan, V. 2018. The increasing burden of diabetes and variations among the states of India: The global burden of disease study 1990–2016. *Lancet Global Health*, 6(12), e1352–1362.

Thakur, P., Kumar, K. 2019. Nutritional importance and processing aspects of pseudocereals. *Journal of Agricultural Engineering and Food Technology*, 6(2), 155–160.

Thompson, L.U. 1993. Potential health benefits and problems associated with antinutrients in foods. *Food Research International Journal*, 26(2), 131–149.

Ti, H., Zhang, R., Zhang, M., Wei, Z., Chi, J., Deng, Y., Zhang, Y. 2015. Effect of extrusion on phytochemical profiles in milled fractions of black rice. *Food Chemistry*, 178, 186–94.

Topping, D.L., Clifton, P.M. 2001. Short-chain fatty acids and human colonic function: roles of resistant starch and nonstarch polysaccharides. *Physiological Reviews* 81, 1031–1064.

Tripathi, V., Singh, A., Ashraf, M.T. 2018. Avenanthramides of oats: medicinal importance and future perspectives. *Pharmacognosy Reviews*, 12(23), 66–71.

Tucker, L.A., Thomas, K.S. 2009. Increasing total fibre intake reduces risk of weight and fat gains in women. *Journal of Nutrition*, 139(3), 576–581.

Ullah, R., Nadeem, M., Khalique, A., Imran, M., Mehmood, S., Javid, A., Hussain, J. 2016. Nutritional and therapeutic perspectives of Chia (*Salvia hispanica* L.): A review. *Journal of Food Science and Technology*, 53(4), 1750–1758.

Upadhyay, A., Karn, S.K. 2018. Brown rice: Nutritional composition and health benefits. *Journal of Food Science and Technology Nepal*, 10, 47–52.

USDA. 2019. *National nutrient database for standard reference legacy release: Full report* (all nutrients) 20067, sorghum grain. Retrieved from https://ndb.nal.usda.gov/ndb/foods/show/20067?n1=%7BQv%3 D1%7D&fgcd=&man=&lfacet=&count=&max=25&sort=default&qlookup=sorghum&offset=&format =Full&new=&measureby=&Qv=1&ds=&qt=&qp=&qa=&qn=&q=&ing=

Van Rensberg, S.J. 1981. Epidemiological and dietary evidence for a specific nutritional predisposition to esophageal cancer. *Journal of the National Cancer Institute*, 67(2), 243–251.

Vanegas, S.M., Meydani, M., Barnett, J.B., Goldin, B., Kane, A., Rasmussen, H., Brown, C., Vangay, P., Knights, D., Jonnalagadda, S., Koecher, K., Karl, J.P., Thomas, M., Dolnikowski, G., Li, L., Saltzman, E., Wu, D., Meydani, S.N. 2017. Substituting whole grains for refined grains in a 6-wk randomized trial has a modest effect on gut microbiota and immune and inflammatory markers of healthy adults. *American Journal of Clinical Nutrition*, 105(3), 635–650.

Verma, B., Huel, P., Chhibbar, R.N. 2009. Phenolic acid composition, antioxidant capacity, and alkali hydro-lyzed wheat bran fractions. *Food Chemistry*, 116(4), 947–954.

Vitaglione, P., Mennella, I., Ferracane, R., Rivellese, A.A., Giacco, R., Ercolini, D., Gibbons, S.M., La Storia, A., Gilbert, J.A., Jonnalagadda, S., Thielecke, F., Gallo, M.A., Scalfi, L., Fogliano, V. 2015. Whole-grain wheat consumption reduces inflammation in a randomized controlled trial on overweight and obese sub-jects with unhealthy dietary and lifestyle behaviours: role of polyphenols bound to cereal dietary fiber. *American Journal of Clinical Nutrition*, 101(2), 251–261.

Vuholm, S., Nielsen, D.S., Iversen, K.N., Suhr, J., Westermann, P., Krych, L., Andersen, J.R., Kristensen, M. 2017. Whole-grain rye and wheat affect some markers of gut health without altering the fecal microbiota in healthy overweight adults: A 6-week randomized trial. *Journal of Nutrition*, 147(11), 2067–2075.

Vuksan, V., Whitham, D., Sievenpiper, J.L. 2007. Supplementation of conventional therapy with the novel grain salba (*Salvia hispanica L.*) improves major and emerging cardiovascular risk factors in type 2 diabetes: results of a randomized controlled trial. *Diabetes Care*, 30(11), 2804–2810.

Wang, T., He, F., Chen, G. 2014. Improving bioaccessibility and bioavailability of phenolic compounds in cereal grains through processing technologies: a concise review. *Journal of Functional Foods*, 7, 101–111.

Wiseman, M. 2008. *The Second World Cancer Research Fund/American Institute for cancer research Expert Report. Food, Nutrition, Physical Activity, and the Prevention of Cancer: A Global Perspective.* Nutrition Society and BAPEN Medical Symposium on 'Nutrition Support and Cancer Therapy,' Proceedings of the Nutrition Society, 67(3), 253–256.

World Cancer Research Fund/American Institute for Cancer Research. 2018. *Food, Nutrition, Physical Activity, and the Prevention of Cancer: A Global Perspective.* World Cancer Research Fund/American Institute for Cancer Research; Washington. DC: USA.

Young, G.P., Hu, Y., Le Leu, R.K., Nyskohus, L. 2005. Dietary fibre and colorectal cancer: A model for envi-ronment–gene interactions. *Mol. Nutr. Food Res.* 49, 571–584

Zettel, V, Hitzmann, B. 2018. Applications of Chia (*Salvia hispania L.*) in food products. *Trends in Food Science and Technology*, 80, 43–50.

Zhu, F. 2018. Anthocyanins in cereals: composition and health effects. *Food Research International*, 109, 232–249.

Zielińska D., Zieliński H. 2009. Low molecular weight antioxidants and other biologically active components of buckwheat seeds. *The European Journal of Plant Science and Biotechnology*, 3, 29–38.

CHAPTER 6

Post-Harvest Losses of Cereals

Samreen Ahsan, Atif Liaqat, Muhammad Farhan Jahangir Chughai, Adnan Khaliq,
Tariq Mehmood, Ayesha Siddiqa, Ayesha Ali, Kanza Saeed, Shoaib Fayyaz, Nimra Sameed,
Syed Junaid-ur-Rehman and Aqib Saeed

CONTENTS

6.1 INTRODUCTION

Post-harvest losses (PHLs) of cereal are losses that occur throughout the supply chain of grains from harvesting to end consumption. Most post-harvest losses result during handling practices and at the farm level (Gustavsson et al., 2011). These losses can occur in various forms like weight reduction that happens due to spoilage, viable potential, quality and nutritional value (Aulakh et al., 2013). The Food and Agriculture Organization (FAO) reported that various forms of wastage together caused about 19% of total losses of grain cereals (FAO, 2014; Lipinski et al., 2013). The cereal supply chain

DOI: 10.1201/9781003252023-6

starts right from harvesting and further involves many operations like threshing, cleaning of grains, washing, drying, storage, transportation and processing operations. Therefore, most of the losses occur at these stages due to improper handling operations, poor storage facilities, inadequate transportation and inefficient processing units. One of the major reasons for these cereal grains losses is insect/pest infestation and microbial spoilage that deteriorate stored grains and represent one of the major threats to food security (Chaboud & Daviron, 2017).

In the cereal supply chain, handling is one of the primary and most critical stages that play important roles in the overall control of quality and quantity of cereal grains. Previously harvesting was performed manually by sickles, knives, cutters and so on, which caused more quantitative losses (Sawicka, 2019). With advancements in agricultural mechanization, now harvesting of various cereal grains is performed by combined harvesters that help to control the losses due to poor labor practices. But still, losses at harvesting largely depend on crop harvest time, moisture content and harvest methods. Types of losses at the harvesting stage are edible proportion left due to poor handling, harvesting practice, plowing into the soil, insect/pest or bird attack on the mature crop and method and time of harvesting of crop being unsuitable (Olorunfemi & Kayode, 2021). Early-stage harvest causes losses as grains have high moisture content and are more susceptible to microbial spoilage, especially mold growth, and demand more drying cost. The resulting yield also decreases by a significant level due to a higher moisture percentage in grains. The optimal harvesting stage is about a 20–30% moisture level (De Lucia & Assennato, 2006). Late-harvest grains also cause losses due to shattering and are more susceptible to attack by birds/rodents. Natural climacteric disasters also cause substantial losses in late harvest conditions. According to reported studies, post-harvest losses of wheat due to late harvesting increased to 67% (Grover & Singh, 2013), while a 10% higher loss of paddy rice resulted from poor harvesting machinery (Elumalai et al., 2015). After harvesting grains, threshing and cleaning through screening or winnowing is performed to remove grains and separate whole grains or foreign matter. Losses at this stage are about 4% due to winnowing or inadequate treatment that promotes infestations, resulting in loss of sensory characteristics (Sarkar et al., 2013). Drying is performed to reduce the grain's moisture content to <13% to improve shelf life, as it controls microbial and insect spoilage. But poor drying practices like open-sun drying result in grains losses of about 3.5–4.5% as birds feed on it and dust and dirt also contaminate it (Abass et al., 2014; Kebede et al., 2019).

Cereal grains are usually stored for longer period, as produce is harvested in a particular period and used throughout the year. Therefore, post-harvest losses of all cereals are observed to be maximum in storage, particularly in traditional storage structures. In developing countries, due to the traditional type of storage systems, losses are about 40 to 50% (Grover & Singh, 2013). Maize losses were observed to be the highest, at ~60% in a traditional storage system in only a 3-month period (Costa, 2014). Transportation also causes losses due to inadequate transport facilities of about 2 to 10% (Alavi, 2011; Kitinoja & Tokala, 2018). Milling is one of the critical steps of cereal grain processing, as after that step, it proceeds to the final processing step for consumption purposes. Evidence confirmed that milling losses are quantitatively highest, about 70%, mostly due to milling practices (Alavi, 2011). Milling machines also contribute to quality loss, as in traditional mills that incorporate a foreign matrix (Qu et al., 2021).

6.2 STAGES OF POST-HARVEST LOSSES

Post-harvest loss is the degradation of quantity and quality of a food product from harvesting until consumption. Quality losses of food include detrimental changes in its caloric/nutrient values, consumer acceptance and edibility. Such losses are more prevalent in developing countries. Quantity losses of food include degradation in the amount of the food. According to a report by the FAO, the magnitude of lost and degraded food is higher in downstream phases along the food chain as compared to high-income regions (Hengsdijk & De Boer, 2017; Vilariño et al., 2017).

6.2.1 Harvesting

Crop harvesting is regarded as the initial stage of the food supply chain and serves an important part in determining the overall quality of a cereal crop. Crop harvesting is mainly performed by manual means in developing countries using various tools including sickles and cutters. Among others, harvesting time and method (manual vs machinery) are considered important factors for assessing post-harvest losses during various operations. Significant losses occur during the harvesting operations of cereal crops if not performed optimally and if the physiological characteristics (maturity, moisture content, grain size and shape) of crops are neglected (Šotnar et al., 2018).

Harvesting a cereal crop at early stages with high moisture content renders it vulnerable to microbial growth, insect attack and high drying cost and results in broken grains and reduced yield (Schmidt et al., 2018b). However, delayed harvesting of a mature cereal crop leads to shattered losses, exposing the crop to bird and rodent attack. Such crops are also damaged by natural calamities (rain, wind and hailstorms). Developed countries mostly employ manual means of harvesting, which require more labor and are time consuming as well. A severe labor shortage is commonly faced by most countries during peak harvesting seasons, resulting in delayed harvesting of crops and more losses. Shattering losses for wheat were found to be 67% in Punjab, India, due to delayed harvesting (Alexander et al., 2017; Manandhar et al., 2018).

6.2.2 Threshing and Cleaning

Threshing operations for cereal crops are performed to separate grain from the panicles. It can be done by rubbing, pounding, stripping or a combination of these actions. Both manual (trampling and beating) and mechanical means can be used for this purpose. Developed countries mostly use manual means of threshing. Major reasons for post-harvest losses of cereals include incomplete removal of grain from chaff, grain breakage while pounding and grain spillage (Belay & Fetene, 2021; Dumitru et al., 2020). Delayed threshing after harvesting results in significant losses of cereal crops both in terms of qualitative and quantitative aspects, as the crop becomes susceptible to atmospheric factors, rodents and insect and bird attacks. Lack of mechanization and unavailability of farm equipment contribute to delayed harvesting of cereal crops. High moisture content may also lead to mold growth by providing favorable conditions for growth (Paulsen et al., 2015).

Followed by threshing, a cleaning operation is performed in order to isolate compact grains from damaged grains and foreign particles, including stones, chaff and straw. Winnowing is widely employed by developing countries for this purpose. Screening is another alternative to achieve this goal and can be done by both manual and mechanical means. Insufficient cleaning of grains can result in insect attack and microbial growth, impart unpleasant taste and disturb appearance while in storage and can also affect the processing machinery. A significant quantity of cereal grains are wasted at this step due to spillage and post-harvest losses, while screening may reach up to 4% of the total grain production (Kumar & Kalita, 2017).

6.2.3 Drying

The harvesting of cereal crops at the farm level is usually done at high moisture of the crop to alleviate the losses occurring due to shattering. Usually moisture below 13% is considered safe for storage over a longer period of time in most cases for cereal crops. Improper drying of grains can contribute to mold growth and result in significant post-harvest losses during storage and processing operations. That's why drying after harvesting is considered an important step in order to maintain quality of the crop, reduce losses during storage and minimize transportation cost. A drying effect can be achieved both by natural or mechanical means, such as sun drying and drying by using hot air ovens. Natural drying techniques including sun drying are regarded as economical and conventional means for drying cereal crops. Sometimes the whole crop is left in the field without performing threshing operations to dry under sunlight (Bradford et al., 2020; Kumar & Kalita, 2017).

For instance, after wheat harvesting, the crop is tied in the form of bundles and left in the field to dry under sunlight. Sun drying relies mainly on the weather conditions, requires more labor, is time consuming and also results in more losses. Furthermore, grains left in an open field are susceptible to bird and insect damage and also get contaminated with stones, dust and various foreign particles present in the environment. Sudden rains and harsh weather conditions may limit proper drying operation, and the cereal crop is stored with high moisture content, which leads to mold growth and more losses. Around 4.5% and 5.5% post-harvest losses were observed during drying of maize in various regions of Zambia and Zimbabwe. The application of plastic sheets is a common practice by farmers for drying grains, which reduces the chances of contamination by dirt and eases the process of grain collection (Affognon et al., 2015; Kumari et al., 2015). Mechanical drying, on the other hand, is useful and addresses certain concerns of natural drying, including reduction in handling losses, temperature regulation and space utilization. However, mechanical drying requires high maintenance cost, sufficient size availability, proper training and knowledge and optimum processing conditions. Due to such restrictions, mechanical dryers are mostly used by smallholders in developed countries (Gao et al., 2016; Kumar & Kalita, 2017).

6.2.4 Milling

Milling operations vary according to the nature of crops and grains. Concerning rice crops, the objective of the milling operation is to separate the husk and bran covering the paddy and get whole white rice kernels, which are later used for human consumption. The milling operation can be done both by manual and mechanical means. Milling machines are widely used by food processing facilities for this purpose. Traditionally, milling is done by continuous pounding of cereal crops in rural areas (Bartholomeu et al., 2017). The yield of such crops depends largely on the milling technique being employed, expertise of the operator and conditions of crop during the milling operation. Milling of adulterated rice grains leads to the production of cracked and damaged kernels. Such kernels can also affect milling machines. Improper operation of milling machines can also damage kernels and affect yield. According to a survey, post-harvest milling losses are highest in five countries: Indonesia, Philippines, China, Vietnam and Thailand. The yields from rural areas were reported to be 55% due to small scale, inexperienced workers operating the milling machine, poor calibration of milling equipment and maintenance. Cereal crops with high moisture content and improperly cleaned paddy lead to lower yields (Cardoen et al., 2015; Kumari et al., 2015).

6.2.5 Transportation

Transportation of cereals and grains is a critical step in the grain value chain, as products need to be transported from one place to the other such as farm to processing sites, farm to storage sites and processing sites to market. The unavailability of proper transport facilities causes damage to food commodities through bruising and spillage. Transportation losses are lower in developing countries around the world due to proper road infrastructure and engineered sites on the farm and industrial level. Food commodities in such regions are loaded and unloaded with minimum damage. At farm sites, crops are mainly transported via carts or containers in Asian countries. Grains for domestic use are usually transported in plastic sheets or bags from farm to processing sites. Poor roads and improper modes of transportation lead to larger spills and increased contamination. Continuous transport of crop from one place to another is also a major factor for increased losses during transportation (Amentae et al., 2016; Bradford et al., 2020).

In Pakistan and India, harvested and bagged wheat is transported multiple times from trucks before it is processed for milling. During this operation, various wheat grains are lost due to spillage. As compared to bulk handling systems employed in developing countries, loading and unloading from vehicles at processing facilities is done manually and results in more losses. Inferior-quality jute bags are usually used by processing facilities for transportation and storage of grains (Chegere, 2018). These bags cause more spillage due to leakage, and since large quantities of grains are stored in bags, the losses are huge. Furthermore, hooks used to carry these bags result in tears and lead to more spillage. Trucks are commonly used to carry cereals in most countries. The improper handling and unsuitability of trucks to carry cereals and oil seed crops causes losses during processing and transportation in Asian countries (Dumitru et al., 2020).

6.2.6 Storage

Cereal storage is an important step in the food supply chain, and various studies have shown that most post-harvest losses occur at this step. In various regions, cereal crops are grown according to the season and after harvesting are kept for a certain period of time as food reserves for later use and as seeds for the coming season crop. In countries like Pakistan, it is a common practice to store grains in traditional structures like kanaja, earthen containers and kacheri at the domestic level for household consumption. Grain storage containers are made from local materials such as grass, wood or mud without any particular scientific design, and such structures also do not protect grains against pests (Kumar & Kalita, 2017; Swai et al., 2019).

6.3 POST-HARVEST LOSSES OF CEREAL GRAINS

Major cereal grains in growing countries are wheat, maize and rice, and all others come after these three. Developing countries export a large number of grains to foreign countries other than national export countries. The quality of grains should be maintained, as good-quality grains produced as exports raise demand, and those with moderate quality result in losses not only of quality and yield of cereal grains but also of health and the economy. Bangladesh was fourth in the list of largest rice-producing countries, and it fulfills its dietary demand of rice consumption by importing by controlling post-harvest losses (Kumar & Kalita, 2017). According to World Statistics data, about 8.6% of total cereal grains and pulses undergo post-harvest catastrophe from the farm level to consumer. It is predicted that the cereal loss that occurs in India annually could overcome the demand of increasing population even though it was second in rice production worldwide. Lack of proper handling and processing, storage conditions and mechanization are the major reasons for post-harvest losses, as well as lack of knowledge of Good Manufacturing Practices (GMPs) and Good Agricultural Practices (GAPs), resulting in wide damage at a small scale or at the farm level. Even in developed countries like Sweden, most of the cereal undergoes PHLs, as maize losses are about 11.7% due to inadequate post-harvest practices (Chegere, 2018). According to a report published by the World Bank, it is estimated that around 10% of grain losses occur during post-harvest operations, 3–4% during distribution operations in market and 5–10% due to improper storage conditions. It was also reported that around 10 to 15 million tons of cereal grains are discarded every year, which could sufficiently meet the food demands of around one-third of the population. However, regardless of the seriousness of the issue, the continuous and consistent availability of post-harvest losses is still a challenging scenario. Various research institutes are making efforts to study and assess the actual scenario of post-harvest losses along food supply chains in numerous regions across the world (Alam et al., 2018; Brown et al., 2020). Reducing post-harvest losses by optimizing processing operations from farm to fork can be helpful in coping with food demands and reducing the burden on the economy (Alhassan & Kumah, 2018; Stathers et al., 2018).

6.3.1 Wheat

Wheat is a major cereal crop and staple food for most countries including Europe, Asia, America, China and Russia. It is greatly influenced by storage, environmental conditions or insect/pest attack, resulting in various quality and yield losses of produce (Kumar & Kalita, 2017). According to a report published by the US National Academy of Sciences, post-harvest losses of wheat in Sudan were reported to be around 5–20%. Among various post-harvest operations, storage losses were found to be the greatest for wheat despite the fact that wheat has a short storage period (Majumder et al., 2016).

Wheat losses are observed to be greatest when storage conditions are not proper, the structure of the commodity or area in which the wheat is stored is not adequate and pest or rodent entry is not blocked properly due to weather conditions and field damage. Among post-harvest wheat losses, storage losses are highest, at about 41% even at short storage periods (Bala et al., 2010). In 2020–21, the annual production of wheat was 772.64 million metric tons (USDA, 2021), while 10–30% of production was lost at supply chain stages like harvesting, storage, transport and processing.

In a survey conducted in India, data were compiled from 50 farmers, 20 wholesalers, 15 retailers and 20 food processors from the major crop producer of each district in Karnataka state. The overall post-harvest losses in the whole wheat supply chain from harvesting to retailer were found to be 4%. Farm operations resulted in 75% of the total post-harvest losses. The losses were found to be greatest during storage operations because of the unavailability of storage facilities, poor infrastructure, rodent attack and improper drainage facilities (Amentae et al., 2017).

6.3.2 Maize

Maize, scientifically known as *Zea mays*, is the staple food of Brazil, Mexico, the United States and other countries. The United States is the largest producer of corn/maize, with an annual production of 360.25 million metric tons in 2020–21. After the United States, China and Brazil are the largest producers. In Brazil the annual loss of maize production is approximately 9.34%. Recent evidence reported post-harvest losses in Brazil due to soybean and maize to combine production of 12–16% of annual yield, approximately 3.75 million tons individually for each of the two crops (Hampf et al., 2021). On average 24% of the maize stored is flossed by combined factors of storage, rodents, pests and by rotting in Ethiopia. The main weevil that catalyzes maize damage is *Sitophilus zeamais*. Studies showed that in developing countries, most cereal grains are stored by traditional methods, mostly at small scale for daily consumption, and thus lack the proper conditions needed for ideal storage. If maize is stored traditionally in granny bags or polystyrene bags, it results in a high percentage of loss, about 59% after a 90-day storage period (Costa, 2014).

6.3.3 Rice

Rice is the most-grown crop in the world and is extensively consumed in developing and developed countries. Rice accounts for 20% of the global calorific value and is among the high-energy caloric foods. Moreover, it is principally a staple food consumed in more than half of countries, including China, Bangladesh, India, Indonesia and so on. The extent of post-harvest losses in the rice supply chain vary and rely on the economic conditions, agricultural and ergonomic practices and climatic factors of the region (Sadiya & Hassan, 2018). Food and agriculture organizations reported that about 10 to 35% losses of rice occur in south Asian regions (Alavi, 2011).

In regions such as Bangladesh, rice production accounts for more than 75% of the total food production and around 75% of calorie intake. In West African regions, Nigeria is among the major producers of rice, with an annual production of about 3 million tons. Regardless of the massive production and imports, rice is still unable to meet the nutritional needs of people in Nigeria. Bangladesh is among the top four rice producers across the globe, but it is still food deficient and meets the requirements of its population by importing over a million tons of rice each year (Alhassan et al., 2018; Stathers et al., 2018). Post-harvest losses in rice have been reported to be minimum at 3% to maximum at 24% in India and Nigeria, respectively. Based on the 24% losses in Nigeria, the worth of total grain loss in the rice supply chain was around 55 billion Nigerian naira. The post-harvest losses occurring in the rice value chain from producer to retailer were around 11% in Bangladesh. Most of the post-harvest losses were reported to occur at the farm level and farm processing operations including harvesting, threshing, screening and cleaning. In a study provided by the FAO, it was found that 10–20% of total post-harvest losses in rice occur in Southeast Asia, and 5–25% losses were reported in China (Gao et al., 2016; Nath et al., 2016).

Post-harvest losses of rice usually occur due to poor harvest practices and during the transportation of rice. Combine harvesting is usually employed in rice harvesting and is more efficient than segment harvesting, because during segment harvesting, there are pile up and bunching losses along the cutting and bundle losses. Similarly, during the threshing and flailing of rice, there are approximately equal losses, and leftover rice obtained after beating and threshing is not used for commercial sale purposes, as it become loaded with dust. Losses during the winnowing of rice contribute a large part of the losses, as the forced air pushes away the lighter straw and rice. Inappropriate packing, bags and stacking of bags cause significant losses during transportation (Gummert et al., 2020). About 30 to 50% of post-harvest losses of rice can be controlled through the application of mechanized machinery and training staff, but it carries a high operational cost and is costly for small stakeholders (Gummert et al., 2020). Most of PHLs occur at the farm level, resulting in a maximum loss of rice of >85%, followed by storage losses going up to 40% (Kumar & Kalita, 2017).

6.4 FACTORS CONTRIBUTING TO POST-HARVEST LOSSES

Huge losses occur during storage of crops due to inadequate infrastructure. Two categories are suggested to categorize storage losses:

- Direct losses: These are because of products that are physically lost
- Indirect losses: These are because of losses in the quality and nutritional status of maize

Both loss during storage and insect damage are considered important factors rather than weight loss for stored grain loss. "Damage" can be related to physical signs of spoilage, such as bean punctures. Grain quality is mainly affected by this. The total vanishing of the food that can be calculated quantitatively is called "loss." The loss of quality leads to a reduction of the value of the products and occasionally even to a complete refusal of the products. The refusal rate is dependent on the cultural background and economic situation of the individual. For example, a farmer belonging to a poor family may consume spoiled food to a small extent, while wealthy families reject foods when food is even slightly spoiled. Some losses occur because of spills from leaking bags that can be seen when bags are stored in empty places and grain can be seen on the floor (Kiaya, 2014; Schmidt et al., 2018a).

Various factors can affect storage losses, and they can be categorized into two main classes:

- Living factors (insects, pests, fungi, rodents)
- Non-living factors (humidity, rainfall, temperature)

The most crucial factors that can affect storage life are temperature and moisture content. Feasible requirements for mold growth in storage conditions are that most storage molds grow rapidly at a relative humidity of more than 70% and temperature range of 20–40°C. Mold growth is limited by low humidity levels, which keep relative humidity below 70%. In conventional storage facilities, climatic changes lead to temperature fluctuations and can cause moisture to collect at every part of the grain mass, and it depends on the air convection direction. It can be minimized by limiting the difference in temperature between the outside and inside of the storage facility. Kernels are required to be dried to around a moisture content of up to 13% to reduce losses before storage. At a moisture content of more than 16%, the safe storage period for rice is only a few weeks. Another crucial factor for grain storage is grain quality, which can also cause losses during storage. Threshing and harvesting lead to mechanical damage on affected areas of the grain, which can contribute as a focus of infection and lead to spoilage. The cruciality of all the factors depends on the conditions in which grains are being stored (Hodges & Maritime, 2012; Kiaya, 2014; Manandhar et al., 2018).

6.4.1 Insect Infestation

Among all living factors, pests and insect are evaluated as the most crucial and lead to huge grain losses (30–40%). Researchers in Ghana reported that insect infestation contributes to maize losses of as much as 50%, if all quality, quantity and economic losses because of early sale estimation are taken into account. It was reported that pests and insects cause losses up to 80–90% of cereals that are being stored. *Callosobruchus maculatus* (F.), a legume weevil, has shown to cause up to 24% of stored legume losses in Nigeria (Ekeh *et al.*, 2013). Losses by pest insects in corn storage of about 12 to 44% have been reported in the western highlands of Cameroon. The grain borer (LGB) and corn weevil (*Sitophilus zeamais* and *Prostephanus truncatus*) are the major pests of corn. Losses of around 23% were seen in maize grains that have been stored for only 6 months, and the main cause was observed by an infestation of LGB in Benin and corn weevil. LGB was found in Central America and was accidentally introduced to Africa in the late 1970s. Today, it is observed in almost all parts of Africa and is thought to be the most devastating pest, which can lead to a significant loss in a very short time. The spore-forming nature of LGB even makes it difficult to handle because it does not attack all the stores in the same vicinity and it is not certain that it will appear every year. At the farm storage level, a weight loss of more than 30% has been suspected in maize because of pest attack. In Ghana, a few studies of loss in maize crops have estimated a loss in market value of around 5 to 10% because of infestation by *Sitophuilus* spp. In general, in this area, family income loss was estimated at about 5% (Ahmed, 1983; Alonso-Amelot & Avila-Núñez, 2011; Dobie, 1974; Hodges & Maritime, 2012; Keskin & Ozkaya, 2015).

6.4.2 Mycotoxins

Contamination by mycotoxins is also a huge challenge, chiefly in corn, which makes the feed unfit for utilization by humans and animals. A huge quantity of cereal grains about (25–40%) are contaminated with mycotoxins, which are produced by storage fungi worldwide. Mycotoxins and molds lead to loss in quality and dry matter as well as inducing risk in the food supply chain. The most important mycotoxins commonly found in maize species include fumonisins, aflatoxins, ochratoxin and deoxynivalenol. *Aspergillus flavus* and *A. parasiticus* aflatoxins are two fungal species that are responsible for producing secondary metabolites, aflatoxins, and are supposed to be the most devastating group of mycotoxins, increasing the chances of liver cancer and affecting the growth of adolescents. In developing countries, aflatoxins are responsible for food contamination, and about 4.5 billion people suffer from it. Aflatoxicosis is caused by high concentrations of aflatoxins and leads to serious illness and can even prove fatal. *Penicillium verrucosum* (ochratoxin), an important mycotoxigenic mold, is usually found in cool, humid temperature ranges (e.g. northern Europe), and in temperate and tropical conditions, *Aspergillus flavus* is commonly observed.

Grain damage and germination are reduced by molds that affect them during storage. They also degrade grain quality because of the musty smell. They also result in reduced sugar and starch contents and enhanced fatty acid contents. Food spoilage is also caused by lipid peroxidation, which also imparts flavor and smell and can lead to adverse impacts on human health. Oilseed varieties with more oil content require special observation while in storage, as higher levels of humidity deteriorate vegetable oil, result in high levels of fatty acids and occasionally may lead to self-heating.

In storage on farms in developing countries, rodents are also responsible for crop damage, while fungi also contribute to causing spoilage during storage at high relative humidity. The utilization of science-based structures for storage and effective handling of grains can reduce losses during storage to less than about 1%. Losses can also be reduced by physically preventing rodent and insect entry by sustaining environmental factors that prevent microorganisms from growing. If control points are known during harvesting and drying before storage, it can help reduce grain storage losses. Taking control measures for living factors in time can be very efficient in lowering losses while storing grains (Mesterházy et al., 2020; Wilson et al., 2021).

6.5 STRATEGIES TO REDUCE POST-HARVEST LOSSES ALONG THE GRAIN VALUE CHAIN

It is a great challenge for the food supply chain to reduce PHLs along the grain chain by controlling insects, pests, and microbial infestation. Post-harvest losses can occur at any stage during harvesting, threshing, drying, storage and transportation. Reducing grain post-harvest losses generates more revenue than improving yield by genetic variation (Mesterházy et al., 2020).

Biodeterioration of cereal grains by attack of insects, pests, rodents, birds, mold and fungi also leads to post-harvest losses and quality deterioration. Moreover, inadequate storage of grains causes insect infestation and facilitates the exposure of grains to moisture. Adoption of better post-harvest practices is the most promising way to prevent post-harvest losses. This involves the careful handling of grains during all stages of the grain value chain, improving current post-harvest technologies and introducing novel technologies. So the first and foremost priority is to prevent post-harvest losses. If insects, pests and microbes gain entry into the grain value chain, then disinfection techniques are employed for their inactivation, and adequate storage techniques must be adopted to prevent recontamination (Hodges et al., 2013).

6.5.1 Strategies to Reduce Field (Pre-Harvest) Losses

The key challenge in cereal production is to reduce loss in crop yield initiated by microbial, insect and pest attack. It was reported that the application of plant protection techniques improved crop theoretical yield from 42% in 1665 to 70% in 1990. But 30% of the theoretical yield is still lost to inadequate pest management practices. The absence of effective pest management practices leads to a loss of crop yield up to 70% by pest attack (Mesterházy et al., 2020). Therefore, the following protection systems must be implemented on-farm to reduce losses:

- Integrated crop management (ICM)
- Integrated pest management (IPM)

Integrated crop management is a universal economically feasible approach that utilizes modern techniques to enhance crop production and prevent insect and pest attacks. The main aim of ICM is to produce high-quality food along with a reduction in pesticide usage and its cost. Along with ICM, an integrated pest management system must also be implemented on-farm. IPM involves physical, chemical and biological practices to control insect and pest infestation in the agriculture system. It also utilizes insect predators or parasites for effective pest control (Bagheri et al., 2019; Hagstrum & Flinn, 2018).

6.5.2 Strategies to Reduce Post-Harvest Losses

After cleaning and drying, grain is stored in various storage structures that are an integral part of the grain supply chain. Losses during storage include direct (loss in the physical structure of the grain) and indirect losses (loss in nutritional quality). There are many biotic and abiotic factors responsible for these losses during storage. Temperature and moisture content are two crucial factors that influence the storage life of cereals (Mesterházy et al., 2020). Most cereal crops are stored in various storage structures after harvest for later use across the season or next season. Various storage methods are adopted, including on-farm storage in traditional structures in developing countries and storage in metal silos in developed countries. The latter is the better method and leads to a reported reduction in grain losses of up to 1–2% for grain stored in storage silos in comparison to 20–50% losses in the grain storage in traditional storage structures and on-farm storage. Conventional storage structures are not scientifically designed and are constructed of wood, mud or grass that does not protect against pests for a longer period of time (Jayas, 2012; Kumar & Kalita, 2017). It was reported that 59% of maize is lost during 90 days of storage in indigenous storage structures made from polypropylene bags and granaries (Costa, 2014). Moreover, biotic factors such as insects, pests and rodents also account for about 10–20% loss of stored grains. On average, about 420 metric tons (mt) of grains are lost annually during storage (Phillips & Throne, 2010). Therefore, storage methods must be adopted to provide maximum protection and reduce storage losses. Grains must be cleaned before storage and contaminants such as weeds, seeds, straws, insects and chaff must be eliminated. Toxins must be tested, and grains with varying levels of toxins must be stored separately. Moreover, temperature and relative humidity must be monitored accurately within storage structures, as high moisture and heat result in an increased probability of fungal and insect attack (Mesterházy et al., 2020). Despite the enormous challenges, the adoption of advanced storage infrastructures, good storage practices and good storage technology can reduce cereal grain storage losses (Kumar & Kalita, 2017). Therefore, various disinfection techniques and storage strategies or technology interventions are applied to reduce waste during storage.

6.5.2.1 Disinfection Techniques

Grain infestation by insects, rodents and molds can be reduced by chemical methods (chemical fumigation) and non-chemical methods (infrared and microwave heating), as well as non-thermal methods (ozonation and ultraviolet heating) (Srivastava & Mishra, 2021).

6.5.2.1.1 Chemical Fumigation

Chemical disinfection is an extensively employed method around the globe for the control of insects and pests. Fumigation is the best method and is cheap and easy to use and has greater penetration power. Ethyl formate, phosphine, sulfuryl fluoride and methyl bromide are commonly employed fumigants, especially in developing countries. Moreover, other fumigants such as propylene oxide, carbonyl sulfide and allyl isothiocyanate are also used for the disinfection of grains (Paul et al., 2020). Phostoxin is also employed to control infestations of larger grain borer in dry maize grains. Actellic super is also used by farmers to control pest attacks in grain stored for months in polypropylene bags (De Groote et al., 2013).

Despite their effectiveness, the application of many fumigants has been limited or banned by authorities in many parts of the world due to their toxicity, associated health risks, adverse impacts on environments and probability of development of resistance in treated pests. The application of fumigants on grain stored in conventional storage structures is a challenging task, as there is a chance of reinfestation. Another major challenge is lack of knowledge about application techniques, time and dosage of the chemicals that may reduce the effectiveness of treatment and result in high losses during storage (Paul et al., 2020).

6.5.2.1.2 Ozone Treatment

Ozone (O_3) is highly reactive triatomic oxygen that is generated by the combination of oxygen molecules with a free radical of oxygen. It has been generally recognized as safe (GRAS) and has also been approved by the FDA for its direct application on foods as antimicrobial agent. After application, ozone is converted into molecular oxygen, leaving no toxic residue on the product. It has potent bactericidal effect against a wide range of microbes. The mode of action involves the oxidation of proteins, amino acids and peptides, ultimately leading to cell death and microbial inactivation. It has been widely applied on cereals for microbial decontamination, especially as a fungicide (Conte et al., 2020). The application of gaseous ozone was reported to reduce *Aspergillus parasiticus* contamination up to 63% in corn (Kells et al., 2001). Moreover, ozone in gaseous form was also found to be effective for its application on wheat, as it reduced fungal spores up to 96% (Wu et al., 2006). Mycotoxin contamination in wheat and barley can be inactivated by the application of ozone. This treatment exerts no adverse impact on the grain's physical and nutritional parameters. Ozone treatment has also been reported to improve the storage characteristics of various cereal grains such as rice, corn and wheat (Sujayasree et al., 2021).

6.5.2.1.3 Thermal Treatments

Pasteurization and sterilization are the most widely employed thermal techniques, especially in the food industry. Thermal treatments are also applied in various forms for post-harvest disinfection of grains such as superheated steam, hot water dips and dry heat with varied time/temperature combinations (Schmidt et al., 2018a).

- Dry heat treatment

Dry heat treatment is considered a potent alternative to the chemical disinfection method. It is also found to be effective in controlling fungal infestation with minimum effect on grain viability.

It was reported that the application of dry heat with a time/temperature combination of 60°C for 15 days, 70°C for 5 days and 80°C for 2 days was found to be effective to obstruct the germination of microbial spores and fungus in wheat grains with no adverse impact on grain viability. However, the application of a higher temperature of 90°C for 5 days led to substantial loss in grain viability (Gilbert et al., 2005). Thus, it is recommended to apply dry heat treatment before storage of grain for reduction of moisture content and inactivation of microbes, thus reducing the post-harvest spoilage of the grain and ultimately post-harvest losses. However, dry heat treatments are time and energy consuming and can take many days. Dry heat treatment alone is also not effective against mycotoxin degradation due to their greater heat resistance. Another limitation of hot water treatment is the particle size of the treated sample, as it is more effective for milled grain in comparison to whole grain due to reduced particle size (Chang et al., 2015).

- Wet heat treatment

Wet heat treatment to control microbial and fungal infestation in grains has emerged as a more effective method in comparison to dry heat treatment (Syamaladevi et al., 2016). Wet heat treatment includes both applications of superheated steam with temperatures up to 250°C along with saturated water (up to 100°C). Superheated steam is gaining more popularity due to higher enthalpy and increased heat transfer rate, leading to a rapid rise in the temperature in contrast to saturated steam (Jin et al., 2021). Many researchers have reported that the application of superheated steam even for a period of less than 30 sec is enough to control the vegetative form of bacterial and fungal species in cereal grains, and no adverse impact on sensory attributes and nutritional quality of the product was reported. Fungal and bacterial spores are more resistant to heat in comparison the vegetative form, thus posing a major challenge (Ban & Kang, 2016; Chang et al., 2015; Hu et al., 2016). Nevertheless, superheated steam was reported to be effective in inhibiting or destroying the spores of *Geobacillus stearothermophilus* during an exposure time of 20 min. Another technique, ultra-superheated steam, that uses temperatures of 400–500°C has attracted the attention of researchers.

Application of ultra-superheated steam (210–250°C) to wheat, rye and barley for up to 15 sec inhibited fungal spoilage the most, though few data are available on the application and effectiveness of ultra-superheated steam (Bari et al., 2015).

6.5.2.1.4 Ionizing Radiation

Ionizing radiation arising from nuclear and electromagnetic sources has found potential applications in food processing and preservation. This radiation offers great penetration power irrespective of the size of the package (Paul et al., 2020). This is radiation with shorter wavelengths and higher energy, including:

1. Gamma irradiation
2. Electron beam irradiation
3. X-rays

Gamma irradiation is widely applied to agricultural products to prevent sprouting, delay the ripening process and control pest infestation. Along with gamma radiation, electron beam radiation and X-rays are also applied to control pest infestation. The mechanisms of inhibition involve the rise in temperature by increasing the irradiation dose, leading to a delay in development, ceasing reproduction and halting respiration in insects, ultimately leading to cell death. Irradiation also causes disruption of the cell cycle, cell lesions and mutation in DNA, thus killing insects and pests. Irradiation was approved by the FDA in 1963 for application on wheat and wheat flour as an insect disinfection method. The WHO has recognized irradiated food as safe when treated with irradiation doses up to 10 kGy in 1981. Electron beam irradiation has low penetration power in comparison to gamma rays and X-rays. Thus, it should be employed in synergy with other techniques or on free-flowing grains to attain maximum inactivation. Despite the advantage, the application of irradiation may lead to vitamin loss up to 15%, off-flavor development and textural changes in the product. Moreover, due to high operational costs, it cannot be utilized by small farmers (Paul et al., 2020).

6.5.2.1.5 Non-Ionizing Radiation

Non-ionizing radiation is also excessively applied to cereal for decontamination and control of insect, pest and fungal infestation. Due to its non-ionizing nature, it has no adverse effects on product quality. Moreover, non-ionizing disinfection techniques have greater consumer acceptability in comparison to ionizing radiation. Non-ionizing radiation techniques involve:

1. Microwave treatment
2. Ultra-sonication
3. Infrared treatment

Microwave treatment is considered the best substitute for other chemical methods that are employed to control microbial, fungal and insect infestation in grains. It is a safe, environment-friendly, highly efficient method but may affect product quality. Microwaves are electromagnetic radiation with a frequency ranging from 0.3 GHz to 300 Hz and wavelength between 1 mm and 1 m. The movement of molecules in the pulsating electromagnetic field leads to an increase in the internal pressure of the product, thus inhibiting microbes. It may also result in nucleic acid, protein and enzyme denaturation. The inhibition of enzymes may also affect the germination ability of the grains in downstream processing. Nevertheless, optimizing microwave treatment operational parameters can entirely control the microbial counts in cereal without affecting grain enzymatic activity (Yadav et al., 2014). A number of studies have suggested that application of microwaves reduced *F. graminearum* (seed-borne infection) in wheat up to 7% without affecting the seed germination level. It may affect the seed viability adversely (Los et al., 2018). Moreover, microwaves with a frequency of 2450 MHz and energy output of 1.25 kW applied for 120 sec would be effective for fungal inhibition, thus substantially lowering the production of mycotoxin in cereals (Schmidt et al., 2018a). The use of microwaves in combination with other techniques such as irradiation, high temperature and chemical disinfectants can effectively disinfect cereal grain by reducing the radiation dose and keeping product quality intact. Conclusively, this will become a major technology to reduce post-harvest losses of cereal grains (Yadav et al., 2014). Limitations of microwave treatment include uneven heat distribution and damage to the grains if applied for a longer time.

Ultrasound is a sound wave with a frequency higher than the human audible threshold and can be classified as high-frequency ultrasound (2–20 MHz) and power ultrasound (20–100 kHz). High-power ultrasound has been considered a novel technique that is used extensively for microbial disinfection in foods. However, there are limitations of the application of this method on cereals, as ultrasound waves are generated in a liquid medium and only applied to liquid foods. It is not clear that ultrasonic waves can produce cavitation through cereal crops powerful enough to inactivate microbes. No research is available on the application of ultrasound on solid food products (Schmidt et al., 2018a).

Infrared (IR) treatment or infrared heating is a novel technique also widely utilized for drying as well as disinfection of cereal grains due to its ability to promote fast and uniform heating. IR heating leads to the denaturation of protein and nucleic acid of microbes, thus inactivation fungal growth in cereals.

IR with selected wavelengths trailed by tempering is applied to crops for inactivation of mycotoxigenic fungi. The application of IR with a wavelength of 3.3 for 60 sec lowered *Aspergillus* contamination in corn up to 22%. Therefore, selected IR systems can be scaled up for application on a larger scale for the inactivation of fungal species in corn and various other cereals (Wilson et al., 2021).

6.5.2.1.6 High Hydrostatic Pressure

One recent measure to control post-harvest losses of cereals is high pressure treatment. It is mostly employed to control microbial spoilage either through the inactivation of microorganisms or of fungal spores (Heinz & Buckow, 2010). It is an emergent approach that is recognized as safe and widely adopted because it does not cause any adverse effect on sensory characteristics or nutritional value of food products (Marti et al., 2014). Mycotoxin spoilage is one of the major issues in cereal storage life. Now high hydrostatic pressure (HHP) is widely employed for microbial decontamination of cereal grains by applying high pressure of about 100–800 MPa for a few seconds to minutes. High hydrostatic pressure treatment for cereal grain decontamination is about 400 MPa and effectively inhibits fungal spoilage without quality damage. Heat-resistant fungal spores are difficult to destroy by a single HHP treatment; therefore, it is combined with temperature (20 to 50°C) to kill spores (Black et al., 2007). Thus, HHP treatment combined with temperature is a promising, efficient approach for cereal post-harvest loss control, especially at the storage step.

6.5.2.2 Storage Techniques

There are many promising storage techniques that have the potential to prevent insect and pest infestation and ultimately reduce post-harvest losses. The choice of appropriate storage techniques depends on the type of the crop, farmers' economic condition, scale of production and affordability (Elik et al., 2019). Some storage techniques that can reduce post-harvest losses of cereal grains are discussed in the following:

- Cold or refrigerated storage
- Modified atmosphere storage technique
- Warehouse storage

- On-farm and community storage structures:
  - Self-build silos
  - Hermetic storage
  - Metallic silos
  - Hermetic bagging technology

6.5.2.2.1 Cold or Refrigerated Storage

Cold or refrigerated storage is economical and the most common method employed for storage of cereal grains and seed conservation. The technique is suitable to use because it avoids post-harvest losses in cereals by preventing insect infestation, mold growth, heat buildup, fermentation, toxin production and grain metabolism without the use of chemicals and insecticides. The main objective of refrigeration storage in hot climates is to achieve a temperature less than 18°C to prevent insect infestation (Navarro et al., 2016). In this technique, the temperature of the ambient air is reduced and passed over bulk grains with an aeration system. The initial installation cost of this system is higher, but along with the dehumidified air technique, this system provides significant possibilities for safe commercial storage, especially in tropical regions (Pekmez, 2016).

6.5.2.2.2 Modified Atmosphere Storage Technique

Modified atmosphere storage has successfully replaced the traditional disinfection technique of fumigation (Navarro et al., 2016). Cold atmosphere storage and modified atmosphere storage not only reduce post-harvest losses by controlling insects but also preserve the quality of produce without any residue. These techniques are an alternative to fumigation chemicals used for controlling pests in oilseeds, cereals and grains (Pekmez, 2016). Integrated pest management can be used along with aeration, chilling with refrigerated air storage and bio-generated modified air for ensuring insect and quality control. In controlled atmospheres, storage disinfection is achieved by altering the nitrogen, carbon dioxide and oxygen content in storage areas and silos. Insects, being aerobic in nature, due to desiccation cannot survive this atmosphere of reduced oxygen. Modified atmospheres include hermetic storage, vacuum or high-pressure carbon dioxide treatment (Navarro, 2012). The degree of modification is high in controlled as compare to modified atmospheres. Both of these techniques are non-toxic and preserve the quality of cereal grains without posing a threat to environment. They are more effective when applied in combination with other techniques such as hurdle technology and other disinfection techniques. Modified atmosphere packaging to control or prevent deterioration in fresh produce is also a widely studied and practiced technique.

6.5.2.2.3 Warehouse Storage

Warehouses are a common grain storage practice in developing countries. Cover and plinth structures or warehouses are common housed storage practices to protect cereal grains (Manandhar et al., 2018). Community-run or government-based warehouses are common in developing countries where small-scale farmers can store their grains in bags by paying a fee. Cover and plinth storage refers to the storage of cereal bags by stacking on wooden pallets. These bags are then covered with waterproof low-density polyethylene sheets on four sides. Grains can be stored for 6 to 12 months by employing this method. The warehouse or cover and plinth structures allow storage at large scale and provide protection against elements but are not suitable to protect grains from insects, pests and molds. In such systems, pest and insects control is achieved by using pesticides and fumigants. Poison baits are used for rodent control.

6.5.2.2.4 On-Farm and Community Storage Structures

On-farm storage bins or silos provide farmers with better control of produce. In comparison to self-build silos, tower silo storage bins have a much higher capacity, allowing grain drying to desired moisture content before storage. Grains are stored at a safer moisture level and protected from harsh weather, insects and pest and rodent attack. The use of fumigants and chemicals is required for further protection. The challenge for small farmers is the investment for building these storage structures at the farm level, especially at the individual level, unless they invest in groups (Edwards, 2018). Condominium space for storage provided by commercial elevators manages quality storage throughout a period but requires investment and strong technical skills.

6.5.2.2.5 Self-Build Silos

These silos are made up of corrugated iron roof material (galvanized), or high-density polyethylene sheets act as insulation between the environment and stored grains. These storage structures are popular in India and Africa, particularly for small farmers due to their low cost. They are called pusa bins (Manandhar et al., 2018).

6.5.2.2.6 Hermetic Storage

In developing countries, "sealed storage" or hermetic storage for pulses, beans and cereals is becoming popular. This is also termed "airtight storage." The reason for the popularity of this technique is its chemical additive– and pesticide-free nature, which make it safer for the end user. A hermetic seal creates an atmosphere of high carbon dioxide inside the package, thereby creating a modified atmosphere environment in sealed packaging. Aerobic organisms such as insects and microorganisms are unable to grow in this airtight packaging where carbon dioxide is increased and oxygen depleted, thus creating an inhibitory environment (self-regulated modified atmosphere packaging). The amount of carbon dioxide inside the package is used as an index of biological activity of grains (Cardoso et al., 2008). It has been reported that at such high concentrations of carbon dioxide, aflatoxin production from *Aspergillus flavus* is also reduced (Tefera et al., 2011). For international shipments over long distances, storage losses can be decreased by employing hermetic storage (<1%) (Villers et al., 2010). Hermetic storage offers advantages such as safety due to elimination of pesticide usage, installation ease and cost-effective and simple infrastructure. In a trial study conducted by the World Food Programme, it was found that proper hermetic sealing units were found to be effective themselves for killing insect pests without the use of fumigation (Costa, 2014). Some hermetic storage options developed in the last years include Purdue improved cowpea storage (PICS), metallic silos and SuperGrain bags. These effective, less costly and practical storage techniques are practiced in a few countries today (Zeigler et al., 2014).

- *Metallic Silos*

A metallic silo is a hermetically sealed cylindrical structure constructed from galvanized iron sheet and has been used widely for the safe storage of harvested cereal grains from the attack of rodents and insects. In some countries metal silos made of painted aluminum sheets are also gaining popularity due to their corrosion-free nature and improved appearance. A metal silo is an airtight container that eliminates oxygen inside and kills insects that may gain entry. Hermetic metal silos discourage mold formation by providing a barrier to moisture exchange between the outer environment and inside (Chigoverah et al., 2016). Thus metal silos are an effective post-harvest technology to ensure food safety and conserve food losses. As compared to traditionally used chemicals and storage methods, metal silos are found to be most effective for cereal grain storage (Manandhar et al., 2018). However, for safe storage of cereal in metal silos without mold growth, the moisture content of grains must be less than 14% (Gitonga et al., 2013). A survey was conducted in which a group of farmers who adopted metal silos for grain storage for a period of 1.8–2.4 months were compared to non-adopters. It was found that farmers who adopted metal silos for grain storage lost only 3 Kg, with an average worth of $2, compared to non-adopters, who lost 198 Kg worth $132 (Manandhar et al., 2018). In developing countries, with 5 to 6 average family members, the most suitable storage capacities of metal silos are 820 and 550 Kg (Tefera et al., 2011). Further, in order to maintain storage effectiveness, control of oxygen level, maintenance of the hermetic seal and pressure decay tests are required for efficient storage.

- **Hermetic bagging technology**

Hermetic bagging technology includes Purdue improved cowpea storage and SuperGrain bags. In this technology, two-layer high density polyethylene bags (HDPE) along with a polypropylene layer are used for grain storage. A triple-bagging hermetic storage technique is employed in PICS bags; these bags are common in America and Latin America (De Bruin et al., 2012). The bag consists of two layers of HDPE of thickness 80 microns to limit oxygen permeability, and one layer of woven polypropylene ensures physical durability and acts as the casing of two HDPE inner layers (Murdock & Baoua, 2014). SuperGrain bags consist of one HDPE layer that ensures hermetic sealing with an oxygen barrier and propylene bag for protection. As compared to traditional woven plastic bags, PICS and SuperGrain bags are found to be effective for safe grain storage under similar conditions (Baoua et al., 2013). A major disadvantage of using hermetic sealing bags is their susceptibility to mechanical damage, perforations from insect damage and abrasion. Additionally these bags can rupture while shifting or in transport from one place to another (De Groote et al., 2013).

6.6 CONCLUSION

Post-harvest loss of cereals throughout the grain value chain during harvesting, drying, milling, transportation and storage is multifaceted problem. Bio-deterioration of cereal caused by insect, pest and fungal infestation is a major reason for post-harvest losses, along with inadequate storage techniques. Therefore, many disinfection techniques and storage technologies are adopted to reduce post-harvest losses. Many conventional (chemical fumigation, ozonation, dry heat treatment) and novel disinfection techniques (ionizing radiation, non-ionizing radiation, high hydrostatic pressure) are extensively employed to control insect, pest and microbial activity. Optimal results are reported by applying technologies in synergy. Moreover, many emerging storage technologies such as hermetically sealed bags, metallic silos and multilayer systems have also contributed to a substantial reduction in post-harvest losses.

REFERENCES

Abass, A. B., Ndunguru, G., Mamiro, P., Alenkhe, B., Mlingi, N., and Bekunda, M. 2014. Post-harvest food losses in a maize-based farming system of semi-arid savannah area of Tanzania. *Journal of Stored Products Research* 57: 49–57.

Affognon, H., Mutungi, C., Sanginga, P., and Borgemeister, C. 2015. Unpacking postharvest losses in sub-Saharan Africa: a meta-analysis. *World Development* 66: 49–68.

Ahmed, H. 1983. Losses incurred in stored food grains by insect pests-a review. *Pakistan journal of Agriculture Research* 4(3), 198–207.

Alam, M., Ahmed, K., Sultana, A., Firoj, S., Hasan, I. M. 2018. Ensure food security of Bangladesh: Analysis of post-harvest losses of maize and its pest management in stored condition. *Journal of Agricultural Engineering and Food Technology* 5(1): 26–32.

Alavi, H. R. 2011. *Trusting trade and the private sector for food security in Southeast Asia.* London: World Bank Publications.

Alexander, P., Brown, C., Arneth, A., Finnigan, J., Moran, D., and Rounsevell, M. D. J. a. S. 2017. Losses, inefficiencies and waste in the global food system. *Agricultural systems* 153: 190–200.

Alhassan, N. F., Kumah, P. 2018. Determination of postharvest losses in maize production in the upper West region of Ghana. *American Scientific Research Journal for Engineering, Technology, and Sciences* 44(1): 1–18.

Alonso-Amelot, M. E., and Avila-Núñez, J. L. 2011. Comparison of seven methods for stored cereal losses to insects for their application in rural conditions. *Journal of stored products research* 47(2): 82–87.

Amentae, T. K., Hamo, T. K., Gebresenbet, G., and Ljungberg, D. 2017. Exploring wheat value chain focusing on market performance, post-harvest loss, and supply chain management in Ethiopia: The case of Arsi to Finfinnee market chain. *Journal of Agricultural Science* 9(8): 22.

Amentae, T. K., Tura, E. G., Gebresenbet, G., and Ljungberg, D. 2016. Exploring value chain and post-harvest losses of teff in Bacho and Dawo districts of central Ethiopia.

Aulakh, J., Regmi, A., Fulton, J. R., and Alexander, C. E. 2013. *Estimating post-harvest food losses: Developing a consistent global estimation framework.* 2013 Annual Meeting, August 4–6, 2013, Washington, DC. Agricultural and Applied Economics Association.

Bagheri, A., Bondori, A., and Damalas, C. A. 2019. Modeling cereal farmers' intended and actual adoption of integrated crop management (ICM) practices. *Journal of Rural Studies* 70: 58–65.

Bala, B., Haque, M., Hossain, M. A., and Majumdar, S. 2010. Post harvest loss and technical efficiency of rice, wheat and maize production system: Assessment and measures for strengthening food security. *Final Report CF* 6(08).

Ban, G.-H., and Kang, D.-H. 2016. Effectiveness of superheated steam for inactivation of *Escherichia coli* O157: H7, *Salmonella typhimurium, Salmonella enteritidis* phage type 30, and *Listeria monocytogenes* on almonds and pistachios. *International Journal of Food Microbiology* 220: 19–25.

Bari, L., Ohki, H., Nagakura, K., and Ukai, M. 2015. Application of ultra superheated steam technology (USST) to food grain preservation at ambient temperature for extended periods of time. *Advances in Food Science and Nutritional Sciences* 1: 14–21.

Bartholomeu, D. B., da Rocha, F. V., Péra, T. G. and Vicente, J. 2017. Postharvest losses in the wheat logistics chain: A Brazilian case study. *Journal of Agricultural Science and Technology* 6: 321–329.

Belay, D., and Fetene, M. 2021. The effect of moisture content on the performance of Melkassa multicrop thresher in some cereal crops. *Bioprocess Engineering* 5(1): 1–10.

Black, E. P., Setlow, P., Hocking, A. D., Stewart, C. M., Kelly, A. L., and Hoover, D. G. 2007. Response of spores to high-pressure processing. *Comprehensive reviews in food science and food safety* 6(4): 103–119.

Bradford, K. J., Dahal, P., Van Asbrouck, J., Kunusoth, K., Bello, P., Thompson, J., and Wu, F. 2020. The dry chain: Reducing postharvest losses and improving food safety in humid climates. *Trends in Food Science & Technology* 71: 84–93.

Brown, P. R., Singleton, G. R., Belmain, S. R., Htwe, M. M., Mulungu, L., and Mdangi, M. 2020. Advances in understanding rodent pests affecting cereal grains. In: Maier, Dirk E., (ed.) *Advances in postharvest management of cereals and grains. Burleigh Dodds series in agricultural science (88).* Cambridge: Burleigh Dodds Science Publishing Limited, pp. 1–30.

Cardoen, D., Joshi, P., Diels, L., Sarma, P. M., Pant, D. 2015. Agriculture biomass in India: Part 2. Post-harvest losses, cost and environmental impacts. *Resources, Conservation and Recycling* 101: 143–153.

Chaboud, G., and Daviron, B. 2017. Food losses and waste: navigating the inconsistencies. *Global Food Security* 12: 1–7.

Chang, Y., Li, X. P., Liu, L., Ma, Z., Hu, X. Z., Zhao, W. Q., and Gao, G. T. 2015. Effect of processing in superheated steam on surface microbes and enzyme activity of naked oats. *Journal of Food Processing and Preservation* 39(6): 2753–2761.

Chegere, M. J. 2018. Post-harvest losses reduction by small-scale maize farmers: The role of handling practices. *Food Policy* 77: 103–115.

Chegere, M. J. J. F. P. 2018. Post-harvest losses reduction by small-scale maize farmers: The role of handling practices. *Food Policy* 77: 103–115.

Conte, G., Fontanelli, M., Galli, F., Cotrozzi, L., Pagni, L., and Pellegrini, E. 2020. Mycotoxins in feed and food and the role of ozone in their detoxification and degradation: An update. *Toxins* 12(8), 486.

Costa, S. J. 2014. Reducing food losses in sub-Saharan Africa. *An 'Action research' evaluation trial from Uganda and Burkina Faso.* https://documents.wfp.org/stellent/groups/public/documents/special_initiatives/WFP265205.pdf

De Groote, H., Kimenju, S. C., Likhayo, P., Kanampiu, F., Tefera, T., and Hellin, J. 2013. Effectiveness of hermetic systems in controlling maize storage pests in Kenya. *Journal of stored products research* 53: 27–36.

De Lucia, M., and Assennato, D. 2006. Agricultural engineering in development: post-harvest operations and management of foodgrains. *FAO Agricultural Services Bulletin (FAO).* https://agris.fao.org/agris-search/search.do?recordID=XF2006446709. Agricultural engineering in development: post-harvest operations and management of foodgrains (fao.org)

Dobie, P. 1974. The laboratory assessment of the inherent susceptibility of maize varieties to post-harvest infestation by *Sitophilus zeamais* Motsch. (Coleoptera, Curculionidae). *Journal of Stored Products Research* 10(3–4), 183–197.

Dumitru, O.-M., Iorga, S., Vlădut, N.-V., and Brăcăcescu, C. 2020. Food losses in primary cereal production. A review. *INMATEH-Agricultural Engineering* 62(3): 133–142.

Ekeh, F.N., Onah, I.E., Atama, C.I., Ivoke, N. and Eyo, J.E., 2013. Effectiveness of botanical powders against Callosobruchus maculatus (Coleoptera: Bruchidae) in some stored leguminous grains under laboratory conditions. *African Journal of Biotechnology, 12*(12).

Elik, A., Yanik, D. K., Istanbullu, Y., Guzelsoy, N. A., Yavuz, A., and Gogus, F. 2019. Strategies to reduce post-harvest losses for fruits and vegetables. *Strategies* 5(3): 29–39.

Elumalai, K., Pramod, K., Kedar, V., and Abraham, H. 2015. Assessment of pre and post harvest losses of rice and red gram in Karnataka. *Agricultural Situation in India 72*(9): 101–105.

Gao, L., Xu, S., Li, Z., Cheng, S., Yu, W., Zhang, Y., . . . Wu, C. J. 2016. Main grain crop postharvest losses and its reducing potential in China. *Transactions of the Chinese Society of Agricultural Engineering* 32(23): 1–11.

Gilbert, J., Woods, S., Turkington, T., and Tekauz, A. 2005. Effect of heat treatment to control *Fusarium graminearum* in wheat seed. *Canadian Journal of Plant Pathology* 27(3): 448–452.

Grover, D., and Singh, J. 2013. Post-harvest losses in wheat crop in Punjab: Past and present. *Agricultural Economics Research Review* 26(2): 293–297.

Gummert, M., Cabardo, C., Quilloy, R., Aung, Y. L., Thant, A. M., Kyaw, M. A., . . . Singleton, G. R. 2020. Assessment of post-harvest losses and carbon footprint in intensive lowland rice production in Myanmar. *Scientific reports* 10(1): 1–13.

Gustavsson, J., Cederberg, C., Sonesson, U., Van Otterdijk, R., and Meybeck, A. 2011. *Global food losses and food waste.* Rome: FAO.

Hagstrum, D. W., and Flinn, P. W. 2018. Integrated pest management *Integrated management of insects in stored products.* London: CRC Press, pp. 399–407

Hampf, A. C., Nendel, C., Strey, S., and Strey, R. 2021. Biotic yield losses in the Southern Amazon, Brazil: Making use of smartphone-assisted plant disease diagnosis data. *Frontiers in plant science* 12: 548.

Heinz, V., and Buckow, R. 2010. Food preservation by high pressure. *Journal für Verbraucherschutz und Lebensmittelsicherheit* 5(1): 73–81.

Hengsdijk, H., and De Boer, W. J. 2017. Post-harvest management and post-harvest losses of cereals in Ethiopia. *Food Security* 9(5): 945–958.

Hodges, R., Bennett, B., Bernard, M., and Rembold, F. 2013. Tackling post-harvest cereal losses in sub-Saharan Africa. *Rural 21: The International Journal for Rural Development,* 47(1), 16–18.

Hodges, R. J., and Maritime, C. 2012. Postharvest quality losses of cereal grains in sub-Saharan Africa. *Afr. Postharvest Losses Information System* 22: 7–22.

Hu, Y., Nie, W., Hu, X., and Li, Z. 2016. Microbial decontamination of wheat grain with superheated steam. *Food Control* 62: 264–269.

Jayas, D. S. 2012. Storing grains for food security and sustainability. *Agricultural Research* 1(1): 21–24.

JIN, C., GUO, J., ZHU, H., and WEN, J. 2021. Optimization of superheated steam treatment conditions for wheat aleurone layer flour. *Food Science and Technology* 1–8.

Kebede, L., Getnet, B., Lema, Y., Alebachew, M., and Ageze, M. 2019. Post-harvest processes and advances to introduce loss-reducing technologies for rice. *Advances in Rice Research and Development in Ethiopia* 185.

Kells, S. A., Mason, L. J., Maier, D. E., and Woloshuk, C. P. 2001. Efficacy and fumigation characteristics of ozone in stored maize. *Journal of Stored Products Research* 37(4): 371–382.

Keskin, S., and Ozkaya, H. 2015. Effect of storage and insect infestation on the technological properties of wheat. *CyTA-Journal of Food* 13(1): 134–139.

Kiaya, V. 2014. Post-harvest losses and strategies to reduce them. *Technical Paper on Postharvest Losses, Action Contre la Faim (ACF)* 25.

Kitinoja, L., and Tokala, V. 2018. Brondy. A. 2018. A review of global postharvest loss assessments in plant-based food crops: Recent findings and measurement gaps. *Journal of Postharvest Technology* 6(4): 1–15.

Kumar, D., and Kalita, P. 2017. Reducing postharvest losses during storage of grain crops to strengthen food security in developing countries. *Foods* 6(1): 8.

Kumari, A., Pankaj, P. P., and Baskarm, P. 2015. Post-harvest losses of agricultural products: management and future challenges in India. In: *Recent trends in post harvest management. First Edn.* New Delhi: Mangalam Publishers, pp. 141–153.

Lipinski, B., Hanson, C., Waite, R., Searchinger, T., Lomax, J., and Kitinoja, L. 2013. Reducing food loss and waste. *World Research Institute Working Paper* 40.

Los, A., Ziuzina, D., and Bourke, P. 2018. Current and future technologies for microbiological decontamination of cereal grains. *Journal of Food Science* 83(6), 1484–1493.

Majumder, S., Bala, B., Arshad, F. M., Haque, M., and Hossain, M. J. 2016. Food security through increasing technical efficiency and reducing postharvest losses of rice production systems in Bangladesh. *Food Security* 8(2): 361–374.

Manandhar, A., Milindi, P., and Shah, A. 2018. An overview of the post-harvest grain storage practices of smallholder farmers in developing countries. *Agriculture* 8(4): 57.

Marti, A., Barbiroli, A., Bonomi, F., Brutti, A., Iametti, S., Marengo, M., and Pagani, M. A. 2014. Effect of high-pressure processing on the features of wheat milling by-products. *Cereal Chemistry* 91(4): 318–320.

Mesterházy, Á., Oláh, J., and Popp, J. 2020. Losses in the grain supply chain: Causes and solutions. *Sustainability* 12(6): 2342.

Nath, B., Hossen, M., Islam, A., Huda, M., Paul, S., and Rahman, M. J. 2016. Postharvest loss assessment of rice at selected areas of Gazipur district. *Bangladesh Rice Journal* 20(1): 23–32.

Olorunfemi, B. J., and Kayode, S. E. 2021. Post-harvest loss and grain storage technology—A review. *Turkish Journal of Agriculture-Food Science and Technology* 9(1): 75–83.

Paul, A., Radhakrishnan, M., Anandakumar, S., Shanmugasundaram, S., and Anandharamakrishnan, C. 2020. Disinfestation techniques for major cereals: A status report. *Comprehensive Reviews in Food Science and Food Safety* 19(3): 1125–1155.

Paulsen, M. R., Kalita, P. K., and Rausch, K. D. 2015. *Postharvest losses due to harvesting operations in developing countries: A review.* Paper presented at the 2015 ASABE Annual International Meeting. American Society of Agricultural and Biological Engineers.

Phillips, T. W., and Throne, J. E. 2010. Biorational approaches to managing stored-product insects. *Annual Review of Entomology* 55: 375–397.

Qu, X., Kojima, D., Wu, L., and Ando, M. 2021. The losses in the rice harvest process: A review. *Sustainability* 13(17): 9627.

Sadiya, S., and Hassan, I. 2018. Post-harvest loss in rice: causes, stages, estimates and policy implications. *Agriculture Research and Technology: Open Access Journal* 15(4).

Sarkar, D., Datta, V., and Chattopadhyay, K. S. 2013. Assessment of pre and post harvest losses in rice and wheat in West Bengal. In: *Agro-Economic Research Centre, Visva-Bharati, Santiniketan*, New Delhi: Santiniketan, pp. 6–66.

Sawicka, B. 2019. Post-harvest losses of agricultural produce. *Sustainable Development* 1: 1–16.

Schmidt, M., Zannini, E., and Arendt, E. K. 2018a. Recent advances in physical post-harvest treatments for shelf-life extension of cereal crops. *Foods* 7(4): 45.

Šotnar, M., Pospíšil, J., Mareček, J., Dokukilová, T., and Novotný, V. 2018. Influence of the combine harvester parameter settings on harvest losses. *Acta technologica agriculturae* 21(3): 105–108.

Srivastava, S., and Mishra, H. N. 2021. Ecofriendly nonchemical/nonthermal methods for disinfestation and control of pest/fungal infestation during storage of major important cereal grains: A review. *Food Frontiers* 2(1), 93–105.

Stathers, T., Ognakossan, K., Priebe, J., Mvumi, B., and Tran, B. J. J.-K.-A. 2018. Counting losses to cut losses: Quantifying legume postharvest losses to help achieve food and nutrition security. 8–18. https://gala.gre.ac.uk/id/eprint/24062/. Greenwich Academic Literature Archive - Counting losses to cut losses: quantifying legume postharvest losses to help achieve food and nutrition security)

Sujayasree, O., Chaitanya, A., Bhoite, R., Pandiselvam, R., Kothakota, A., Gavahian, M., and Mousavi Khaneghah, A. 2021. Ozone: An advanced oxidation technology to enhance sustainable food consumption through mycotoxin degradation. *Ozone: Science & Engineering* 1–21.

Swai, J., Mbega, E. R., Mushongi, A., Ndakidemi, P. A. 2019. Post-harvest losses in maize store-time and marketing model perspectives in sub-Saharan Africa. *Journal of Stored Products and Postharvest Research* 10(1): 1–12.

Syamaladevi, R. M., Tang, J., Villa-Rojas, R., Sablani, S., Carter, B., and Campbell, G. 2016. Influence of water activity on thermal resistance of microorganisms in low-moisture foods: A review. *Comprehensive Reviews in Food Science and Food Safety* 15(2): 353–370.

Vilariño, M. V., Franco, C., and Quarrington, C. 2017. Food loss and waste reduction as an integral part of a circular economy. *Frontiers in Environmental Science* 5: 21.

Wilson, S. A., Mohammadi Shad, Z., Oduola, A. A., Zhou, Z., Jiang, H., Carbonero, F., and Atungulu, G. G. 2021. Decontamination of mycotoxigenic fungi on shelled corn using selective infrared heating technique. *Cereal Chemistry* 98(1): 31–43.

Wu, J., Doan, H., and Cuenca, M. A. 2006. Investigation of gaseous ozone as an anti-fungal fumigant for stored wheat. *Journal of Chemical Technology & Biotechnology: International Research in Process, Environmental & Clean Technology* 81(7): 1288–1293.

Yadav, D. N., Anand, T., Sharma, M., and Gupta, R. 2014. Microwave technology for disinfestation of cereals and pulses: An overview. *Journal of Food Science and Technology* 51(12): 3568–3576.

CHAPTER 7

Industrial Applications of Cereals

Nazmul Sarwar, Taslima Ahmed, Nahidur Rahman, Gulzar Ahmad Nayik and Saeme Asgari

CONTENTS

DOI: 10.1201/9781003252023-7

7.1 INTRODUCTION

Cereals are edible grains consisting of seeds, endosperms, and germs from the *Poaceae* or *Gramineae* family that have been cultivated for hundreds to thousands of years. Corn, wheat, rye, barley, rice, sorghum, oats, and millets are globally valued cereals (Hill & Li, 2016). However, among cereals, rice and wheat are the dominant products in both Asian and Western cuisines (Samota et al., 2017). Cereal-based food products are widely consumed throughout the world, as they are proven to provide energy; carbohydrates; proteins; B-vitamins, particularly B1 (thiamine), B2 (riboflavin), and B3 (niacin); and minerals such as calcium, phosphorus, and iron (Huang & Miskelly, 2016). However, cereals are widely used as a staple food for adults and weaning food for infants in many developing countries (Wrigley, 2010). These foods are mostly made from cereal grains, with a variety of formulations containing wheat, maize, or rice, which account for two-thirds of total human food consumption (Day, 2013).

Cereals are typically eaten whole or ground to make bread, ready-to-eat breakfast cereals, noodles, and pasta. The rapid expansion of the bakery industry is regarded as an important step in the development of modern countries (Cauvain, 2012). Technological advancements have sparked the production of food products with strong cultural foundations using cereals and legumes as raw materials (McSweeney & Day, 2016). White bread, for example, has been surpassed by a variety of breads and baked goods consumed by elite peoples of the early 19th century. Some other bread types such as naan, whole meal, pita bread, focaccia and rye bread, are commercially produced nowadays and have become a part of our regular diet. Meanwhile, breakfast cereals in variety of forms or shapes, including whole grain (higher fiber), ready-to-eat, wheat-based biscuits, porridge, oats, and those containing other ingredients aimed to improve the nutritional profile of the food consumed as the day's first meal (McSweeney & Day, 2016).

Some other snacks items such as muffins, cakes, noodles, pasta, bulgur, popcorn, and rice cakes have occupied a larger share in the human diet and can be produced using grains or milled flours. However, cereals present several nutritional problems, especially during cooking, such as starch swelling, limited bioavailability of mineral contents, lower protein fraction, and limited amino acid profile (Nout, 2009). Despite the challenges, the many beneficial effects of grains can be used in a variety of ways, resulting in widespread industrial applications for functional foods or grain ingredients.

7.2 BREAD

Bread is a well-known baked food item that is particularly made with flour and is kneaded, moistened, and frequently fermented for an extended period of time. It has been the major food stuff since prehistoric periods and can be made with diverse raw materials and methods. Historically, the very first bread was made from a mixture of coarsely ground grain and water in the Neolithic period, nearly 12,000 years ago, and the resulting dough was further baked on heated stones. Furthermore, the Egyptians invented the baking oven after discovering that when wheat dough was fermented to produce gas, it also produced light expanded bread (Britannica, 2021).

Modern industrial baking processes are highly mechanized and require the best-quality flours since it affects the process variables and the quality of end products. Besides, flour quality also regulates water absorption by dough, optimum mixing time, and the final bread characteristics. However, the general sequential steps in bread production are dry ingredient mixing, dough kneading, fermentation, panning/molding, and baking. Most yeast-leavened breads are made from refined hard wheat flours known as durum flour (Serna-Saldivar, 2016).

7.2.1 Mechanism of Dough Development

The development of dough is an important step in the bread-making process. As the flour, water, and other ingredients combine to form dough, a series of chemical interactions take place. Rheological properties show that dough is typically viscoelastic. Bread improvers typically oxidize cysteine sulfhydryl or thiol (—SH) groups in wheat gluten, resulting in fewer or no exchange reactions with disulfide (—S—S—) bonds. However, at the molecular level, oxidation of the sulfhydryl or thiol (—SH) groups can result in the formation of newly identical disulfide (—S—S—) bonds and later on cross-links between the protein matrices. Development of dough entails the simultaneous breakdown and formation of newer bonds. These bonds usually keep the proteins in their initial configuration. These proteins, on the other hand, form a three-dimensional network while breaking down previously molded bonds to form dough for bread making (Edwards, 2007). However, in both the sponge batter and bulk fermentation processes, the bonds are usually broken by flour improvers or by the action of enzymes found in flours. Furthermore, intense mechanical pressures and the actions of flour improvers frequently break down the existing bonds in the Chorleywood dough-making method. Similarly, in the activated dough development (ADD) process, slow-acting oxidation followed by reduced cysteine level accompanied by mild mechanical action produces similar effects during dough formation (Shukla, 2001).

7.2.2 Bread-Making Procedures

Many technological procedures, such as dough mixing, dough fermentation, shaping or molding, proofing, pan baking, rapid cooling, and packaging, are widely considered essential for the production of high-quality bread (Figure 7.1). Each of the processes is important in the development of dough and, ultimately, in the formulation of bread. As a result, in order to ensure premium-quality finished products, all the procedures must be handled with care within the stipulated scopes.

7.2.2.1 Mixing

In the bread-making process, mixing aims for an even distribution of ingredients as well as the formation of gluten. However, the mixing operation starts with hydrating the flours along with other formula ingredients. When the flours are hydrated, it creates a gliadin–glutelin network within the dough, and gluten starts to interact by forming hydrophobic and disulfide bonds. As a result, the dough system loses its moisture, becomes more compact, and eventually reaches a point of rigid consistency or decreased mobility. To get the best-quality bread, the gluten forms a continuous sheet or film at this point.

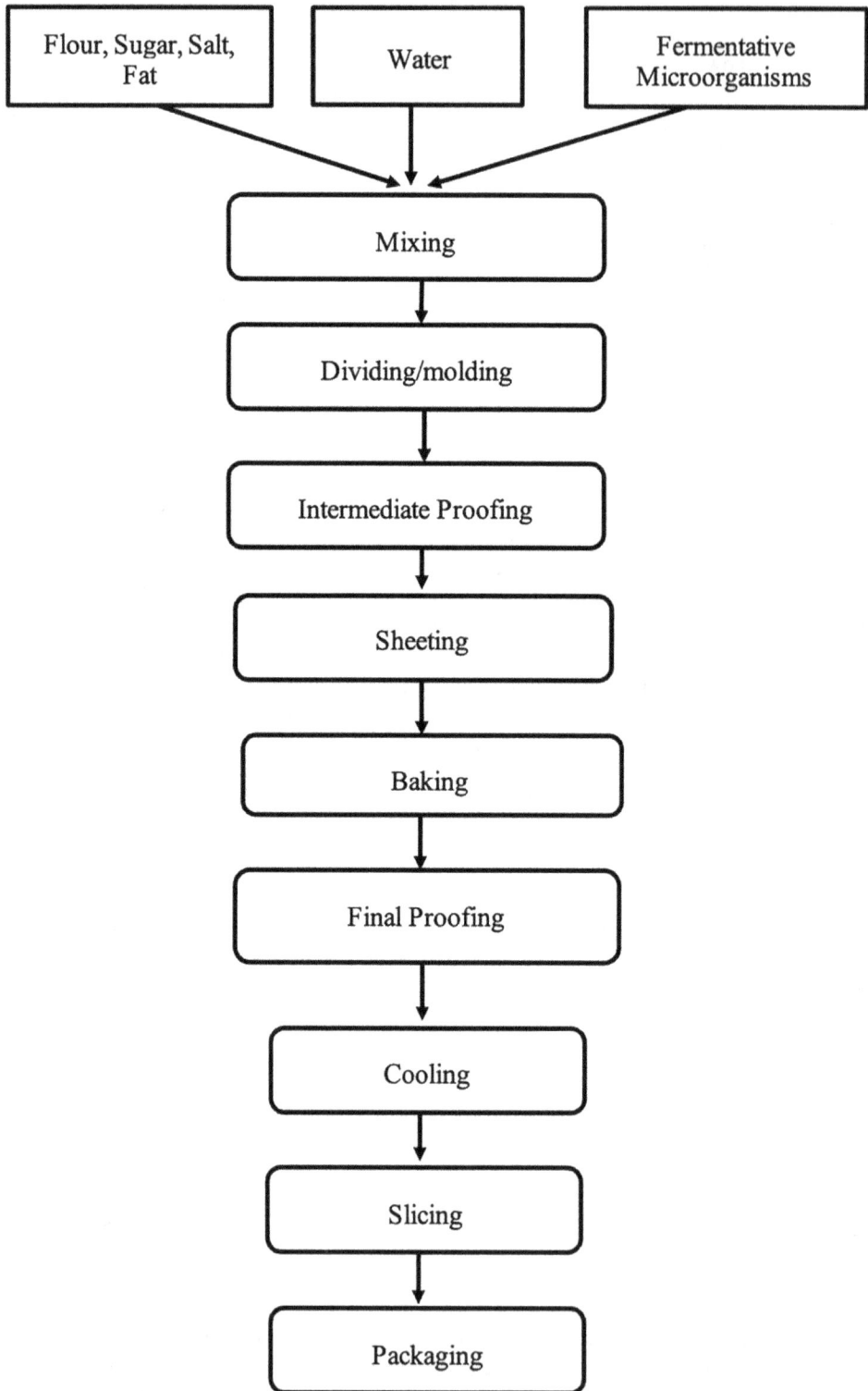

Figure 7.1 Bread-making procedures.

(*Source*: Zhou et al., 2014)

There are three main control parameters in the critical stage of mixing or kneading the dough: the moisture absorption rate of the flour, the mixing time, and the mixing temperature. The temperature of the dough significantly influences the rheological properties and workability of the bread. Some industrial mixers are equipped with a water jacket to ensure optimum kneading temperature. Moreover, with optimal kneading time, the dough will have a smooth, shiny texture and will tend to retain the maximum amount of gas generated during fermentation. Good-quality bread flours have a farinograph dough development time of 4 to 8 min and more than 12 min on a commercial low speed mixer. High-protein flours, in general, are more stable or less prone to over-mixing compared to low-protein flours (Cauvain & Young, 1998; Kulp & Ponte, 2000; Sluimer, 2005).

7.2.2.2 Dough Dividing and Rounding

Dough division is usually performed with a volumetric divider (Bhardwaj et al., 2015). The dough is first fed into a divider hopper. It flows down and enters the adjustable chamber by gravity and suction. A piston works in the chamber in a cycle where the dough is cut and released from a moving belt. These distributors are mechanically or hydraulically controlled. Another common divider is a rotary divider equipped with an extruder that carries the dough cut with a rotating knife. The speed of rotation of the knife determines the weight of the dough. Splitting should be done as soon as possible to ensure the same size and weight.

When dough is discharged from the divider, the piece is rough and sticky and requires rounding in order to seal the outer surface of the dough piece so as to minimize gas diffusion and enhance the formation of new gas vesicles known as loci (Serna-Saldivar, 2016). Rounding also enhances the formation of smooth and non-sticky dough that facilitates further handling in subsequent mechanical steps. There are various types of commercial rounders. The most common rounders are the bowl type, the umbrella type, and the cone type (Serna-Saldivar, 2016).

7.2.2.3 Fermentation

Fermentation starts when yeast cells (*Saccaromyces cerevisiae*) and flour are hydrated. Dry yeast requires a longer activation time than compressed yeast. Activity of yeasts is affected by the temperature and relative humidity of the processing zone, which is responsible for producing gas and, more importantly, developing flavors. Fermentation is usually carried out in cabinets or continuous proofers at 20–40°C under high humidity (≥85%). However, the optimum pH range for yeast activation ranges between 4 and 6 (Bhatia, 2017). Higher relative humidity prevents surface dehydration and dough crust, which affects the quality and yield of the final product. Activated yeast cells break down available substrates (sucrose, glucose, fructose, maltose, etc.) to form carbon dioxide, organic acids, aldehydes, ketones, ethanol, and other alcohols. Produced gases, however, leaven the dough into foams, which are further separated and stabled. Fermented dough owing to baking operation changes its foamy structure to a spongy texture and improves the aerating structure of the finished breads.

7.2.2.4 Degassing/Punching and Molding

The purpose of dough punching is to remove carbon dioxide gas trapped in the gluten and create a new cell or locus. Punching reactivates yeast by removing high concentrations of carbon dioxide trapped in the dough (Serna-Saldivar, 2016). However, fermented dough is usually degassed by molding. This operation involves three consecutive steps: sheeting, curling, and scaling (Bhatia, 2017). Pre-formed dough is passed through 2–3 closely spaced rolls to degas the dough and then conveyed to curling section using a belt conveyor. Finally, the rolled dough sheets are passed under a pressurized disk to remove and seal the gas pockets inside (Bhatia, 2017). During this mechanical operation, trapped gases are released and small new air bubbles are formed. The properly degassed dough is formed and, in some instances, placed inside baking pans in preparation for proofing. The preformed piece of dough is finally proofed for a given amount of time prior to baking.

7.2.2.5 Baking

The preformed, fermented, and proofed dough is baked at 200–300°C for 12–25 min, varying by the size and type of the finished bread (Kulp, 1988). During baking, heat is conveyed inside the oven through conduction, convection, radiation, condensation, and evaporation of steam and water (Serna-Saldivar, 2016). Yeast cells die about 8 min after baking at temperatures of approximately 220°C. However, during baking, hydrated starch granules gelatinize and eventually acquire a strong water-holding capacity. In fact, the bread develops its characteristic crust color due to Maillard and caramelization reactions at this stage. Right after baking, the bread crust is hard and, upon cooling, attains the typical soft consistency. This is because water gradually migrates or equilibrates from the internal crumb to the more dehydrated surface or crust. During maximum baking schedules, there is sufficient time to destroy all microorganisms and spores (Serna-Saldivar, 2016). Thus, breads exiting from the oven are practically sterile. Thus, it is crucial to design the layout of good sanitation and cooling procedures so as to avoid cross-contamination that would compromise shelf life.

7.2.2.6 Cooling and Slicing

Breads are allowed to cool down in cooling racks or through a series of open tiers that discharge the breads into slicing and bagging areas. The shelf life of baked goods is greatly dependent on the effectiveness of this operation. Some cooling chambers are furnished with fans to speed up heat transfer. Also, cooling rooms are best equipped with ultra-violet light and microbiological filters and usually have restricted entrance to prevent cross-contamination. However, the cooling rate depends on the size and characteristics of bread, cooling room temperature, and whether fans are used. Most breads are subjected to cooling schedules of at least 20–30 min. Inadequate cooling causes moisture condensation or sweating inside the packaging, loss of bread texture, and microbial growth (Serna-Saldivar, 2016). The final step in pan bread production is the slicing of the loaf of bread. There are basically two types of slicers: reciprocating and belt type. Slicers are equipped with saws. The latter is more suitable for softer, bulky breads. The thickness of the slices varies, typically 1–1.3 cm. Slicers should be kept clean in order to minimize cross-contamination and optimize shelf life.

7.2.2.7 Packaging

The purpose of packaging the bread is to keep it fresh by preventing it from drying out too quickly and losing its texture. Packaging also prevents cross-contamination with spores and prolongs the microbial shelf life. The most commonly used packaging material is low density polyethylene (LDPE), glazed with imitation parchment paper and paraffin wax. The ends of the LDPE bag are further wrapped around and twisted with adhesive tape (Robertson, 1993). Certain types of Italian and French breads are packaged in perforated bags to allow moistness while retaining their characteristic crispness (Serna-Saldivar, 2016).

7.3 BISCUITS

Biscuits are flour-based baked goods that are popular all over the world. The term biscuit first originated from the Latin term *panis biscoctus*, and later on the French word *bescuit*, meaning "twice-baked bread," and was referred to as ship's biscuits (Edwards, 2007). The name biscuit now encompasses a wide variety of products and is accepted by consumers of all ages owing to different colors, shapes, and fillings. Biscuit production is a major sector of the food industry. However, several types of biscuits are manufactured in bakeries as well as in supermarkets, while some biscuits are also prepared in households. Biscuit manufacturers therefore must have a good understanding of chemistry to maintain quality and attract consumers (Chavan et al., 2016).

7.3.1 Biscuit Production Process

The major ingredients for biscuit production are flour, water, sugar, baking powder, and fats. During biscuit production, all these major or minor ingredients are mixed homogenously in dough mixers (Misra & Tiwari, 2014). This dough is further processed using a roller to obtain sheets of predefined thickness. Rolled sheets are then conveyed towards the baking oven. Meanwhile, the shapes of the biscuits are given according to mold cuttings. After baking, the biscuits are allowed to cool before being packaged (Misra & Tiwari, 2014). However, this describes the commercial biscuit production procedure. A flow chart of the biscuit production process is given in Figure 7.2.

7.3.1.1 Mixing

Mixing influences the formation of gluten and the rheological properties of dough used for biscuit production; hence the overall quality of the finished product is also dependent on optimal mixing (Manohar et al., 1999). Mixing is primarily done to form uniform dough and involves sequential steps: blending the ingredients, dissolving the solid materials in water, continuous kneading and spreading, and unloading the mixture on a cart or conveyor belt for rolling. The mixing time varies according to flour type and characteristics, formula ingredients, and dough temperature and typically ranges between 15 and 25 min (Chavan et al., 2016). Both batch mixers and continuous mixers are widely used in large-scale production of biscuits. However, these mixers are programmable and automated; therefore, their variable-speed can be modified accordingly (Caballero et al., 2015). Biscuit dough is typically divided into four types: hard doughs, semi-sweet doughs, short doughs, and batters.

1. **Hard dough:** Hard dough is commonly used for crackers and semi-sweet biscuits. The fat and sugar content is relatively low. Cracker dough involves mixing in an all-in-one process where pre-formed dough is kneaded until achieving the final temperature (26–30°C). This can be accomplished by lowering the internal energy (Edwards, 2007). The obtained dough is then left aside for fermentation and other subsequent production steps.
2. **Semi-sweet dough:** This recipe contains more sugar and fat than crackers. Usually, semi-sweet dough involves only one stage of kneading. All of the baking ingredients are charged inside a mixer and stirred continuously. Then, pre-formed dough should be kneaded to a certain final temperature, and the kneading time should be at least 4 min (Edwards, 2007).
3. **Short dough:** Gluten formation is undesirable in this dough. High levels of fat and sugar inhibit gluten hydration (Edwards, 2007). Therefore, even distribution of fats and water in dough is quite a challenging task. The sugar content of these foods is too high to be completely soluble in water. However, this dough is mixed in two stages. In the first stage, when an emulsion of fat-in-water is formed, flour is added to promote dispersion, while a very small amount of flour along with a portion of sugar is added in the second mixing stage to prevent the formation of gluten.
4. **Batters:** Some biscuits are so soft that they become really doughy. These types of foods usually contain eggs or egg powder. However, batter-type products are very similar to muffins, which are typically prepared in a cake-type mixer (Edwards, 2007).

7.3.1.2 Sheeting and Molding

All the individual biscuit types are formed in a variety of ways, including sheeting (semi-sweet), sheeting and lamination (crackers), molding (short dough), depositing (wafers), and extrusion

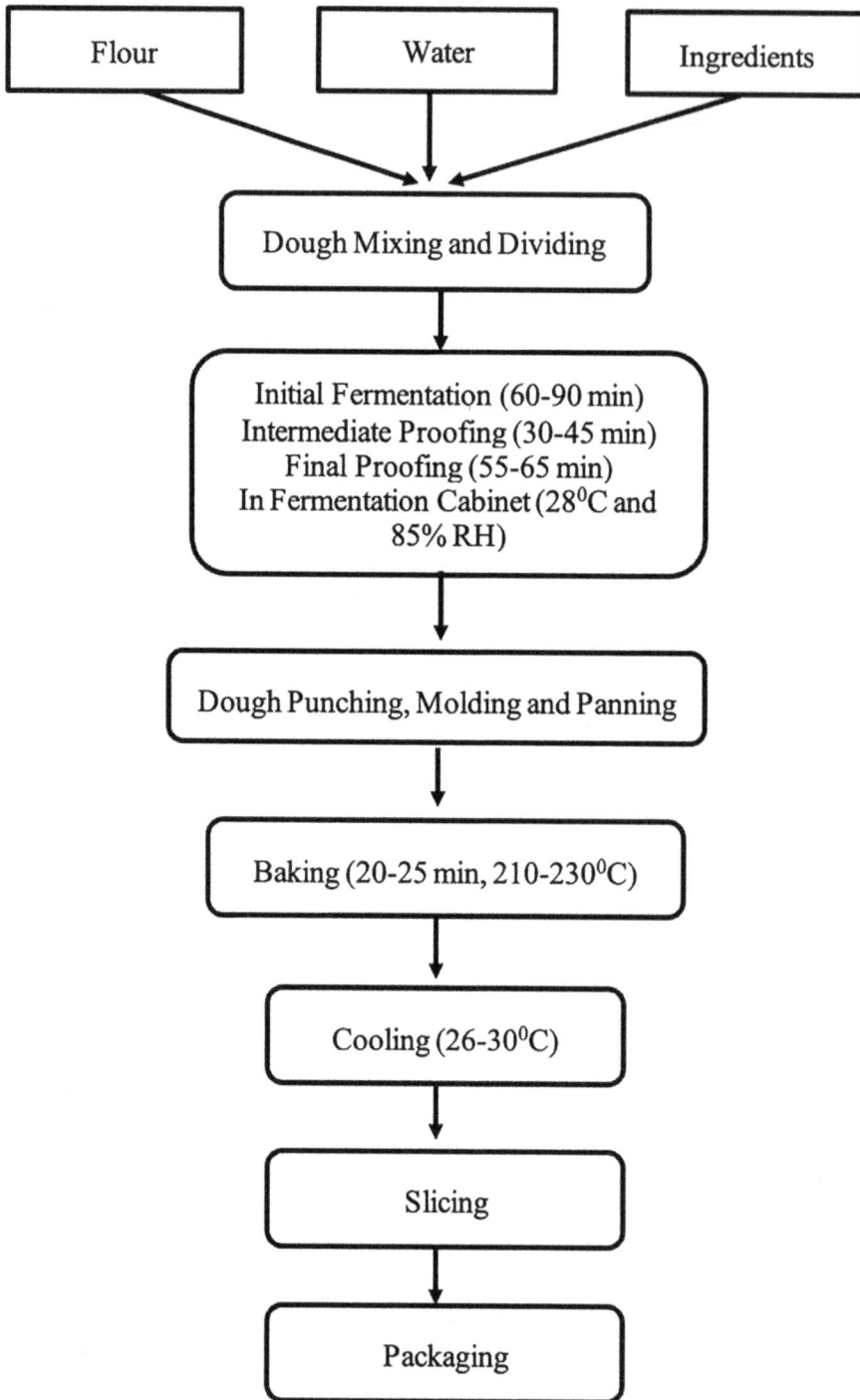

Figure 7.2 Biscuit manufacturing process flow chart.

(*Source:* Misra & Tiwari, 2014)

(rout-pressing), although the mixing procedure is same for all types (Cauvain & Young, 2006). Sheeting is accomplished by passing the dough through a series of "gauge rolls," which reduce the thickness as specified in a controlled way. The gauge rolls, consisting of a series of cylindrical rollers, continuously revolve in the opposite direction (Oliver et al., 1996). The dough sheets with the desired thickness are then conveyed through a cutting roller, usually made of stainless steel. A dough cutter contains an engraved knot that cuts the incoming dough according to the desired shape. However, to maintain an ideal size and shape, the dough plate should be moved at a programmed speed beneath the cutter such that the blade falls on the dough plate and moves along the dough. Then it returns to its primary position and continue the processes. After cutting, the remaining dough can be re-added to a horizontal mixer while kneading a new batch for biscuit production (Misra & Tiwari, 2014).

7.3.1.3 Baking

Baking is a complex process in which mass (water) and heat are transferred simultaneously inside the oven (Broyart & Trystram, 2002). The quality and shelf life of the finished product depend on the efficiency of the oven in which it is baked. During baking, the density of the dough pieces changes, the surface becomes hard, and the structure becomes porous (Chavan et al., 2016). However, biscuits can be baked in any type of oven, including travelling ovens, deck ovens, and rack ovens. Traveling or band ovens are widely used at the industrial scale, while smaller bakery industries typically rely on static ovens. One of the advantages is that these ovens can be arranged in multiple zones, allowing the food to enter the hottest part of the oven and move towards the cooling zone as cooking proceeds. In addition, baking profiles vary according to biscuit type, ingredient formulation, and the rheology of texture. Therefore, baking time differs from 3 to 12 min: crackers tend to have the shortest baking time, whereas cookies tend to have the longest baking time. The temperature range also varies from 140 to 240°C, depending on the processing zone and product specification (Chavan et al., 2016).

7.3.1.4 Cooling

Freshly baked biscuits must be cooled prior to packaging to settle the smooth texture. The temperature of the biscuit coming out of the baking oven is about 100°C. If baked biscuits are too soft, they may not withstand the packaging process, and water vapor may condense inside the packaged product, causing the packaging material to shrink and reduce product quality. In industrial production, stripper (wire mesh structured) conveyors are often installed along with the oven to cool down baked biscuits immediately and avoid contact with hot surfaces. Biscuits are further transported to the packaging area. In general, foods are cooled to 40–45°C, while foods requiring secondary processing, such as cream filling, are cooled to 18–26°C (Chavan et al., 2016). However, owing to reduced moistness in biscuits, they confer microbial safety along with guaranteed shelf life.

7.3.1.5 Packaging

Biscuits are highly hygroscopic. Therefore, keeping the quality of biscuits largely depends on the intrinsic properties of the biscuits themselves, methods of distribution and storage facilities, and most importantly the functional and barrier properties of the packaging materials (Robertson, 2016). According to functional requirements, a combination of primary, secondary, and tertiary packaging is recommended for the overall packaging of biscuits. Primary packaging is usually made up of flow wraps, slugs, sachets, displays, tubes, and shrink wrappings, typically polypropylene films, laminated paper, and metal cans (Chavan et al., 2016). In addition, secondary packaging is typically used for the packing of wholesale or bulk packages. Tertiary packaging often uses corrugated or grooved cardboard boxes and containers (Chavan et al., 2016). Biscuits are first arranged and distributed in rows by a stacking machine before being subjected to any of the packing machines, which include rim type, cream sandwich type, and pile type. Furthermore, horizontal flow wrap packing machines are frequently used for flexible packaging (Misra & Tiwari, 2014).

7.3.2 Chemical and Physical Changes during Baking

When dough is kept inside the baking oven, several physico-chemical changes may occur. However, some of the changes are desirable, while others may tend to degrade the quality of the finished product.

1. **Oven-spring:** The dough expands rapidly in the first few minutes of baking. This sudden rise is referred to as oven-spring. Oven-spring is caused by a number of factors. The presence of gases such as air and carbon dioxide, as well as heat and water volume, causes an increase in the internal pressure of the dough, causing the dough to rise rapidly during the first stage of baking.
2. **Crust formation:** When dough is heated in the oven, moisture evaporates quickly from the surface, resulting in the formation of a crust. The developed crust, on the other hand, provides the physical strength of the biscuits.
3. **Yeast activity:** Yeast activity is dependent on baking temperature. Yeast activity increases rapidly when the dough is placed in the oven, but it becomes inactive after 55°C (Bhatia, 2017).
4. **Starch gelatinization:** Starch begins to gelatinize at around 60°C. The dough contains limited water to completely gelatinize the starch. This limited gelatinization of the dough aids in gas retention and texture setting.
5. **Gluten coagulation:** Gluten coagulation is related to water removal. The gluten matrix that surrounds the individual cells is transformed into a semi-rigid film structure during this process. As a result, redistribution of water from the gluten phase to the starch phase occurs.
6. **Enzyme activity:** Amylase's action on starch increases with temperature. Heat inactivation, on the other hand, denatures both α- and ß-amylase. When amylase activity is too low, it restricts the loaf volume; thus the starch becomes rigid.
7. **Browning reaction:** The Maillard reaction occurs between proteins and reducing sugars. This reaction will spread to the interior of the biscuit as it heats up through conduction, forming colored compounds such as melanoidins. This reaction imparts color and flavor to the biscuits.

7.3.3 Methods of Bread Making

7.3.3.1 Straight-Dough System

The kneading procedure for a straight-dough system consists of mixing the flour with the remaining dry ingredients and water until a proper dough is formed. The resulting dough is manually or mechanically separated and placed in a fermentation cabinet or resistant vessel with strict temperature control (25–30°C) and relative humidity (approximately 85%) for about 2 h. The fermented dough is further punched, formed, panned, and proofed for 50–70 min until the desired calibration heights are reached. It is baked in a loaf of bread for 20–25 min at temperatures of 210–230°C. It is further cooled for about 30 min, sliced, and packaged in bags, preferably moisture-proof plastic bags (Serna-Saldivar, 2016). A straight-dough bread-making system is illustrated in Figure 7.3.

7.3.3.2 Sponge-Dough System

The sponge-dough bread-making method is widely used in industrial systems and occasionally in noncommercial baking. The dough requires more handling and longer fermentation time compared to straight-dough system but is more tolerant of variations, especially of fermentation time and temperature (Penfield & Campbell, 1990). It is a semi-continuous process since the sponge and dough mixing steps are batch, whereas the rest of the processes are continuous. Half of the flour (60–70%), water, and yeasts are mixed and kneaded. The sponge is then placed in troughs in large fermentation rooms at 28°C under 85% relative humidity and fermented for about 4–6 h (Serna-Saldivar, 2016).

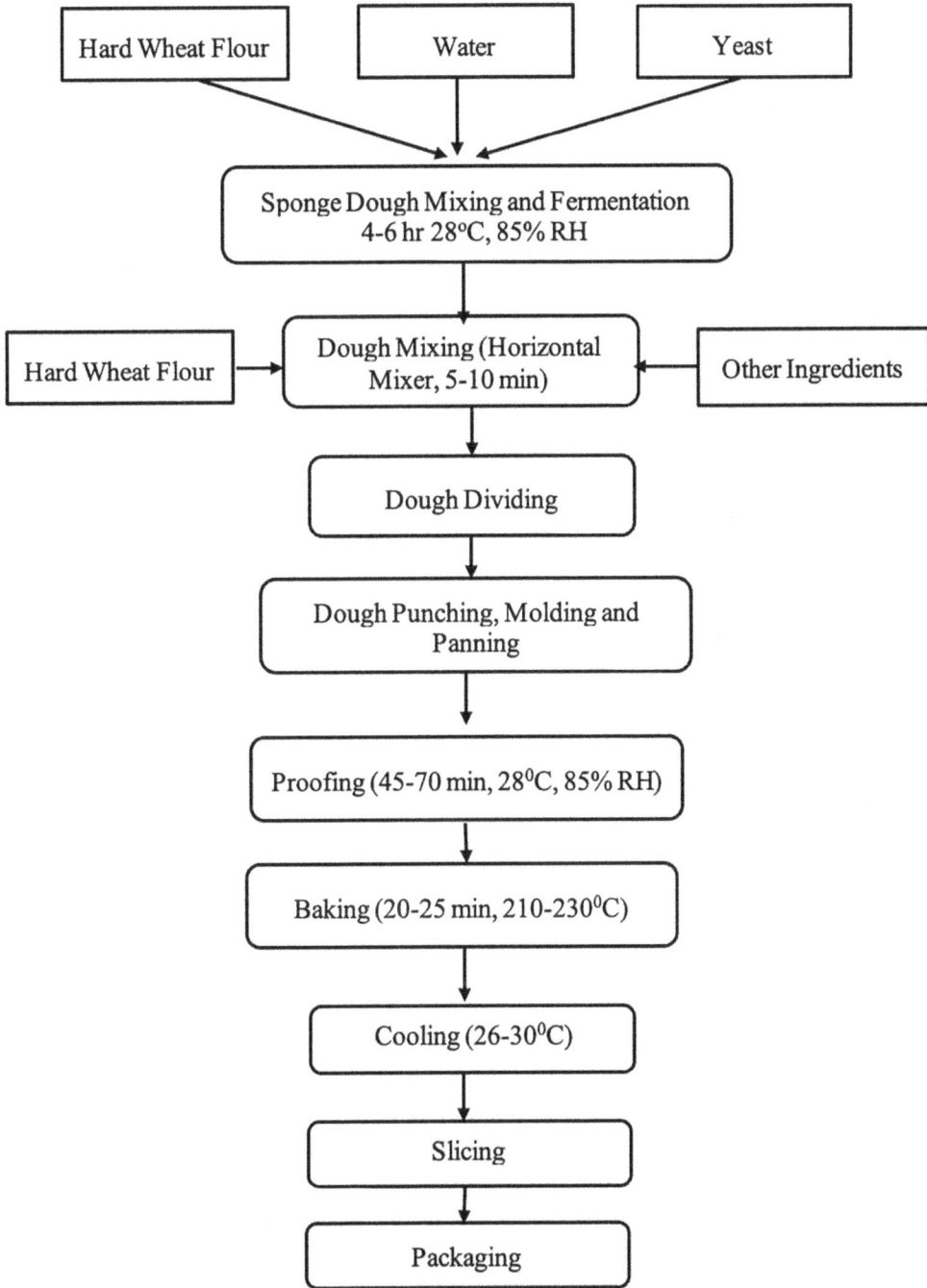

Figure 7.3 Straight-dough bread-making systems (pan bread and related products).

(*Source:* Serna-Saldivar, 2016)

During rising, the sponge becomes light and frothy and sometimes falls back because viscosity is too low for gas retention. After the sponge has risen, the remaining ingredients are added and the dough is handled like straight dough, with bulk fermentation, scaling, molding, and proofing. The rising time of the dough is rather short because of the longer sponge fermentation. However, the method derives its name from the sponge-like appearance of this mixture. The sponge method

Chorleywood Process   **Sourdough Process**

Figure 7.4 depicts two parallel bread-making process flow diagrams.

Chorleywood Process (left column):
- All ingredients
- Tweedy mixer 3-5 min
- Divider
- Rounder
- Intermediate proofer 2-15 min
- Sheeter
- Moulder
- Proofer 50-60 min
- Oven 16-30 min

Sourdough Process (middle and right columns):
- Mother dough, sourdough
- Flour, water
- Mixer 1st Refreshment
- Bulk fermentation 90 min
- Flour, water
- Mixer 2nd Refreshment
- Bulk fermentation 90 min
- Flour, water
- Mixer 3rd Refreshment
- Bulk fermentation 90 min
- Flour, water, salt
- Mixer
- Bulk fermentation 90 min
- Final dough
- Weighing 1st Shaping
- Resting 30 min
- Weighing 2nd Shaping
- Resting 30 min
- Oven

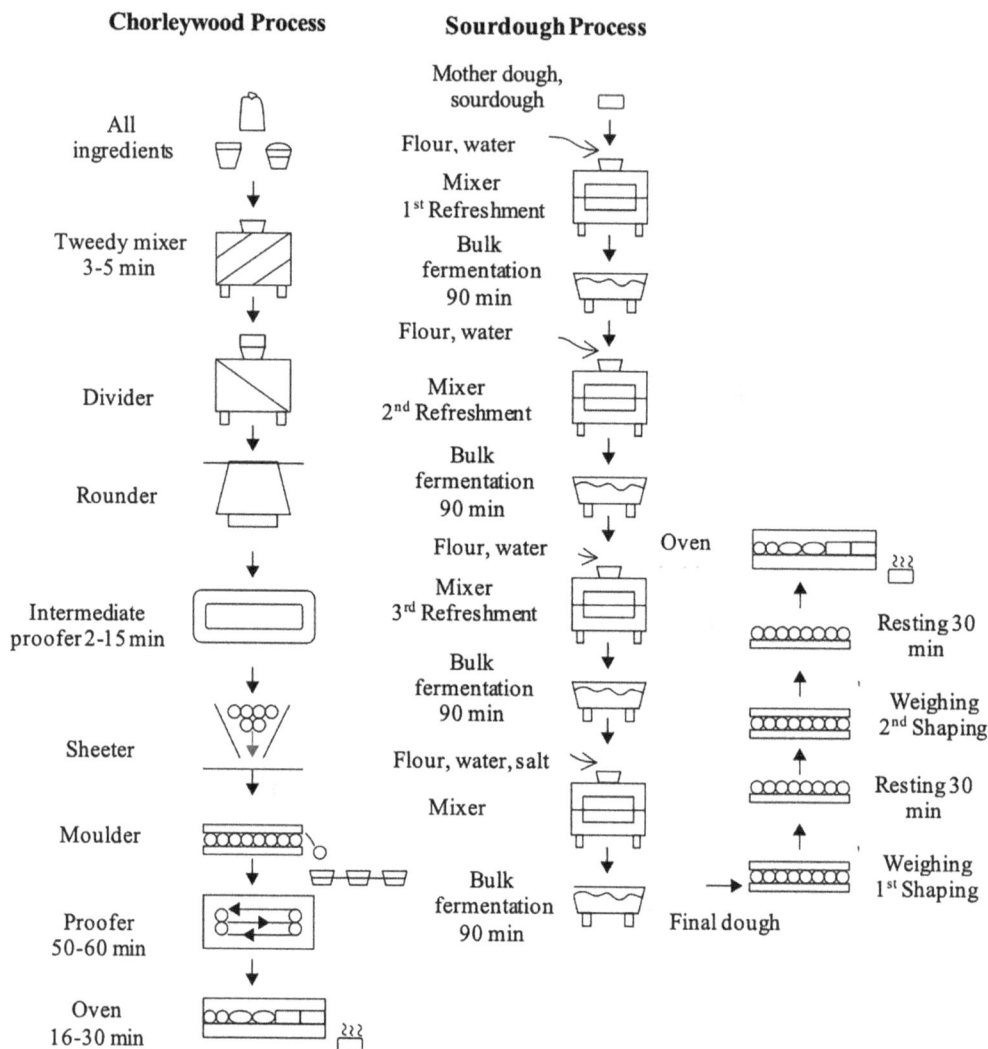

Figure 7.4 Sponge-dough bread-making systems (pan bread and related products).
(*Source:* Serna-Saldivar, 2016)

is not desirable with weak flour because of the long total fermentation period. See Figure 7.4 for the details steps of a sponge-dough system.

7.3.3.3 Chorleywood Bread-Making System

The Chorleywood bread-making system uses low-protein flour to shorten processing times and produce soft, fluffy bread. It was first introduced by the British Bakery Research Association in 1961 (Hui, 2006). The quality of bread largely depends on the pre-processing of dough while mixing the flour and other baking ingredients. However, superior-quality bread can be obtained from the optimal level of work input, that is, stirring at 40 J/g (11 W-h/kg) for approximately 2–4 min (Hui, 2006). The Chorleywood bread-making system traps air in the dough during mixing to form cells, and mechanical agitation separates the air into bubbles. High-speed mixing usually generates higher temperatures, which can be cooled further by using advanced mixers equipped with cooling jackets or chilled water. The Chorleywood bread-making system requires more yeast, doubled as compared to the straight-dough system, while yielding similar effects during fermentation (Hui et al., 2008). The dough also requires more water due to the better hydration of the flour resulting from intensive mixing.

7.3.3.4 Sourdoughs

Sourdough uses an ancient method of leavening bread. Rather than using baker's yeast, it leavens the dough with a combination of lactic acid bacteria and wild yeasts found naturally in wheat flour (Rizzello et al., 2019). Sourdough is similar to a sponge, except that it is fermented until it becomes extremely acidic. While formulating dough, part of the starter is retained and combined with additional flour and water for another batch. However, bacterial fermentation produces predominantly organic acids, whereas yeast fermentation produces primarily carbon dioxide and alcohol. These organic acids hydrolyze the aromatic esters during baking and thereby produce the characteristic flavors in the bread. Hetero-fermentative lactic acid bacteria create the most desired organic acids. The type of acid produced is dependent on the fermentation type and temperature. Low fermentation temperature, in general, favors the development of beneficial organic acids, including lactic, fumaric, and citric acids, whereas higher fermentation temperatures appear to favor the synthesis of long-chain fatty acids such as butyric acid, resulting in unpleasant tastes (Hui et al., 2008). The processing steps of a sourdough system are illustrated in Figure 7.5.

7.4 CAKES

Cake is one kind of sweet baked good that is usually prepared with flour, sugar, and other ingredients. Cakes' popularity is due not only from their sweetness and richness but also from their versatile uses. However, cakes can be represented in multiple ways, including simple sheeted cakes in restaurants and decorated cakes for special occasions such as birthdays, weddings, graduations, and so on. Cakes contain a lot of sugar and fat compared to other cereal-based bakery products. Therefore, using a well-balanced formulation followed by appropriate mixing is required for the production of superior-quality cake.

7.4.1 Raw Ingredients for Cake Formulation

To produce a high-quality cake, the major ingredients are combined in precise proportions. Each of the ingredients thereby influences the different quality characteristics such as porous texture, moist surface, and so on (Conforti, 2006). However, the most common ingredients for cake production are flours, water, butter or oil, sugar, eggs, and leavening agents. Some other minor ingredients and flavorings are fresh, dried, or candied fruits; cocoa butter; nuts; and vanilla extracts. Cakes can be further decorated with fruit fillings, preserves, and other icings.

7.4.2 Types of Cakes

Cakes are frequently classified into three categories based on the attributes, functionality, processing conditions, and variation in the formulations (Gisslen, 2021; Labensky et al., 2009).

1. **Batter type:** Chemically leavened and aerated oil-in-water emulsion that relies on flour, milk, and eggs for the spongy texture. Baking powder contributes significantly to the volume of the finished product. Pound cakes and layer cakes are examples of batter-type cakes.
2. **Foam type:** These types of cakes rely primarily on the denaturation and extension of egg proteins. They can also be called cakes without shortening owing to the absence of oil-in-water emulsion. Both angel food cakes and sponge cakes are examples of foam-type cakes where differences are reported due to the egg fraction used during cake production.
3. **Chiffon type:** Modified "foam-like" grainy texture that uses both formulations of batter- and foam-type cakes.

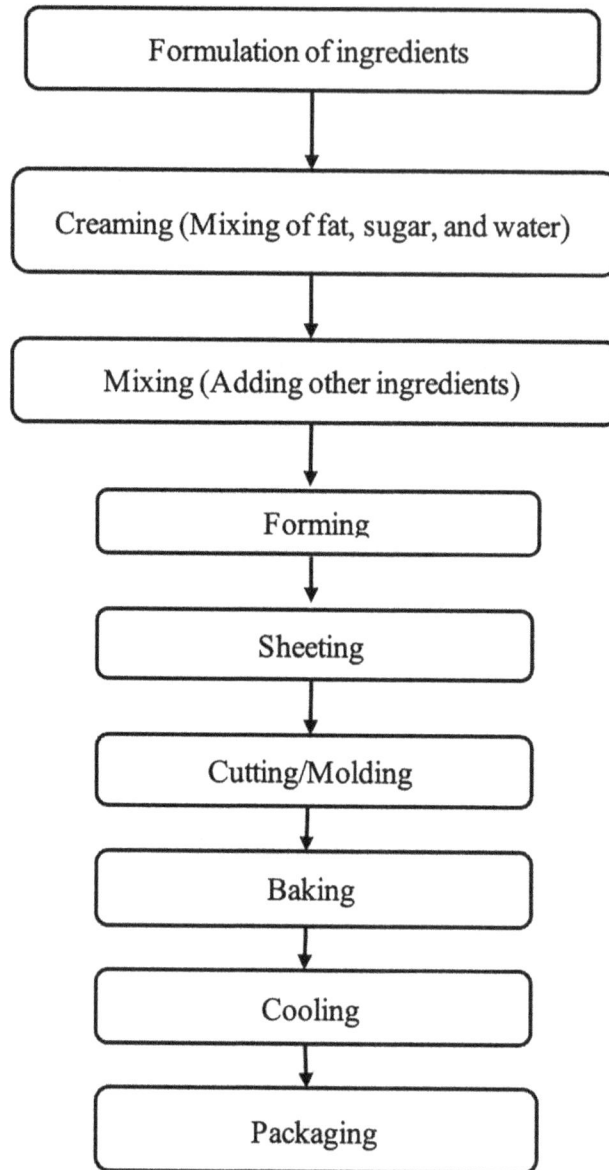

Figure 7.5 Chorleywood and sourdough bread-making systems.

(*Source:* Arendt & Zannini, 2013)

7.4.3 Cake Production Procedures

Commercial bakery industries use largely automated processes to ensure that finished products are uniform in size, shape, and appearance. However, commercial mixers, filling machines, baking ovens, and so on work together to yield a larger volume of end products in the most consistent way.

7.4.3.1 Mixing

A commercial mixer is used to combine the batter. After mixing all of the ingredients, the mixer beats the mixture for another 10 min to trap air within the protein matrix and allow the cake to rise properly. The trapped air, combined with the added leavening agent, causes the cake to expand in size during baking. After the batter has been thoroughly mixed, it is poured into a mold and baked (Delcour & Hoseney, 2010).

7.4.3.2 Filling up the Mold

A thin layer of oil emulsion is sprayed on an automated mold in order to prevent batter from sticking as baking proceeds. Emulsified molds are further filled with batter using an automated dosing system that simultaneously monitors the batter's weight, shape, and level prior to baking.

7.4.3.3 Baking

Molds filled with batter are transferred to automated ovens using conveyor belts. In order to produce the best-quality cakes, the oven's temperature as well as relative humidity should be carefully controlled. However, a very small amount of water is sprayed over the crust to reduce case hardening during the initial baking stage. The crust is then allowed to develop a sharp texture. The formation of an extensive protein network during baking, on the other hand, affects the strength of this structure, which ultimately contributes to the quality of the finished products (Wilderjans et al., 2013).

7.4.3.4 Cooling

Soon after baking, the cakes must be cooled thoroughly so as to prevent the crust collapsing. As cooling proceeds, it releases moisture, which is immediately removed from the cooling station by using powerful suction pumps to prevent condensation.

7.4.3.5 Stripping and Packaging

After the completion of cooling, the cakes are removed from the mold and transported to the packaging section. Condensation of moisture would damage the finished product if it is not completely cooled. However, commercial cakes are typically wrapped in plastic bags with a cardboard tray as secondary packaging material.

7.5 PASTA AND NOODLES

7.5.1 Pasta

Since pasta is made from semolina flour, it is one of the simplest products in terms of raw materials, with salt and eggs added in some cases. It is a versatile food loved by people of all ages. The widespread popularity of pasta can be attributed to several factors, including easier preparation, convenience, and being a rich source of complex carbohydrates (Bugialli, 1988). The origin of pasta is still debated. Some believe that Marco Polo brought pasta to Italy from China in the early 1300s. However, recent research suggests that ancient Etruscans ate spaghetti-like products. Meanwhile, the early Romans developed the first pasta and lasagna machines. But the Italians are proclaimed as the originators of pasta. In fact, the word *pasta* is derived from an Italian phrase *paste* (dough) *alimentari*, meaning related to nourishment.

7.5.1.1 Raw Ingredients for Pasta Making

Pasta is usually made from a mixture of semolina flour and water. Semolina flour is usually made from the endosperm of durum wheat, which is rich in protein and typically grown for pasta production. It contains lower carbohydrate contents than all-purpose flour, thereby making it easier to digest. Similarly, another coarse grain, farina flour, is also used by some manufacturers to make pasta. However, semolina and farina flour are sometimes fortified with B-vitamins and iron before being shipped to a pasta processing factory. Eggs; vegetable juices such as tomatoes, spinach, carrots, and beets; and herbs and spices such as basil, garlic, and thyme are often added for additional flavors, color, and aroma.

7.5.1.2 Pasta-Making Processes

The production of most pasta products consists of three major steps: hydration of the semolina, mixing—forming, and drying (Pagani et al., 2007). Most pasta are dried to decrease the moisture content to 10% and water activity to 0.5. Pasta products are usually made by means of mixing the major and minor ingredients. Formed dough is further transferred to a continuous, high-capacity extruder, equipped with varying dies and shapes. The pasta is finally dried and packaged prior to market supplying (see Figure 7.6).

7.5.1.2.1 Mixing

Semolina flours are transferred to a mixer machine fitted with rotating blades. Water from the warm water–jacketed pipeline is also supplied to the mixer. Eggs and additional ingredients can also be added. Then the mixture is kneaded until it forms a lumpy consistency. During the mixing process, small air bubbles may form due to the presence of air. This can reduce the mechanical strength of the pasta and yields a finished product with a chalky-white appearance. Therefore, modern pasta presses are fitted with a vacuum chamber to remove the air bubbles prior to extrusion.

7.5.1.2.2 Extrusion

The dough is transported to the extruder once it is mixed completely. The extrusion auger impacts the overall quality characteristics of the finished pasta by controlling the production rate. Incoming dough is cut or rolled into dies depending on the type of pasta. Rotating blades cut string-style and ribbon-like pasta such as spaghetti, fettuccine, linguini, and capellini. In addition, doughs in the extruder are pressed through a metal die to form tubular or shell-shaped pastas like rigatoni, ziti, elbow pasta, and fusilli. To make capellini and vermicelli pasta, the dough is passed through holes (diameter 0.8–0.5 mm) and then cut into pre-defined lengths (250 mm) and finally twisted into curls. However, spaghetti pasta ranges from 1.5 to 2.5 mm in diameter and is left straight (Rosentrater & Evers, 2018). Tortellini pasta is made by cutting dough into small circles and then layering pre-measured ricotta cheese on top of each circle. The dough is then folded in half and connected at the ends to form round pasta. To prepare ravioli pasta, pre-specified portions of cheese fillings are placed on a sheet at certain intervals. As it passes along the conveyer belt, another sheet of pasta is laid on top of it. The two layers are then fed through a perforating machine, which cuts the pasta into pre-measured squares (Pollini et al., 2012).

7.5.1.2.3 Drying

Most pasta processing industries use a pre-dryer immediately after extrusion to prevent the pasta from sticking together. It aims to harden the exterior surface while retaining softness and flexibility of the interior surface. The final dryer is further used to remove most of the accessible moisture from pasta. Drying time varies depending on the type of pasta. It ranges from 3 h for elbow pasta and egg noodles to 12 h for spaghetti (Pollini et al., 2012). Drying time is critical because pasta is more likely to break if dried too quickly and become spoiled if dried too slowly. Therefore, it is important to tailor the drying cycle as recommended for each product type.

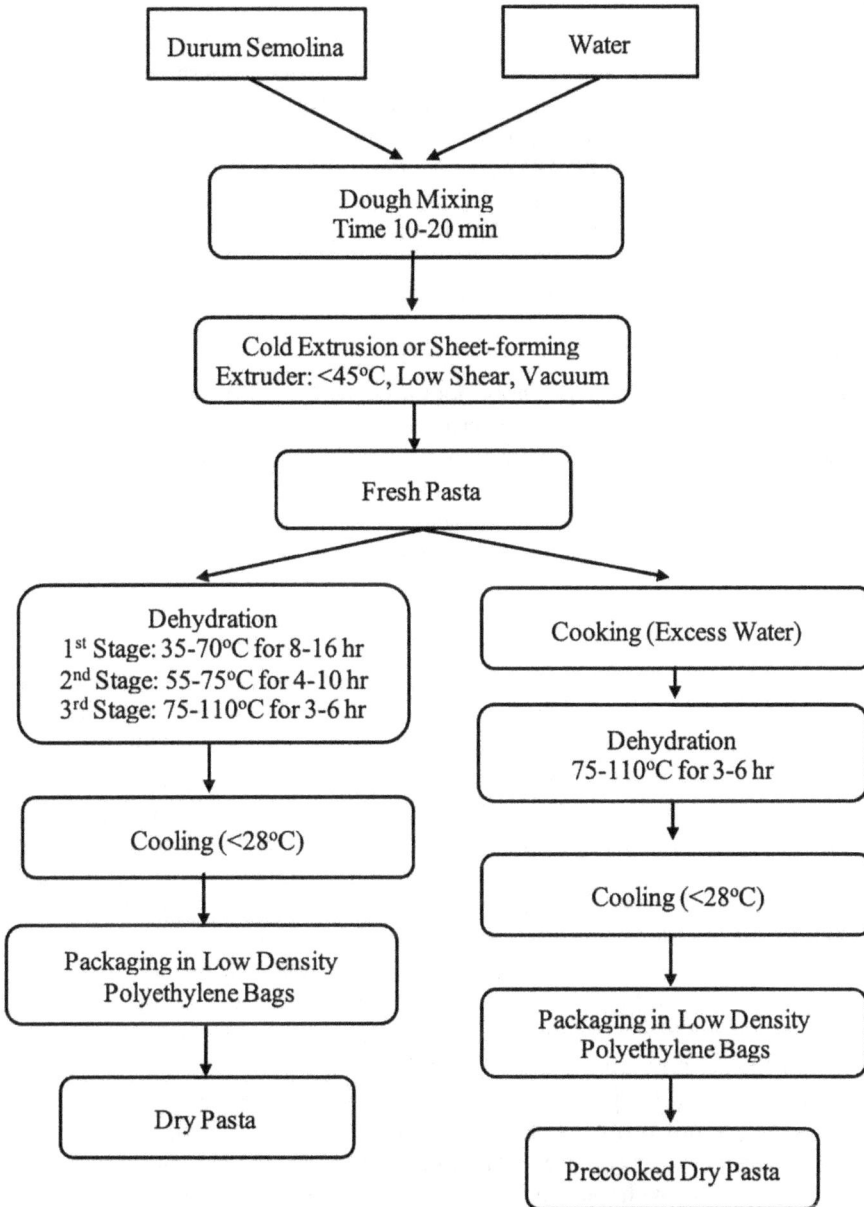

Figure 7.6 Industrial pasta (fresh, dried, and pre-cooked) making processes.

(*Source:* Serna-Saldivar, 2016)

7.5.1.2.4 Packaging

Packaging protects pasta from physical damage and external contamination during transportation and promotes convenient handling during storage. The main packaging material for pasta is the cellophane bag, which is easy to use on automated packaging machines. Fresh dried pasta is folded together and placed in clear plastic containers. A plastic sheet covers each container as it moves along the conveyer belt and is further sealed with a hot press. Simultaneously, a small tube draws air from the container and replaces it with a mixture of nitrogen and carbon dioxide to extend the shelf life of the pasta. However, many food processors prefer boxes rather than bags to wrap pasta.

7.5.2 Noodles

Noodles have been considered a staple food in most Asian and Western countries since ancient times. The most probable origin of noodles is China (Hou, 2020). They are usually made from rice, wheat, and other cereals like buckwheat, as well as starches from potatoes and legumes (Arendt & Zannini, 2013). Noodles can be eaten in a variety of forms, such as steamed, boiled, or fried. Types of noodles vary according to culinary properties and processing methods (Hou, 2020). However, noodles are diverse components of various dishes and often served with sauce or soup.

7.5.2.1 Raw Ingredients for Noodle Production

The primary ingredient of wheat noodles is plain wheat flour. The protein content of wheat flour should be in the range of 10–14%, which is responsible for the elastic texture in noodles (Wieser et al., 2020). The swelling and gelling properties of starch determine whether the external surface of the noodles is sticky after steaming or cooking. Some other primary ingredients are water and salt. The level of water should be 30–38% of the flour, while the salt level is recommended as 1–3%. However, the addition of salt strengthens the structure of the dough, followed by improved rolling and elasticity (Nip, 2007). Additional additives, such as egg solids, improve the color of the end product (Wieser et al., 2020).

7.5.2.2 Noodle Production Process

The production process of noodles is comparatively simple compared to other cereal-based baked products. Flour, salt, and other minor ingredients are mixed and kneaded with water for about 5–10 min. The mixer is then allowed for the even distribution of the ingredients (Hou, 2020). The homogenously mixed dough is then pressed between two to three rotating rolls to form a single sheet of noodle with uniform thickness. However, the thickness of the sheet as well as the sheeting speed significantly influences the quality of noodles. Finally, the dough sheet is passed through a slitter equipped with different cutting rolls, resulting in noodles of the desired size and shape (Serna-Saldivar, 2016).

Noodles come in a wide range of shapes, which adds to their appeal and diversification. Long thin strips, waves, helix strings, shells, and tubes are the most common types of noodles. Non-cooked foods have a shelf life of about 2–3 days when refrigerated. Dried noodles, on the other hand, can maintain their quality for about a year. In Western countries, dried noodles are consumed most, while frying is preferred in Asian countries (Hatcher, 2001). To fix the texture and make the noodles less fragile, freshly prepared noodles are steamed at 100°C for about 1–5 min (Hou, 2020). Accordingly, hot air drying is also carried out at 70–90°C for about 30–40 min, resulting in dried noodles. Furthermore, to obtain fried noodles, freshly dried and steamed noodles are fried in plant oil at 140–150°C for about 0.5–2 min (Wieser et al., 2020). However, both fresh and dried noodles must be cooked for 10–15 min in boiling salty water along with spices or vegetable chunks prior to serving.

Instant-precooked noodles can be prepared by steaming followed by drying or frying of fresh noodles (Wieser et al., 2020). Instant fried noodles usually have a firm texture, non-sticky surface, and relatively strong bite. Both Chinese and Japanese instant noodles are packaged in Styrofoam cups or bowls with dried soup base, vegetables, shrimp, and/or meat and are simply prepared and consumed instantly by adding hot water (Kruger et al., 1996). A flow chart of the noodle production process is shown in Figure 7.7.

7.6 BREAKFAST CEREALS

Breakfast cereals or ready-to-eat (RTE) cereals are typical grains, such as wheat, corn, oats, rice, and so on that have been treated with additional flavors, vitamins, or minerals. Since their invention, they have offered good nutrition as the first meal of the day and increasingly on other occasions, such as snacking (Caldwell et al., 2016). They are convenient to prepare and usually have a long shelf life.

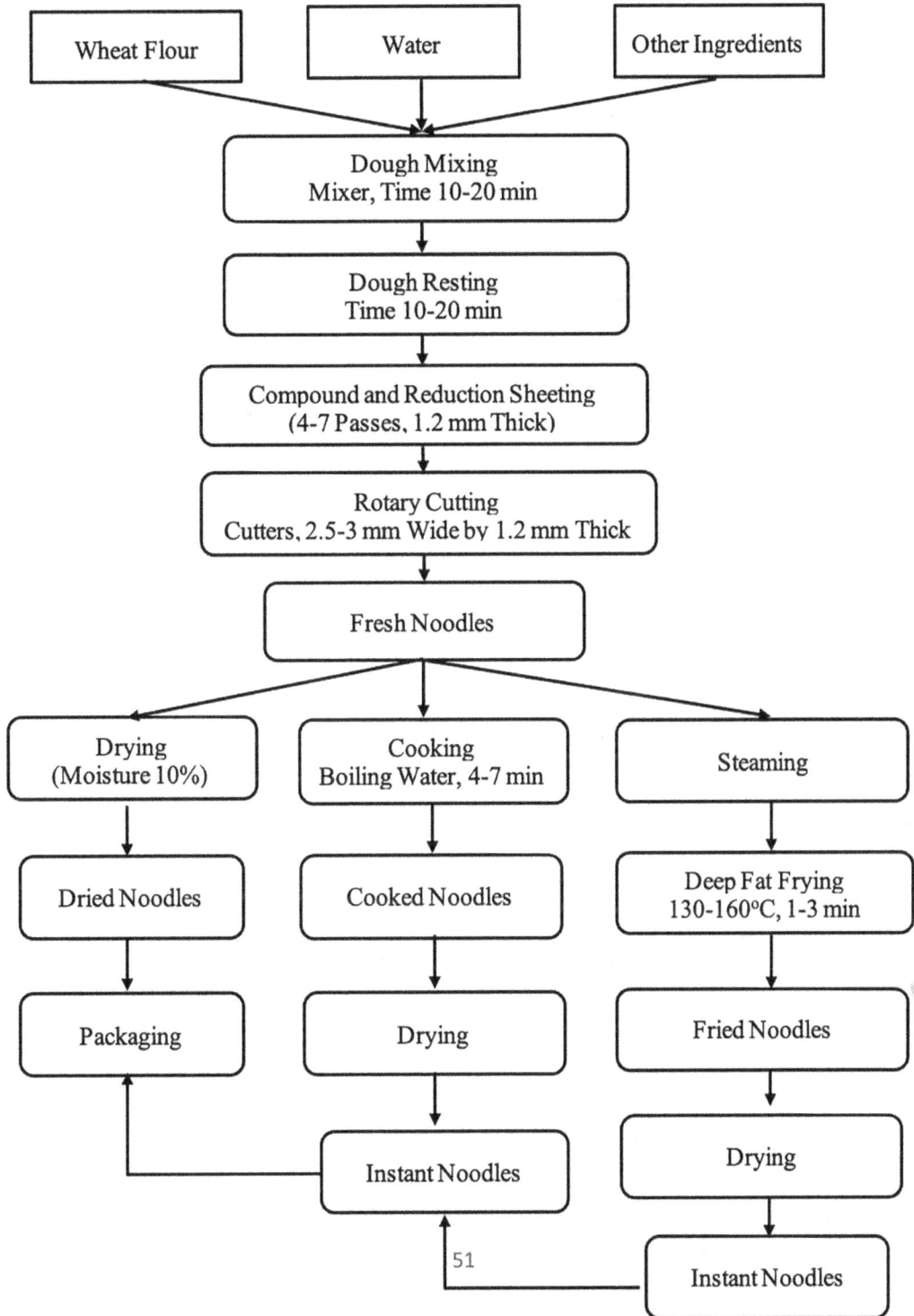

Figure 7.7 Industrial production processes of noodles (fresh, dried, and precooked).

(*Source:* (Serna-Saldivar, 2016)

Breakfast cereals are primarily of two types: hot cereals (e.g., porridge), which require further cooking to gelatinize the starches, and cold ready-to-eat cereals (Tribelhorn, 1991). Cold cereals are eaten right away after being mixed with milk, yogurt, or fruit. RTE cereals, on the other hand, are made through a series of cooking, shape-forming, drying, particulate additions, and packaging (Fast et al., 2020).

7.6.1 Corn Flakes

Corn flakes are a popular breakfast cereal, usually made by toasting flakes of corn (maize). It can be prepared by using either a traditional steam cooking method or continuous extrusion method. The primary ingredient for traditionally cooked corn flake is dry milled corn. Dry milling usually removes both the germ and bran from the kernel, leaving only chunks of endosperm, which are known as corn flaking grits (Caldwell et al., 2016). Grits from de-germinated yellow maize or corn are further cooked with flavorings such as sugar, salt, and malt before being dried and tempered to a firm texture, flaked by passing through rotating rolls, and finally toasted or dried.

The purpose of cooking is to achieve proper gelatinization of the starch and development of characteristic color and flavor. Cooked grits, however, tend to stick together and are fed to a lumping machine that separates the coagulated grits. The cooked grits are then conveyed to a counter-current dryer operating at a temperature of approximately 65°C and further equilibrated inside a tempering vessel for about 6 to 24 h. The hard, dark-cooked grits are further crushed by a series of rollers rotating in opposite directions. Hollow rolls are equipped with water to dissipate frictional heat. On the other hand, soft and floppy flakes are baked in cylindrical rotary ovens. Toasting times vary from 50 s to 3 min, depending on the temperature profile. Flakes are typically toasted at 288–302°C (Fast et al., 2020). Toasting dehydrates the flakes and is responsible for their crisp texture, distinctive flavor, and color. The golden color in corn flakes is primarily due to the Maillard reaction or non-enzymatic browning. After toasting, the flakes are immediately coated with flavorings, sweeteners, and fortifying mixtures (Serna-Saldivar & Carrillo, 2019). The flakes are finally cooled to prepare them for packaging. The optimal moisture content to maintain crispiness should be lower than 2% (Caldwell et al., 2016). The processing of corn flakes is illustrated in Figure 7.8.

7.6.2 Rice Flakes

Whole grain (head rice) or broken pieces of whole kernels (second heads) can be used to make rice flakes. However, broken rice is preferred for flaking because it is less expensive. For oven-puffed rice cereals, whole grains are preferred. Rice is first cooked with other flaking ingredients such as sugar, malt syrup, salt, and a small amount of water to increase the moisture content to 28%. The mixture is then pressure-cooked for 60 min at 103–124 kPa. The subsequent operations, lump breaking, drying, cooling, tempering, flaking, and toasting, are performed in a manner similar to that of corn flakes. The optimal moistures for flaking and toasted flakes are 17% and 2%, respectively (Fast & Caldwell, 2000).

7.6.3 Wheat Flakes

Wheat flake processing differs from corn flake processing to some extent due to differences in grain structure. The primary ingredient of wheat flakes is a whole wheat grain containing germ, bran, and endosperm with intact seed. Production of wheat flakes includes pre-processing, cooking, lump breaking, drying, cooling, tempering, and flaking similar to that of corn flakes. However, the moisture content of finished toasted wheat flakes should be 1–3% to extend the storage stability (Fast & Caldwell, 2000).

7.6.4 Extruded Flakes

Extruded flakes differ from traditional flakes in that the grit is formed by extruding the mixed flour ingredients through a die hole and cutting off pellets of dough according to the desired size.

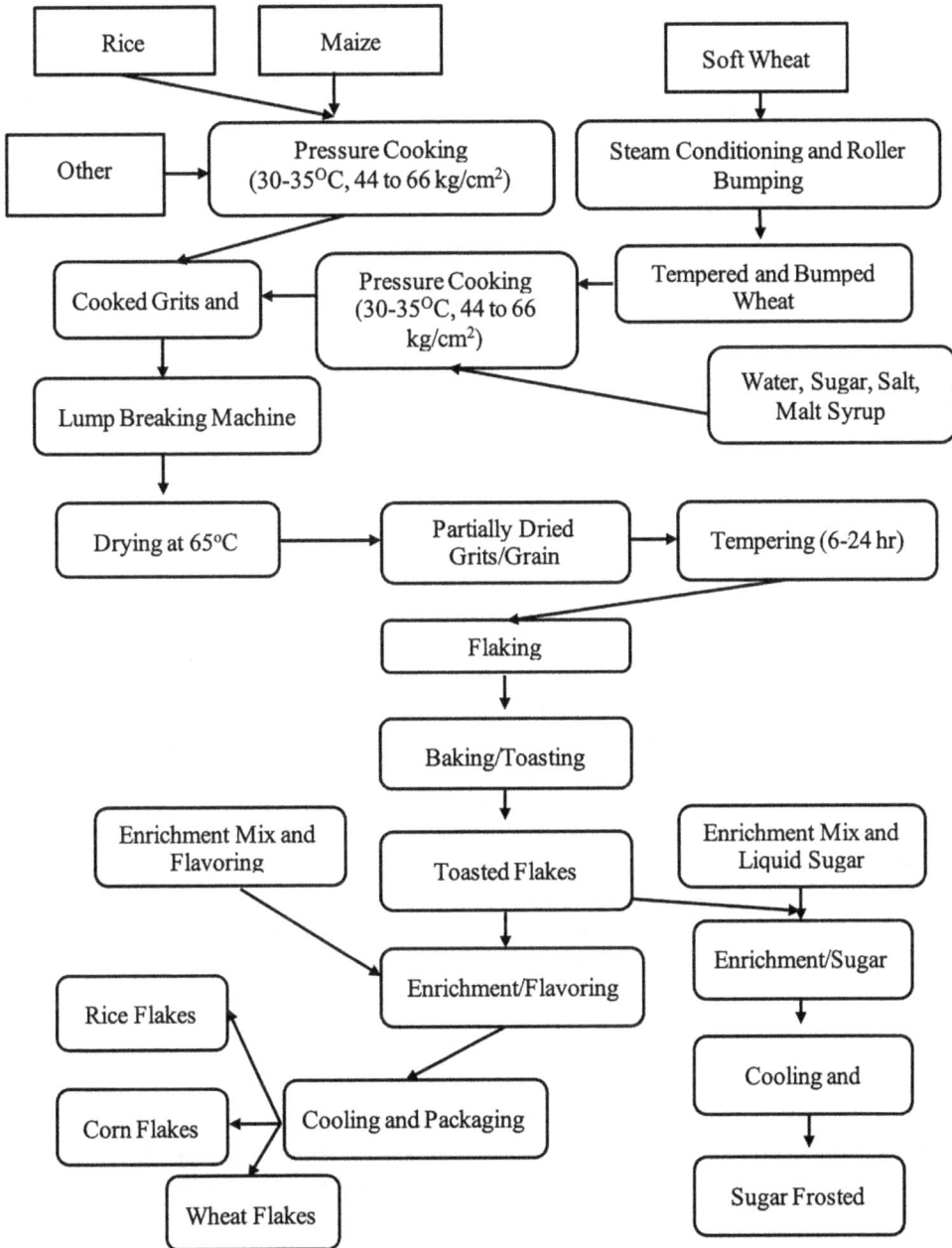

Figure 7.8 Traditional processes to manufacture corn, rice, and wheat flakes.

(*Source:* Serna-Saldivar, 2016)

Extruders can be of varying types, including single screw, twin screw, low shear, high shear, or a combination. Temperature, pressure, shear in the extruder barrel, and perforations in the die and the design of the cut-off knives influences the shape, size, and texture of the final product. To enhance sensorial appeal, the extruded products are further coated with sweetening and flavoring agents and then oven-dried to reduce the moisture levels (Caldwell et al., 2016).

7.6.5 Puffed Cereals

Wheat and rice are the most common grains used to make puffed cereals. Puffed cereals can be made either by using whole grain kernels or using flour blends only. Both traditional or extrusion cooking methods can be optimized to obtain these types of ingredients. Dried cooked kernels or extruded pellets are puffed in a toaster oven or puffing machine to produce puffed cereals (Perdon et al., 2020). Another popular method is direct extrusion, which involves cooking and puffing a flour mixture in a single step.

7.6.6 Granola Cereals

Regular or thin-flaked rolled oats are the primary grains used to make granola cereals. Other ingredients such as nut pieces, corn syrup, brown sugar, cocoa nibs, malt extract, honey, dried fruits, vegetable oils, and milk are combined to prepare this cereal product. Meanwhile, liquid and dry ingredients are mixed and further spread in a uniform layer on the band of a continuous oven and baked at 150–220°C (Caldwell et al., 2016). After cooling, the flakes are broken or cut into the desired shape and packaged to preserve for further use.

7.7 CONCLUSION

The increased prevalence of obesity-related problems in developed countries necessitates special attention in the maintenance of normal and sound health through various formulations and activities, particularly delivering soluble fibers to consumers through various foods such as cereals and cereal-based products. Cereals such as maize, wheat, oats, and rice are now used in the preparation of foods that resemble traditional foods and are consumed in a similar manner but have the additional benefits of assisting physiological functions as well as providing nutrition. Eating habits can drastically reduce healthcare expenditure if people can modify their diets based on their nutritional knowledge. In today's world, cereal-based products have demonstrated significant potential for improving human health and enabling functional food innovations. The concept of cereal-based industry can be expanded towards developing newer technologies for cereal processing in order to improve nutritional value and acceptance by end users.

REFERENCES

Arendt, E.K. and Zannini, E., 2013. *Cereal grains for the food and beverage industries.* Elsevier.

Bhardwaj, A., Tayal, A. and Goyal, A., 2015. System modeling & behavioural analysis of bread manufacturing unit—Case study. *International Journal on Theoretical and Applied Research in Mechanical Engineering (IJTARME), 4*(3), pp. 48–53.

Bhatia, S.C., 2017. Food Biotechnology: A review. *Food Biotechnology*, pp. 3–17.

Britannica, E., 2021. The Editors of Encyclopedia. "Bread". *Encyclopedia Britannica*, www.britannica.com/topic/bread. Accessed on: January 5, 2022.

Broyart, B. and Trystram, G., 2002. Modelling heat and mass transfer during the continuous baking of biscuits. *Journal of Food Engineering, 51*(1), pp. 47–57.

Bugialli, G., 1988. *Bugialli on Pasta.* Simon & Schuster.

Caballero, B., Finglas, P. and Toldrá, F., 2015. *Encyclopedia of food and health.* Academic Press.

Caldwell, E.F., McKeehen, J.D. and Kadan, R.S. 2016. Breakfast cereals. In *Reference module in food science.* Elsevier.

Cauvain, S.P. ed., 2012. *Breadmaking: improving quality.* Elsevier.

Cauvain, S.P. and Young, L.S., 1998. *Technology of bread making.* Blackie Academic & Professional.

Cauvain, S.P. and Young, L.S., 2006. Key characteristics of existing bakery-product groups and typical variations within such groups. *Baked products science, technology and practice*, pp. 14–34.

Chavan, R.S., Sandeep, K., Basu, S. and Bhatt, S., 2016. *Biscuits, cookies, and crackers: chemistry and manufacture*. Elsevier.

Conforti, F.D., 2006. Cake manufacture. *Bakery products: Science and Technology*, *22*, pp. 393–410.

Day, L., 2013. Proteins from land plants—potential resources for human nutrition and food security. *Trends in Food Science & Technology*, *32*(1), pp. 25–42.

Delcour, J. and Hoseney, R.C., 2010. *Principles of cereal science and technology*. AACC International Press.

Edwards, W.P., 2007. *The science of bakery products*. Royal Society of chemistry.

Fast, R.B. and Caldwell, E.F., 2000. *Breakfast cereals and how they are made* (ed. 2). American Association of Cereal Chemists.

Fast, R.B., Perdon, A.A. and Schonauer, S.L., 2020. Breakfast—Forms, ingredients, and process flow. In *Breakfast cereals and how they are made* (pp. 5–35). AACC International Press.

Gisslen, W., 2021. *Professional baking*. John Wiley & Sons.

Hatcher, D.W., 2001. Asian noodle processing. In *Cereals processing technology* (pp. 131–157). Woodhead Publishing.

Hill, C.B. and Li, C., 2016. Genetic architecture of flowering phenology in cereals and opportunities for crop improvement. *Frontiers in Plant Science*, *7*, p. 1906.

Hou, G.G. ed., 2020. *Asian noodle manufacturing: Ingredients, technology, and quality*. Elsevier.

Huang, S. and Miskelly, D., 2016. *Steamed breads: Ingredients, processing and quality*. Woodhead Publishing.

Hui, Y.H. ed., 2006. *Handbook of food science, technology, and engineering* (Vol. 149). CRC press.

Hui, Y.H., Corke, H., De Leyn, I., Nip, W.K. and Cross, N.A. eds., 2008. *Bakery products: science and technology*. John Wiley & Sons.

Kruger, J.E., Matsuo, R.B. and Dick, J.W., 1996. *Pasta and noodle technology*. American Association of Cereal Chemists.

Kulp, K. ed., 2000. *Handbook of Cereal Science and Technology, revised and expanded*. CRC Press.

Kulp, K., 1988. Bread industry and processes. *Wheat chemistry and technology*, *2*, pp. 371–406.

Labensky, S.R., Martel, P. and Van Damme, E., 2009. *On baking: A textbook of baking and pastry fundamentals*. Pearson/Prentice Hall.

Manohar, R.S., Rao, P.H., Manohar, R.S. and Rao, P.H., 1999. Effect of mixing method on the rheological characteristics of biscuit dough and the quality of biscuits. *European Food Research and Technology*, *210*(1), pp. 43–48.

McSweeney, P.L.H. and Day, L., 2016. *Food products and ingredients*. Elsevier.

Misra, N.N. and Tiwari, B.K., 2014. Biscuits. In *Bakery products science and technology* (pp. 585–601). John Wiley & Sons, Ltd.

Nip, W.K., 2007. Asian (Oriental) Noodles and their manufacture. In *Handbook of food products manufacturing* (pp. 539–564). John Wiley & Sons, Ltd.

Nout, M.R., 2009. Rich nutrition from the poorest—Cereal fermentations in Africa and Asia. *Food Microbiology*, *26*(7), pp. 685–692.

Oliver, G., Wheeler, R.J. and Thacker, D., 1996. Semi-sweet biscuits: 2. Alternatives to the use of sodium metabisulphite in semi-sweet biscuit production. *Journal of the Science of Food and Agriculture*, *71*(3), pp. 337–344.

Pagani, M.A., Lucisano, M. and Mariotti, M., 2007. Traditional Italian products from wheat and other starchy flours. In *Handbook of food products manufacturing* (p. 2). John Wiley & Sons, Ltd.

Penfield, M.P. and Campbell, A.M., 1990. Yeast Breads: Flat Breads, Experimental Food Science. *Academic Press*, *3*, pp. 438–441.

Perdon, A.A., Schonauer, S.L. and Poutanen, K. eds., 2020. *Breakfast cereals and how they are made: raw materials, processing, and production*. Elsevier.

Pollini, C.M., Pantò, F., Nespoli, A., Sissons, M. and Abecassis, J., 2012. Manufacture of pasta products. In *Durum wheat: Chemistry and technology*. AACC International Press.

Rizzello, C.G., Portincasa, P., Montemurro, M., Di Palo, D.M., Lorusso, M.P., De Angelis, M., Bonfrate, L., Genot, B. and Gobbetti, M., 2019. Sourdough fermented breads are more digestible than those started with baker's yeast alone: An in vivo challenge dissecting distinct gastrointestinal responses. *Nutrients*, *11*(12), p. 2954.

Robertson, G.L., 1993. Packaging of cereal and snack foods. In *Food packaging. Principles and practice*. Marcel Dekker, Inc.

Robertson, G.L., 2016. *Food packaging: principles and practice*. CRC Press.

Rosentrater, K.A. and Evers, A.D. 2018. *Extrusion processing of pasta and other products. Kent's Technology of Cereals*, Fifth Edition. Woodhead Publishing,

Samota, M.K., Sasi, M., Awana, M., Yadav, O.P., Amitha Mithra, S.V., Tyagi, A., Kumar, S. and Singh, A., 2017. Elicitor-induced biochemical and molecular manifestations to improve drought tolerance in rice (*Oryza sativa* L.) through seed-priming. *Frontiers in Plant Science, 8*, p. 934.

Serna-Saldivar, S.O. 2016. *Cereal grains: Properties, processing, and nutritional attributes.* CRC press.

Serna-Saldivar, S.O. and Carrillo, E.P., 2019. Food uses of whole corn and dry-milled fractions. In *Corn* (pp. 435–467). AACC International Press.

Shukla, T.P., 2001. Chemistry of dough development. *Cereal Foods World, 46*(7), pp. 337–339.

Sluimer, P., 2005. *Principles of breadmaking: functionality of raw materials and process steps.* AACC International Press.

Tribelhorn, R.E., 1991. Breakfast cereals. In *Handbook of cereal science and technology* (pp. 741–762). Marcel Dekker, Inc.

Wieser, H., Koehler, P. and Scherf, K.A., 2020. *Wheat-an exceptional crop: botanical features, chemistry, utilization, nutritional and health aspects.* Elsevier.

Wilderjans, E., Luyts, A., Brijs, K. and Delcour, J.A., 2013. Ingredient functionality in batter type cake making. *Trends in Food Science & Technology, 30*(1), pp. 6–15.

Wrigley, C.W., 2010. An introduction to the cereal grains: major providers for mankind's food needs. In *Cereal grains* (pp. 3–23). Woodhead Publishing.

Zhou, W. and Therdthai, N., 2014. Manufacture. In *Bakery products science and technology* (pp. 473–488). John Wiley & Sons, Ltd.

CHAPTER 8

Cereals and Their By-Products

Jagbir Rehal, Kulwinder Kaur and Preetinder Kaur

CONTENTS

ABBREVIATIONS

Expanded form	Abbreviation
Alcohol by volume	ABV
Dried distillers' grains and solubles	DDGS
Dietary fiber	DF
Dry matter	DM
Deoiled rice bran	DRB
Defatted rice germ	DRG

(Continued)

Expanded form	Abbreviation
Defatted wheat germ	DWG
Free fatty acids	FFA
Food and Agriculture Organization	FAO
Food grade distillers' dried grains	FDDG
γ-aminobutyric acid	GABA
High-density lipoprotein	HDL
Insoluble dietary fiber	IDF
Low-density lipoprotein	LDL
Nanoparticle rice bran oil	NPRBO
Oil binding capacity	OBC
Polyunsaturated fatty acid	PUFA
Rice bran oil	RBO
Rice germ	RG
Rice hull	RH
United States Department of Agriculture	USDA
Un-stabilized defatted wheat germ	USDWG
United Nations Environment programme	UNEP
Water binding capacity	WBC
Wheat bran	WB
Wheat bran fiber	WBF
Wheat germ	WG
Wheat germ oil	WGO

8.1 INTRODUCTION

Bringing about a world without hunger and with high-quality and microbiologically safe food for everyone is one of the greatest challenges we face. The food service industry, including agriculture and food production, retail, and household consumption, generates a total of 931 million tons of food loss and waste every year (UNEP, 2021). Losses are associated with the food (and feed) value chain, which starts at harvest time and continues with post-harvest handling and processing, marketing, and distribution until consumer consumption. The losses associated with food processing also include by-products generated during the production of target products across all food sectors. Cereal by-products, on the other hand, are an unutilized source of numerous components or fractions with high nutritive quality that might be used as a novel resource not only for feed products but also for food commodities.

Cereals exist in a variety of shapes and sizes, yet they all have basic characteristics. These are cultivated in massive quantities owing to their economic value as a commodity that provides more food and energy than any other commodity on the globe. Cereal grains are hence regarded as staple crops. Cereal processing is not only a major crucial element of the food production chain, but it also provides a variety of necessary nutrients to the masses. Cereal grains are easy to store if their enzymatic activity is under control, and they may be utilized in making a variety of foods (Amadou et al., 2013; Yishak G & Shimelis, 2015). Corn (1148), rice (paddy, 755), wheat (768), barley (159), sorghum (58), oat (23), and rye (13) were the primary cereals produced around the globe in 2019–20, ranked by tonnage (in million tons) (FAO, 2022).

Cereal is a grass belonging to the family of *Gramineae* that is grown for the edible ingredients of its grain or kernel. In general, it is a caryopsis in the strictest sense, as it is made up of a pericarp and a seed. The pericarp clings firmly to the seed coat, enclosing the germ and endosperm in the remainder of the seed. Next to the pericarp is the aleurone layer. Protein and minerals abound in this stratum. The endosperm is the vast central section of the kernel that is primarily starch, while the germ, also known as embryo, is smaller in size and located at the bottom of the kernel that contains mostly oil (Delcour & Hoseney, 2010).

8.2 GRAIN STRUCTURE AND ITS IMPORTANCE

Most cereal grains comprise four major components: the hull or husk, the outermost layer (bran), starchy endosperm, and the oil containing the embryo or germ. The husk is a dry and membranous outer covering in which grain is enclosed. It is not present in all grains. Bran is the outermost tissue layer enriched with dietary fiber, grain protein, and lipid in addition to phytochemical compounds. During milling, bran is rubbed off through either abrasion or friction action. The endosperm is the starchy component of grain surrounding the embryo. It is consumed in the form of flour and processed food (bread, buns, snacks, cookies, etc.). The germ is the oil-containing component, which has to be separated from the grain to enhance its storage life. It can be a major product or by-product depending upon the end use. Even though the major cereal grains, wheat, rice, and maize, exhibit similar anatomy, the proportions of major components such as protein, fiber, carbohydrate, and lipid vary depending upon the grain structure. The anatomy of cereal grain also plays crucial role in understanding the type of processing (wet, dry, or dry-grind milling) and need of unit operations (debranning, dehulling, degerming) required for that particular cereal grain. Wet milling, for example, separates primary grain components, whereas dry milling is often a size reduction technique. Paddy contains a low-value husk that is not edible for human consumption and has to be removed by a dehuller. The structures of the major grains (wheat, rice, maize, and sorghum) are shown in Figure 8.1.

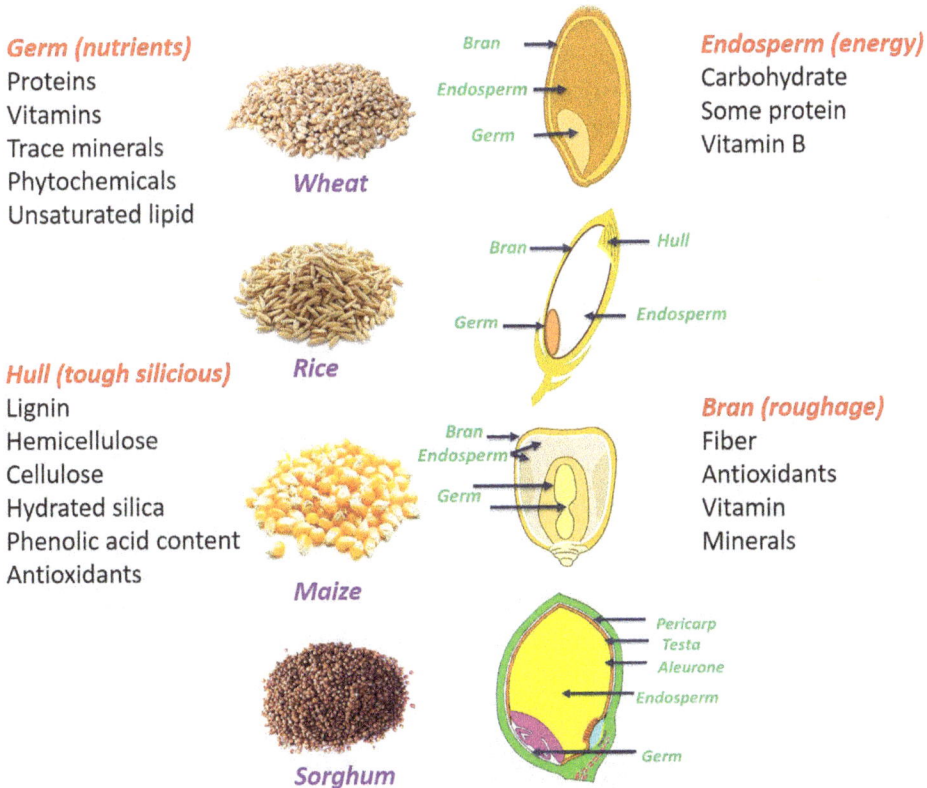

Germ (nutrients)
Proteins
Vitamins
Trace minerals
Phytochemicals
Unsaturated lipid

Wheat

Endosperm (energy)
Carbohydrate
Some protein
Vitamin B

Rice

Hull (tough silicious)
Lignin
Hemicellulose
Cellulose
Hydrated silica
Phenolic acid content
Antioxidants

Maize

Bran (roughage)
Fiber
Antioxidants
Vitamin
Minerals

Sorghum

Figure 8.1 Structural composition of the cereal grains, wheat, rice, maize, and sorghum.

8.3 MILLING AND ITS FRACTIONS

Cereal by-products are primarily supplied by the milling sector. Milling is the most important technique in the grain sector, and it is divided into two types: dry and wet, each with its unique set of features. The outermost bran layer and germ, which are regarded as by-products of the cereal grains, are separated during the debranning and degerming unit operations of dry milling. Dry milling is also termed pearling, an abrasive and frictional process to be carried out to remove only the seed coat (testa and pericarp), thereby allowing the nutritious part (aleurone and subaleurone layers) to remain in the intact grains. The steps taken prior to milling produce a by-product known as "grain screenings," which primarily contain shriveled grains; other cereals; pest-damaged grains; grains with discolored germs; germinated grains; and other unacceptable material such as broken grains; decayed grains; and other undesirable material like extraneous seeds, husks, and dead insects (Union, 2010). The technique of conditioning or tempering involves moistening the kernels with a regulated amount of water so that the inner endosperm softens and the bran hardens. This method is intended to reduce bran breakdown, aids in progressive detachment during milling, and increases sieving efficiency. The conjunction of time and temperature for hydration of grains prior to milling/grinding depends on the family of grain, its variety, initial moisture content, and so on. Conditioning for hard or translucent grains, such as durum, is done in dual stages. Hard wheat usually requires more time for conditioning in order to reach an even higher level of final moisture content than that of soft wheat. On the other hand, cereals such as rye and triticale, owing to their softer endosperm, can be conditioned to lower moisture content than durum wheat.

In maize or corn dry milling, tempering of grains is carried out in three stages in order to attain the optimum moisture content (10–18%) for further milling (Jung et al., 2018). Size and shape of grain, intensity of adherence between outer layers and endosperm, and kernel hardness are key parameters for an efficient cereal milling process.

Wet milling, nevertheless, is mostly used to produce starch and gluten as major products along with steep solids (those high in nutrients used in the pharmaceutical sector), germ, and bran as coproducts. Wet milling is primarily used in maize and wheat crops to yield maximum starch for its commercial and industrial applications such as pharmaceutical, packaging material, ethanol production, and so on. It can also be utilized in food industries to develop syrups (such as glucose, maltodextrins, and other starch derivatives), food additives, bakery and confectionery items, soups, baby foods, and alcoholic beverages. This procedure might also be effectively used in other cereals such as barley, sorghum, and oats depending upon availability of equipment or processing facilities. Wet milling is composed of cleaning of grains and steeping, followed by milling/grinding of steep grains and sifting into different fractions. The unit operations involved in milling and the resulting fractions of major products as well as by-products are shown in Figure 8.2.

8.4 CEREAL PROCESSING BY-PRODUCTS

During cereal milling, the husk, bran, embryo, and aleurone layers are separated from the inner starchy endosperm. Biofunctional chemicals, fiber, vitamins, minerals, lignans, phenolic substances, and phytoestrogen compounds abound in these grain fractions. The nutritional content of the by-products varies substantially depending on the milling procedure adopted. Apart from usage as feed, the fractions produced during dry, dry-grind, or wet milling can be used in a variety of food and non-food items.

8.4.1 Wheat

Wheat, which belongs to farinaceous grass, is also known as *Triticum* spp. It is a staple food crop consumed by more than 33% of the world's population due to its high energy content (calories). Wheat is a rich source of major nutrients like proteins, vitamins B, minerals, and dietary fiber (Kanojia et al., 2018). Wheat is mainly produced from two species, *Triticum aestivum* vulgare (bread wheat)

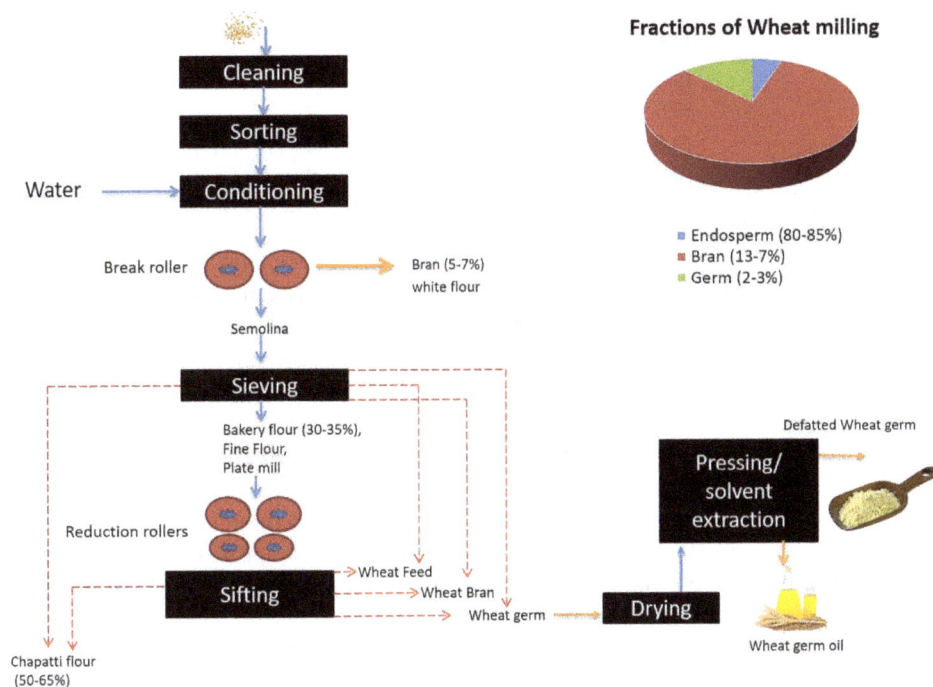

Figure 8.2 Major products and by-products obtained during milling of wheat.

and *Triticum turgidum* durum (hard wheat), and finds wide applications in bakery and confection-ary products (bread, biscuits, cakes, muffins, noodles, and vital wheat gluten). The grain is usually oval shaped but may also be round; nearly spherical; or long, narrow, and flattened. Wheat grains are usually between 5 and 9 mm long, between 35 and 50 mg in weight, and have a crease on one side where the grain was linked to the wheat flower. In wheat grains, 2–3% of the grain is germ, 13–17% is bran, and 80–85% of the grain is mealy endosperm (Sramkova et al., 2009). A process flow diagram of wheat with fractions of major by-products during the milling operation is shown in Figure 8.2. The primary by-products of the wheat milling industry include bran, shorts, and germ, along with a lower fraction of secondary by-products, including wheat middlings, red dog, and feed flour. More than 25–30% of wheat is converted into by-products (Huang et al., 2014). Wheat feed may or may not be mixed with wheat bran. It is sold in different fractions depending upon fiber content: low (<5%), intermediate (<7%), and higher fiber content (<9%). Many feed wheat products are surplus to human requirements or are low quality (low test weight or damaged wheat) and therefore inappropriate for human consumption (Blair, 2008).

According to nutritional analysis, wheat bran contains more fiber than any other wheat mill feed. In addition to wheat bran, wheat midds or middlings are another typical by-product of wheat milling. This is finely ground components of wheat bran, wheat germ, wheat shorts, and wheat flour, along with offal from the "mill tail." Additionally, it may include undesirable materials such as weed seeds, which have to be removed before milling. Wheat middlings are commonly used in commercial feeds both as a source of energy and as a means to improve pellet quality (Blair, 2008). The appearance of wheat shorts is similar to that of wheat middlings. Red dog is a minor millfeed. In comparison with original grain, red dog is higher in protein. However, the by-products still remain deficient in the essential amino acids. Most of the water-soluble vitamins can be found in the by-products. Among minerals, feeds made from by-products are fairly high in phosphorus, magnesium, and manganese. By-products of wheat milling are fed to poultry, cattle, pigs, and swine to provide energy and amino acids (Huang et al., 2014). However, these can be further processed

Table 8.1 Proximate Composition (% of Dry Matter) of Wheat Grain and Its By-Products

Major Nutrients	Protein	Starch	Lipid/Fat	Fiber	Ash
Wheat grain	13.5	68	1.9	2.6	1.8
Wheat bran	15.6	22.3	4.7	12.3	5.6
Wheat germ	28.5	–	8.8	3.5	4.9
Middlings	16.9	–	4.4	7.8	4.7

(*Source:* Hertrampf & Piedad-Pascual, 2000)

to improve the nutritional value of the feed if used appropriately. Due to differences in varieties of wheat and environmental factors, wheat by-products have a chemical composition that varies with processing techniques and differences in environmental factors (Huang et al., 2014). The major composition of wheat grain and its by-products is given in Table 8.1.

8.4.1.1 Nutritional and Functional Characteristics of Wheat By-Products

Bran is a good source of dietary fiber with a low lipid content. The bran contains about 4.7% lipids, 15.6% protein, 22.3% starch, and 12.3% fiber content (Hertrampf & Piedad-Pascual, 2000). In addition to primary components, it also contains phytonutrients such as vitamins and phenolic compounds (Skendi et al., 2011). Wheat bran, however, has been found to reduce the dough rheology and handling capabilities of cereal-based products, as well as their quality and sensory parameters. Different solutions are being tested to offset these negative effects on cereal-based goods. One method of compensating for the loss of quality in baked items with a high bran content is to incorporate essential gluten into the flour to improve the gluten content (Koletta et al., 2014). Nonetheless, bran processing (either mechanically, thermally, or enzymatically) modulates the functional properties of wheat bran as well as its effect on food products.

It is estimated that wheat germ accounts for 2 to 3% of the kernel weight. The germ contains about 8.8% lipids, 28.5% protein, 3.5% fiber, and about 4.9% ash (Hertrampf & Piedad-Pascual, 2000). In addition to this, bioactive compounds such as phytosterols (24–50 mg/kg), tocopherols (300–750 mg/kg DM), policosanols (10 mg/kg), riboflavin (6–10 mg/kg), thiamin (15–23 mg/kg), and carotenoids (4–23%) are also present in significant amounts within this product (Brandolini & Hidalgo, 2011). Numerous enzymes exhibiting an embryonic nature are also associated with wheat germ. Germ removal during processing improves the shelf life of flour, allowing it to be milled as feed for mills and sold without enrichment. It has been estimated that 25 million tons of germ are separated from grain each year during the milling process in the world (Rizzello et al., 2010).

Wheat germ has about 10–15% oil (Dunford & Zhang, 2003). Mechanical pressing and solvent extraction are used to recover oil and recover about 50% or 90% of total lipids, respectively. Supercritical CO_2 extraction is an innovative method that yields 92% of solvent-extracted oil without using toxic solvents (Panfili et al., 2003). The minerals phosphorus (1.4 g/kg) and iron (0.5 g/kg) figure prominently in wheat germ oil (WGO) (Wang & Johnson, 2001). The tocopherol and phytosterol (mainly sitosterol and campesterol) content of WGO are predominant among commonly used vegetable oils. Additionally, it has large amounts of polyunsaturated fatty acids, with 80% linolenic and linoleic acid. It is vital for human health to consume these fatty acids since they are considered essential (cannot be synthesized by the body).

A defatted wheat germ meal is the principal by-product of the oil extraction process. Zhu et al. (2006) showed that deffatted germ meal has globulins (15.6%) and a balanced amino acid profile (high in lysine and vital amino acids) in addition to high protein content (approx. 35%), with albumins accounting for about 34.5%, thus showing it to be one of the most productive and attractive sources of vegetable proteins. Furthermore, defatted germs are also a source of carbohydrate sources such as sugars (around 20%, 58.5% sucrose and 41.5% raffinose), starch, pentosans, and fiber, along with a low amount of carotenoids (3 mg/kg) and flavonoids (0.35 g rutin per kg). Its mineral composition consists of potassium> magnesium> calcium> zinc> manganese.

Wheat shorts, which are a mixture of endosperm, bran, and germ, are an important source of protein, vitamins, minerals, and other nutrients. These are regularly used as animal feed when mixed with bran. Shorts, if used immediately after production, can, however, be used in food for humans.

8.4.1.2 Utilization of Wheat By-Products

The primary by-products of the wheat milling industry include bran, shorts, and germ, along with a lower fraction of secondary by-products, including wheat middlings, red dog, and feed flour. A detailed summary of utilization of wheat by-products is listed in Table 8.2.

Table 8.2 Utilization of Wheat By-Products in Different Foods

By-Product	Application	Salient Findings	References
Wheat germ	Pasta	Increase in nutritional value	Aktaş et al. (2014)
Wheat germ	Macaroni	Increase in protein content up to 17%	Pınarlı et al. (2004)
Wheat germ	Bread dough	Smaller volume, less elasticity and cohesiveness, firmer, and darker in color	Noori and Sabir (2019)
Wheat germ	Biscuits	Increased crude fiber; vitamins C, A, and E; and folic acid	Youssef (2015)
Wheat germ	Cookie	High protein, fat, fiber, and mineral contents	Petrovic et al. (2017)
Wheat germ	Cake	Increased viscosity and consistency of batter	Aalami et al. (2018)
Wheat germ	Bread	Reduction in phytic acid content, improved textural and antioxidant attributes of the product	Ebrahimi et al. (2021)
Wheat bran	Pasta	Higher resistance to overcooking and good sensory attributes	Sobota set al. (2015)
Wheat germ	Fried crackers	Increase in protein, fiber, and mineral contents	Meriles et al. (2021)
Wheat bran	Biscuits	Increase in dietary fiber by 3.25 times than that of control	Nandeesh et al. (2010)
Wheat germ	Idli	Accelerated fermentation rate along with improved protein, fat content, texture, and sensorial attributes	Hemavathi et al. (2017)
Wheat germ	Buttermilk	Increased antioxidant activity and apparent viscosity	Abbas et al. (2015)
Defatted Wheat germ flour	noodles	Enhanced amino acid composition, minerals, and vitamin B	Yiqiang Ge et al. (2001)
Defatted wheat germ	Cookie dough	No effect on the rheological and textural properties up to 15% inclusion	Petrović et al. (2015)
Wheat bran	Biscuits	Increase in antioxidant activity, phenolic concentration	Tiwari and Mishra (2019)
Wheat bran	Dough	Increased functional attributes and soluble dietary fiber (SDF) content of bran	Lee et al. (2021)
Wheat bran	Biscuits	High protein and soluble dietary fiber (DF) contents and low arabinoxylan and insoluble DF contents	Ma et al. (2021)
	Pancakes	High protein content	Ma et al. (2021)
	Bread	Reduced luminosity and specific volume	Alán et al. (2014)
	Noodles	Increase in tensile strength and water absorption index	Song et al. (2013)
	Doughnuts	Increased volume and darker-colored doughnuts, reduced fat content	Kim et al. (2012)
	Alkaline noodles	Increased fiber content	Wardhana and Banawi (2020)
Wheat germ oil	Noodles	Increased hardness, chewiness, and swelling power	Yan and Lu (2021)
	Noodles	Improvement in protein, in vitro protein digestibility, total phenolic content, phytic acid content, and total dietary fiber content	Demir et al. (2021)

Wheat bran enhances the amount of dietary fiber in bread, muffins, and cookies by replacing some of the flour (Mi et al., 2007). The in vitro starch hydrolysis index of biscuits by incorporating coarse wheat bran up to 30% improved its nutritional profile without affecting its nutritional value (Sozer et al., 2014). In contrast, when coarse bran addition was increased to 30%, which equals 15% of dietary fiber in cookies, a larger hydrolysis index was obtained as a result of improved gelatinization of the starch. As part of the same study, a lower particle size of wheat bran (68 μm) enhanced the elastic modulus as well as hardness of biscuits when added to flour at 15%. The sensory parameters showed that higher levels of coarse bran incorporation (above 5%) lead to rougher and more damaged biscuits in the mouth. Biscuits with wheat bran added had less freshness, odor, and color values than those without, and the size of the bran particles had no significant effect on these characteristics. Steaming and roasting bran gave the lowest hardness values among the treated brans, and biscuit dough hardened more after adding bran. By adding glycerol monostearate to the dough along with steamed and roasted bran, the textural properties of the biscuits improved in terms of breaking strength and spread ratio. Compared to the control, this biscuit scored higher on quality (87%) and had a 3.25 times higher fiber content.

The presence of OH groups in the structure of wheat bran fiber seems to interfere with hydration of protein and dilute the network of gluten in the dough, which in turn affects the resulting bread by increasing its water absorption capacity (Coda et al., 2014). As a result, dough stability is affected and dough development time is extended. Fried cereal products containing wheat bran (WB) have been shown to have lower oil contents. WB can be blended with wheat flour through manual mixing or through hybridization techniques. Wheat bran and soya bean hulls have been combined to produce composites that have been utilized to develop fried doughnuts with less oil content (Kim et al., 2012). Wheat bran with a high content of dietary fiber is a rich by-product of wheat flour; however, the majority of dietary fiber found in wheat bran is insoluble. Fiber-fortified products are adversely affected by the addition of insoluble dietary fiber, resulting in an undesirable texture and a drop in consumption (Lei et al., 2021). However, higher concentrations of insoluble dietary fiber (above 4%) can lead to tighter structures, which negatively affect the formation of gluten networks.

Wheat germ enhanced the nutritional quality of pasta in terms of protein, fat, ash, and mineral content (Aktaş et al., 2014; Pınarlı et al., 2004). Raw and microwave treated WG added in macaroni had a similar in vitro protein digestibility as the conventional product (Pınarlı et al., 2004). Biscuits supplemented with 5–15% wheat germ had higher protein, fiber, fat, and mineral content (Petrovic et al., 2017). Likewise, cakes with a similar concentration of wheat germ also had a high mineral content (Ca, Fe, Cu, P, K, Mn, and Zn). However, muffins showed a significant increase in Mg, as reported by Levent and Bilgiçli (2013). Biscuits enriched with 20% wheat germ had a high crude fiber content as well as vitamins C, A, and E and folic acid. The addition of WG increased the water requirement for making biscuits, as fiber and protein were contained in WG (Bansal & Ml, 2011; Petrović et al., 2015). This would result in the formation of a strong structure with high ductility and durability (Bansal & Ml, 2011).

The color of crust and crumb of cakes is significantly influenced by the incorporation of WG (Levent & Bilgiçli, 2013), indicating a darker cake crust as more and more WG is added. Such a result could be attributed to the increase in caramelization of sugar and Maillard browning reactions (Arshad et al., 2007). Undamaged stabilized debitterized wheat germ (USDWG) can be utilized in a regular diet via traditional fermented food products like idli, being an excellent source of high-quality protein, polyunsaturated fatty acids, vitamin E, vitamin B, polyphenols, healthy fiber, and vital minerals, as well as having potent anticancer properties (Hemavathi et al., 2017). Fermented products are becoming popular owing to increasing awareness about healthy or functional foods. As evidenced by currently available supplements fortified with WG that have prebiotic properties, WG-based fermented products are predicted to increase the market for functional beverages (Coda et al., 2011; Mueller et al., 2011; Mueller & Voigt, 2011; Judson et al., 2012; Otto et al., 2016). However, further research is needed to explore stability of these beverages during storage for consumer acceptance.

In recent years, consumers' awareness of diet and health has increased, as well as their need for foods rich in saturated and monounsaturated fats (Jalal et al., 2013). Among all the plant sources of vitamin E, wheat germ oil is the richest, as it contains about 10–15% phytosterols that help lower fat content (M. S. Arshad et al., 2013). Arshad et al. (2007) formulated cookies out of wheat germ oil at levels of 0–100%. Results indicated that rats fed diets containing WGO had significantly lower serum cholesterol levels and lower concentrations of low-density lipoproteins (LDLs) than their control group counterparts. In another study, Khalid et al. (2021) used wheat germ oil and wheat bran fiber (WBF) to develop low-fat hamburger patties and observed superior quality, stability, and low cholesterol content in the patties treated with 4.5% WGO plus 3% WBF in both raw and cooked form. Also, WGO exhibits natural antioxidants and bioactives as well as antimicrobial compounds. The diverse biological activities and health properties of this compound make it a potent ingredient to be used in formulations of nutraceuticals, functional foods, and other food products. The lipid profile of rats fed cookies containing WGO was monitored in a study by Arshad et al. (2007), and it was observed to reduce lipid peroxidation, a health factor associated with heart disease. WG and WGO can also be utilized in germ-enriched bakery products like cakes, bread, snacks, and break-fast cereals (Yazicioglu et al., 2015). Gumus et al. (2015) prepared black seed oil and WGO using an infusion of calendula flowers for experiments focusing on wound healing and radioprotection. This study showed that WGO could be a therapeutic ingredient for food supplements.

Nutritional properties of noodles were enhanced by the addition of 15% DWG flour with respect to amino acid composition, minerals, and vitamin B (Ge et al., 2001). According to Zhu et al. (2011), the total phenols from defatted germ have a high level of antioxidant activity. These results suggest that phenols could be utilized in the formulation of nutraceuticals that can reduce oxidative stress (Zhu et al., 2011). Furthermore, their addition to cereal-based foods is an effective supplementation method for nutritional enhancement (Arshad et al., 2007). The proteins found in defatted wheat germ meal have similar emulsifying properties and comparable stability to bovine serum albumin, in addition to good foaming capabilities and good water retention (Y Ge et al., 2000).

8.4.2 Rice

Rice (*Oryza sativa* L.) is the most common staple food crop, with total production of approximately 755 million tons throughout the world (Anon, 2022). Rice is a rich source of carbohydrates with a moderate amount of protein; fat; and vitamin-B complex including riboflavin, thiamin, and niacin (Fresco, 2005). Starch is the main carbohydrate present in rice and is composed of polysaccharides molecules such as amylose and amylopectin. The milling of rice grains facilitates retention of more nutritional value, fast cooking, and better taste (Dhankhar & Hissar, 2014). Depending on the variety of rice and method used, up to 40% is lost as the by-products are dumped in landfills or utilized as animal feed (Linscombe, 2006).

It is reported that utilization of these by-products in food or alternative industries could improve the yield and sustainability of rice production (Sanchez et al., 2019). During rice processing, low-value by-products are generated, such as husks (20% of the total weight of whole rice) and bran (7% of the total weight of whole rice) that protect rice seeds during growth (Butsat & Siriamornpun, 2010). The major nutrient composition of rice grain and its by-products is given in Table 8.3. The nutritional composition and health benefits of by-products generated during rice milling and their further applications in the food industries are discussed in the following section.

Table 8.3 Proximate Analysis of Parts of the Rice Grain (g per 100 g) at 14% Moisture

Major Nutrients	Protein	Carbohydrates	Lipid/Fat	Fiber	Ash
Brown rice	7.1–8.3	73–87	1.6–3.1	2.9–4.4	1–1.5
Rice husk	2.0–2.8	22–34	0.3–0.8	66–74	13–21
Rice bran	11–14.9	34–64	15–19.7	19–29	6.6–9.9
De-oiled rice bran	16.9	58.5	1.5	9.6	–
Milled rice	6.3–7.1	77–89	0.3–0.7	0.7–2.7	0.3–0.8

(*Source:* Juliano & Tuaño, 2019 and Sunphorka et al., 2012)

8.4.2.1 Nutritional and Functional Characteristics of Rice By-Products

Rice grain is composed of rice husk (20%), rice bran (7%), and starchy endosperm (70%) (Dhankhar & Hissar, 2014). During harvesting, mature grain (brown rice or caryopsis) is usually covered with rough and tough siliceous hull, and this whole assembly is known as the paddy (Juliano & Tuaño, 2019). The rice hull acts as a protective cover to the caryopsis and also prevents fungi infestation of the grain. Inside the hull, there is a starchy endosperm and embryo/germ of the mature rice grain covered by three outer layers: the pericarp, seed coat (tegmen), and aleurone layer. During caryopsis development, the pericarp undergoes extensive degeneration as the mature ovary wall ripens. The pericarp is followed by a single layer of crushed cells called the tegmen, which protects the seed. There is a layer of aleurone that completely encircles the rice grain and the outside surface of the embryo/germ. It is closely wrapped around the underlying starchy endosperm cells and most of the germ. At the base of the rice grain is the embryo. The starchy endosperm is primarily divided into two regions: (1) the subaleurone layer, which consists of the two outermost cells surrounding the aleurone layer, and (2) the middle region, which comprises the rest of the endosperm.

Rice that has been milled is slightly shorter than brown rice. The outer surface of brown rice is smooth, non-glistening, and waxy white. Brown rice has a slightly higher protein content than milled rice due to presence of a protein-rich bran layer (Table 8.3). In addition, brown rice contains also more crude fat, dietary fiber, and ash. Endosperms or milled rice grains are utilized for direct consumption or for further processing, while the remaining 30% yield is considered waste by-products. The by-products of the milling process include rice husks, bran, germ, and fine broken

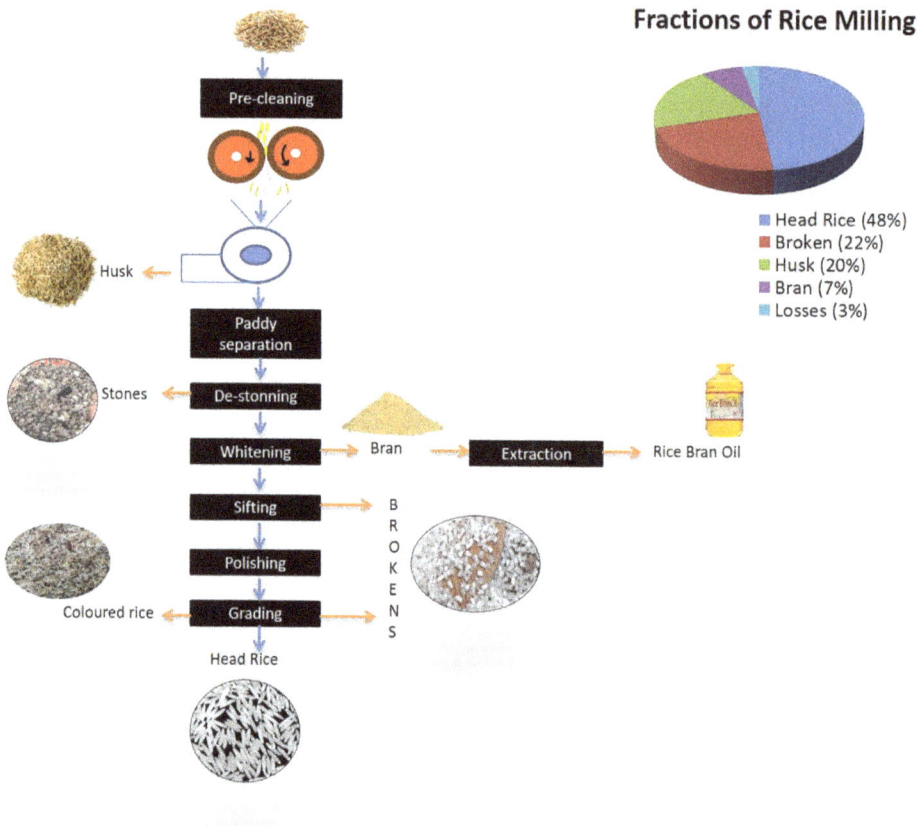

Figure 8.3 Major products and by-products obtained during milling of rice.

rice (Kennedy & Burlingame, 2003b), as shown in Figure 8.3. Despite high nutritional value, these products have usually been discarded as waste into landfills (Sharif et al., 2014). According to Rohman et al. (2014), waste by-products have the potential to be utilized in the food industry for formulation of functional foods or to extract bioactive compounds for commercial application. In the following subsection, the primary by-products associated with the milling process are discussed.

Rice husk is an important waste product produced during the rice milling process. Rice hull, also known as husk, primarily consists of lignin, hemicellulose, cellulose, and hydrated silica, which are difficult to digest by humans. This is the reason rice husk often ends up either in landfills or burned in the open, leading to environmental pollution (Alshatwi et al., 2015). Chemical analysis of hulls indicated that they are rich in crude fiber (66–74%), with moderate values of crude protein (2.0–2.8%), crude fat (0.3–0.8%), ash (13–21%), and moisture (14%) in addition to phenolic acid (vanillic acid and p-coumaric acid), γ-oryzanol and tocopherol content, and antioxidant capacity (Table 8.3). The effects of consumption include the reduction of blood glucose levels and lipid concentrations, as well as the enhancement of gastrointestinal content viscosity (Dikeman et al., 2006). Therefore, rice husk can be utilized in the food sector by extracting bioactive compounds and utilizing them in development of functional foods.

Rice bran, which is extracted from the outer layer of the rice grain, contains an aleurone layer, along with some proportions of endosperm and germ (10%) of the rice grain (Justo et al., 2013). Rice bran accounts for about 65% of all the nutrients of the entire rice grain with 11–14.9% of proteins, 15–19.7% of oil, 19–29% of fiber, 14% of moisture, and 6.6–9.9% of ash. It is also rich in micronutrients, such as vitamins, and in minerals, such as Al, Ca, Cl, Fe, Mg, and Mn (Begum et al., 2015) (Table 8.3). Besides being high in nutritional value, rice bran has high dietary fiber and polyunsaturated fatty acids, and bioactive compounds such as γ-oryzanol, tocotrienol, and tocopherols (Abdul-Hamid et al., 2007). These substances play a vital role in the prevention and treatment of chronic diseases of the cardiovascular system, as well as boosting thyroid function (Renuka Devi & Arumughan, 2007)

On average, rice bran contains about 10–23% rice bran oil (RBO) (Kennedy & Burlingame, 2003b). There are various types of unsaturated fatty acids present in rice bran oil, such as linoleic and oleic fatty acids, tocopherols, phytosterols, and tocotrienols, that have been supplemented in food to enhance heart health (Gul et al., 2015). Following bran stabilization, it must be further processed to yield edible oil (Rafe et al., 2017). Commonly used techniques for producing edible oil out of rice bran include hydraulic pressing, solvent extraction, and X-M milling (Nagendra Prasad MN et al., 2011). Extraction by solvent is the most widely used method for oil extraction. This is due to the relatively high yield of 0.549 g RBO/g bran and the ease of removing the solvent (Chiou et al., 2013). Rice bran oil is separated into two types: crude bran oil and defatted rice bran oil. Crude bran oil is composed of about 4% unsaponifiable material (fat, oil, and wax), 4% FFA, and 90% lipids, which requires refining (Chandrasekar et al., 2014).

Rice germ contains the highest amount of nutrients. The amount of vitamin E (α-tocopherol) in rice germ is five times higher than that in rice bran (γ-tocopherol). Rice germ also contains substantial amounts of fiber, vitamins (B_1, B_2, and B_6), and amino acid (GABA), a neurotransmitter that has been documented to lower blood pressure, improve cognition, and reduce blood glucose. However, there was five times less γ-oryzanol in rice germ than in rice bran (Yu et al., 2007).

8.4.2.2 Utilization of Rice By-Products

The primary by-products of the rice milling industry include rice hull, rice bran, and rice germ, along with a major fraction of secondary by-products, including broken rice. A detailed summary of the utilization of rice by-products is given in Table 8.4.

In rice hulls, there are higher levels of dietary fiber (>90% DWB with 52% cellulose and 23% arabinoxylan) and lower levels of ash (40–50% total dietary fiber and 15–21% ash). Furthermore, it is light in color, bland in taste, has small particles, and exhibits moderate water retention. The natural hydrating qualities of fiber make it suitable for use in many varieties of foods, such as snacks, bakery products, cereals, and nutritional supplements, especially those in which it does not compete with other ingredients for water. Using rice hull fiber can lower calories in finished products,

Table 8.4 Utilization of Rice By-Products in Different Foods

By-Product	Application	Salient Findings	References
Rice hull	Cookie, cracker, and cake dough	Improve the texture and sensory acceptability	Podolske et al. (2013)
Rice bran	Wheat noodles	Reduction in cohesiveness and improvement in polyphenols, flavonoids, anthocyanins, and antioxidative activity	Kong et al. (2012)
	Sweet potato pasta	Increase in protein and dietary fiber content	Krishnan et al. (2012)
	Wheat pasta	Increase in storage stability up to 4 months	Kaur et al. (2012b)
	Rice pasta	Improved textural and cooking characteristics	L. Wang et al. (2017)
	Bread	Inclusion of 10 and 20% bran increased hardiness, chewiness, and gumminess	Lima et al. (2002)
Defatted rice bran	Bread	Improvement in antioxidant potential and storage stability up to 5–10% bran inclusion	Sairam et al. (2011)
Rice bran	Pork frankfurter	Improvement in textural consistency and reduction in gelling capacity with inclusion of 2.5% rice bran into a vegetable-oil emulsion	Álvarez et al. (2012)
	Corn flakes and tortilla chips	An organoleptic and rheological amelioration of the final product	Al-Okbi et al. (2014)
	Cookies	WBC, OBC, emulsifying properties improved with long-term stability	Younas et al. (2011)
Defatted rice bran	Breakfast cereal	Reduced overall expansion and increased hardness with retention of desired crispness for longer time	Charunuch et al. (2014)
Rice bran oil	Fried food	Better taste and flavor, less cooking time, and saving in energy due to 15% less absorption of oil	Sharma (2002)
	Bread	Enhanced bread making and organoleptic quality of breads with replacement of bakery shortening with rice bran oil at 50% level	Kaur et al. (2012a)
	Salad dressing	Highest powder yield (42.70%), solubility (98.04%), and stability (100%) of spray-dried powder of salad dressing prepared with of 0.5% rice bran oil	Garba (2020)
	Mayonnaise	Desired color, optimum spreadability, and excellent emulsion stability	Chetana et al. (2019)
	Yoghurt	More compact and dense structure of frozen yoghurt prepared with NERBO of size 150–300 nm	Alfaro et al. (2015)
	Emulsion	Greater emulsion stability but more oxidation if prepared with brown rice bran oil as compared to purple rice bran oil	Alfaro et al. (2016)
Broken rice flour with mung flour	Extruded snacks	Increased protein content, protein quality, dietary fiber, and mineral content	C. Sharma et al. (2017)
Broken rice flour	Bread	Highest specific volume and sensorial properties with low glycemic index	Feizollahi et al. (2018)
Broken rice flour	Gluten-free rolled papers	Increased starch digestibility, improvement in freshness and rollability	M. H. Abdel-Haleem (2016)
Broken grains	Gluten-free beer	Volatile compounds comparable with a barley malt bottom-fermented beer	Mayer et al. (2016)

enhanced dietary fiber levels to support nutrient content claims, as well as to manage texture and moisture. To impart rich fiber content into formulated products, direct expanded snack puffs using 8% RH fiber have been produced successfully (Podolske et al., 2013). There have also been cookie, cracker, and cake formulas developed with 5 to 8% RH fiber to meet the definition of an excellent source of fiber (and provide more than 2.5 grams each). It is also acceptable to use much higher levels of RH fiber due to its physical properties. As an example, reduced-calorie bread containing 20% RH fiber (flour basis) had acceptable dough development, bread loaf volume, cellular structure, and sensory attributes.

Similar to other cereal-based insoluble fibers, RH fiber can be incorporated into bread and cracker doughs by adjusting water content. For most baked goods, 1–3 g of additional water is required for every gram of fiber (Podolske et al., 2013). Gluten-free diets are usually low in several essential nutrients, including fiber. RH fiber offers a fiber enrichment solution for gluten-free foods due to their rising demand. Several studies have shown that addition of RH fiber improves the flavor and texture of gluten-free sponge cake at standard level (2.5 g/55 g serving or 5% usage rate). Adding rice hulls to food can also serve as an anti-clumping agent, as suggested by Kennedy & Burlingame, (2003a).

In rice bran, 15–20% of the protein is of high quality. The unique hypoallergenic and anticancer properties of RB proteins make them more effective than cereal proteins, although they seem to be unavailable commercially. Kong et al. (2012) added black rice bran at levels of 2, 6, 10, and 15% to wheat-based noodles and evaluated for textural and antioxidant properties. Comparing the noodles to control noodles, the RB-enriched noodle had a lower cohesiveness, along with a high level of polyphenols, flavonoids, anthocyanins, and antioxidative activity. Rice bran was also evaluated as an ingredient in various other products, such as Asian noodles (Izydorczyk et al., 2005), sweet potato pasta (Krishnan et al., 2012), wheat pasta (Kaur et al., 2012b), and rice pasta (Wang et al., 2017) and was found highly acceptable with respect to quality. A study to compare the effects of defatted and full-fat rice brans yielded from longer-, medium-, and shorter-grain rice on quality of bread was carried out (Lima et al., 2002). Bread volume increased substantially with the inclusion of 10% and 20% bran into wheat flour, with the medium-grain bran showing the biggest increase, due to lower fiber and greater starch content among the three kinds. Breads with greater hardness, chewiness, and gumminess had no significant changes in springiness and cohesiveness but did have enhanced chewiness, hardness, and gumminess.

Overall, full-fat rice bran outperformed defatted rice bran in terms of bread texture qualities. Sairam et al. (2011) evaluated the effect of various levels of commercial DRB on bread quality, considering that rice bran is a good source of dietary fiber and minerals. The sensory evaluation revealed that bread containing 5–10% bran was acceptable, but bread containing 15% bran was acceptable only if it was combined with bread improvers. Rice bran could be used as an additive to increase the antioxidant potential and storage stability of high-fiber bread. Using rice bran in vegetable oil emulsions improved texture consistency and reduced gelling capacity in frankfurters, indicating interactivity between the rice bran and the proteins and fats in the meat (Álvarez et al., 2012). Rice bran has some interesting technological properties that can be enhanced using specific alcoholic solvents for the food industry (Capellini et al., 2020). Another study by Liu et al. (2019) showed that rice bran stabilized using microwave was marginally better than dry heated bran in terms of the water-binding and oil-binding capacity, emulsifying properties, and shelf life stability. In addition, rice bran protein fractions (12–16%), have the potential to form gel, emulsion, and foam (Sharif et al., 2014).

A study performed by (Al-Okbi et al., 2014) assessed the effectiveness of rice bran when added to corn flakes and tortilla chips, reporting that a reduction in protein content improved the final product's organoleptic and rheological properties. Younas et al. (2011) developed cookies with both heat- and acid-stabilized RB and optimized the recipe with a 10% substitution of RB. RB solid exhaust is a low–fat content ingredient due to the high added value of the RB oil. For example, deoiled rice bran was utilized at a different levels for the preparation of extruded breakfast cereal (Charunuch et al., 2014) and Indian flat bread chapati (Yadav et al., 2012). There are few studies demonstrating that pretreated rice bran can be used for food preparation via physical, thermal, or bioprocessing, probably due to the scarcity of industrial- and pilot-scale studies. Rice bran protein is, therefore, a good source of protein for the food industry due to its easy availability. Considering that rice bran proteins are hypoallergenic, they have specific functions. This makes this field a valuable asset for the food industry, since these substances are bioactive and have functional properties at the same time.

Rice bran oil is considered a "wonder oil" in Japan, China, and India due to its health benefits for humans. RBO, owing to its hypoallergenic properties, can be used as a substitute for regular cooking oils by persons who are allergic to them (Nayik et al., 2015). It also produces a better coating for fried dishes since rice bran oil is used in lower amounts in cooking, resulting in fewer polymers and healthy taste in the food due to the lower fat uptake during frying (Zarei et al., 2017).

The rise in consumer demand for natural products, coupled with the scarcity of olive oil, has led to RBO becoming more and more important in the economic realm (Sohail et al., 2017). There is evidence that rice bran oil provides poly-unsaturated fatty acids (PUFAs), which may lower cardiovascular risks. RBO reduces cholesterol more effectively than corn, sunflower, and safflower oil (Kaur et al., 2012a). In the manufacture of mayonnaise (Chetana et al., 2019) and salad dressings (Garba, 2020), RBO is considered superior. In addition to its superior nutritional properties, rice bran oil is more stable at higher temperatures, contains more micronutrients, and gives food items better flavor. It also saves energy because frying takes less time and is more economical because rice bran oil is 15% less expensive than palm oil (Sharma 2002). Alfaro et al. (2015) proposed that frozen yoghurt might be enriched with a nano-emulsion containing purple rice bran oil (NPRBO) to develop a dairy product with unique marketing potential. In another study, purple and brown rice bran oils were used to produce an oil in water emulsion with good stability and oxidative protection (Alfaro et al., 2016). The sonication process could be used to produce emulsions containing little oxidation from purple rice bran oil. Functional ingredients could be incorporated into these emulsions for use as food additives.

Broken rice is also a by-product of the rice milling process that is present at a high percentage in poor nations due to the usage of outdated equipment and inexperienced manpower for rice milling and polishing. It could be a good option for the formulation of a broad range of products. A higher percentage of broken rice is obtained while milling raw rice in comparison with parboiled rice (22–30 and 15%, respectively) (C. Sharma et al., 2017). Because of nutritional benefits for humans, broken rice is reduced into flour and used as a food ingredient (Kim et al., 2012; Feizollahi et al., 2018), whereas gluten is usually found in everyday meals such as cereals, snacks, baked goods, and pasta.

Rice flour is gluten free, making it a viable option for making gluten-free foods (Quiñones et al., 2015). Marcoa and Rosell (2008) claimed that rice flour is hypoallergenic. Rice flour is preferred over other flours for infant food, porridge, and many other foodstuffs because it poses a lower risk to persons with food sensitivities (Gujral & Rosell, 2004). As a result, grinding broken rice to make flour for such uses has become even more economically feasible (Qian & Zhang, 2013). Broken rice has gained in popularity in the pet food industry in the United States (Buff et al., 2014).

A broken rice dish called banh da nem is a popular product in Vietnam. Rice paper can be made from broken rice kernels mixed with water to create a gluten-free traditional spring roll film (Abdel-Haleem, 2016; Nagano et al., 2000). Any kernels that may fracture during the milling process (30%) are regarded as undesirable and are thus sold to the brewing industry at a lower cost. Brewer's rice gets its name from the fact that it's frequently used in the brewing of beer. This type of rice gives beer its flavor, fragrance, color, and mouth feel (Marconi, 2017). Brewer's rice can also be used as raw materials to ferment and produce alcohol. Ceppi and Brenna (2010) used the Italian rice variety *loto* and a mashing-in liquor-to-grist ratio of 1:3.5 to produce a gluten-free beer (3.5–4.5% ABV).

Similarly, an all-rice malt beer (4.4–4.8% ABV) was produced with the Italian rice variety *centauro* using 1:4 liquor-to-grist ratio (Mayer et al., 2016).

8.4.3 Maize

Maize (*Zea mays*), also referred to as corn, a monocotyledonous cereal, has its origin traced back to Mexico and is now grown globally, primarily in tropical and temperate climate areas of the world. Maize production in 2021 was estimated to be 1136.3 million metric tons (Zhang et al., 2021). It finds prime application as livestock feed, amounting to 60 to 70%, and the rest, 30 to 40%, is processed for human consumption. It ranks third as a staple food after wheat and rice. The kernel of maize has a general composition of 82–84% endosperm, 10–12% germ, 5–6% pericarp, and 1% tip cap (on a dry basis).

Processing is required to render the cereal edible or separate the different components of the grain. Milling is the primary procedure the cereals are subjected to and can either be wet or dry depending on the end use. Maize undergoes both wet and dry milling as well as dry-grind milling based on the availability of equipment or the desired products. Primarily, wet and dry-grind milling are done to get ethanol as the major product, while dry milling helps to remove the outer bran layer as well as the germ, which is achieved by pearling and involves abrading the grain to obtain an end product free of seed coat and germ. A complete process for different types of milling of maize is shown in Figures 8.4–8.6. Various by-products are generated in these processes that have high nutritive value and phytochemical content.

8.4.3.1 Wet Milling

Wet milling of maize helps in separating the chemical constituents of the grains like starch, proteins, oil, and fiber. It is done primarily to obtain starch as the major product, which has varied food, pharmaceutical, and industrial applications. Maize wet milling aims to isolate the starch from the other fractions of the kernel. For this, the grains are cleaned and then steeped in a warm sulfurous acid solution for 30–48 hours to soften the endosperm and solubilize the corn solids. The swollen kernel needs less mechanical strength during milling, and the rubbery germ is also separated easily. Soluble nutrients of maize also come in the steep water, which is later evaporated to obtain condensed corn fermented extractives, or corn steep liquor. The milling process separates the germ from the kernel, which is removed by floatation and is used to obtain maize germ oil and results in maize germ meal as a by-product. The remaining portion of the kernel is screened to remove the bran from the starch and gluten protein, which pass through screens to centrifugal separators that separate the lighter gluten from the heavier starch. The bran is another by-product that is combined with a steep water co-stream to get corn gluten feed. Gluten is dried to get corn gluten meal, while the starch obtained is utilized for production of various sweeteners, ethanol, and so on, as shown in Figure 8.4.

8.4.3.2 Dry-Grind Milling

The process is carried out to generate ethanol, but unlike wet milling, the kernel components are not separated; the whole kernel is crushed, initially by hammer mills, and then by roller mills, to reduce the particle size. Water is then added to this milled corn to get a slurry that is subjected to cooking. The enzyme amylase is added to the slurry to allow liquefaction, followed by the addition of glucoamylase and yeast to promote fermentation of the mash. The "beer" hence obtained is then removed to the distillation column to strip off the ethanol, and the solids that are non-fermentable are left behind. This is the only by-product generated by the dry grinding process and consists of protein, fiber, and the fat component of the kernel, which is then combined with stillage water and dried to get what is known as dried distillers' grains and solubles (DDGS), as shown in Figure 8.5.

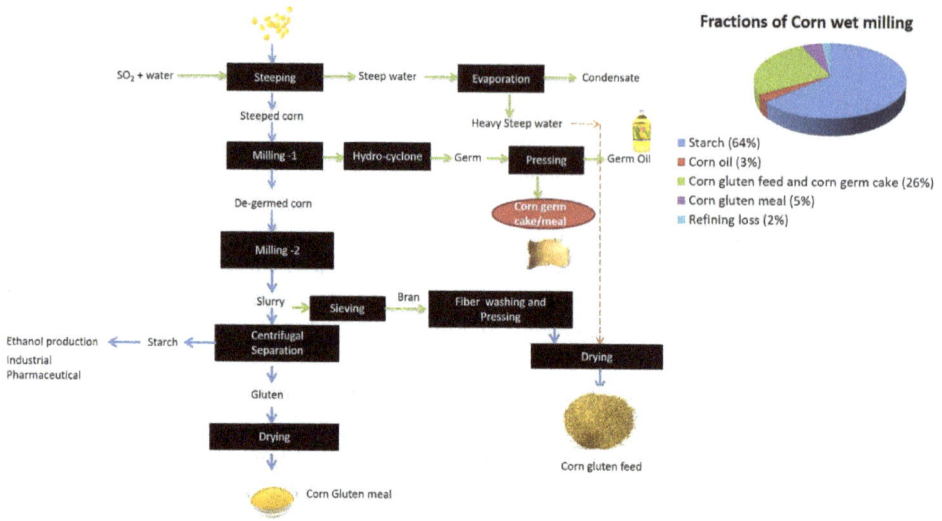

Figure 8.4   Major products and by-products obtained during wet milling of maize.

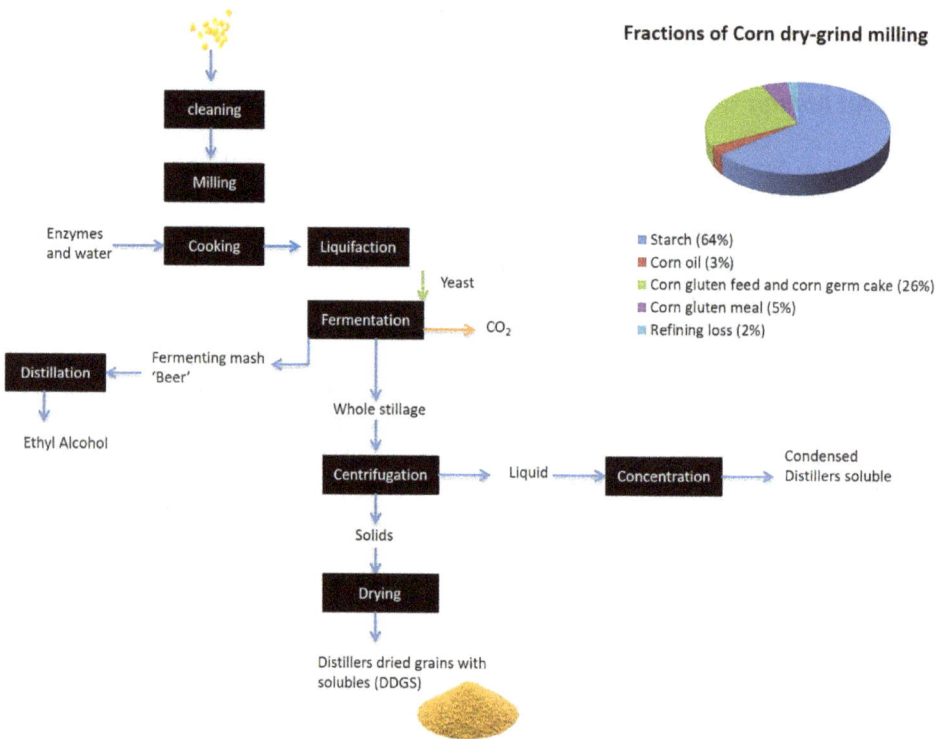

Figure 8.5   Major products and by-products obtained during dry-grind milling of maize.

8.4.3.3 Dry Milling

Dry milling of maize is done primarily to generate products for human consumption. It involves the removal of the germ, tip cap, and pericarp from the endosperm by the abrasion process of the tempered corn kernels. Aspiration is done to remove the pericarp portion, and gravity tables are employed to separate the germ from the endosperm. The germ is used to extract the corn oil from

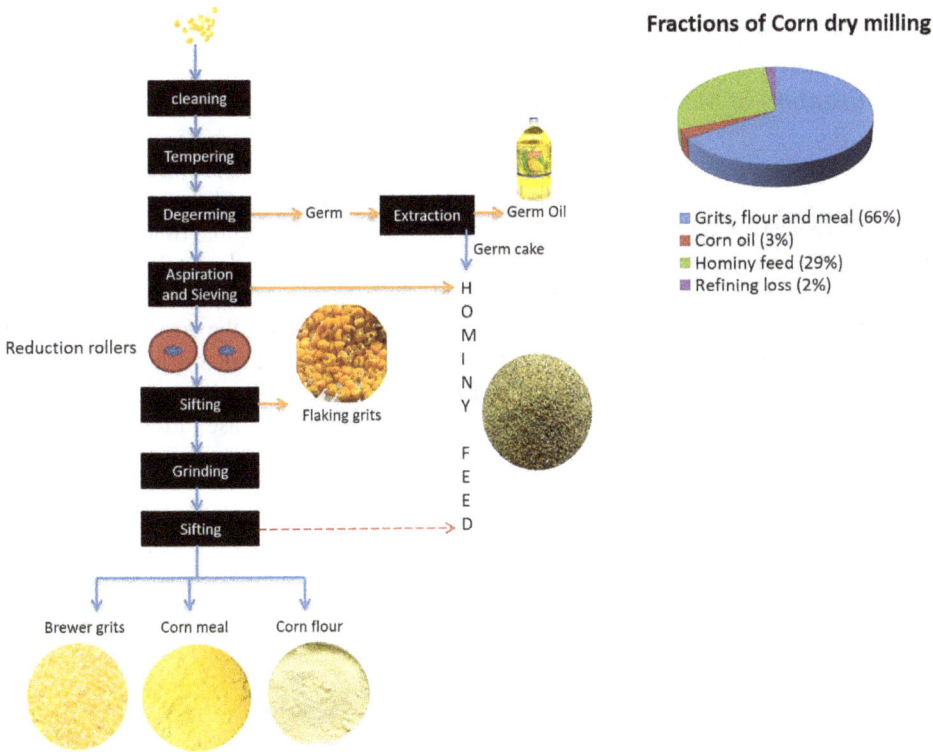

Figure 8.6 Major products and by-products obtained during dry milling of maize.

it, and germ cake is generated as a by-product. The endosperm fraction has varied streams used for various food applications like flaking, baking, and so on. The bran, germ cake, and waste from endosperm fractions are combined to get hominy feed as the other major by-product of dry milling (Figure 8.6).

8.4.3.4 Nutritional and Functional Characteristics of Maize By-Products

Hominy feed accounts for nearly 35% of starting corn (Papageorgiou & Skendi, 2018). Maize hominy contains approximately 57% starch, 25% fiber, 11% protein, and 5% fat (Sharma et al., 2008). The composition of DDGS ranged from 26.1–28.3% for protein, 2.4–4.4% for ash, and 7.7–10.9% for fiber (Tsen et al., 1982). An inclusion of up to 20% of maize germ obtained by dry milling was done in short dough biscuits by Paraskevopoulou et al. (2019), who found that this addition influenced the product characteristics like protein, fat, spread ratio, and breaking strength only to a limited extent while fortifying biscuits with fiber. The deoiled maize germ cake contains 8.83% moisture, 15.51% fat, 19.94% protein, 4.69% ash, 9.84% crude fiber, and 41.64% carbohydrate (Barnwal, 2013), while Nasir et al. (2010) reported that defatted maize germ flour contains 27.6% protein, 5.1% fat, 13.1% fiber, 7.5% ash, and 46.7% nitrogen free extract. Compositional analysis of low-fat corn germ flour shows that it contains 2.57% moisture, 1.94% ash, 19.77% protein, 1.89% fat, and 9.09% fiber (Masoumikhah & Zargari, 2013). Corn bran contains 3.54% ash, 12.02% proteins, 17.74% fat, 28.90% total dietary fiber, 23.70% insoluble fiber, and 3.54% soluble fiber (Ribeiro et al., 2018).

Wet milling produces four major co-products from the isolated steep water, bran, germ meal, and gluten. Together these represent about 25 to 30% of corn processed (Davis, 2001). According to Ramírez et al. (2009), the by-product yield calculated on a dry weight basis for the corn wet milling process was dry germ 7.7%, gluten feed (soak water solids plus fiber) 19.4%, and gluten meal 6.2%. Table 8.5 lists the nutritional profile of corn milling by-products.

8.4.3.5 Utilization of Maize By-Products

Table 8.6 lists the utilization of maize by-products in different foods. The fermentation process results in converting the starch present in the grain in alcohol and carbon dioxide and the remaining components like protein, fat, fiber, minerals, and vitamins reach up to three times concentration in the DDGS. Hence, it is nutrient dense and can be utilized in human food as well. DDGS utilization for human food is possible by first rendering it suitable for human consumption—food grade distillers' dried grains (FDDG). Owing to its high protein and fiber values, the enriched food products will help to deal with coronary heart diseases, diabetes, obesity, and colon cancer (Kaczmarczyk et al., 2012). Pourafshar et al. (2018) utilized DDGS at three levels with other additives to study its effect on *barbari* bread and found that it resulted in a significant increase in the fiber and protein content of the bread in a cost-effective way.

Table 8.5   Nutritional Profile of Corn/Maize Milling By-Products

By-Product	Protein (%)	Fat (%)	Fiber (%)
Corn steep liquor	25	–	–
Corn germ meal	20	2	9.5
Corn gluten feed	21	2.5	8
Corn gluten meal	60	2.5	1
DDGS	27	11	9

(*Source:* Davis, 2001)

Table 8.6   Utilization of Corn or Maize By-Products in Different Foods

By-Product	Application	Salient Findings	References
Corn bran	Extruded snack	Increase in dietary fiber	Ogunmuyiwa et al. (2017)
	Low-calorie snack bar	40% inclusion	Sousa et al. (2019)
	Short dough biscuits	20% inclusion	Paraskevopoulou et al. (2019)
	White bread	Good volume, texture, and microstructure	Ortiz de Erive et al. (2020)
	Chicken nuggets	Increased firmness, toughness	Pathera et al. (2018)
	Chicken sausages	Increased firmness, toughness, dietary fiber, storability, and acceptance	Yadav et al. (2016)
Defatted germ flour	Biscuits	Increased protein, better texture	Barnwal (2013)
	Buns	Increased fiber, yield, hardness	Arora and Saini (2016)
	Wheat flour	Increase in all functional properties of flour	Siddiq et al. (2009)
	Cookies	Increased sensory scores, thickness, and firmness	Nasir et al. (2010)
	Bread	Increased acceptability and nutritional composition	Pauaucean and Man (2013)
	Macaroni	Increased rheological, baking, and sensory qualities	Masoumikhah and Zargari (2013)
	Crackers		Kuchtová et al. (2016)

Liu et al. (2011) developed cornbread with the addition of DDGS and reported that up to 25 g/100 g can be added in the formulation without any adverse effects on the quality and processability while enhancing the nutritional values and lowering the glycemic index. Singha et al. (2018) added DDGS to garbanzo flour and corn grits in varying ratios to produce extruded snacks and found that they can be included at a 20% level to obtain a healthier snack alternative with low fat, high protein, and dietary fiber. Tsen et al. (1982) added DDGS in wheat flour at 15 and 25% levels for making sugar cookies and found it suitable for enriching the protein and fiber content of cookies while adversely affecting the color of the product.

Tsen et al. (1983) replaced wheat flour with 10% of DDGS, which resulted in breads that were superior in term of loaf volume, crumb grain, and color as compared to whole wheat bread. Abbott et al. (1991) evaluated muffins and rolls prepared by the addition of DDGS of different particle sizes and found that the products had good acceptability scores with good product quality traits. Bread with 5 to 10% DDGS addition had acceptable sensory attributes (Brochetti et al., 1991).

Extruded snacks were produced by adding corn bran to blends of Bambara groundnut flour and cassava starch in varying proportions ranging from 0–50%, and it was found that corn bran had a significant effect on the total dietary and ash content (Ogunmuyiwa et al., 2017). Sousa et al. (2019) found that corn bran up to an extent of 40% was well accepted in a low-calorie snack bar, which also enhanced ash content and total dietary fiber content as compared to the control. Ortiz de Erive et al. (2020) achieved equivalent beneficial results by addition of lower amounts of micro-fluidized corn bran in bread that yielded loaves with similar micro-structure and specific loaf volume and texture as that of control bread with up to 22% inclusion of micro-fluidized corn bran.

Emulsion-based meat products like chicken nuggets can be enriched with dietary fiber utilizing corn bran without any adverse effects on the sensory and textural quality of the nuggets, where corn bran can be added up to 11.74% in the formulation (Pathera et al., 2018). Dietary fiber–enriched chicken sausage had very good acceptability by incorporating corn bran at a 3% level, resulting in a product with increased cooking yields and emulsion stability (Yadav et al., 2016). Ribeiro et al. (2018) observed that partial substitution of wheat flour with corn bran to prepare tagliarini pasta resulted in increased cooking time, enhanced volume, reduced cooking loss, and absorption up to a level of 14.53% when added after thermal treatment of corn bran.

The addition of partially deoiled maize germ cake flour up to 10% replacement in wheat flour was found acceptable for biscuit preparation by Barnwal (2013). Wheat buns prepared by Arora and Saini (2016) with the inclusion of deoiled maize germ cake flour up to a 25% level showed increased percentage of protein, fat, ash, and crude fiber in the buns, while buns prepared with 10% incorporation of deoiled maize germ flour had the highest scores for overall appearance. Siddiq et al. (2009) also added defatted maize germ to wheat flour for making bread and found that breads with defatted maize germ flour up to 15 g/100 g flour showed no difference for crumb color, cell uniformity, aroma, firmness, mouth feel, or off flavor, and hence acceptable bread can be made with this formulation.

Wheat flour was replaced with defatted maize germ flour to prepare cookies, and it was reported that cookies made with up to 15% defatted maize germ flour exhibited sensory scores within an acceptable range (Nasir et al., 2010). To improve the nutritional value of wheat bread, defatted maize germ flour, when added up to a level of 15% by Pauaucean and Man (2013), was best to obtain bread with good acceptance without any negative quality attributes. Kuchtová et al. (2016) added corn germ to wheat flour at a level of up to 15% to prepare crackers, which resulted in a product with lower specific volume, volume index, width, and thickness as compared to control, but the crackers were acceptable, as shown by their sensory scores. Macaroni prepared by 10% low fat corn germ flour incorporation exhibited higher quality due to better rheological, baking, and sensory characteristics as compared to higher incorporation levels (Masoumikhah & Zargari, 2013).

8.4.4 Sorghum

Sorghum (*Sorghum bicolor* (L.) Moench) is an indigenous crop of Africa with the United States holding the first place in its production with 11,967 thousand metric tons, while India takes the sixth position in the world, producing 4,400,000 metric tons for the year 2021 (USDA, 2021). It is widely grown in diverse climatic conditions and on marginal lands with limited use of fertilizer, pesticide, or irrigation.

Sorghum can be classified into two types: brown sorghum, which is high in tannins due to colored testa, and white sorghum, which has no tannins due to the absence of pigmented testa. There are four parts to the grain of sorghum: pericarp, testa, endosperm, and germ. Unlike other cereal grains, sorghum grains contain a testa, which lies between the pericarp and endosperm. Sorghum grain has a pigmented pericarp (i.e., red, black, yellow, and brown) as well as a white pericarp (i.e., non-pigmented) (see Figure 8.1).

8.4.4.1 Nutritional and Functional Characteristics of Sorghum By-Products

The composition of sorghum flour shows that it contains 9.79% moisture, 10.43% protein, 3.61% fat, 2.84% fiber, and 3.38% ash and a very good elemental composition, with potassium 363.87 mg/100 g, calcium 220.56 mg/100 g, magnesium 165.87 mg/100 g, and phosphorus 32.5 mg/100 g (Volkova et al., 2020). A 100% flour extraction of sorghum had a protein content of 8.47%, fat 3.04%, ash 1.34%, fiber 1.50%, moisture 8.49%, and total starch 79.70% (Trappey et al., 2015). The presence of phytochemicals like tannins, phenolic acids, anthocyanin, and phytosterols make it an ideal grain for lending various benefits like reduction in cardiovascular diseases (CVDs), cancer, diabetes, and obesity (de Morais Cardoso et al., 2017).

8.4.4.2 Utilization of Sorghum By-Products

Like corn, sorghum is also processed using both wet and dry processes (Figures 8.7 and 8.8). Owing to the presence of starch, it is used to produce ethanol (Do et al., 2016). Wet milling of sorghum grains resembles corn wet milling (Belhadi et al., 2013), and the resulting starches have same physical and

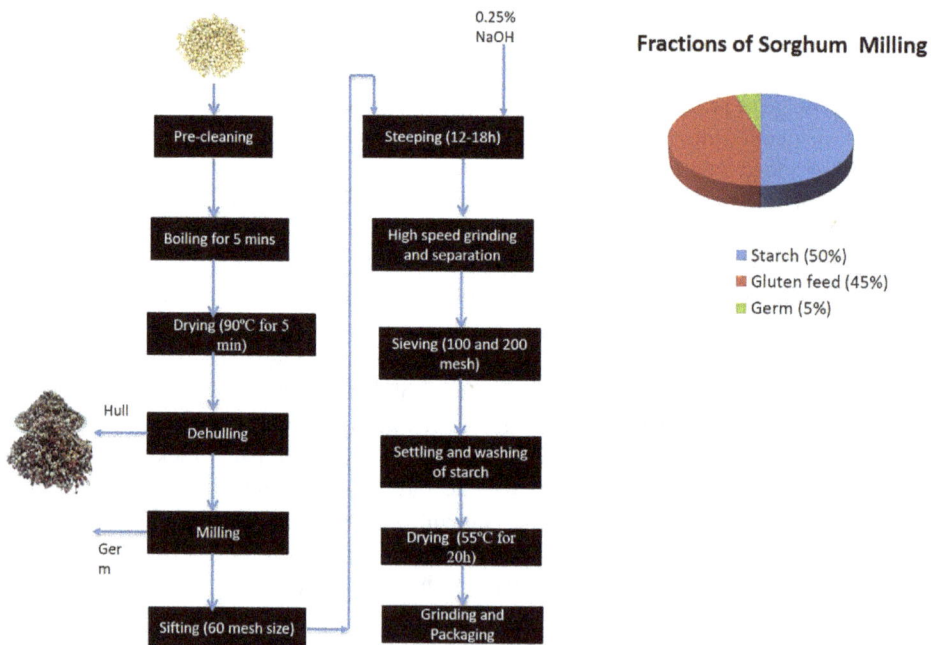

Figure 8.7 Major products and by-products obtained during wet milling of sorghum.

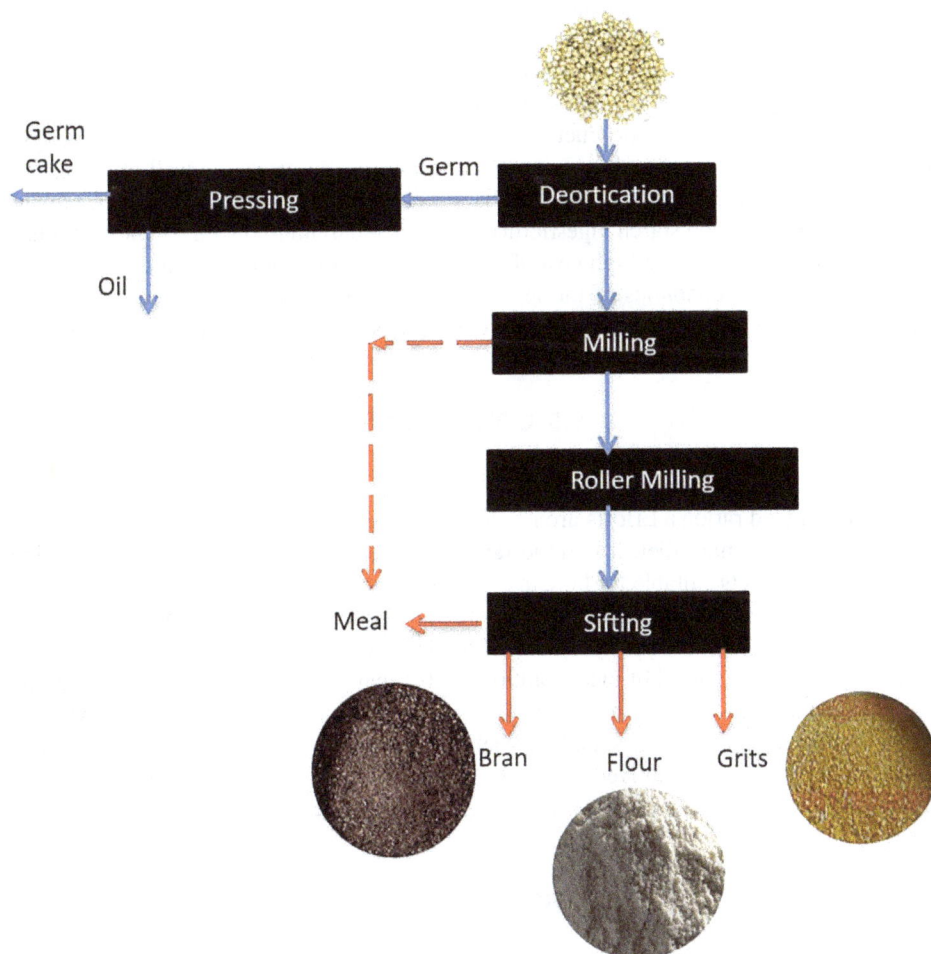

Figure 8.8 Major products and by-products obtained during dry milling of sorghum.

chemical properties, but the recovery of starch from sorghum is comparatively difficult owing to the structure of the sorghum kernel and the fragile nature of the pericarp of sorghum, whose particles impede the separation of starch and protein (F. C. Wang et al., 2000). Belhadi et al. (2013) isolated starch from red and white sorghum with the recovery ranging from 64.8 to 80.4% with a purity of 92.61 to 99.29%.

Sorghum grains that are intended for human consumption must be dehulled to remove the seed coat, as it contains anti-nutrients like tannins and phytic acids. Moreover, kafirns, which are the main prolamin present in the dark-colored varieties of sorghum, form complexes with tannins and result in low protein digestibility (Tobias et al., 2018). Hence, appropriate processing methods like soaking, dry heating, germination, and fermentation are employed to enhance digestibility and to increase the palatability, availability, and shelf life of sorghum-based foods.

Sorghum is processed at the domestic level using conventional means or small grain mills. Sorghum can be milled to produce flour and grits (semolina) which can be utilized in various traditional recipes. The flour characteristics are influenced by the extraction rate, particle size, and starch damage, especially in the production of baked goods like breads and so on. The consumption of sorghum is in the form of flour or semolina; as flakes; in extruded and baked products; and as popped grains, porridges, alcoholic beverages, instant mixes, malted drinks, and so on (Ratnavathi, 2016).

Sorghum is a gluten-free grain and has been a grain of interest in many food products like breads, tortilla chips, noodles, and biscuits (Trappey et al., 2015). Dordoni et al. (2015) optimized the utilization of white sorghum and its by-product for development of a sustainable cereal bar named SO crock that used sorghum grain, sorghum syrup (obtained by enzymatic hydrolysis of grains), and sorghum fiber (obtained as a by-product during syrup production). Popped sorghum is a popular snack in India and Africa and is becoming popular around the world as well (Kent & Rooney, 2021) due to its neutral flavor and reduced hulls as compared to popcorn. Moreover, popped sorghum has higher protein and starch digestibility as compared to non-popped sorghum (Nathakattur Saravanabavan et al., 2013). The high cost of maize in feed formulations as an energy source puts immense strain on the economics of the feed, and so sorghum and maize spent grains obtained from the beer-making industry are being explored for incorporation in feed rations (Fasuyi, 2005).

8.5 CONCLUSION

The by-products of cereal processing are rich sources of nutrients and phytochemicals and are utilized primarily in feed rations. Efforts are needed to devise technologies and processes to increase their adoption in the human diet. It is important to develop appropriate processing techniques to render cereal by-products suitable for human consumption as well as to improve their quality attributes, as they have anti-nutrients and anti-functional components present in them. Various research studies have reported that the supplementation of food products with cereal by-products shows good acceptance up to a certain level of incorporation, with enhanced nutritional and functional properties with good health benefits. Optimal utilization of cereal by-products and industrial waste is very important, as they are nutrient dense and will help to achieve food security and environmental sustainability. Future research initiatives need to be focused on the development of cereal by-product–based products with enhanced shelf life stability and consumer acceptance.

REFERENCES

Aalami, M., Rahbari, M., and Avazsufiyan, A. 2018. Rheological properties of sponge cake based on the rice and wheat germ flour. *Electronic Journal of Food Processing and Preservation* 10(1#T00971):17–32. www.sid.ir/en/journal/ViewPaper.aspx?id=804263

Abbas, H. M., Hussein, A. M. S., Seleet, F. L., Bayoumi, H. M., and Abd El-Aziz, M. 2015. Quality of some dairy by-products supplemented with wheat germ as functional beverages. *International Journal of Dairy Science* 10(6): 266–277. https://doi.org/10.3923/ijds.2015.266.277

Abbott, J., O'Palka, J., and McGuire, C. F. 1991. Dried distillers' grains with solubles: partaicle size effects on volume and accetability of baked products. *Journal of Food Science* 56(5): 1323–1326. https://doi.org/10.1111/j.1365-2621.1991.tb04763.x

Abdul Hamid, A., Sulaiman, R. R., Osman, A., and Saari, N. 2007. Preliminary study of the chemical composition of rice milling fractions stabilized by microwave heating. *Journal of Food Composition and Analysis* 20: 627–637. https://doi.org/10.1016/j.jfca.2007.01.005

Nayik, G.A., Majid, I., Gull, A., and Muzaffar, K. 2015. Rice bran oil, the future edible oil of India: A mini review. *Journal of Rice Research* 3(4): 4–6. https://doi.org/10.4172/2375-4338.1000151

Aktaş, K., Bilgiçli, N., and Levent, H. 2014. Influence of wheat germ and β-glucan on some chemical and sensory properties of Turkish noodle. *Journal of Food Science and Technology* 52. https://doi.org/10.1007/s13197-014-1677-z

Alán, P.-A., Rouzaud-Sández, O., Romero-Baranzini, A., Vidal-Quintanar, R., and Salazar-García, M. 2014. Relationships between chemical composition and quality-related characteristics in bread making with wheat flour—Fine bran blends. *Journal of Food Quality* 38. https://doi.org/10.1111/jfq.12103

Alfaro, L., Hayes, D., Boeneke, C., Xu, Z., Bankston, D., Bechtel, P., and Sathivel, S. 2015. Physical properties of a frozen yogurt fortified with a nano-emulsion containing purple rice bran oil. *LWT—Food Science and Technology* 62. https://doi.org/10.1016/j.lwt.2015.01.055

Alfaro, L., Zhang, J., Chouljenko, A., Scott, R., Xu, Z., Bankston, D., Bechtel, P., and Sathivel, S. 2016. Development and characterization of emulsions containing purple rice bran and brown rice oils. *Journal of Food Processing and Preservation* 41. https://doi.org/10.1111/jfpp.13149

Al-Okbi, S., Hussein, A., Hamed, I., Mohamed, D., and Helal, A. 2014. Chemical, rheological, sensorial and functional properties of gelatinized corn-rice bran flour composite corn flakes and tortilla chips. *Journal of Food Processing & Preservation* 38. https://doi.org/10.1111/j.1745-4549.2012.00747.x

Alshatwi, A. A., Athinarayanan, J., and Periasamy, V. S. 2015. Biocompatibility assessment of rice husk-derived biogenic silica nanoparticles for biomedical applications. *Materials Science & Engineering. C, Materials for Biological Applications* 47: 8–16. https://doi.org/10.1016/j.msec.2014.11.005

Álvarez, D., Xiong, Y. L., Castillo, M., Payne, F. A., and Garrido, M. D. 2012. Textural and viscoelastic properties of pork frankfurters containing canola-olive oils, rice bran, and walnut. *Meat Science* 92(1): 8–15. https://doi.org/10.1016/j.meatsci.2012.03.012

Amadou, I., Gounga, M. E., and Le, G. W. 2013. Millets: Nutritional composition, some health benefits and processing—A review. *Emirates Journal of Food and Agriculture* 25(7): 501–508. https://doi.org/10.9755/ejfa.v25i7.12045

Anon. 2022. https://ricetoday.irri.org/global-demand-for-rice-genetic-resources/#:~:text=In%202019%2C%20over%20755%20million,over%20the%20last%20several%20decades. (Accessed on October 14, 2022)

Arora, A., and Saini, C. S. 2016. Development of bun from wheat flour fortified with de-oiled maize germ. *Cogent Food and Agriculture* 2(1). https://doi.org/10.1080/23311932.2016.1183252

Arshad, M. S., Anjum, F. M., Khan, M. I., Shahid, M., Akhtar, S., and Sohaib, M. 2013. Wheat germ oil enrichment in broiler feed with α-lipoic acid to enhance the antioxidant potential and lipid stability of meat. *Lipids in Health and Disease* 12: 164. https://doi.org/10.1186/1476-511X-12-164

Arshad, M. U., Anjum, F. M., and Zahoor, T. 2007. Nutritional assessment of cookies supplemented with defatted wheat germ. *Food Chemistry* 102(1): 123–128. https://doi.org/10.1016/j.foodchem.2006.04.040

Bansal, S., and Ml, S. 2011. Nutritional, microstructural, rheological and quality characteristics of biscuits using processed wheat germ. *International Journal of Food Sciences and Nutrition* 62: 474–479. https://doi.org/10.3109/09637486.2010.549116

Barnwal, P. 2013. Effect of partially de-oiled maize germ cake flour on physico-chemical and organoleptic properties of biscuits. *Journal of Food Processing & Technology* 04(04). https://doi.org/10.4172/2157-7110.1000221

Begum, A., Sarma, J., Borah, P., Moni Bhuyan, P., Saikia, R., Hussain Ahmed, T., Karki, S., Rupa Dowarah, J., Gupta, P., R Khanna, R., Rai, L., Lalhmangaihzuali, J., Goswami, A., and Chowdhury, P. 2015. Microwave (MW) energy in enzyme deactivation: Stabilization of rice bran from few widely consumed indigenous rice cultivars (*Oryza sativa* L.) from eastern Himalayan range. *Current Nutrition & Food Science* 11(3): 240–245. https://doi.org/10.2174/1573401311666150521233113

Belhadi, B., Djabali, D., Souilah, R., Yousfi, M., and Nadjemi, B. 2013. Three small-scale laboratory steeping and wet-milling procedures for isolation of starch from sorghum grains cultivated in Sahara of Algeria. *Food and Bioproducts Processing* 91(3): 225–232. https://doi.org/10.1016/j.fbp.2012.09.008

Blair, R. 2008. Approved ingredients for organic diets. In *Nutrition And Feeding of Organic Poultry* (pp. 66–207). Cromwell Press, Trowbridge.

Brandolini, A., and Hidalgo, A. 2011. Wheat germ: Not only a by-product. *International Journal of Food Sciences and Nutrition* 63(Suppl 1), 71–74. https://doi.org/10.3109/09637486.2011.633898

Brochetti, D., Penfield, M. P., and Heim-Edelman, M. F. 1991. Yeast bread containing distillers' dried grain: Dough development and bread quality. *Journal of Food Quality* 14: 331–344.

Buff, P. R., Carter, R. A., Bauer, J. E., and Kersey, J. H. 2014. Natural pet food: A review of natural diets and their impact on canine and feline physiology. *Journal of Animal Science* 92(9): 3781–3791. https://doi.org/10.2527/jas.2014-7789

Butsat, S., and Siriamornpun, S. 2010. Antioxidant capacities and phenolic compounds of the husk, bran and endosperm of Thai rice. *Food Chemistry* 119: 606–613. https://doi.org/10.1016/j.foodchem.2009.07.001

Capellini, M. C., Novais, J. S., Monteiro, R. F., Veiga, B. Q., Osiro, D., and Rodrigues, C. E. C. 2020. Thermal, structural and functional properties of rice bran defatted with alcoholic solvents. *Journal of Cereal Science* 95: 103067. https://doi.org/https://doi.org/10.1016/j.jcs.2020.103067

Ceppi, E. L. M., and Brenna, O. V. 2010. Brewing with rice malt—A gluten-free alternative. *Journal of the Institute of Brewing* 116(3): 275–279. https://doi.org/10.1002/j.2050-0416.2010.tb00431.x

Chandrasekar, V., Sampath, C., Belur, P., and Iyyaswami, R. 2014. Refining of edible oils: A critical appraisal of current and potential technologies. *International Journal of Food Science & Technology* 50. https://doi.org/10.1111/ijfs.12657

Charunuch, C., Limsangouan, N., Prasert, W., and Wongkrajang, K. 2014. Optimization of extrusion conditions for ready-to-eat breakfast cereal enhanced with defatted rice bran. *International Food Research Journal* 21: 713–722.

Chetana, R., Bhavana, K. P., Babylatha, R., Geetha, V., and Suresh Kumar, G. 2019. Studies on eggless mayonnaise from rice bran and sesame oils. In *Journal of Food Science and Technology* 56(6): 3117–3125. https://doi.org/10.1007/s13197-019-03819-1

Chiou, T.-Y., Ogino, A., Kobayashi, T., and Adachi, S. 2013. Characteristics and antioxidative ability of defatted rice bran extracts obtained using several extractants under subcritical conditions. *Journal of Oleo Science* 62: 1–8. https://doi.org/10.5650/jos.62.1

Coda, R., Cassone, A., Rizzello, C. G., Nionelli, L., Cardinali, G., and Gobbetti, M. 2011. Antifungal activity of *Wickerhamomyces anomalus* and *Lactobacillus plantarum* during sourdough fermentation: Identification of novel compounds and long-term effect during storage of wheat bread. *Applied and Environmental Microbiology* 77(10): 3484–3492. https://doi.org/10.1128/AEM.02669-10

Coda, R., Kärki, I., Nordlund, E., Heiniö, R.-L., Poutanen, K., and Katina, K. 2014. Influence of particle size on bioprocess induced changes on technological functionality of wheat bran. *Food Microbiolog*, 37: 69–77. https://doi.org/10.1016/j.fm.2013.05.011

Davis, K. S. 2001. *Corn Milling, Processing and Generation of Co-products*. Minnesota Nutrition Conference-Technical Symposium, Minnesota, USA, Minnesota Corn Growers Association.

de Morais Cardoso, L., Pinheiro, S. S., Martino, H. S. D., and Pinheiro-Sant'Ana, H. M. 2017. Sorghum (*Sorghum bicolor* L.): Nutrients, bioactive compounds, and potential impact on human health. *Critical Reviews in Food Science and Nutrition* 57(2): 372–390. https://doi.org/10.1080/10408398.2014.887057

Delcour, J., and Hoseney, R. 2010. Principles of cereal science and technology. *AACC Internationa* 229–289.

Demir, M., Bilgiçli, N., Türker, S., and Demir, B. 2021. Enriched Turkish noodles (erişte) with stabilized wheat germ: Chemical, nutritional and cooking properties. *LWT* 149: 111819. https://doi.org/10.1016/j.lwt.2021.111819

Dhankhar, P., and Hissar, T. 2014. Rice milling. *IOSR Journal of Engineering* 4(5), 34–42.

Dikeman, C. L., Murphy, M. R., & Fahey, G. C. J. 2006. Dietary fibers affect viscosity of solutions and simulated human gastric and small intestinal digesta. *The Journal of Nutrition* 136(4): 913–919. https://doi.org/10.1093/jn/136.4.913

Do, T. K., Dann, E. K., and Stanley, R. 2016. Screening of sorghum grain biorefinery by-products for growth inhibition of two common postharvest pathogens of mango. *Acta Horticulturae* 1120: 219–224. https://doi.org/10.17660/ActaHortic.2016.1120.33

Dordoni, R., De Cesare, S., Dioni, M., Mastrofilippo, T., Quadrelli, D., Rizzi, R., Corrado, S., and De Faveri, D. M. 2015. Design and development of an eco-innovative sorghum snack. *Chemical Engineering Transactions* 43(Table 1): 187–192. https://doi.org/10.3303/CET1543032

Dunford, N., and Zhang, M. 2003. Pressurized solvent extraction of wheat germ oil. *Food Research International* 36: 905–909. https://doi.org/10.1016/S0963-9969(03)00099-1

Ebrahimi, M., Sadeghi, A., Sarani, A., and Purabdolah, H. 2021. Enhancement of technological functionality of white wheat bread using wheat germ sourdough along with dehydrated spinach puree. *Journal of Agricultural Science and Technology* 23(4): 839–851.

FAO. 2022. Food and agriculture organization of the United Nations. "FAOSTAT". www.fao.org. Archived from the original on 6 January 2022. (Accessed on October 14, 2022)

Fasuyi, A. O. 2005. Maize-sorghum based brewery by-products as a energy substitute in broiler starter: Effect on performenace, carcass characteristics, organs and muscle growth. *International Journal of Poultry Science* 4(5): 334–338.

Feizollahi, E., Mirmoghtadaie, L., Mohammadifar, M. A., Jazaeri, S., Hadaegh, H., Nazari, B., and Lalegani, S. 2018. Sensory, digestion, and texture quality of commercial gluten-free bread: Impact of broken rice flour type. *Journal of Texture Studies* 49(4): 395–403. https://doi.org/10.1111/jtxs.12326

Fresco, L. 2005. Rice is life. *Journal of Food Composition and Analysis* 18(4): 249–253. https://doi.org/10.1016/j.jfca.2004.09.006

Garba, U. 2020. Preparing spray-dried cholesterol free salad dressing emulsified with enzymatically synthesized mixed mono- and diglycerides from rice bran oil and glycerol. *Journal of Food Science and Technology* 58. https://doi.org/10.1007/s13197-020-04611-2

Ge, Y, Sun, A., Ni, Y., and Cai, T. 2000. Some nutritional and functional properties of defatted wheat germ protein. *Journal of Agricultural and Food Chemistry* 48(12): 6215–6218. https://doi.org/10.1021/jf000478m

Ge, Yiqiang, Sun, A., Ni, Y., and Cai, T. 2001. Study and development of a defatted wheat germ nutritive noodle. *European Food Research and Technology* 212: 344–348. https://doi.org/10.1007/s002170000253

Gujral, H., and Rosell, C. 2004. Improvement of the breadmaking quality of rice by glucose oxidase. *Food Research International* 37: 75–81. https://doi.org/10.1016/j.foodres.2003.08.001

Gul, K., Yousuf, B., Singh, A. K., Singh, P., and Wani, A. A. 2015. Rice bran: Nutritional values and its emerging potential for development of functional food—A review. *Bioactive Carbohydrates and Dietary Fibre* 6: 24–30. https://doi.org/10.1016/j.bcdf.2015.06.002

Gumus, Z. P., Guler, E., Demir, B., Barlas, F. B., Yavuz, M., Colpankan, D., Senisik, A. M., Teksoz, S., Unak, P., Coskunol, H., and Timur, S. 2015. Herbal infusions of black seed and wheat germ oil: Their chemical profiles, in vitro bio-investigations and effective formulations as phyto-nanoemulsions. *Colloids and Surfaces. B, Biointerfaces* 133: 73–80. https://doi.org/10.1016/j.colsurfb.2015.05.044

Hemavathi, K., Abhilash, H. S., Senthil, A., and Kumar, S. 2017. Physico-chemical chemical and nutritional evaluation of wheat germ idli. *International Journal of Current Research* 9(6): 52953–52959.

Hertrampf, J. W., and Piedad-Pascual, F. 2000. *Wheat and wheat by-products BT—Handbook on ingredients for aquaculture feeds* (Ed. J. W. Hertrampf & F. Piedad-Pascual., pp. 531–542). Springer, Dordrecht.

Huang, Q., Shi, C. X., Su, Y. B., Li, D. F., Liu, L., Huang, C., and Piao, X. 2014. Prediction of the digestible and metabolizable energy content of wheat milling by-products for growing pigs from chemical composition. *Animal Feed Science and Technology* 196. https://doi.org/10.1016/j.anifeedsci.2014.06.009

Izydorczyk, M., Lagassé, S. L., Hatcher, D., Dexter, J. E., and Rossnagel, B. G. 2005. The enrichment of Asian noodles with fiber-rich fractions derived from roller milling of hull-less barley. *Journal of the Science of Food and Agriculture* 85: 2094–2104. https://doi.org/10.1002/jsfa.2242

Jalal, H., Salahuddin, M., Wani, S., Sofi, H., Pal, M., and Rather, F. 2013. Development of low fat meat products. *International Journal of Food Nutrition and Safety* 4(3): 98–107.

Judson, P. L., Al Sawah, E., Marchion, D. C., Xiong, Y., Bicaku, E., Bou Zgheib, N., Chon, H. S., Stickles, X. B., Hakam, A., Wenham, R. M., Apte, S. M., Gonzalez-Bosquet, J., Chen, D.-T., and Lancaster, J. M. 2012. Characterizing the efficacy of fermented wheat germ extract against ovarian cancer and defining the genomic basis of its activity. *International Journal of Gynecological Cancer* 22(6): 960–967. https://doi.org/10.1097/IGC.0b013e318258509d

Juliano, B. O., and Tuaño, A. P. P. 2019. *2—Gross structure and composition of the rice grain* (J. B. T.-R. (Ed. Fourth E. Bao., pp. 31–53). AACC International Press, London. https://doi.org/https://doi.org/10.1016/B978-0-12-811508-4.00002-2

Jung, H., Lee, Y. J., and Yoon, W. B. 2018. Effect of moisture content on the grinding process and powder properties in food: A review. *Processes* 6(6): 6–10. https://doi.org/10.3390/pr6060069

Justo, M. L., Rodriguez-Rodriguez, R., Claro, C. M., Alvarez de Sotomayor, M., Parrado, J., and Herrera, M. D. 2013. Water-soluble rice bran enzymatic extract attenuates dyslipidemia, hypertension and insulin resistance in obese Zucker rats. *European Journal of Nutrition* 52(2): 789–797. https://doi.org/10.1007/s00394-012-0385-6

Kaczmarczyk, M. M., Miller, M. J., and Freund, G. G. 2012. The health benefits of dietary fiber: Beyond the usual suspects of type 2 diabetes mellitus, cardiovascular disease and colon cancer. *Metabolism: Clinical and Experimental* 61(8): 1058–1066. https://doi.org/10.1016/j.metabol.2012.01.017

Kanojia, V., Kushwaha, N., Reshi, M., Rouf, A., and Muzaffar, H. 2018. Products and byproducts of wheat milling process. *IJCS* 6(4): 990–993. www.researchgate.net/profile/Monica_Reshi2/publication/330171167_Products_and_byproducts_of_wheat_milling_process/data/5c311f0592851c22a35eca9a/6-4-365-256.pdf

Kaur, A., Jassal, V., Thind, S. S., and Aggarwal, P. 2012a. Rice bran oil an alternate bakery shortening. *Journal of Food Science and Technology* 49(1): 110–114. https://doi.org/10.1007/s13197-011-0259-6

Kaur, G., Sharma, S., Nagi, H., and Dar, B. 2012b. Functional properties of pasta enriched with variable cereal brans. *Journal of Food Science and Technology* 49: 467–474. https://doi.org/10.1007/s13197-011-0294-3

Kennedy, G., and Burlingame, B. 2003a. Analysis of food composition data on rice from a plant genetic resources perspective. *Food Chemistry* 80(4): 589–596. https://doi.org/10.1016/S0308-8146(02)00507-1

Kennedy, G., and Burlingame, B. 2003b. Analysis of food composition data on rice from a plant genetic resources perspective. *Food Chemistry* 80: 589–596. https://doi.org/10.1016/S0308-8146(02)00507-1

Kent, M., and Rooney, W. 2021. Effects of field processing of sorghum grain on popping traits. *Agronomy* 11(5): 1–7. https://doi.org/10.3390/agronomy11050839

Khalid, A., Sohaib, M., Nadeem, M. T., Saeed, F., Imran, A., Imran, M., Inam, A. M., Ramzan, S., Nadeem, M., Anjum, F. M., and Arshad, M. S. 2021. Utilization of wheat germ oil and wheat bran fiber as fat replacer for the development of low-fat beef patties. *Food Science and Nutrition* 9: 1271–1281. https://doi.org/10.1002/fsn3.1988 20487177, 2021, 3.1988

Kim, B. K., Chun, Y. G., Cho, A. R., and Park, D. J. 2012. Reduction in fat uptake of doughnut by microparticulated wheat bran. *International Journal of Food Sciences and Nutrition* 63(8): 987–995. https://doi.org/10.3109/09637486.2012.690027

Koletta, P., Irakli, M., Papageorgiou, M., and Skendi, A. 2014. Physicochemical and technological properties of highly enriched wheat breads with whole grain non wheat flours. *Journal of Cereal Science 60*. https://doi.org/10.1016/j.jcs.2014.08.003

Kong, S., Kim, D.-J., Oh, S.-K., Choi, I.-S., Jeong, H.-S., and Lee, J. 2012. Black rice bran as an ingredient in noodles: chemical and functional evaluation. *Journal of Food Science* 77(3): C303–7. https://doi.org/10.1111/j.1750-3841.2011.02590.x

Krishnan, J. G., Menon, R., Padmaja, G., Sajeev, M. S., and Moorthy, S. N. 2012. Evaluation of nutritional and physico-mechanical characteristics of dietary fiber-enriched sweet potato pasta. *European Food Research and Technology* 234(3): 467–476. https://doi.org/10.1007/s00217-011-1657-8

Kuchtová, V., Minarovičová, L., Kohajdová, Z., and Karovičová, J. 2016. Effect of wheat and corn germs addition on the physical properties and sensory quality of crackers. *Potravinarstvo* 10(1): 543–549. https://doi.org/10.5219/598

Lee, Y., Ma, F., Byars, J., Felker, F., Liu, S., Mosier, N., Lee, J., Kenar, J., and Baik, B. 2021. Influences of hydrothermal and pressure treatments on compositional and hydration properties of wheat bran and dough mixing properties of whole wheat meal. *Cereal Chemistry* 98. https://doi.org/10.1002/cche.10411

Lei, M., Huang, J., Tian, X., Zhou, P., Zhu, Q., Li, L., Li, L., Ma, S., and Wang, X. 2021. Effects of insoluble dietary fiber from wheat bran on noodle quality. *Grain & Oil Science and Technology* 4(1): 1–9. https://doi.org/10.1016/j.gaost.2020.11.002

Levent, H., and Bilgiçli, N. 2013. Quality evaluation of wheat germ cake prepared with different emulsifiers. *Journal of Food Quality 36*. https://doi.org/10.1111/jfq.12042

Lima, I., Guraya, H., and Champagne, E. 2002. The functional effectiveness of reprocessed rice bran as an ingredient in bakery products. *Die Nahrung* 46:112–117. https://doi.org/10.1002/1521-3803(20020301)46:2<112::AID-FOOD112>3.0.CO;2-N

Linscombe, S. 2006. *Rice quality determines payment*. https://www.lsuagcenter.com/portals/our_offices/research_stations/rice/features/publications/rice-quality-determines-payment (Accessed on October 14, 2022)

Liu, S. X., Singh, M., and Inglett, G. 2011. Effect of incorporation of distillers' dried grain with solubles (DDGS) on quality of cornbread. *LWT—Food Science and Technology* 44(3): 713–718. https://doi.org/10.1016/j.lwt.2010.10.001

Liu, Y. Q., Strappe, P., Zhou, Z. K., and Blanchard, C. 2019. Impact on the nutritional attributes of rice bran following various stabilization procedures. *Critical Reviews in Food Science and Nutrition* 59(15): 2458–2466. https://doi.org/10.1080/10408398.2018.1455638

M. H. Abdel-Haleem, A. 2016. Production of gluten-free rolled paper from broken rice by using different hydrothermal treatments. *International Journal of Nutrition and Food Sciences* 5(4): 255. https://doi.org/10.11648/j.ijnfs.20160504.14

M. Noori, L. Y., and Sabir, D. A. 2019. Effect of wheat germ on quality of wheat bread dough. *Kurdistan Journal of Applied Research* 4(2): 102–109. https://doi.org/10.24017/science.2019.2.10

Ma, F., Lee, Y. Y., Park, E., Luo, Y., Delwiche, S., and Baik, B. K. 2021. Influences of hydrothermal and pressure treatments of wheat bran on the quality and sensory attributes of whole wheat Chinese steamed bread and pancakes. *Journal of Cereal Science* 102(July): 103356. https://doi.org/10.1016/j.jcs.2021.103356

Marcoa, C., and Rosell, C. M. 2008. Effect of different protein isolates and transglutaminase on rice flour properties. *Journal of Food Engineering* 84(1): 132–139. https://doi.org/https://doi.org/10.1016/j.jfoodeng.2007.05.003

Marconi, O. 2017. *The use of rice in brewing* (Ed. V. Sileoni; Ch. 4). IntechOpen, London. https://doi.org/10.5772/66450

Masoumikhah, Z., and Zargari, K. 2013. Effects of additional low fatted corn germ flour on rheological properties and sensory of macaroni. *Annals of Biological Research* 4(10): 61–66.

Mayer, H., Ceccaroni, D., Marconi, O., Sileoni, V., Perretti, G., and Fantozzi, P. 2016. Development of an all rice malt beer: A gluten free alternative. *LWT—Food Science and Technology* 67:67–73. https://doi.org/10.1016/j.lwt.2015.11.037

Meriles, S. P., Steffolani, M. E., Penci, M. C., Curet, S., Boillereaux, L., & Ribotta, P. D. 2021. Effects of low-temperature microwave treatment of wheat germ. *Journal of the Science of Food and Agriculture.* https://doi.org/10.1002/jsfa.11595

Ml, S., Vetrimani, R., and Krishnarau, L. 2007. Influence of fibre from different cereals on the rheological characteristics of wheat flour dough on biscuit quality. *Food Chemistry* 100: 1365–1370. https://doi.org/10.1016/j.foodchem. 2005.12.013

Mueller, T., and Voigt, W. 2011. Fermented wheat germ extract—Nutritional supplement or anticancer drug? *Nutrition Journal* 10(1): 89. https://doi.org/10.1186/1475-2891-10-89

Mueller, T., Jordan, K., and Voigt, W. 2011. Promising cytotoxic activity profile of fermented wheat germ extract (Avemar®) in human cancer cell lines. *Journal of Experimental & Clinical Cancer Research : CR* 30(1): 42. https://doi.org/10.1186/1756-9966-30-42

Nagano, H., Shoji, Z., Tamura, A., Kato, M., Omori, M., To, K. A., Dang, T. T., and Le, V. N. 2000. Some characteristics of rice paper of Vietnamese traditional food (Vietnamese spring rolls). *Food Science and Technology Research* 6(2): 102–105. https://doi.org/10.3136/fstr.6.102

Nagendra Prasad, M. N., and Khatokar M, S. 2011. Health benefits of rice bran—A review. *Journal of Nutrition & Food Sciences* 1(3):1–7 https://doi.org/10.4172/2155-9600.1000108

Nandeesh, K., Jyotsna, R., and Rao, G. 2010. Effect of differently treated wheat bran on rheology, microstructure and quality characteristics of soft dough biscuits. *Journal of Food Processing and Preservation* 35: 179–200. https://doi.org/10.1111/j.1745-4549.2009.00470.x

Nasir, M., Siddiq, M., Ravi, R., Harte, J. B., Dolan, K. D., and Butt, M. S. 2010. Physical quality characteristics and sensory evaluation of cookies made with added defatted maize germ flour. *Journal of Food Quality* 33(1): 72–84. https://doi.org/10.1111/j.1745-4557.2009.00291.x

Nathakattur Saravanabavan, S., Manchanahally Shivanna, M., and Bhattacharya, S. 2013. Effect of popping on sorghum starch digestibility and predicted glycemic index. *Journal of Food Science and Technology* 50(2): 387–392. https://doi.org/10.1007/s13197-011-0336-x

Ogunmuyiwa, O. H., Adebowale, A. A., Sobukola, O. P., Onabanjo, O. O., Obadina, A. O., Adegunwa, M. O., Kajihausa, O. E., Sanni, L. O., and Keith, T. 2017. Production and quality evaluation of extruded snack from blends of bambara groundnut flour, cassava starch, and corn bran flour. *Journal of Food Processing and Preservation* e13183. https://doi.org/10.1111/jfpp.13183

Ortiz de Erive, M., Wang, T., He, F., and Chen, G. 2020. Development of high-fiber wheat bread using microfluidized corn bran. *Food Chemistry* 310. https://doi.org/10.1016/j.foodchem.2019.125921

Otto, C., Hahlbrock, T., Eich, K., Karaaslan, F., Jürgens, C., Germer, C.-T., Wiegering, A., and Kämmerer, U. 2016. Antiproliferative and antimetabolic effects behind the anticancer property of fermented wheat germ extract. *BMC Complementary and Alternative Medicine* 16(1): 160. https://doi.org/10.1186/s12906-016-1138-5

Panfili, G., Cinquanta, L., Fratianni, A., and Cubadda, R. 2003. Extraction of wheat germ oil by supercritical CO_2: Oil and defatted cake characterization. *Journal of the American Oil Chemists' Society* 80(2):157–161. https://doi.org/10.1007/s11746-003-0669-1

Papageorgiou, M., and Skendi, A. 2018. Introduction to cereal processing and by-products. *Sustainable Recovery and Reutilization of Cereal Processing By-Products April*: 1–25. https://doi.org/10.1016/B978-0-08-102162-0.00001-0

Paraskevopoulou, A., Rizou, T., and Kiosseoglou, V. 2019. Biscuits enriched with dietary fibre powder obtained from the water-extraction residue of maize milling by-product. *Plant Foods for Human Nutrition* 74(3): 391–398. https://doi.org/10.1007/s11130-019-00752-8

Pathera, A. K., Riar, C. S., Yadav, S., and Singh, P. K. 2018. Effect of egg albumen, vegetable oil, corn bran, and cooking methods on quality characteristics of chicken nuggets using response surface methodology. *Korean Journal for Food Science of Animal Resources* 38(5): 901–911. https://doi.org/10.5851/kosfa.2018.e23

Pauaucean, A., and Man, S. 2013. Influence of defatted maize germ flour addition in wheat: maize bread formulations. *Journal of Agroalimentary Processes and Technologies* 19(3): 298–304.

Petrović, J., Fišteš, A., Rakić, D., Pajin, B., Lončarević, I., and Šubarić, D. 2015. Effect of defatted wheat germ content and its particle size on the rheological and textural properties of the cookie dough. *Journal of Texture Studies* 46(5): 374–384. https://doi.org/10.1111/jtxs.12137

Petrovic, J., Rakić, D., Fistes, A., Pajin, B., Lončarević, I., Tomovic, V., and Zarić, D. 2017. Defatted wheat germ application: Influence on cookies' properties with regard to its particle size and dough moisture content. *Food Science and Technology International* 23:108201321771310. https://doi.org/10.1177/1082013217713101

Pınarlı, İ., Öner, M., and İbanoğlu, Ş. 2004. Effect of wheat germ addition on the microbiological quality, in vitro protein digestibility, and gelatinization behavior of macaroni. *European Food Research and Technology* 219: 52–59. https://doi.org/10.1007/s00217-004-0932-3

Podolske, J., Cho, S. S., Gonzalez, R., Lee, A. W., and Peterson, C. 2013. Rice hull fiber: Food applications, physiological benefits, and safety. *Cereal Foods World* 58(3): 127–131. https://doi.org/10.1094/CFW-58-3-0127

Pourafshar, S., Rosentrater, K. A., and Krishnan, P. G. 2018. Production of barbari bread (traditional Iranian bread) using different levels of distillers dried grains with solubles (DDGS) and sodium stearoyl lactate (SSL). *Foods* 7(3):1–16. https://doi.org/10.3390/foods7030031

Qian, H., and Zhang, H. 2013. 22—Rice flour and related products. In B. Bhandari, N. Bansal, M. Zhang, & P. B. T.-H. of F. P. Schuck (Eds.), *Woodhead publishing series in food science, technology and nutrition* (pp. 553–575). Woodhead Publishing, London.

Quiñones, R. S., Macachor, C. P., and Quiñones, H. G. 2015. Development of gluten-free composite flour blends. *Tropical Technology Journal* 19(1): 3–6. https://doi.org/10.7603/s40934-015-0003-3

Rafe, A., Sadeghian, A., and Hoseini-Yazdi, S. Z. 2017. Physicochemical, functional, and nutritional characteristics of stabilized rice bran form tarom cultivar. *Food Science and Nutrition* 5(3): 407–414. https://doi.org/10.1002/fsn3.407

Ramírez, E. C., Johnston, D. B., McAloon, A. J., and Singh, V. 2009. Enzymatic corn wet milling: Engineering process and cost model. *Biotechnology for Biofuels* 2: 1–9. https://doi.org/10.1186/1754-6834-2-2

Ratnavathi, C. V. 2016. Sorghum processing and utilization. In *Sorghum biochemistry: An industrial perspective*. Elsevier Inc, London.

Renuka Devi, R., and Arumughan, C. 2007. Phytochemical characterization of defatted rice bran and optimization of a process for their extraction and enrichment. *Bioresource Technology* 98(16): 3037–3043. https://doi.org/10.1016/j.biortech.2006.10.009

Ribeiro, G. O., Carolina, E., Brito, L., Camilloto, G. P., and Cruz, R. S. 2018. Effect of corn bran addition on technological properties of tagliarini pasta. *Journal of Food and Nutrition Research* 6(2): 130–136. https://doi.org/10.12691/jfnr-6-2-10

Rizzello, C., Nionelli, L., Coda, R., De Angelis, M., and Gobbetti, M. 2010. Effect of sourdough fermentation on stabilisation, and chemical and nutritional characteristics of wheat germ. *Food Chemistry* 119: 1079–1089. https://doi.org/10.1016/j.foodchem.2009.08.016

Rohman, A., Helmiyati, S., Penggalih, M., and Setyaningrum, D. 2014. Rice in health and nutrition. *International Food Research Journal* 21: 13–24.

Sairam, S., Gopala Krishna, A. G., and Urooj, A. 2011. Physico-chemical characteristics of defatted rice bran and its utilization in a bakery product. *Journal of Food Science and Technology* 48(4): 478–483. https://doi.org/10.1007/s13197-011-0262-y

Sanchez, J., Thanabalan, A., Khanal, T., Patterson, R., Slominski, B. A., and Kiarie, E. 2019. Growth performance, gastrointestinal weight, microbial metabolites and apparent retention of components in broiler chickens fed up to 11% rice bran in a corn-soybean meal diet without or with a multi-enzyme supplement. *Animal Nutrition* 5(1):41–48. https://doi.org/10.1016/j.aninu.2018.12.001

Sharif, M. K., Butt, M. S., Anjum, F. M., and Khan, S. H. 2014. Rice bran: A novel functional ingredient. *Critical Reviews in Food Science and Nutrition* 54(6): 807–816. https://doi.org/10.1080/10408398.2011.608586

Sharma, A. R. 2002. *Edible rice bran oil—Consumer awareness programme*. Mumbai: Rice bran oil promotion committee. Solvent Extractors Association of India.

Sharma, C., Singh, B., Hussain, S. Z., and Sharma, S. 2017. Investigation of process and product parameters for physicochemical properties of rice and mung bean (*Vigna radiata*) flour based extruded snacks. *Journal of Food Science and Technology* 54 (6): 1711–1720. https://doi.org/10.1007/s13197-017-2606-8

Sharma, V., Moreau, R. A., and Singh, V. 2008. Increasing the value of hominy feed as a coproduct by fermentation. *Applied Biochemistry and Biotechnology* 149(2): 145–153. https://doi.org/10.1007/s12010-007-8110-2

Siddiq, M., Nasir, M., Ravi, R., Dolan, K. D., and Butt, M. S. 2009. Effect of defatted maize germ addition on the functional and textural properties of wheat flour. *International Journal of Food Properties* 12(4): 860–870. https://doi.org/10.1080/10942910802103028

Singha, P., Singh, S. K., Muthukumarappan, K., and Krishnan, P. 2018. Physicochemical and nutritional properties of extrudates from food grade distiller's dried grains, garbanzo flour, and corn grits. *Food Science and Nutrition* 6(7): 1914–1926. https://doi.org/10.1002/fsn3.769

Skendi, A., Biliaderisa, C. G., Izydorczyk, M. S., Zervou, M., and Zoumpoulakis, P. 2011. Structural variation and rheological properties of water-extractable arabinoxylans from six Greek wheat cultivars. *Food Chemistry* 126(2): 526–536. https://doi.org/10.1016/j.foodchem.2010.11.038

Sobota, A., Rzedzicki, Z., Zarzycki, P., and Kuzawińska, E. 2015. Application of common wheat bran for the industrial production of high-fibre pasta. *International Journal of Food Science and Technology* 50(1): 111–119. https://doi.org/10.1111/ijfs.12641

Sohail, M., Rakha, A., Butt, M. S., Iqbal, M. J., and Summer, R. 2017. Rice bran nutraceutics: A comprehensive review. *Critical Reviews in Food Science and Nutrition* 57(17): 3771–3780. https://doi.org/10.1080/10408398.2016.1164120

Song, X., Zhu, W., Pei, Y., Ai, Z., and Chen, J. 2013. Effects of wheat bran with different colors on the qualities of dry noodles. *Journal of Cereal Science* 58: 400–407. https://doi.org/10.1016/j.jcs.2013.08.005

Sousa, M. F. de, Guimarães, R. M., Araújo, M. de O., Barcelos, K. R., Carneiro, N. S., Lima, D. S., Santos, D. C. Dos, Batista, K. de A., Fernandes, K. F., Lima, M. C. P. M., and Egea, M. B. 2019. Characterization of corn (*Zea mays* L.) bran as a new food ingredient for snack bars. *LWT-Food Science and Technology* 101: 812–818. https://doi.org/10.1016/j.lwt.2018.11.088

Sozer, N., Cicerelli, L., Heiniö, R.-L., and Poutanen, K. 2014. Effect of wheat bran addition on in vitro starch digestibility, physico-mechanical and sensory properties of biscuits. *Journal of Cereal Science* 60. https://doi.org/10.1016/j.jcs.2014.01.022

Sramkova, Z., Gregová, E., and Šturdík, E. 2009. Chemical composition and nutritional quality of wheat grain. *Acta Chimica Slovaca* 2: 115–138.

Sunphorka, S., Warinthorn, C., Yoshito, O., and Somkiat, N. 2012. Protein and sugar extraction from rice bran and de-oiled rice bran using subcritical water in a semi-continuous reactor: Optimization by response surface methodology. *International Journal of Food Engineering* 8(3): 1–22. https://doi.org/10.1515/1556-3758.2262

Tiwari, A., and Mishra, S. 2019. Sensory Evaluation of wheat bran biscuits mixed with flaxseed. *Asian Journal of Advanced Research and Reports* 6(3): 1–5. https://doi.org/10.9734/ajarr/2019/v6i330155

Tobias, J. R., Castro, I. J. L., Peñarubia, O. R., Adona, C. E., and Castante, R. B. 2018. Physicochemical and functional properties determination of flour, unmodified starch and acid-modified starch of Philippine-grown sorghum (*Sorghum bicolor* L. Moench). *International Food Research Journal* 25(6): 2640–2649.

Trappey, E. F., Khouryieh, H., Aramouni, F., and Herald, T. 2015. Effect of sorghum flour composition and particle size on quality properties of gluten-free bread. *Food Science and Technology International* 21(3): 188–202. https://doi.org/10.1177/1082013214523632

Tsen, C. C., Eyestone, W., and Weber, J. L. 1982. Evaluation of the quality of cookies supplemented with distillers' dried grain flours. *Journal of Food Science* 47(2): 684–685. https://doi.org/10.1111/j.1365-2621.1982.tb10156.x

Tsen, C. C., Weber, J. L., and Eyestone, W. 1983. Evaluation of distillers' dried grain flour as a bread ingredient. *Cereal Chemistry* 60(4): 295–297.

Union, E. 2010. Commission Regulation (EC) No 165/2010 of 26 February 2010, amending Regulation (EC) No 1881/2006 setting maximum levels for certain contaminants in foodstuffs as regards aflatoxins. *Official Journal of the European Union* 50(2009): 8–12.

United Nations Environment Programme. 2021. *Food waste index report 2021.* Nairobi: United Nations Environment Programme

USDA, 2021. https://ipad.fas.usda.gov/cropexplorer/cropview/commodityView.aspx? cropid=0459200. (Accessed on December, 2021).

Volkova, A. V., Kazarina, A. V., Antimonova, O. N., Nikonorova, Y. Y., & Atakova, E. A. 2020. Use of by-products of millet, amaranth and sorghum grains in bakery production. *BIO Web of Conferences* 17: 00047. https://doi.org/10.1051/bioconf/20201700047

Wang, F. C., Chung, D. S., Seib, P. A., and Kim, Y. S. 2000. Optimum steeping process for wet milling of sorghum. *Cereal Chemistry* 77(4): 478–483. https://doi.org/10.1094/CCHEM.2000.77.4.478

Wang, L., Duan, W., Zhou, S., Qian, H., Zhang, H., and qi, X. 2017. Effect of rice bran fibre on the quality of rice pasta. *International Journal of Food Science & Technology* 53. https://doi.org/10.1111/ijfs.13556

Wang, T., and Johnson, L. 2001. Refining high-free fatty acid wheat germ oil. *Journal of Oil & Fat Industries* 78: 71–76. https://doi.org/10.1007/s11746-001-0222-2

Wardhana, Y. R., and Banawi, L. S. 2020. Fortification of yellow alkaline noodles with wheat bran and the impact on physical and sensorial properties. *Journal of Nutritional Science and Vitaminology* 66: S190—S195. https://doi.org/10.3177/jnsv.66.S190

Yadav, D. N., Singh, K. K., and Rehal, J. 2012. Studies on fortification of wheat flour with defatted rice bran for chapati making. *Journal of Food Science and Technology* 49(1): 96–102. https://doi.org/10.1007/s13197-011-0264-9

Yadav, S., Malik, A., Pathera, A., Islam, R. U., and Sharma, D. 2016. Development of dietary fibre enriched chicken sausages by incorporating corn bran, dried apple pomace and dried tomato pomace. *Nutrition and Food Science* 46(1): 16–29. https://doi.org/10.1108/NFS-05-2015-0049

Yan, H., and Lu, Q. 2021. Physicochemical properties of starch-wheat germ oil complex and its effects on water distribution and hardness of noodles. *LWT—Food Science and Technology* 135: 110211. https://doi.org/10.1016/j.lwt.2020.110211

Yazicioglu, B., Sahin, S., and Sumnu, G. 2015. Microencapsulation of wheat germ oil. *Journal of Food Science and Technology* 52(6): 3590–3597. https://doi.org/10.1007/s13197-014-1428-1

Yishak G, Y., and Shimelis, T. 2015. Development of value added products from byproducts of Ethiopian wheat milling industries. *Journal of Food Processing & Technology* 6(8). https://doi.org/10.4172/2157-7110.1000474

Younas, A., Bhatti, M. S., Ahmed, A., and Randhawa, M. A. 2011. Effect of rice bran supplementation on cookie baking quality. *Pakistan Journal of Agricultural Sciences*, 48(2): 129–134.

Youssef, H. 2015. Assessment of gross chemical composition, mineral composition, vitamin composition and amino acids composition of wheat biscuits and wheat germ fortified biscuits. *Food and Nutrition Sciences* 6: 845–853. https://doi.org/10.4236/fns.2015.610088

Yu, S., Nehus, Z. T., Badger, T. M., and Fang, N. 2007. Quantification of vitamin E and gamma-oryzanol components in rice germ and bran. *Journal of Agricultural and Food Chemistry* 55(18): 7308–7313. https://doi.org/10.1021/jf071957p

Zarei, I., Brown, D. G., Nealon, N. J., and Ryan, E. P. 2017. Rice bran metabolome contains amino acids, vitamins & cofactors, and phytochemicals with medicinal and nutritional properties. *Rice* 10(1): 1–21. https://doi.org/10.1186/s12284-017-0157-2

Zhang, R., Ma, S., Li, L., Zhang, M., Tian, S., Wang, D., Liu, K., Liu, H., Zhu, W., and Wang, X. 2021. Comprehensive utilization of corn starch processing by-products: A review. *Grain & Oil Science and Technology* 4(3): 89–107. https://doi.org/10.1016/j.gaost.2021.08.003

Zhu, K.-X., Lian, C.-X., Guo, X.-N., Peng, W., and Zhou, H. M. 2011. Antioxidant activities and total phenolic contents of various extracts from defatted wheat germ. *Food Chemistry* 126: 1122–1126. https://doi.org/10.1016/j.foodchem.2010.11.144

Zhu, K.-X., Zhou, H.-M., and Qian, H. F. 2006. Proteins extracted from defatted wheat germ: Nutritional and structural properties. *Cereal Chemistry* 83: 69–75. https://doi.org/10.1094/CC-83-0069

CHAPTER 9

Phytochemical Profile of Cereal Grains

Ghulam Mustafa Kamal, Arfa Liaquat, Ayesha Noreen, Asma Sabir, Muhammad Saqib,
Muhammad Khalid and Rukhsana Iqbal

CONTENTS

9.1 INTRODUCTION

Grains are the world's most important crops. Since the beginning of civilization, grains have been used as staple foods for both human consumption and animal feed (Arzani & Ashraf, 2017; Ragaee et al., 2013). Cereals are harvested for their nutritionally fruitful seeds known as grains belonging to the family *Poaceae*, which are used indirectly for livestock as well as for human consumption (Burak & Imen, 1999; Smýkal et al., 2015). Cereals like rice, wheat, barley, and maize are significant foods for humans on earth. Three major agricultural cereal grains, rice, wheat, and maize, together constitute at least 75% of world's grain production (Harlan, 1992; Van Hung, 2016). In addition, 50% of the protein and 56% of food energy is obtained from eight types of cereal grains: barley, wheat, rice, maize, rye, millet, sorghum, and oat (Stoskopf, 1985; Van Hung, 2016).

Phytochemicals present in cereals including flavonoids, anthocyanins, phenolics, and so on have antioxidant activity (Harlan, 1992; Martín-Sánchez et al., 2014; Wu et al., 2018). Small quantities of bioactive phytochemicals like carotenoids, dietary fibers, phytosterols, and phenolic compounds are

DOI: 10.1201/9781003252023-9

extra-nutritional constituents present in plants (Kris-Etherton et al., 2002; Xiong et al., 2022). The major cereals like wheat, oats, barley, maize, rice, millet, rye, and sorghum consumed worldwide are not only important foods but also exhibit several components that are beneficial for health (Niu & Hou, 2020; Slavin, 2003). Cereals are staple foods for the world's population, providing proteins, carbohydrates, minerals, and vitamin B. Plant bioactive substances or phytochemicals present in cereals have health-promoting effects. Cereal germs and grains contain several phytochemicals like flavones, phytic acid, phenolic acids, terpenes, coumarins, and flavonoids. Cereals rich in several bioactive phytochemicals with health-promoting effects are an important part of food for humans and provide a wide source of energy, carbohydrates, vitamin B, and minerals (Ragaee et al., 2013). Carotenoids are important phytochemicals in cereal grains and cause the yellow color of the endosperm, giving aesthetic and nutritional quality to cereal-based products (Trono, 2019). Whole-grain cereals contain commonly found carotenoids like lutein, beta-cryptoxanthin, and zeaxanthin (Sofi et al., 2019).

9.2 PHYTOCHEMICALS

Major chronic disease risks are reduced using phytochemicals, which are non-nutrient bioactive plant components found in vegetables, fruits, grains, and other foods obtained from plants. Though a substantial percentage remain unknown, more than 5000 phytochemicals have been identified (Jayashri et al., 2019; Liu, 2013; Shahidi & Naczk, 1995). A phenolic compound contains one or more hydroxyl groups and aromatic rings with vitamin E (Siddiqui et al., 2019). Carotenoids in whole-cereal phytochemicals have antioxidant activity and the ability to kill free radicals that can biologically oxidize useful molecules (Liu, 2007). Whole-grain barley contains phytochemicals, which include tocols, phenolic acids, flavonoids, lignans, folate, and phytosterols (Figure 9.1). These phytochemicals show strong anti-proliferative, cholesterol-lowering, and antioxidant abilities that are beneficial in reducing the risk of various diseases (Badr et al., 2000; Fratianni et al., 2016).

Bioactive phytochemicals such as phenolic compounds, phytosterols, carotenoids, and dietary fibers are additional nutrients that are often present in small amounts in plants (Kris-Etherton et al., 2002). These are bioactive compounds based on plant nutrients that have antimicrobial and food

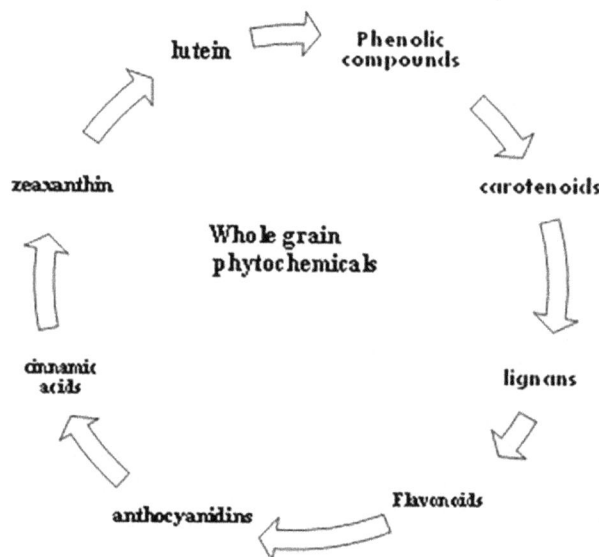

Figure 9.1 Whole-grain phytochemicals.

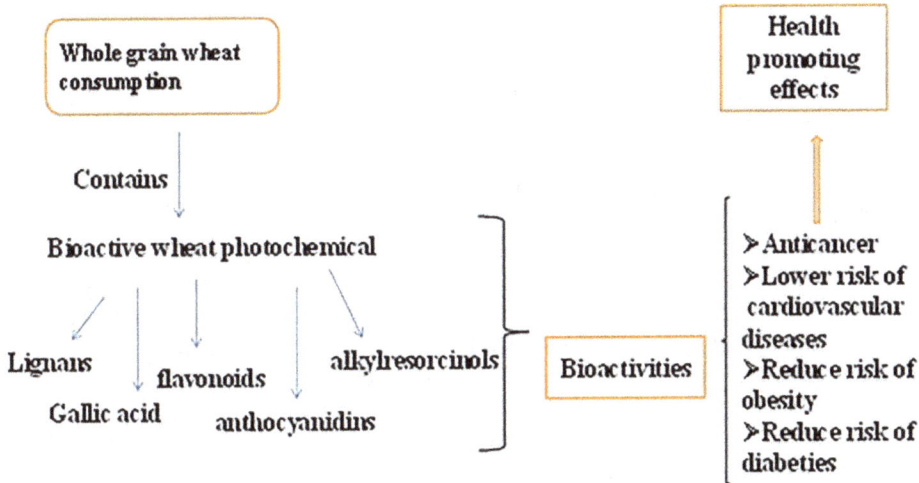

Figure 9.2 Association between grain consumption and risk of chronic diseases.
(*Source:* (Zhu & Sang, 2017)

nutritional properties that often play certain roles in the functioning of the human body, including antioxidant and anti-inflammatory properties, hormonal regulation, immune development (Xiong et al., 2019b), immunity-potentiating, dietary fiber, detoxifying, anticancer, and other neuro-pharmacological activities (Singh et al., 2020), as shown in Figure 9.2. The chemicals in phytochemicals, either individually or combined, have great therapeutic potential in treating various diseases. Dietary phytochemicals primarily promote health benefits and protect the body from diseases such as cancer; high blood pressure; diabetes; heart disease; inflammation; parasitic, microbial, and viral diseases; spasmodic conditions; mental illness; ulcers; osteoporosis and related disorders; and so on

These bio-active components derived from plants have recently attracted great interest due to their various uses. Medicinal plants are rich biological resources for parts of traditional medicine processes, modern medicine, nutraceuticals, dietary supplements, traditional medicine, pharmaceutical intermediates, and synthetic chemical associations (Ncube et al., 2008). Economically important foods, such as grains, vegetables, fruits, and nuts, contain phytochemicals that are used as active foods, soft drinks, and many other nutritious foods (Singh et al., 2020). Some of the major phytochemical properties are listed in the following (Thakur et al., 2020).

- They are non-nutritive plant chemicals that can protect from or prevent diseases.
- There are more than 1000 known phytochemicals.
- In the early days, plants used to make these chemicals for their own protection, but nowadays they can be used to protect humans from various diseases.
- Some of the most well-known phytochemicals are isoflavone in soy, lycopene in tomatoes, and flavonoids in fruit.
- They are not essential nutrients and are not needed by the human body for survival.

Phytosterols are types of phytochemicals present in vegetables, fruits, seeds, and nuts. These have antioxidant activity to suppress the growth of cell lines of various tumors with the onset of apoptosis and the corresponding binding of cells in the G1 cell cycle (Dillard & German, 2000; Elliott, 2005; John et al., 2007). Limonoids are also phytochemicals found in citrus fruits, which reduce phase II enzymes, inhibit phase I enzymes, detoxify enzymes in the liver, and give protection to lung tissues by releasing enzymes (Lam et al., 1994; Ozaki et al., 1995; Willcox et al., 2004).

9.3 PHENOLIC COMPOUNDS

The most complex and diverse class of phytochemicals in grains are phenolic compounds (Adom et al., 2005; Okarter & Liu, 2010), which are major contributors to antioxidant activities in whole grains (Jonnalagadda et al., 2011). Compounds containing a benzene ring with more hydroxyl groups are called phenolic compounds (e.g., condensed tannins, coumarins, phenolic acids, flavonoids, and alkyl-resorcinols). Phenolic compounds or polyphenols are large groups of chemicals, which include one or more benzene rings containing hydroxyl groups such as phenolic acid, coumarins, flavonoids, tannins, and stilbenes (Blanca Hernandez, 2013). These compounds are found everywhere in plants and are therefore a necessary part of the human diet (Dykes & Rooney, 2007; Ragaee et al., 2013). Phenolic compounds have many physiological functions including anti-inflammatory, antioxidant, anti-thrombotic, antimicrobial, anti-allergenic, and anti-carcinogenic effects, which promote health benefits by decreasing the risk of chronic diseases (Dai & Mumper, 2010; Fardet et al., 2008; Scalbert et al., 2005).

They play a vital role in promoting health due to their high anti-cancer contents (Van Hung, 2016), and the second group of phenolic metabolites counteracts oxidative stress in humans while maintaining a balance between oxidants and antioxidants. Phenolic acids are found in plants derived from cinnamic and benzoic acids (Figure 9.4) (Dykes & Rooney, 2007; Van Hung, 2016). Common hydroxycinnamic, ferulic, caffeic, and p-coumaric acids (Figure 9.4) usually occur in foods as esters containing glucose and quinic acid.

The antioxidant effect has been extensively studied due to its health benefits and is considered an important property of phenolic compounds. Numerous phenolic compounds have been shown to have powerful antioxidant activity in cell culture, in vivo, and in vitro studies (Fernandez-Panchon et al., 2008; Masisi et al., 2016; Wan et al., 2015; Wolfe & Liu, 2008). The antioxidant mechanisms of phenolic compounds are complex.

It has been suggested that phenolics exert antioxidant effects via the following three main actions:

- direct destruction of free radicals,
- inhibiting the formation of free radicals by inhibiting oxidant enzymes or digestion of certain metals, and
- regulation of antioxidants and detoxifying enzymes (López-Alarcón & Denicola, 2013; Masisi et al., 2016).

They contain derivatives of cinnamic and benzoic acids with flavonols, flavonoids, flavones, avenanthramides, lignans, alkyl-resorcinols, and anthocyanidins. The most common forms of phenolic compounds in wheat grains are flavonoids and phenolic acids. Due to their strong antioxidant effect, these compounds possess diabetes alleviation properties, anti-carcinogenic activity, and anti-inflammatory properties, which could be linked with aging control and cardiovascular and obesity disease prevention. Phenolic acids obtained from cinnamic acids and benzoic acids are divided into hydroxycinnamic acids (e.g., ferulic, p-coumaric, caffeic, and sinapic acids) and hydroxybenzoic (e.g., gallic acids, p-hydroxybenzoic, vanillic, protocatechuic, and syringic) (Duodu, 2011; Fereidoon Shahidi et al., 2019), as shown in Figure 9.3.

Secondary plant metabolites including phenolic compounds have an initial function of protecting against pathogen attack, ultraviolet light, and oxidative stress (Shahidi & Naczk, 1995). The antioxidant activity of phenolic compounds can protect from reactive oxygen species such as peroxy radicals, superoxide anions, and hydroxyl radicals, which are involved in various degenerative diseases (i.e., cancer and heart disease) (Harborne & Williams, 2000). Phenolic compounds are important due to their protective role against degenerative diseases like heart disease and cancer and antioxidant properties (Harlan, 1992). Through several signaling pathways, free radicals could be quenched with the help of strong antioxidant activities of phenolic phytochemicals. By modulating several enzymes and proteins at all levels of glucose homeostasis, these compounds can

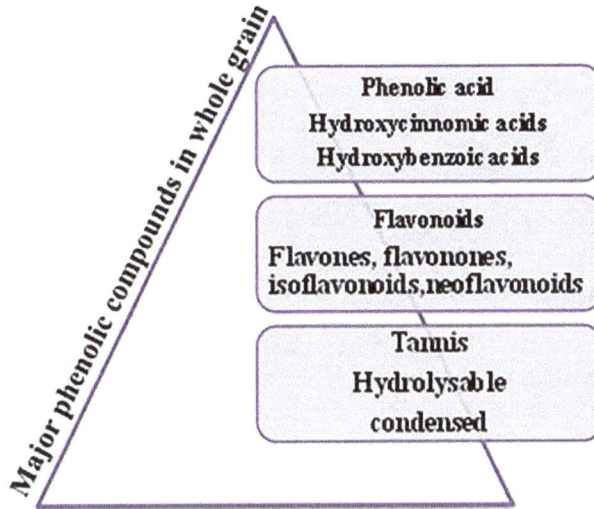

Figure 9.3 Phenolic compounds.

(*Source:* (Adebo & Gabriela Medina-Meza, 2020)

cause reverse diabetic regulation, including epigenetic modulations, protein expressions, enzyme activities, and gene expressions. Diabetic potential is also found in various phenolic phytochemicals (Hoda et al., 2019).

9.3.1 Alkyl-Resorcinols

Alkyl-resorcinols (ARs) with an odd-number n-alkyl side chain at C-5 on the benzene ring are 1,3-dihydroxybenzene derivatives. These compounds are found in barley, triticale, rye, and the bran of wheat. Barley, wheat, and rye contain 8, 339–759, and 575–1008 µg/g of alkyl-resorcinols, respectively (Chen et al., 2004; Mattila et al., 2005; Ross et al., 2001). Some studies have reported 108, 758–4, and 211–3 µg/g and 225 µg/g of alkyl-resorcinols in rye brains and wheat, respectively (Chen et al., 2004; Mattila et al., 2005). Cereal grains such as rye (720–761 mg/g) and wheat (489–642 mg/g) have also been reported to contain abundant alkyl-resorcinols, while these AR compounds are present in lower quantities in maize, barley, and millet (Ross et al., 2003). ARs also possess antimicrobial and antioxidant activities. In vitro results reported the anti-carcinogenic activities of this group. Distal colon cancer risk is reduced around 52–66% when extremely low concentrations of ARs are present in plasma. ARs show high cytotoxicity toward cancerous cell lines (Kruk et al., 2017; Fu et al., 2018). ARs have been indicated as biomarkers due to their high absorption rates in humans to correlate health benefits to whole-grain cereal consumption (Ross et al., 2003).

In vitro alkyl-resorcinol has anti-fungal, antioxidant, and antibacterial activities (Ross et al., 2004). Whole cereal grain intake as biomarkers could help to elucidate the connection between health and whole grains (Ross et al., 2004). Phenolic lipids derived from plants contain ARs, which are particularly found in cereals. The maximum amount of ARs is found in rye, which can be twice that in wheat (Ross et al., 2003). ARs are 1, 3-dihyydroxy benzene derivatives with an alkyl chain that gives amphiphilic properties. Mutagenic activity of indirect mutagens is reduced by ARs (Gsiorowski et al., 1996). ARs become biologically active when they are incorporated into biological membranes and possess stabilizing and antioxidative properties (Kozubek & Tyman, 1999; Ross et al., 2004).

Alkyl-resorcinol concentrations in wheat and grain products have been measured to vary between 142 and1784 mg/g DM (whole meal wheat bread, wheat bran-based products) (Ross et al., 2003) and 202–353 mg/g DM in whole grain wheat soft bread (Chen et al., 2004). In vitro ARs themselves can form liposomal membranes that are incorporated into biological membranes (Kozubek, 1987). They influence liposome size, increase bilayer rigidity, and increase membrane permeability and stability (Gubernator et al., 1999; Kozubek, 1987; Kozubek & Demel, 1980). They may have anti-carcinogenic and anti-mutagenic activities (Gsiorowski et al., 1996). Alkyl-resorcinols can have cyclo-oxygenase-2 (COX-2) promoter activity as another compound with a resorcinol moiety and therefore have protection ability in colon carcinogenesis (Mutoh et al., 2000).

9.3.2 Anthocyanidins

The most-studied flavonoids in cereals are a group of water-soluble pigments (red to purple, blue, orange). Pelargonidin, petunidin, peonidin, malvidin, cyanidin, and delphinidin are six anthocy-anidins, as shown in Figure 9.4. These substances have been mentioned in the case of pigmented varieties of wheat, rice, maize, rye, and barley. Polymerized flavanol units present in procyanidins or proantho-cyanidins called condensed tannins are helpful in food astringency. Barley and sorghum with red finger millets and a pigmented test layer contain these compounds (Dykes & Rooney, 2006; Goupy et al., 1999). Almost 69% of total anthocyanidins are made by major aglycone in blue wheat delphinidin. An anthranilic acid derivative related to derivatives of hydroxy-cinnamic acid are avenanthramides. Three major avenanthramides, B, C, and A, reported in oat are also known as oat avenanthramides1, 3, and 4 (Collins, 1989; Gani et al., 2012; Meydani, 2009). These are bioavailable and have shown anti-atherogenic, anti-inflammatory, and antioxidant properties (Bratt et al., 2003; Chen et al., 2004; Emmons & Peterson, 1999; Liu, 2013; Peterson, 2001). After smoking cessation, β-cryptoxanthin acts as a protective agent against lung cancer. β-cryptoxanthin has a

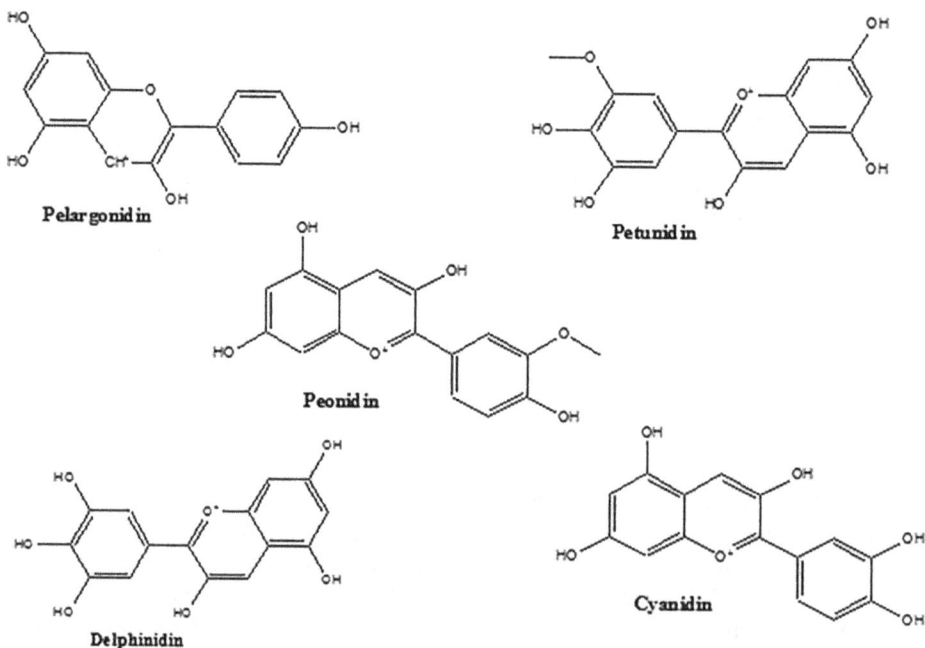

Figure 9.4 Chemical structures of some anthocyanidins present in cereal grains.

direct effect on bone formation and inhibits bone regeneration. β-cryptoxanthin was found in lower amounts in women with osteoporosis than in postmenopausal women due to overuse. Specific poly-phenols from oats are avenanthramides. There are 25 distinct entities, and they are substituted for cinnamic acid amides of anthranilic acids. Avenanthramides are present more in oat flakes (26–27 µg/g) than oat bran (13 µg/g).

9.3.3 Flavonoids

Flavonoids are naturally extruded products of plants and are found in distinct parts of the plant. These are secondary metabolites with a poly-phenolic structure, which is found in certain drinks, vegetables, and fruits. They possess antioxidant and biochemical effects linked with various diseases such as atherosclerosis, cancer, and Alzheimer's disease (AD) (Burak & Imen, 1999; Castañeda-Ovando et al., 2009; Lee et al., 2009). They are also responsible for inhabitation of various enzymes (e.g., phosphoinositide 3-kinase, xanthine oxidase [XO], cyclo-oxygenase [COX], and lipoxygenase (Metodiewa et al., 1997; Walker et al., 2000).

Vegetables use flavonoids in their growth to protect against plaque (Havsteen, 2002). They belong to the low-molecular-weight group of phenolic compounds that are still commonly distributed in the plant kingdom. In higher plants, they are responsible for forming one of the most prominent classes of compounds. For numerous angiosperm families, flavonoids are identified as flowering pigments. Flavonoids plays a key role in biological functions of bacteria, plants, and animals. In plants, flavo-noids are responsible for the smell and color of flowers, and the fruit attract pollen, and as a result, the fruit is dispersed to help the seeds and seed germination, development, and growth of seedlings (Griesbach, 2010).

Flavonoids constitute a huge family of phenolic compounds in nature and grain fields, focusing on the grain component (Dykes & Rooney, 2007). The main flavonoids (subclasses) reported in grains include flavonols, anthocyanidins, flavanones, flavones, and flavan-3-ols (Figure 9.4). Among the larger grains, sorghum grains have the most varied and abundant fla-vonoids (Xiong et al., 2022). Anthocyanidins, a collection of water-soluble dyes (blue, orange, and red to purple) are the most widely studied flavonoid in grain. Anthocyanidin content varies between grains. High concentrations are found in grains with colored pericarp such as purple-blue to orange-red or black wheat, rice, barley, and sorghum (Idehen et al., 2017; Rouf Shah et al., 2016; Xiong et al., 2022). Sorghum grains contain a rare form of anthocyanidin called 3-deoxyanthocyanidin, which does not have an OH group in place of C-3. 3-deoxyanthocyani-dins are more stable than anthocyanidins and have been suggested as potential natural pigment compounds (Xiong et al., 2019a, 2019b). Some grain flavonoids are also found in fruits and vegetables. Antioxidant, anti-inflammatory, anti-cancer, and anti-mutagenic effects, as well as modification of certain cellular enzymes, have been suggested as the main health benefits of flavonoids (Panche et al., 2016).

9.3.3.1 Flavonols

Flavonols are the building blocks of proanthocyanins and a type of flavonoids with a ketone group. These are found abundantly in different vegetables and fruits. The most common fla-vonols are myricetin, fisetin, kaempferol, and quercetin (Figure 9.5). Tea, red wine, lettuce, onions, apples, berries, kale, tomatoes, and grapes are major sources of flavonols. Flavonol supplementation has been found to be linked with a diversity of health benefits like reduced risk of cardiovascular disease and antioxidant potential. In comparison to flavones, flavonols are quite different from hydroxylation, methylation, and glycosylation patterns, which are the most ordinary subgroup of flavonoids present in vegetables, fruits (e.g., quercetin), and several plant foods (Iwashina, 2013).

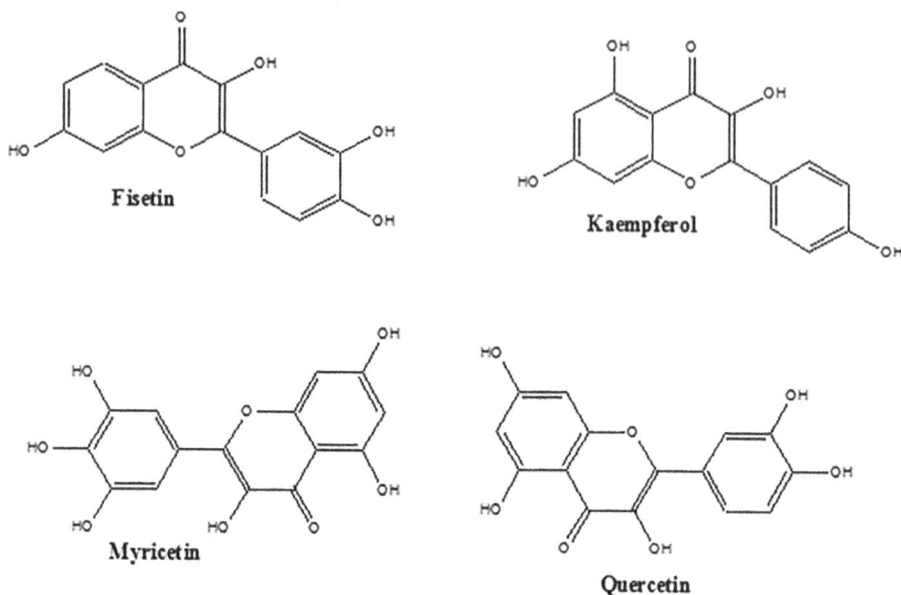

Figure 9.5 Chemical structures of some flavonols.

9.3.3.2 Flavones

Flavones are the most important components of flavonoids. These are found in flowers, fruits, and leaves and as glucosides. Edible vegetables with succulent branches, chamomile, parsley, mint, red pepper, and ginkgo are the main sources of flavone. Apigenin, tangeritin, and luteolin are part of this class of flavonoids. The citrus fruits are rich in tangeritin, polymethoxylated flavones, nobiletin, and sinensetin (Manach et al., 2004). Most fruit and vegetable flavones contain the hydroxyl group at the fifth place of the A-ring, whereas hydroxylation is at other positions (Iwashina, 2013).

9.3.3.3 Flavanones

Flavanones are another vital component of all citrus fruits and is responsible for the bitter taste of citrus fruits, such as oranges, lemons, and grapes. Naringenin, eriodictyol, and hesperitin are examples of flavanones. These give various health benefits due to their free radical detoxification properties. Citrus flavonoids have various medicinal effects such as blood lipid-ejaculation, antioxidant, cholesterol-lowering, and anti-inflammatory properties. Flavanones are also known as dihydroflavones and have a full C-ring; but due to the presence of a saturated double bond between positions 3 and 2, two flavonoid groups are differentiated.

9.3.4 Isoflavonoids

Isoflavonoids are a very important and diverse group of flavonoids. They have extremely limited distribution in plants and are mostly found in legume and other soybean crops. Other isoflavonoids have been reported in microbes (Matthies et al., 2008). They are also responsible for the development of phytoalexins and act as a precursor during plant-specific interactions (Aoki et al., 2000; Dixon & Ferreira, 2002). Isoflavonoids show great immunity against several diseases. Due to the estrogenic activity of isoflavones in certain animal models, they are considered phyto-estrogens, such as daidzein and genistein.

9.4 CAROTENOIDS

Carotenoids are a group of more than 600 living organisms that occur naturally in tetraterpenoids composed of algae, photosynthetic bacteria, and plants commonly found in mutations. The molecules that are the source of colors such as orange, red, and yellow have gained attention because they act as antioxidants and pro-vitamin supplements (Britton, 1995). These compounds are categorized into two classes: xanthophylls (zeaxanthin, astaxanthin, lutein) and carotenes (α, β, γ-carotene and β-cryptoxanthin). Carotenoids are found in whole-grain cereals (Adom et al., 2005), whereas vegetables and fruits are the main sources of carotenoids such as carrots, parsley, tomatoes, green and orange leafy vegetables, fenugreek, chenopods, spinach, radish, cabbage, and turnips. Their antioxidant activities protect against cervical, lung, prostate, and digestive cancers and colorectal diseases (Britton, 1995; Elliott, 2005; Paiva & Russell, 1999; Prakash & Sharma, 2014; Ribaya-Mercado & Blumberg, 2004), while their deficiency can lead to many diseases such as keratomalacia, corneal ulceration, night blindness, irreversible blindness, scarring, and xerophthalmia (Vucenik & Shamsuddin, 2006).

Carotenoids can prevent liver cancer (Nishino et al., 2009), sunburn, and retinal deterioration (Stahl & Sies, 2005) and improve the immune system. In addition, they exhibit specific anti-inflammatory and antiapoptotic properties and participate in the display of intracellular signals by influencing enrollment factors (Kaulmann & Bohn, 2014).

In grain, carotenoids occur naturally in esterified or free forms (especially linoleic and palmitic acid) and depend upon cereal genotype (Mellado & Hornero, 2012). Barley (*Hordeum vulgare* L) is an old berry grown in a cool environment and well-used in brewing. In grains, barley contains an exceedingly small amount of total carotenoids as compared to regular bread wheat. Yellow barley is the most common grain type, containing 2.25 µg/g of total carotenoid, and its purple variation contains twice as much (4.54 µg/g) (Ndolo & Beta, 2013).

Phytochemicals found in cereal grains have the ability to give protection against diabetes, including various carotenoids (e.g., α- and β-carotene), lutein, zeaxanthin, and β-cryptoxanthin. All of these are found in the germ and bran fractions (Adom et al., 2005).

9.4.1 α-β-Carotene and Other Carotenoids

Carotenoids are natural pigments and the most key component of phytochemicals. They are produced by microorganisms and plants but not by animals. These include zeaxanthin, α, β carotene, lutein, lycopene, cryptoxanthin, and astaxanthin. Lycopene is the most efficient extract of singlet oxygen species, while zeaxanthin and lutein are scavengers of radical oxygen species (Blanca Hernandez, 2013).

Carotenoids are classified into several types:

1. Carotenoid hydrocarbons are called carotenes and have some end groups, such as lycopene, that contain two acyclic end groups. β-carotene has two groups of end-type cyclohexene.
2. Oxygen-rich carotenoids are called xanthophylls. These include lutein (hydroxy), spirilloxanthin (methoxy), echinenone (oxo), zeaxanthin, and antheraxanthin (epoxy) (Goodwin, 1980; Rehmanji et al., 2021).

Antioxidant functions of carotenoids depend on their ability to peroxyl radicals and antioxidant properties (Kiokias & Gordon, 2004; Stahl & Sies, 1996). The concentration of each carotenoid in plants and fruits indicates which enzymes may be confined to the biosynthetic cascade, such as a higher level of lycopene in tomatoes suggesting a lack of enzymatic activity to change lycopene to α-carotene (Beecher, 1998).

From phytoene to lycopene, each enzymatic step adds double binding to the molecule, forming lycopene, an symmetrical molecule consisting of 13 double bonds. In a biosynthetic process lycopene involves the enzymatic cycle to form β-carotene (two beta rings) and α-carotene (one beta ring). Formation of xanthophyll takes place due to the addition of oxygen to the molecule.

β-cryptoxanthin and α, β-carotene are precursors of vitamin A and essential nutrients (Zeece, 2020). Unlike α-carotene, β-cryptoxanthin, and β-carotene, zeaxanthin and lutein are not provitamin A carotenoids. On the other hand, they are important for optimal functioning of the retina (Walchuk & Suh, 2020).

9.5 LIGNANS

Lignans are a group of phytoestrogens found in many plants. These are found in whole-grain bran and are high in rye and oat (Durazzo et al., 2013; Rodríguez-García et al., 2019). The lignans that are converted to mammalian lignans (in the colon by microflora) have weak estrogenic and strong antioxidant activities, lower the risk of hormone-related cancers, and enhance colon health (Adlercreutz, 2007; Thompson, 1994). Lignans (natural polyphenols) are extensively distributed in the plant kingdom as natural protective agents. These bioactively behave as phytoestrogens due to their functional and structural similarity to β-estradiol. Lignans have the ability to protect from coronary heart disease due to the presence of a wide range of biological effects, such as antitumor, antioxidant, antibacterial, estrogenic, antiviral, insecticidal, anti-estrogenic, and fungistatic activities (Prasad & Jadhav, 2016; Rhee, 2016; Saarinen et al., 2013).

In rye, the major lignans are pinoresinol, 7-hydroxymatairesinol, syringaresinol, secoisolariciresinol, isolariciresinol, lariciresinol, and mairesinol (Pihlava et al., 2018; Zanella et al., 2017), as shown in Figure 9.6. In addition to the antioxidant properties of lignans, these act as specific nuclear regulators (e.g., liver receptor alpha, estrogen receptors, peroxisome proliferator-activated receptor,

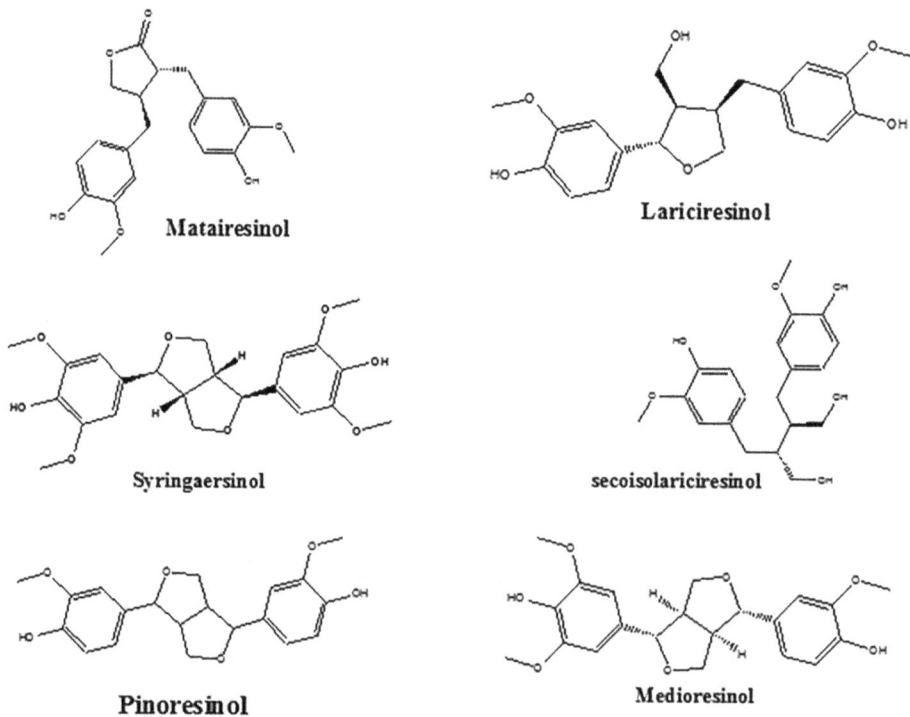

Figure 9.6 Structure of major lignans found in barley.

and gamma), which are involved in insulin action as well as fatty acid and glucose metabolism. This suggests that they may be used effectively for selected body functions (Zanella et al., 2017).

9.6 LUTEIN AND ZEAXANTHIN

Zeaxanthin and lutein are non-provitamins, a structural isomer of carotenoids that can be found in human tissues and blood. The macular pigment of the human retina is made up of xanthophyll identified by zeaxanthin and lutein, but the concentration of lutein and zeaxanthin in the macula is different. Lutein dominates the outer surface of the macula, but zeaxanthin is moderate (Sa, 2013). Along with lutein, xanthophyll also naturally belongs to the carotenoid, β-carotene, and lycopene groups. These are present in large concentrations in green leafy vegetables and certain fruits, for example, kiwi and avocado, and have powerful antioxidant and free-radical effects.

Although β-carotene accumulates in the skin and gives it a golden-yellow color, lutein is stored exclusively in macula lutea, which protects the retina from oxidative damage caused by ultraviolet resorcinol (UVR) (Scarmo et al., 2010). It has antioxidant and anti-inflammatory activity. It is also responsible for image protection, improving skin elasticity and anti-cancer properties (Woodside et al., 2015). The key natural sources of lutein are herbs and vegetables such as lettuce, kale, chicory, spinach, along with legumes, egg yolks, and fresh herbs (Machado et al., 2017). Topical and oral lutein treatment with zeaxanthin provides higher antioxidant activity and better hydration of hard skin than single treatment (Anunciato & da Rocha Filho, 2012).

Animal studies show that it also has hepatoprotective effects against xenobiotics such as carbon tetrachloride, paracetamol, and alcohol. Lutein has been shown to reduce the levels of high serum transfusion of alkaline phosphatase, bilirubin, and amino acids and the levels of conjugated diene, hydroperoxides, and lipid peroxidation in rat liver in treated with ethanol. Lutein treatment in ethanol-controlled mice reduced the levels of hydroxyproline and histopathological abnormalities, which is a symptom of fibrosis (Firdous et al., 2011). Major sources of both zeaxanthin and lutein in diet include green leaves, corn, eggs, and vegetables. Lutein is a leading isomer in the diet, even though some varieties of corn contain substantial amounts of zeaxanthin. Lutein components are commonly found in marigold flowers (Sa, 2013).

9.7 β-CRYPTOXANTHIN

β-cryptoxanthin is a common carotenoid that is found in the human body (Sarungallo et al., 2015). Free β-cryptoxanthin is only type that is found in human tissues and blood because esters of β-cryptoxanthin are enzymatically well hydrolyzed in intestines; however, the enzymes involved in this process are unclear (Mercadante et al., 2017). It has the activity of provitamin A and more biological effects than α-carotene. However, both carotenoids provide the same vitamin A molecule based on composition. Due to β-cryptoxanthin's bipolar nature, various epidemiological investigations have shown that dietary β-cryptoxanthin is related to lower levels of lungs cancer and furthermore improves the human lung function.

β-cryptoxanthin action on bone health requires further research. These compounds have no toxicity or deficiency. β-cryptoxanthin has a number of sources such as papayas, tangerines, peaches, oranges, and yellow corn. In many plants, β-cryptoxanthin, a vital component of carotenoid compounds, is associated with fruits' and vegetables' orange-yellow color. β-cryptoxanthin has both ester and free types of properties.

β-cryptoxanthin and its ester distribution depend on time of ripening and plant type. Sources of β-cryptoxanthin are yellow-orange fruits such as papaya, squash, mango, citrus, apricot, peppers, and persimmons and vegetables. Plants contain both free and ester β-cryptoxanthin. During ripening, β-cryptoxanthin accumulates; however, due to the higher levels, ripening of fruits give features to crypto-cryptoxanthin esters. β-cryptoxanthin esters can be hydrolyzed to crypto-cryptoxanthin in the human body. These have more bioavailability as compared to other major carotenoids, showing that foods rich in cryptoxanthin are a major source of carotenoids (Jiao et al., 2019).

9.8 CONCLUSION

Phytochemicals are found in cereal grains, and their consumption brings various health benefits. Epidemiological, clinical, and preclinical studies reveal that phytochemicals have been effective in preventing and treating several diseases due to their anti-inflammatory and antioxidant activities. Changes in lifestyle and growing health problems have necessitated the use of bioactive compounds in the form of food components. With the increasing use of natural products, there is a need for more scientific investigations supporting the medical and nutritional importance of phytochemicals for the treatment and prevention of various diseases. Nutraceuticals could be obtained from several sources, but nutraceuticals originating from cereal grains (plants) have various health benefits and are considered more valuable. The safety and efficacy of bioactive components is a major public concern. Health claims, labeling, and safety parameters/regulations should be strictly monitored for phytochemical consumption. However, a number of cereal grain–based phytochemicals (bioactive compounds) are still unexploited and underutilized. Therefore, this topic holds exciting opportunities and great future scope. The present chapter is aimed at overviewing the origin, types, structures, and properties of phytochemicals in cereal grains so that their potential can be fully exploited by researchers/consumers.

REFERENCES

Adebo, O. A., & Gabriela Medina-Meza, I. (2020). Impact of fermentation on the phenolic compounds and antioxidant activity of whole cereal grains: A mini review. *Molecules, 25*(4), 927.

Adlercreutz, H. (2007). Lignans and human health. *Critical Reviews in Clinical Laboratory Sciences, 44*(5–6), 483–525.

Adom, K. K., Sorrells, M. E., & Liu, R. H. (2005). Phytochemicals and antioxidant activity of milled fractions of different wheat varieties. *Journal of Agricultural and Food Chemistry, 53*(6), 2297–2306.

Anunciato, T. P., & da Rocha Filho, P. A. (2012). Carotenoids and polyphenols in nutricosmetics, nutraceuticals and cosmeceuticals. *Journal of Cosmetic Dermatology, 11*(1), 51–54.

Aoki, T., Akashi, T., & Ayabe, S.-I. (2000). Flavonoids of leguminous plants: Structure, biological activity and biosynthesis. *Journal of Plant Research, 113*(4), 475.

Arzani, A., & Ashraf, M. (2017). Cultivated ancient wheats (Triticum spp.): A potential source of health-beneficial food products. *Comprehensive Reviews in Food Science and Food Safety, 16*(3), 477–488.

Badr, A., Rabey, H. E., Effgen, S., Ibrahim, H., Pozzi, C., Rohde, W., & Salamini, F. (2000). On the origin and domestication history of barley (Hordeum vulgare). *Molecular Biology and Evolution, 17*(4), 499–510.

Beecher, G. R. (1998). Nutrient content of tomatoes and tomato products. *Proceedings of the Society for Experimental Biology and Medicine, 218*(2), 98–100.

Blanca Hernandez, C.-C. H. O. D. B. L. (2013). Seed components in cancer prevention. In *Nuts and seeds in health and disease prevention* (pp. 101–109). Academic Press, Elsevier.

Bratt, K., Sunnerheim, K., Bryngelsson, S., Fagerlund, A., Engman, L., Andersson, R. E., & Dimberg, L. H. (2003). Avenanthramides in oats (Avena sativa L.) and structure– antioxidant activity relationships. *Journal of Agricultural and Food Chemistry, 51*(3), 594–600.

Britton, G. (1995). Structure and properties of carotenoids in relation to function. *The FASEB Journal, 9*(15), 1551–1558.

Burak, M., & Imen, Y. (1999). Flavonoids and their antioxidant properties. *Turkiye klin tip bil derg, 19*, 296–304.

Castañeda-Ovando, A., de Lourdes Pacheco-Hernández, M., Páez-Hernández, M. E., Rodríguez, J. A., & Galán-Vidal, C. A. (2009). Chemical studies of anthocyanins: A review. *Food Chemistry, 113*(4), 859–871.

Chen, Y., Ross, A. B., Åman, P., & Kamal-Eldin, A. (2004). Alkylresorcinols as markers of whole grain wheat and rye in cereal products. *Journal of Agricultural and Food Chemistry, 52*(26), 8242–8246.

Collins, F. W. (1989). Oat phenolics: Avenanthramides, novel substituted N-cinnamoylanthranilate alkaloids from oat groats and hulls. *Journal of Agricultural and Food Chemistry, 37*(1), 60–66.

Dai, J., & Mumper, R. J. (2010). Plant phenolics: Extraction, analysis and their antioxidant and anticancer properties. *Molecules, 15*(10), 7313–7352.

Dillard, C. J., & German, J. B. (2000). Phytochemicals: Nutraceuticals and human health. *Journal of the Science of Food and Agriculture, 80*(12), 1744–1756.

Dixon, R. A., & Ferreira, D. (2002). Genistein. *Phytochemistry, 60*(3), 205–211.

Duodu, K. G. (2011). Effects of processing on antioxidant phenolics of cereal and legume grains. In *Advances in cereal science: Implications to food processing and health promotion* (pp. 31–54). ACS Publications.

Durazzo, A., Turfani, V., Azzini, E., Maiani, G., & Carcea, M. (2013). Phenols, lignans and antioxidant properties of legume and sweet chestnut flours. *Food Chemistry, 140*(4), 666–671.

Dykes, L., & Rooney, L. W. (2006). Sorghum and millet phenols and antioxidants. *Journal of Cereal Science, 44*(3), 236–251.

Dykes, L., & Rooney, L. W. (2007). Phenolic compounds in cereal grains and their health benefits. *Cereal Foods World, 52*(3), 105–111.

Elliott, R. (2005). Mechanisms of genomic and non-genomic actions of carotenoids. *Biochimica et Biophysica Acta (BBA)-Molecular Basis of Disease, 1740*(2), 147–154.

Emmons, C. L., & Peterson, D. M. (1999). Antioxidant activity and phenolic contents of oat groats and hulls. *Cereal chemistry, 76*(6), 902–906.

Fardet, A., Rock, E., & Rémésy, C. (2008). Is the in vitro antioxidant potential of whole-grain cereals and cereal products well reflected in vivo? *Journal of Cereal Science, 48*(2), 258–276.

Fernandez-Panchon, M., Villano, D., Troncoso, A., & Garcia-Parrilla, M. (2008). Antioxidant activity of phenolic compounds: From in vitro results to in vivo evidence. *Critical Reviews in Food Science and Nutrition, 48*(7), 649–671.

Firdous, A. P., Sindhu, E. R., & Kuttan, R. (2011). Hepato-protective potential of carotenoid meso-zeaxanthin against paracetamol, CCl4 and ethanol induced toxicity. *Indian Journal of Experimental Biology, 49*(1), 44–49.

Fratianni, F., Ombra, M. N., Cozzolino, A., Riccardi, R., Spigno, P., Tremonte, P., . . . Nazzaro, F. (2016). Phenolic constituents, antioxidant, antimicrobial and anti-proliferative activities of different endemic Italian varieties of garlic (Allium sativum L.). *Journal of Functional Foods, 21*, 240–248.

Fu, J., Soroka, D. N., Zhu, Y., & Sang, S. (2018). Induction of apoptosis and cell-cycle arrest in human colon-cancer cells by whole-grain alkylresorcinols via activation of the p53 pathway. *Journal of Agricultural and Food Chemistry, 66*(45), 11935–11942.

Gani, A., Wani, S., Masoodi, F., & Hameed, G. (2012). Whole-grain cereal bioactive compounds and their health benefits: A review. *Journal of Food Process Technology, 3*(3), 146–156.

Gsiorowski, K., Szyba, K., Brokos, B., & Kozubek, A. (1996). Antimutagenic activity of alkylresorcinols from cereal grains. *Cancer Letters, 106*(1), 109–115.

Goodwin, T. (1980). Biosynthesis of carotenoids. In *The biochemistry of the carotenoids* (pp. 33–76). Springer.

Goupy, P., Hugues, M., Boivin, P., & Amiot, M. J. (1999). Antioxidant composition and activity of barley (Hordeum vulgare) and malt extracts and of isolated phenolic compounds. *Journal of the Science of Food and Agriculture, 79*(12), 1625–1634.

Griesbach, R. (2010). Biochemistry and genetics of flower color. *Plant Breeding Reviews, 25*, 89–114.

Gubernator, J., Stasiuk, M., & Kozubek, A. (1999). Dual effect of alkylresorcinols, natural amphiphilic compounds, upon liposomal permeability. *Biochimica et Biophysica Acta (BBA)-Biomembranes, 1418*(2), 253–260.

Harborne, J. B., & Williams, C. A. (2000). Advances in flavonoid research since 1992. *Phytochemistry, 55*(6), 481–504.

Harlan, J. R. (1992). Indigenous African agriculture. In *Crops & man* (pp. 175–191). American Society of Agronomy.

Havsteen, B. H. (2002). The biochemistry and medical significance of the flavonoids. *Pharmacology & Therapeutics, 96*(2–3), 67–202.

Hoda, M., Hemaiswarya, S., & Doble, M. (2019). Role of phenolic phytochemicals in diabetes management. *Role of Phenolic Phytochemicals in Diabetes Management*, 123–143.

Idehen, E., Tang, Y., & Sang, S. (2017). Bioactive phytochemicals in barley. *Journal of Food and Drug Analysis, 25*(1), 148–161.

Iwashina, T. (2013). Flavonoid properties of five families newly incorporated into the order caryophyllales. *Bulletin of the National Museum of Nature and Science, 39*, 25–51.

Jayashri, P., Sharma, S. G., Sharma, M., & Guleria, P. (2019). Influence of naturally occurring phytochemicals on oral health. *Research Journal of Pharmacy and Technology, 12*(8), 3979–3983.

Jiao, Y., Reuss, L., & Wang, Y. (2019). β-Cryptoxanthin: Chemistry, occurrence, and potential health benefits. *Current Pharmacology Reports, 5*(1), 20–34.

John, S., Sorokin, A. V., & Thompson, P. D. (2007). Phytosterols and vascular disease. *Current Opinion in Lipidology, 18*(1), 35–40.

Jonnalagadda, S. S., Harnack, L., Hai Liu, R., McKeown, N., Seal, C., Liu, S., & Fahey, G. C. (2011). Putting the whole grain puzzle together: Health benefits associated with whole grains—summary of American society for nutrition 2010 satellite symposium. *The Journal of Nutrition, 141*(5), 1011–1022.

Kaulmann, A., & Bohn, T. (2014). Carotenoids, inflammation, and oxidative stress—implications of cellular signaling pathways and relation to chronic disease prevention. *Nutrition Research, 34*(11), 907–929.

Kiokias, S., & Gordon, M. H. (2004). Antioxidant properties of carotenoids in vitro and in vivo. *Food Reviews International, 20*(2), 99–121.

Kozubek, A. (1987). The effect of 5-(n-alk(en)yl) resorcinols on membranes. I. Characterization of the permeability increase induced by 5-(n-heptadecenyl) resorcinol. *Acta Biochimica Polonica, 34*(4), 357–367.

Kozubek, A., & Demel, R. (1980). Permeability changes of erythrocytes and liposomes by 5-(n-alk (en) yl) resorcinols from rye. *Biochimica et Biophysica Acta (BBA)-Biomembranes, 603*(2), 220–227.

Kozubek, A., & Tyman, J. H. (1999). Resorcinolic lipids, the natural non-isoprenoid phenolic amphiphiles and their biological activity. *Chemical Reviews, 99*, 1–26.

Kris-Etherton, P. M., Hecker, K. D., Bonanome, A., Coval, S. M., Binkoski, A. E., Hilpert, K. F., . . . Etherton, T. D. (2002). Bioactive compounds in foods: Their role in the prevention of cardiovascular disease and cancer. *The American Journal of Medicine, 113*(9), 71–88.

Kruk, J., Aboul-Enein, B., Bernstein, J., & Marchlewicz, M. (2017). Dietary alkylresorcinols and cancer prevention: A systematic review. *European Food Research and Technology, 243*(10), 1693–1710.

Lam, L., Zhang, J., & Hasegawa, S. (1994). Citrus limonoid reduction of chemically induced tumorigenesis. *Food Technology (Chicago), 48*(11), 104–108.

Lee, Y. K., Yuk, D. Y., Lee, J. W., Lee, S. Y., Ha, T. Y., Oh, K. W., . . . Hong, J. T. (2009). (−)-Epigallocatechin-3-gallate prevents lipopolysaccharide-induced elevation of beta-amyloid generation and memory deficiency. *Brain Research, 1250*, 164–174.

Liu, R. H. (2007). Whole grain phytochemicals and health. *Journal of Cereal Science, 46*(3), 207–219.

Liu, R. H. (2013). Dietary bioactive compounds and their health implications. *Journal of Food Science, 78*(s1), A18–A25.

López-Alarcón, C., & Denicola, A. (2013). Evaluating the antioxidant capacity of natural products: A review on chemical and cellular-based assays. *Analytica Chimica Acta, 763*, 1–10.

Machado, E. C. F. A., Ambrosano, L., Lage, R., Abdalla, B. M. Z., & Costa, A. (2017). Nutraceuticals for healthy skin aging. In *Nutrition and functional foods for healthy aging* (pp. 273–281). Elsevier.

Manach, C., Scalbert, A., Morand, C., Rémésy, C., & Jiménez, L. (2004). Polyphenols: Food sources and bioavailability. *The American Journal of Clinical Nutrition, 79*(5), 727–747.

Martín-Sánchez, A. M., Cherif, S., Ben-Abda, J., Barber-Vallés, X., Pérez-Álvarez, J. Á., & Sayas-Barberá, E. (2014). Phytochemicals in date co-products and their antioxidant activity. *Food Chemistry, 158*, 513–520.

Masisi, K., Beta, T., & Moghadasian, M. H. (2016). Antioxidant properties of diverse cereal grains: A review on in vitro and in vivo studies. *Food Chemistry, 196*, 90–97.

Matthies, A., Clavel, T., Gütschow, M., Engst, W., Haller, D., Blaut, M., & Braune, A. (2008). Conversion of daidzein and genistein by an anaerobic bacterium newly isolated from the mouse intestine. *Applied and Environmental Microbiology, 74*(15), 4847–4852.

Mattila, P., Pihlava, J.-M., & Hellström, J. (2005). Contents of phenolic acids, alkyl-and alkenylresorcinols, and avenanthramides in commercial grain products. *Journal of Agricultural and Food Chemistry, 53*(21), 8290–8295.

Mellado, E., & Hornero, D. (2012). Isolation and identification of lutein esters, including their regioisomers, in tritordeum (× Tritordeum Ascherson et Graebner) grains: Evidence for a preferential xanthophyll acyltransferase activity. *Food Chemistry, 135*(3), 1344–1352.

Mercadante, A. Z., Rodrigues, D. B., Petry, F. C., & Mariutti, L. R. B. (2017). Carotenoid esters in foods-A review and practical directions on analysis and occurrence. *Food Research International, 99*, 830–850.

Metodiewa, D., Kochman, A., & Karolczak, S. (1997). Evidence for antiradical and antioxidant properties of four biologically active N, N-Diethylaminoethyl ethers of flavaone oximes: A comparison with natural polyphenolic flavonoid rutin action. *IUBMB Life, 41*(5), 1067–1075.

Meydani, M. (2009). Potential health benefits of avenanthramides of oats. *Nutrition Reviews, 67*(12), 731–735.

Mutoh, M., Takahashi, M., Fukuda, K., Matsushima-Hibiya, Y., Mutoh, H., Sugimura, T., & Wakabayashi, K. (2000). Suppression of cyclooxygenase-2 promoter-dependent transcriptional activity in colon cancer cells by chemopreventive agents with a resorcin-type structure. *Carcinogenesis, 21*(5), 959–963.

Ncube, N., Afolayan, A., & Okoh, A. (2008). Assessment techniques of antimicrobial properties of natural compounds of plant origin: Current methods and future trends. *African Journal of Biotechnology, 7*(12).

Ndolo, V. U., & Beta, T. (2013). Distribution of carotenoids in endosperm, germ, and aleurone fractions of cereal grain kernels. *Food Chemistry, 139*(1–4), 663–671.

Nishino, H., Murakoshi, M., Tokuda, H., & Satomi, Y. (2009). Cancer prevention by carotenoids. *Archives of Biochemistry and Biophysics, 483*(2), 165–168.

Niu, M., & Hou, G. G. (2020). Whole grain noodles. In *Asian noodle manufacturing* (pp. 95–123). Elsevier.

Okarter, N., & Liu, R. H. (2010). Health benefits of whole grain phytochemicals. *Critical Reviews in Food Science and Nutrition, 50*(3), 193–208.

Ozaki, Y., Ayano, S., Inaba, N., Miyake, M., Berhow, M. A., & Hasegawa, S. (1995). Limonoid glucosides in fruit, juice and processing by-products of Satsuma Mandarin (Chus unshiu Marcov.). *Journal of Food Science, 60*(1), 186–189.

Paiva, S. A., & Russell, R. M. (1999). β-carotene and other carotenoids as antioxidants. *Journal of the American College of Nutrition, 18*(5), 426–433.

Panche, A. N., Diwan, A. D., & Chandra, S. R. (2016). Flavonoids: An overview. *Journal of Nutritional Science, 5*.

Peterson, D. M. (2001). Oat antioxidants. *Journal of Cereal Science, 33*(2), 115–129.

Pihlava, J.-M., Hellström, J., Kurtelius, T., & Mattila, P. (2018). Flavonoids, anthocyanins, phenolamides, benzoxazinoids, lignans and alkylresorcinols in rye (Secale cereale) and some rye products. *Journal of Cereal Science, 79*, 183–192.

Prakash, D., & Sharma, G. (2014). *Phytochemicals of nutraceutical importance.* CABI.

Prasad, K., & Jadhav, A. (2016). Prevention and treatment of atherosclerosis with flaxseed-derived compound secoisolariciresinol diglucoside. *Current Pharmaceutical Design, 22*(2), 214–220.

Ragaee, S., Gamel, T., Seethraman, K., & Abdel-Aal, E. S. M. (2013). Food grains. *Handbook of Plant Food Phytochemicals: Sources, Stability and Extraction*, 138–162.

Rehmanji, M., Suresh, S., Nesamma, A. A., & Jutur, P. P. (2021). Microalgal cell factories, a platform for high-value-added biorenewables to improve the economics of the biorefinery. In *Microbial and natural macromolecules* (pp. 689–731). Elsevier.

Rhee, Y. (2016). Flaxseed secoisolariciresinol diglucoside and enterolactone down-regulated epigenetic modification associated gene expression in murine adipocytes. *Journal of Functional Foods, 23*, 523–531.

Ribaya-Mercado, J. D., & Blumberg, J. B. (2004). Lutein and zeaxanthin and their potential roles in disease prevention. *Journal of the American Aollege of Nutrition, 23*(sup6), 567S–587S.

Rodríguez-García, C., Sánchez-Quesada, C., Toledo, E., Delgado-Rodríguez, M., & Gaforio, J. J. (2019). Naturally lignan-rich foods: A dietary tool for health promotion? *Molecules, 24*(5), 917.

Ross, A. B., Kamal-Eldin, A., & Åman, P. (2004). Dietary alkylresorcinols: Absorption, bioactivities, and possible use as biomarkers of whole-grain wheat–and rye–rich foods. *Nutrition Reviews, 62*(3), 81–95.

Ross, A. B., Kamal-Eldin, A., Jung, C., Shepherd, M. J., & Åman, P. (2001). Gas chromatographic analysis of alkylresorcinols in rye (Secale cereale L) grains. *Journal of the Science of Food and Agriculture, 81*(14), 1405–1411.

Ross, A. B., Kamal-Eldin, A., Lundin, E. A., Zhang, J.-X., Hallmans, G. r., & Åman, P. (2003). Cereal alkylresorcinols are absorbed by humans. *The Journal of Nutrition, 133*(7), 2222–2224.

Rouf Shah, T., Prasad, K., & Kumar, P. (2016). Maize—a potential source of human nutrition and health: A review. *Cogent Food & Agriculture, 2*(1), 1166995.

Sa, T. (2013). Carotenoids: Health effects. In B. Caballero (Ed.), *Encyclopedia of human nutrition* (3rd ed., pp. 292–297). Academic Press.

Saarinen, N., Mäkelä, S., & Santti, R. (2013). *From the institute of biomedicine, department of anatomy.* Paper presented at the Animal Cell Technology: Basic & Applied Aspects.

Sarungallo, Z. L., Hariyadi, P., Andarwulan, N., Purnomo, E. H., & Wada, M. (2015). Analysis of α-cryptoxanthin, β-cryptoxanthin, α-carotene, and β-carotene of Pandanus conoideus oil by high-performance liquid chromatography (HPLC). *Procedia Food Science, 3*, 231–243.

Scalbert, A., Manach, C., Morand, C., Rémésy, C., & Jiménez, L. (2005). Dietary polyphenols and the prevention of diseases. *Critical Reviews in Food Science and Nutrition, 45*(4), 287–306.

Scarmo, S., Cartmel, B., Lin, H., Leffell, D. J., Welch, E., Bhosale, P., . . . Mayne, S. T. (2010). Significant correlations of dermal total carotenoids and dermal lycopene with their respective plasma levels in healthy adults. *Archives of Biochemistry and Biophysics, 504*(1), 34–39.

Shahidi, F., & Naczk, M. (1995). *Foods phenolics: Sources, chemistry, effects, application*. Technomic Publishing.

Shahidi, F., Varatharajan, V., Oh, W. Y., & Peng, H. (2019). Phenolic compounds in agri-food by-products, their bioavailability and health effects. *Journal of Food Bioactives, 5*(1), 57–119.

Siddiqui, A. I., Amjad, F., Majeed, I., & Shah, N. (2019). Polyphenols: The scavengers. *Advanced Food and Nutritional Sciences, 4*, 99–100.

Singh, M. K., Singh, S. K., Sighn, A. V., Verma, H., Sighn, P. P., & Kumar, A. V. (2020). Phytochemicals: Intellectual property rights. In B. Prakash (Ed.), *Functional and preservative properties of phytochemicals* (pp. 363–375). Academic Press.

Slavin, J. (2003). Why whole grains are protective: Biological mechanisms. *Proceedings of the Nutrition Society, 62*(1), 129–134.

Smýkal, P., Coyne, C. J., Ambrose, M. J., Maxted, N., Schaefer, H., Blair, M. W., . . . Besharat, N. (2015). Legume crops phylogeny and genetic diversity for science and breeding. *Critical Reviews in Plant Sciences, 34*(1–3), 43–104.

Sofi, S., Nazir, A., & Ashraf, U. (2019). Cereal bioactive compounds: A review. *International Journal of Agriculture, Environment and Biotechnology, 12*(2), 107–113.

Stahl, W., & Sies, H. (1996). Lycopene: A biologically important carotenoid for humans? *Archives of Biochemistry and Biophysics, 336*(1), 1–9.

Stahl, W., & Sies, H. (2005). Bioactivity and protective effects of natural carotenoids. *Biochimica et Biophysica Acta (BBA)-Molecular Basis of Disease, 1740*(2), 101–107.

Stoskopf, N. C. (1985). *Cereal grain crops*. Reston Publishing Company Inc.

Thakur, M., Singh, K., & Khedkar, R. (2020). Phytochemicals: Extraction process, safety assessment, toxicological evaluations, and regulatory issues. In *Functional and preservative properties of phytochemicals* (pp. 341–361). Elsevier.

Thompson, L. U. (1994). Antioxidants and hormone-mediated health benefits of whole grains. *Critical Reviews in Food Science & Nutrition, 34*(5–6), 473–497.

Trono, D. (2019). Carotenoids in cereal food crops: Composition and retention throughout grain storage and food processing. *Plants, 8*(12), 551.

Van Hung, P. (2016). Phenolic compounds of cereals and their antioxidant capacity. *Critical Reviews in Food Science and Nutrition, 56*(1), 25–35.

Vucenik, I., & Shamsuddin, A. M. (2006). Protection against cancer by dietary IP6 and inositol. *Nutrition and Cancer, 55*(2), 109–125.

Walchuk, C., & Suh, M. (2020). Nutrition and the aging retina: A comprehensive review of the relationship between nutrients and their role in age-related macular degeneration and retina disease prevention. *Advances in Food and Nutrition Research, 93*, 293–332.

Walker, E. H., Pacold, M. E., Perisic, O., Stephens, L., Hawkins, P. T., Wymann, M. P., & Williams, R. L. (2000). Structural determinants of phosphoinositide 3-kinase inhibition by wortmannin, LY294002, quercetin, myricetin, and staurosporine. *Molecular Cell, 6*(4), 909–919.

Wan, H., Liu, D., Yu, X., Sun, H., & Li, Y. (2015). A Caco-2 cell-based quantitative antioxidant activity assay for antioxidants. *Food Chemistry, 175*, 601–608.

Willcox, J. K., Ash, S. L., & Catignani, G. L. (2004). Antioxidants and prevention of chronic disease. *Critical Reviews in Food Science and Nutrition, 44*(4), 275–295.

Wolfe, K. L., & Liu, R. H. (2008). Structure– activity relationships of flavonoids in the cellular antioxidant activity assay. *Journal of Agricultural and Food Chemistry, 56*(18), 8404–8411.

Woodside, J. V., McGrath, A. J., Lyner, N., & McKinley, M. C. (2015). Carotenoids and health in older people. *Maturitas, 80*(1), 63–68.

Wu, N.-N., Li, H.-H., Tan, B., Zhang, M., Xiao, Z.-G., Tian, X.-H., . . . Wang, L.-P. (2018). Free and bound phenolic profiles of the bran from different rice varieties and their antioxidant activity and inhibitory effects on α-amylose and α-glucosidase. *Journal of Cereal Science, 82,* 206–212.

Xiong, Y., Zhang, P., Warner, R. D., & Fang, Z. (2019a). 3-Deoxyanthocyanidin colorant: Nature, health, synthesis, and food applications. *Comprehensive Reviews in Food Science and Food Safety, 18*(5), 1533–1549.

Xiong, Y., Zhang, P., Warner, R. D., & Fang, Z. (2019b). Sorghum grain: From genotype, nutrition, and phenolic profile to its health benefits and food applications. *Comprehensive Reviews in Food Science and Food Safety, 18*(6), 2025–2046.

Xiong, Y., Zhang, P., Warner, R. D., Shen, S., & Fang, Z. (2022). Cereal grain-based functional beverages: From cereal grain bioactive phytochemicals to beverage processing technologies, health benefits and product features. *Critical Reviews in Food Science and Nutrition, 62*(9), 2404–2431.

Zanella, I., Biasiotto, G., Holm, F., & di Lorenzo, D. (2017). Cereal lignans, natural compounds of interest for human health? *Natural Product Communications, 12*(1), 1934578X1701200139.

Zeece, M. (2020). *Introduction to the chemistry of food.* Academic Press.

CHAPTER 10

Processing of Cereals

Tanuva Das, Rahul Thakur, Subhamoy Dhua, Barbara Elisabeth Teixeira-Costa, Menithen Beber Rodrigues, Maristela Martins Pereira, Poonam Mishra and Arun Kumar Gupta

CONTENTS

DOI: 10.1201/9781003252023-10

10.1 INTRODUCTION

Cereals and their products have become a staple in many people's diets around the world. Cereals or edible grains belong to the family *Poaceae* or *Gramineae*. Cereals are the major category in this grass family, with over 10,000 species, and they are widely consumed across the world. Rice, wheat, maize (corn), and sorghum occupy enormous harvesting areas compared to different cereal crops, according to the Food and Agriculture Organization of the United Nations (FAO) Statistics Database (FAOSTAT) on agricultural statistics (FAO, 2020). As a result of their production and consumption levels, they are classed as major cereals rather than minor cereals. Barley and wheat, for example, are significant cereals; on the other hand, einkorn, rye, spelt, and oats are considered minor cereals, per Healthy Minor Cereals (2016).

According to the FAOSTAT statistics, overall cereal grain production increased by 48.4% between 2018 and 2000 for rice (23%), maize (80%), wheat (20%), sorghum (6%), and barley (6%). According to the OCEDFAO Agricultural Outlook (2018–2027), worldwide grain production (such as rice, maize, and wheat) is going to increase by 17.6 million hectares from 2017 to 2027. As a result, production of cereal is critical for nourishing the world's expanding population (estimated to be 6–8.3 billion by 2030), with an expected global consumption rise from 2.6 to 2.9 billion ton (OECD-FAO, 2018). Cereal meals supply a lot of energy in a diet because of their high starch content, and cereals also have other key nutrients such as proteins, dietary fiber, and non-starch carbs, as well as minor nutrients (Gupta et al., 2021; Jha et al., 2021). Basically, cereal grains are consumed in a variety of ways around the world (for example, bakery items and bread, extruded snacks, porridges, cookies, breakfast cereals, and so on) and are partially or totally processed. Cereals, on the other hand, require proper post-harvest management, as well as secondary and/or primary processing, in order to generate good final products. In this era, cereal and cereal product researchers are concentrating their efforts on combining modern processing technologies with conventional approaches to produce healthy and helpful goods based on cereals. Cereal researchers are pursuing a drift of producing end goods that include additional nutrients, have high purposeful characteristics, and have fewer allergens while also increasing the safety of the product through processing processes. This chapter covers cereal properties, the structure of grain and its composition, applications, and various processing methods, including a focus on new techniques for processing.

10.2 TYPES AND PROPERTIES OF CEREALS

Bambusoideae, *Festucoideae*, *Panicoideae*, and *Chloridoideae* were the subfamilies investigated by Shewry and Tatham (1999) (Figure 10.1). Many cereals, like wheat, rye, and barley, are members of the *Pooideae* (also known as *Festucoideae*) subfamily, while oats are members of the *Aveneae* tribe. Rice comes under the subfamily *Bambusoideae*, whereas lesser grains like finger millet (ragi) and teff belong to the *Chloridoideae* subfamily (Shewry & Tatham, 1999).

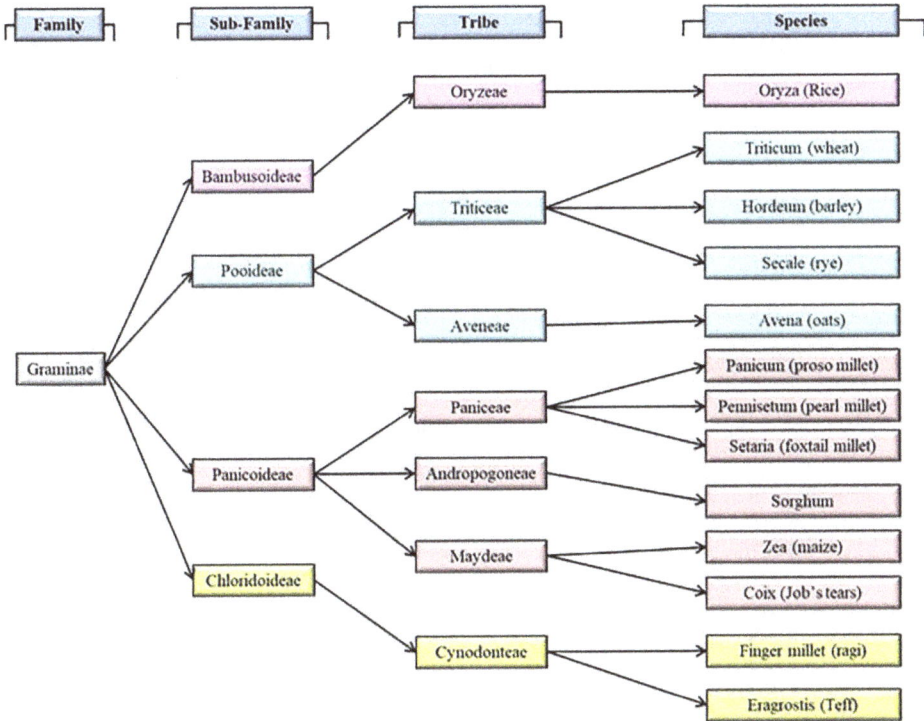

Figure 10.1 Taxonomy relationships of cereals.

The characteristics of cereals can be defined in terms of inflorescences, root system (primary and secondary), stems, leaves, and structure of kernel, but the structure of the kernel is the most important characteristics in determining the processing mode.

10.2.1 Kernel Structures

Seeds originate from the ovule after fertilization and contain the embryo surrounded by outer surfaces such as husk, seed coat (pericarp and testa), layers of aleurone, and endosperm. While wheat, rice, barley, oats, and other monocotyledon grains have seeds with only a single cotyledon or embryonic seed leaf (Hoseney, 1994). The endosperm, which is made up of cells packed with nutrients to keep the embryo alive throughout the process of germination, is the most nutritionally important kernel component. Cereals are well known as a key origin of carbs, protein, several vitamins, phytochemicals, and minerals, all of which help humans meet their energy demands and provide health benefits (Goldberg, 2003). Cereals, in particular, are mostly composed of polysaccharides—principally 56 to 74% starch found in the endosperm and fiber, mainly cellulose, arabinoxylans, and glucans found in the layer of bran with protein extending from 8 to 12% (Koehler & Wieser, 2013). The major nutrients found in cereals are summarized in Table 10.1. The simple anatomy of a kernel is the same in all grains (Alldrick, 2017). Figure 10.1 depicts the architecture of wheat, millet, and rice kernels.

10.2.1.1 Wheat

The embryo, endosperm, and bran (seed coat) are the three primary anatomical sections of the wheat kernel (germ). In general, the germ makes up 2 to 3% of the kernel, the bran 13 to 17%, and the starchy endosperm 83 to 85% of the weight (Pomeranz, 1982). The aleurone layer is the endosperm's outer covering, which is normally linked to the outer coat and ends up in tail end

Table 10.1   Composition of Cereals

Cereals	Crude Protein (%)	Crude Fat (%)	Ash (%)	Crude Fiber (%)	Digestive CHO (%)	Starch (%)	Total Dietary Fiber (%)	Total Phenolics (mg/100 g)
Wheat	10.6	1.9	1.4	1	69.7	64	12.1	20.5
Maize	9.8	4.9	1.4	2	63.6	62.3	12.8	2.91
Brown rice/paddy	7.3	2.2	1.4	0.8	64.3	77.2	3.7	2.51
Barley	11	3.4	1.9	3.7	55.8	58.5	15.4	16.4
Sorghum	8.3	3.9	2.6	4.1	62.9	73.8	11.8	43.1
Pearl millet	11.5	4.7	1.5	1.5	63.4	60.5	7	51.4
Oats	9.3	5.9	2.3	2.3	62.9	52.8	15.4	16.4
Rye	8.7	1.5	1.8	2.2	71.8	68.3	16.1	13.2

Table 10.2   Some Food Applications of Brewer's Spent Grain (BSG) By-Products

Application	Food Product	Amount of BSG or DG	Reference
Bakery	Breadsticks	15%, 25%, and 35% of BSG	Ktenioudaki et al. (2012)
	Dough	15%, 25%, and 35% of BSG	Ktenioudaki et al. (2013)
	Cereal-based snacks	20–25% of BSG	Reis and Abu-Ghannam (2014)
	Breads	15% of BSG (flour basis)	Ktenioudaki et al. (2015)
	Corn extruded snacks	0/100, 5/95, 10/90, and 15/85 of corn grits: BSG	Ačkar et al. (2018)
	Cookies	0, 10, 20, and 30% of BSG	Heredia-Sandoval et al. (2020)
	Muffins	5, 10, and 15% of BSG	Cermeño et al. (2021)
Beverages	Fruit drinks	5% of BSG (aqueous extract)	McCarthy et al. (2013)
	Fermented drinks	0.05–0.25 solid liquid (SL) ratio of BSG	Gupta, Jaiswal, & Abu-Ghannam (2013)
	Lactic acid fermented drinks	14% concentration wort from 72% BSG (unhopped Bavarian pilsner barley malt extract)	Nsogning Dongmo et al. (2017)
Pasta	Fiber-enriched fresh egg pasta	3–25% of BSG	Cappa and Alamprese (2017)
	Semolina dry pasta	5, 10, and 20% of BSG	Nocente et al. (2019)
	Durum wheat pasta	5 and 10% of BSG	Nocente et al. (2021)
	Semolina pasta	60:40 BSG: water ratio	Schettino et al. (2021)
	Pasta	85–98% of BSG	Neylon et al. (2021)

breakdown and reduction flour mill streams or connected to bran particles after grinding and sifting in an industrial roller mill (Pojić et al., 2014). Before manufacturing of refined flour, these layers are separated from the wheat kernel via milling. In general, the inner bran layer has more minerals, fat, and protein, while the external bran layer contains more hemicelluloses and cellulose. Wheat germs are high in vitamins E and B, as well as lysine, unsaturated fatty acids, and minerals.

10.2.1.2 Rice

Generally, rice or paddy is protected by a husk or hull, which accounts for 18 to 28% of the weight of grain, and a caryopsis, which is also known as brown rice, which accounts for 72 to 82% of the weight of grain. The caryopsis is made up of four layers: the starchy endosperm (89–91%), the germ or embryo (2–3%), the aleurone layer or bran (5%), and the outer pericarp layer (1–2%). The aleurone layer, which surrounds the embryo, is made up of one cell to five cell layers, with the surface-shaped dorsal wider than the ventral surface. Short-grain rice has thicker aleurone layers than long-grain rice, and the whitest part of the rice caryopsis is the starchy endosperm (Juliano & Tuao, 2019). Dehusking (removal of the husk) and polishing (separation of the bran) remove the external layer of rice, resulting in the edible endosperm for consumption by humans. The procedure of dehusking also eliminates the various covers of rice, removing fat, carbs, protein, and fiber and so affecting the nutritious content of rice (Fernando, 2013). The further rice grains are polished, the more lipids, thiamine, proteins, and other vitamin-rich components are extracted.

10.2.1.3 Millet

The pericarp, which is made up of 15–21% germ and 70–76% endosperm of millet kernels makes up around 7–10% of the total. Finger millet, pearl millet, proso millet, and foxtail millet are the four principal millet species. The pearl millet's pericarp is firmly connected to the seed (caryopsis), but the pericarps of foxtail, finger, and proso millets are attached to a single point of the seed. Millet has two types of endosperms: the inner floury endosperm and the outer hard endosperm, with the germ accounting for up to a third of the pearl millet caryopsis. The germ and endosperm ratio in millet are roughly 1:4.5, which means the germ makes up about 20% of the overall kernel weight (FAO, 1995). The endosperm contains 60% of the protein content in the pearl millet grain, followed by 31% in the germ and 9% in the pericarp. Pearl millet has a protein level of 8–23%, while proso millet has a protein content of 11–13% (Serna-Saldivar & Rooney, 1995; Lestienne et al., 2007).

10.2.1.4 Barley

The caryopsis is covered by the husk or hull in barley kernels, which are spindle in shape. The hull or husk makes up about 10–13% of the dry weight of the kernel; however, this might change according on the dehulling procedure, which can eliminate up to 20% of the weight. Glucan (70%) makes up the majority of the endosperm cell walls (Tiwari & Cummins, 2009). Depending on the cultivar, the aleurone layer has two or three layers of cells. The pericarp, germ or embryo, seed coat, and starchy endosperm make up the caryopsis, which accounts for 80% of the total weight of the grain. The barley embryo is usually found on the dorsal side of the caryopsis toward the end of the caryopsis. The hull of "hull-less" barley is weakly connected and slips off during harvesting and threshing (separating grains from chaff) (Evers & Millar, 2002).

10.2.1.5 Rye

The rachis is covered by a lemma, a palea, and a glume, and the rye grains are organized in a zigzag pattern. When the grains reach maturity, they easily come off during threshing. The grains are typically greyish yellow in hue and have a shrivelled, rough surface. The starchy endosperm makes up 86.5% of the rye kernel, followed by the bran (10%) and germ (3.5%). The germ and bran of the rye kernel are isolated from the endosperm and ground into flour during the milling operation (Bushuk, 2004).

10.2.1.6 Sorghum

The pericarp, germ, and endosperm are the three major anatomical components of the typical sorghum kernel, accounting for 6, 10, and 84% of its weight, respectively. These percentages, however, change among sorghum cultivars. The endosperm, which makes up the majority of the kernel, has low oil content and minerals. The endosperm is primarily responsible for the kernel's starch (94%), protein (80%), and B-complex vitamin (50–75%) contents, but the germ is responsible for 75% of the oil, 68% of the minerals, and 15% of the protein (FAO, 1995). As a result of processing, the outer pericarp is removed; the relative protein level rises; and the lipid, mineral content, and cellulose of the grain decrease. Alvarenga et al. (2018), for example, illustrated the impacts of milling sorghum into different fractions to provide high-protein animal feed. They discovered that milling fractions had a greater crude protein content (13.4%) than flour (9.68%), highlighting the potential advantages of using the milling fraction for human and animal feed.

10.2.1.7 Oats

The bran, endosperm, and germ make up the caryopses (groats or kernels) of oats, which are comparable to those of barley and wheat. The caryopsis makes up 65–75% of the total kernel, whereas the hull makes up 25–35%. The oat germ is placed on the dorsal side of the caryopsis, where it is partially hidden by the lemma, which consists of two to three plumule leaf shoots and two to three radicle rudimentary roots (Welch, 2012). The bran is made up of tissue layers and aleurone cells found on the groat's exterior layers, whereas the endosperm (55–80%) is made up of lipids, protein, starch, and the highest glucan concentration found inside the groat's wall layers (Tiwari & Cummins, 2009).

10.2.1.8 Maize

The pericarp; embryo or germ; endosperm; bran or hull; and tip cap, a conical arrangement of dead tissue where the kernel joins the cob, make up the maize kernel. The maize kernel contains a bigger germ than other cereals, which is located in the bottom endosperm. The endosperm makes up around 70–86% of the kernel and is mostly made up of protein (8%) and carbohydrates (87.6%). Maize germ has significant quantities of starch (6–21%), protein (12–21%), and lipids (18–41%), but it is also high in tocotrienols, tocopherols, unsaturated fatty acids, and carotenoids, ranging from 7 to 22% of the kernel (Navarro et al., 2016). In monogastric animals that are unable to consume a substantial quantity of minerals, the maize kernel also includes phytate, which acts as an endogenous poisonous chemical and antinutritive component (Humer et al., 2015). The kernel of the corn is flattened, wedge shaped, and wider at the top end than it is at the site of connection of cob. Wet milling produces corn starch and a variety of byproducts such as corn protein meal, germ meal, and corn bran from maize kernels (Papageorgiou & Skendi, 2018). Dry milling produces principal goods such as snack food grits, brewers' grains, and flour.

10.3 DRY MILLING

Dry milling is perhaps the most popular way of preparing grist prior to mashing, and it is used throughout the world. Dry mill operation is quite easy with two smaller-roll mills, but as the number of rolls rises to allow a higher milling capacity and/or more diversity in the milling process, it may become progressively difficult. Two-roll mills work by moving product between two rollers, essentially crushing it to grist before the mashing process. The complexity of four- and six-roll mills increases, but they also have significant advantages. Milling systems with numerous pairs of rollers provide more control over the grain's particle size distribution. With multiple pairs of rolls working in sequence, coarse grist particles can be separated from the loosened husk material, and by keeping the husk substantially intact, the grain is ground further to generate a finer grind, boosting extract potential and supporting good wort separation in run-off.

10.3.1 Corn

In the process of dry milling, corn is cleaned to eliminate kernels that are broken and extraneous material before being conditioned to around 20–24% moisture to liberate the germ and bran from the endosperm. The degerminator, a horizontal cone-shaped drum coated with tiny steel projections that spin within a metal casing, is used to isolate the endosperm from the germ and bran in the first phase of milling. Steel projections adorn the structure, which is additionally screened with one or more perforated metallic screens. The corn moves from the small to the large end of the drum as the germ and bran are separated from the endosperm, with the smaller particles (germ and pericarp) passing through perforated screens (as through stock) and the bigger pieces being released from the drum's end (as tailstock). The larger the percentage of tailstock in the bigger particle sizes, the higher the production of high-valued goods. If the endosperm is extensively fractured before entering the degerminator, a larger proportion will pass through the screens as through stock, lowering tailstock production and, as a result, the yield of big grits.

On both the tailstock and through stock, further grinding, screening, and aspiration take place, separating the endosperm into several size ranges from big flaking grits to flour and meal, as well as separating germ and bran for further processing. Mechanical presses and solvent extraction are used to extract oil from germs. After the oil is removed, the germ cake is milled into corn germ flour or used in hominy feed.

10.3.2 Wheat

Grinding wheat into flour is the oldest continually practiced enterprise in the world, and it is the ancestor of all contemporary industry. The germ, endosperm, and bran are the three primary portions of the wheat kernel, which is usually approximately 4–10 mm in length. As a result, the endosperm is high in carbohydrates and protein, which is needed for energy and building. Roller milling of wheat kernel is an example of a dry milling process. The roller mill's flexibility stems from its ability to change the roll gap, run the rolls at various relative speeds for a different action of cutting, employ different fluting profiles and roll surfaces, and stack subsequent roll pairs in several configurations. The process is divided into two parts: break and reduction systems. To split open the wheat kernel and eliminate the endosperm material from the bran, the brake system employs fluted rolls with increasingly fine fluting. At each stage, some flour is generated, while the bigger endosperm material is transported to the reduction system for additional grinding and bran and germ separation. Each roll is approximately 1 meter long and 250 millimeters in diameter. Break rolls are fluted, whereas reduction rolls are smooth and somewhat frosted. Break rolls have an uneven sawtooth shape and work with a gap between the rolls and a difference of roughly 2.5:1. The energy consumption of the rollers is reduced when they are operated under different speeds (Haque, 1991).

10.4 WET MILLING

In industry, the wet-milling process is used to extract the major components from maize. It involves numerous physical, chemical, biological, and mechanical processes. Wet milling is presently utilized in the industry for maize and wheat, but it may also be used for other cereals like oats, barley, sorghum, or rice. Grain processing; steeping; separation; and recovery of proteins, fiber, germ, and starch were all part of the traditional wet fractionation operation. Over the last 10 years, the use of wet milling to extract components for the production of industrial ethanol has expanded considerably (Awika, 2011).

10.4.1 Corn

Corn wet-milling manufacturing in the United States dates to 1844 when Thomas Kingsford utilized a novel alkali procedure to extract starch from corn at Wm. Colgate & Company in Jersey City, New jersey (CRA, 2013). Initially, the corn business simply ignored the fiber, germ, and protein obtained during the milling process. Non-starch components, on the other hand, have begun to find wider use in a variety of sectors, resulting in changes in wet-milling technology. The traditional procedure included steps such as grain handling; steeping; separation; and recovery of proteins, fiber, germ and starch. A broad diagram of the corn wet-milling process is shown in Figure 10.1.

10.4.2 Rice

Rice flour is used in the preparation of traditional Asian dishes, fermented foods, and celiac-friendly products. Three main types of milling have been practiced for grinding to polish rice and mill it into flour: wet, semi-dry, and dry milling. Chen et al. (1999) found that wet-milling yields the finest flour compared to other milling processes. Wet-milled flour is better than dry-milled flour in the making of baked goods or Thai rice-based snacks. Soaking is the most crucial stage in wet-milling rice because it allows water to spread into the grains, allowing some components to drain out. According to Chiang and Yeh (2002), soaking changes the particle size of rice flour, and increasing the soaking duration (from 3 to 168 hours) results in finer flour with less broken starch. The authors discovered that the resultant starch fraction's small particle size had no effect on its gelatinization, resulting in a high peak viscosity and a low pasting temperature. Heo et al. (2013) compared the quality features of gluten-free dough and noodles made using wet and dry milled rice flours (at 25°C for 4 h). In comparison to wet-milled rice flour, dry-milled rice flour had more starch degradation and more water hydration.

The alkaline protease digestion (digestion period 5–30 h; protease levels of 0.5–1.5%; pH 8.5–10.0) method proposed by Lumdubwong and Seib (2000) could be used to create rice starch from wet-milled rice flour. The protease method recovered roughly 10% more starch than the alkaline (0.05 M NaOH) treatment, according to these researchers. Protease-isolated rice starch was lighter, contained more lipids, and had a poorer consistency after pasting than NaOH-isolated rice starch. On the other hand, the protease procedure required twice as much sodium hydroxide as the NaOH method.

10.4.3 Wheat

Methods for obtaining wheat starch and essential gluten can be traced back to ancient Egypt and Greece. Mostly, starch was found from wheat grains until the latter half of the 18th century. Wheat grains have been investigated as a wet-milling raw material; however, none of the whole grain wheat methods have been commercialized. The yield, protein content, consistency, and color of the produced starch fractions were all impacted by steeping time and acid concentrations in the steering fluid. In the commercial wet-milling method, only wheat flour is now used. When wet-milling wheat, sodium chloride (1–2%) and hemicelluloses (pentosanases or xylanases) may be used to reduce separation medium viscosity and adjust water hardness.

The four most-used industrial wet-milling techniques are the modified Martin, Alfa-Laval/Raisio, hydro cyclone, and Tricanter processes. The key differences are the types of flour-water mixes employed or the fractionation machinery used (centrifuge, hydro cyclone, or screen) (Sayaslan, 2004). The modified Martin technique starts with making hard dough, then utilizing screens to separate the A- and B-starches from the gluten dough, with the starches passing through the screen while the gluten dough is maintained. Using centrifuges or hydro cyclones, the A- and B-starches are separated in the second stage. The dough is allowed to mature for around 30 minutes before being diluted with water, and a succession of hydro-cyclones are used to separate the A-starch from the other ingredients in the hydro cyclone process. There are no moving parts in the hydro cyclones. Because hydro cyclones are so small, each with an internal volume of roughly 5 ml, they are utilized in a series of banks. The suspension is supplied tangentially into the side of the cone at the top of the hydro cyclones. The A-starch granules that escape at the bottom (apex) of the cones are separated from the gluten/B-starch slurry that emerges at the top by the centrifugal force caused by the spinning motion. In the second phase, screens are used to separate the agglomerated gluten from the B-starch. The Alfa-Laval/Raisio method, on the other hand, begins with a thin batter of water and flour. A centrifuge separates the A-starch from the other ingredients, and the batter is then allowed to mature. In the second stage, the B-starch is separated from the gluten by screens. Small gluten agglomerates are formed by centrifugation, which removes the starch from the mixture while keeping the protein suspended. The Tricanter process homogenizes the flour and water mixture with high-pressure water before sending it to a three-phase decanter, which isolates the slurry into B-starch and A-starch gluten and a soluble stream. In the second stage, screening is used to separate gluten and B-starch. The Tricanter procedure has the advantage of removing solubles like pentosans in the primary step, which can obstruct the process of separation.

10.5 MALTING OF CEREALS

Water is essential for biological processes to occur within the cereal. Respiration is triggered by water intake, which allows germination to take place. To germinate, the grain must draw on its own store of nutrients, which are locked within the endosperm and must be accessed. Biological changes occurring in the grain, such as enzyme activation, enzyme synthesis, and metabolic alterations, make these nutrients available. Gibberellic acid comes in a variety of forms, and various types of the hormone are released during grain germination. Alpha-amylase is generated, while beta-amylase, which is already abundant in the endosperm, is released.

Numerous enzymes, such as phosphates, lipases, proteinases, and saccharolytic (xylanases, β-glucanases) enzymes, are also produced and activated and are vital to the grain, delivering energy to the new plant throughout germination and until the roots are formed. The enzymes released during germination cause starch and protein breakdown, resulting in the immediate availability of sugar and amino acids.

The three steps in the malting process are steeping, germination, and drying. However, in reality, the malting process involves more operations, and the distinctions between these stages are not always evident.

In the steeping vessel, kernels are steeped in temperature-controlled water, causing the grain to swell, soften, and prompt living tissues to resume metabolism, while diet, chaff, and broken kernels are removed by washing and floating. The water content of the grain must be at an appropriate level for steeping. The grain is then placed in a germination container. The germination process varies based on the technique utilized; however, most malting procedures are followed for the most part. Grains in circular or rectangular boxes (Saladin boxes) carry out germination.

Germination requires that the grains be distributed to a precise depth and kept in a regulated atmosphere. It takes around 5 days to complete and occurs quickly between 200 and 300°C, with an ideal temperature of 250 to 280°C. When the malt conditions are wet, the germination is then stopped.

The last step in the malting process is kilning. The germinated cereal is exposed to hot air during the kilning process (known as green malt). This prevents germination, reduces moisture in the cereal, and preserves the grains' enzymatic capacity for subsequent processing. The flavor and color components of the malt are affected by kilning. Green malt is dried in many stages, with high temperatures (500°C for 24 hours) being slowly applied. Because high temperatures affect the structure of proteins, kilning regimes induce some loss of enzyme function. Kilned malt, unlike green malt, is brittle and fragile, but due to its decreased microbial count, it is much more stable in storage. The malt is then cooled, cleaned, and stored once it has been kilned.

10.5.1 Current Trends in Malting

Grain yield, germination, malt modification uniformity, and product quality should all be carefully examined via congenital, physiological, and structural factors. Improvements in cereal variety production, insect resistance, extract content, and enzymatic activities are all major arguments for cultural innovation that will have a big impact on size. Currently, cylindro-conical steeps with effective aeration and carbon dioxide extraction are preferred because they provide the finest alternative for ideal sanitary conditions. Alkaline steeping appears to be a potential procedure for the manufacture of gluten-free malt that might be used in the manufacturing of gluten-free beer. New melanoidin-rich malts have recently been released by a number of malt providers. Melano malt is formed through a distinctive germination process followed by a slow drying process up to 130°C, allowing melanoidins to form as part of the kilning process. According to research, adding a specific starter culture to steeping water can prevent the synthesis of toxins from fungi present in barley or malt houses. Tritordeum malt (made by crossing wild barley) has been used as an additive and a source of starch in beer production. It demonstrated high enzymatic activity without the involvement of exogenous enzymes, saving time and money. The use of a barley gene chip array for transcript profiling of transcriptome during malting has aided in the identification of specific malting quality phenotypic traits with high disease-resistant ability.

10.5.2 Nutritional Potential and Application of Malted Cereals

Malting imparts many nutritional properties in malted cereal, resulting in increase of vitamins, minerals, and dietary fibers; enhancement of nutrients and flavors; reduction in anti-nutrients; and easy bioaccessibility of the nutrients.

10.5.2.1 Vitamin and Minerals

Minerals included in barley and malt are essential for nutrient absorption and may also have health advantages. They are produced from the embryo and the aleurone layer in the endosperm and account for 3% of the grain's dry weight and 20% potassium salts (expressed as potassium oxide, K_2O). This procedure employs malted cereal that is high in enzymes and vitamins. The enzymatic power of malts varies greatly depending on the grain variety. Malt has a low calcium content and a high magnesium content, which may help prevent gallstones and kidney stones. Many studies have demonstrated that boosting phytase activity during the malting process can help reduce grain phytate levels and increase iron and zinc availability. Tannins and total phenol levels decrease during germination. Germination boosts vitamin B and C levels substantially. Vitamin E (α-tocopherol) is an antioxidant monophenolic molecule that may be found in barley and malt. Vitamin E's most important activities in the body are antioxidant activity and maintaining membrane integrity (Randhir et al., 2008).

10.5.2.2 Nutrients and Flavors

Germination of cereal grains such as barley, rice, and wheat increase the levels of oligosaccharides and amino acids. Decomposition of high-molecular-weight polymers produces bio-functional chemicals and improves organoleptic properties by softening the texture and increasing the flavor in grains, culminating in a distinct flavor in the resultant products. Germination, in addition to softening the texture of brown rice, also improves flavor and nutrition. The protein efficiency index improves from 1.5 to 1.7, and the relative nutritive value of sprouted sorghum improves from 54.6 to 63%.

According to a study, soaking changes the texture, flavor, scent, and taste of a variety of substances. A large amount of amylases can be found in diastatic malt. The addition of malt to wheat flour low in amylases promotes sugar production via amylase action, yeast activity, and gas production and adds flavor and aroma to the final output. Diastatic malt increases dough sheeting and laminating qualities, as well as crust color and flavor, in crackers. Malt is also high in calcium, iron, thiamin, niacin, and ascorbic acid, which boosts the nutritional value of baked items.

10.5.2.2.1 Enhances Dietary Fiber

The majority of the dietary soluble fiber in a barley grain is created of β-glucans and arabinoxylans. β-glucans in barley food items lower blood cholesterol (Behall et al., 2006) and glycemic index, as well as lowering the chance of cardiovascular disease (Jenkins et al., 2002). Other barley fiber components, most notably arabinoxylans, haven't been studied as thoroughly as β-glucans. In keeping with new research, water-soluble maize, wheat, and rye arabinoxylans boost cereal fermentation, short-chain carboxylic acid synthesis, serum cholesterol reduction, and calcium and magnesium intake.

10.5.2.2.2 Increases Bioavailability of Nutrients

The impact of malting finger millet, wheat, and barley on iron, zinc, calcium, copper, and manganese leads to bioaccessibility. Malting enhances the bioaccessibility of iron in two kinds of finger millet and two strains of wheat by three times. Malting has been suggested as a food-based approach for extracting the most iron and other minerals from cereal grains.

10.5.2.2.3 Reduces Anti-Nutrients

Malting pearl millet minimized the anti-nutrient phytic acid and eradicated the grain's mousy odor, per a study by Pelembe et al. (2003). Phytic acid and oxalate, which are anti-nutrients in sesame seeds, were reduced to 49% and 51%, respectively, in the anti-nutrient composition. Sokrab and colleagues investigated the anti-nutritional elements (phytate and polyphenols) in maize genotype grains and discovered that the phytate concentration of Var-113 and TL-98B-6225 9TL617 genotype grains was 1047.00 and 87.16 mg/100 g, respectively. Within the first two days of germination, the phytic acid content of both genotypes reduced dramatically, indicating that germination minimizes the phytic acid content of the grains over time.

10.5.3 Applications

When cooked, barley and wheat malt is accustomed to boost α-amylase levels in flour as a replacement for fungal amylase or as a source of color and flavor. Malts are utilized in granary-type slices of bread and malt fruit loaves due to their distinct flavor and scent. At levels of 5–10%, germinated cowpea flour enhanced the precise volume marginally, whereas non-germinated flour had no impact. The sensory qualities of grains are modified by germinating and kilning, giving them a definite flavor and aroma that's often thought to be pleasant. Germinating rice boosted the hedonic response to texture, flavor, and appearance when cooked, consistent with research. The inclusion of malted raw materials into goods has also been demonstrated to boost sensory characteristics. Extrudates made up of malted millet and soybean had better flavor and texture than those made up of non-malted millet and soybean, and the malted flavor masks the unpleasant astringent scent of soybeans.

The high amylolytic activities of wheat and barley malts have also been employed in the synthesis of dextrins and breakfast cereals, like Mämmi, a conventional Finnish rye-based Easter dish. Since specified amounts of amylolytic activities are required for optimal baking performance, barley and wheat malts are employed to optimize enzyme levels in baker's flour. Tiwari and colleagues developed a ready-to-use malted cereal pulse mix that will be used as a fast base flour.

10.6 PROCESSING OF CEREAL BY-PRODUCTS

10.6.1 Nutritional Attributes of By-Products and Changes in Composition due to Processing

Wastes generated from food processing industries can be defined as residues, side streams, or by-products derived from processing raw materials into food products (Trigo et al., 2020). The existing disposal and inadequate management of food, as well as agro-industrial by-products, has a great influence on the environment and social-economic sectors globally (Gómez-Garca et al., 2021). Food by-products, on the other hand, have caught the interest of researchers, since they are frequently high in bioactive compounds and have the potential to be employed in food and pharmaceutical products. The utilization of food waste has been recognized as a vital operation to the world since it includes sustainability (use of leftovers), circular economy (production of value-added commodities), research, technology, and health (development of functional things) (Comunian et al., 2021).

By-products are created during grain processing, and they might differ physically and chemically owing to a variety of inherent and operational variables (Papageorgiou & Skendi, 2018). Cereal by-products can be re-utilized and valorized as a valuable source of carbohydrates, proteins, lipids, vitamins (especially B-complex and vitamin E), inorganic compounds, and trace elements (Papageorgiou & Skendi, 2018). However, many of the processing conditions lead to the loss of nutrients and reduced nutritive value of food (Narwal et al., 2020).

Cereal by-products can be utilized to make a variety of chemicals that can be employed as nutraceuticals, nutritional supplements, and functional food formulations (Dapcevic-Hadnadev et al., 2018). The European Commission has approved health claims relating to the consumption of cereal by-products or their constituents (Steiner et al., 2015). Some examples of this are fibers from wheat bran, oat grain, and rye, as well as β-glucans from oat and barley, plant sterols, and so on (Dapčević-Hadnadev et al., 2018). In general, processing residues into products is a challenge to be overcome by food and pharmaceutical industries due to its highly perishable behavior, which may influence the microbiological safety and quality of the final products (Comunian et al., 2021; Lai et al., 2017). Thus, by-products have limited commercial exploitation due to their high water content, high water activity, and high content of carbohydrates, which are significant factors in the growth of pathogens, increased enzymatic activity that can increase spoilage, and a fast auto-oxidation rate (related to a high fat content) (Gobbetti et al., 2020; Trigo et al., 2020; Jahurul et al., 2015). Fast transit of the food by-product and storage in refrigerated settings are tactics for avoiding deterioration in this situation (Comunian et al., 2021).

Additionally, some initial processes aiming to achieve preservation and to improve the safety of food by-products can be performed, such as irradiation and drying (Casarotti et al., 2018; Comunian et al., 2021). When considering the recovery of the bioactive compounds, such as hydrophilic or hydrophobic ones, applying simple steps, such as drying and grinding aiming to obtain small and dried particles, can improve their extraction yield (Lima et al., 2019; Comunian et al., 2021). The physicochemical features of the principal bioactive molecule detected in the by-product have a big role in deciding which extraction procedure to use. Liposoluble chemicals are extracted primarily with organic solvents in this method; however, it can be supplemented with other methods to improve extraction efficiency, such as ultrasonic, pressured liquid, and microwave-assisted extraction (Fierascu et al., 2019). However, it is widely recognized that several organic solvents used in lipid and other material extraction may be hazardous to the environment and dangerous to human health (Trigo et al., 2020).

Thermal processes can degrade bioactive chemicals; thus using environmentally friendly procedures and solvents for the extraction of sensitive components such as phenolics is vital (Gómez-Cruz et al., 2020). Trigo et al. (2020) employed ohmic heating (OH) technology to recover bioactive substances including polyphenols and carotenoids from tomato by-products, suggesting that OH might be an ecologically friendly method. Furthermore, novel technologies such as high hydrostatic pressure, supercritical fluid extraction, pulsed electric fields, cold plasma, and ultrasound can be combined with thermal processes to increase the disruption level of vegetable cells and, as a result, improve the extraction efficiency of valuable substances (Moreira et al., 2019; Wang et al., 2020). Microwave-assisted extraction (MAE) is a novel method that relies on the dissipation of heat to generate pressure inside vegetable cells, rupture, and the release of bioactive chemicals (Chemat et al., 2011; Sonar & Rathod, 2020). The benefits of utilizing this approach include decreased time and solvent consumption, as well as the ability to optimize extraction yields; nevertheless, when volatile or non-polar solvents are utilized, poor extraction efficiency may result (Comunian et al., 2021).

The milling of cereals takes off the bran and germ layers, which house most of the vitamins and provitamins (tocopherols, β-carotene, and zeaxanthin), thus causing their loss and forming by-products, hulls, and polish waste (Fratianni et al., 2015). Milling, sieving, and air classification, as well as the selection of genotypes or species that possess high carotenoid and tocol content of these compounds in the kernel, are steps that can be used to produce by-product fractions with augmented functional substances for food production (Fratianni et al., 2015).

Bran is one of the most important milling by-products, consisting of the gritty outer shell of the seeds, which has a low flour content but a high fiber and protein content (ElMekawy et al., 2013). Corn, rice, and wheat bran are the most frequent forms, although they may also be obtained from oat, barley, and sorghum. Dietary fiber is one of the key added-value components of these by-products (Cassano & Galanakis, 2018). In food products, dietary fibers have key functional qualities (emulsification and gel formation, as well as improved water- and oil-holding capabilities) (Elleuch et al., 2011). Fibers are commonly employed to adjust the texture and sensory characteristics, consistency, and rheological behavior of bread items, dairy, jams, meats, and soups, opening up new possibilities in their industrial applications (Cassano & Galanakis, 2018).

10.6.1.1 Rice By-Products

10.6.1.1.1 Rice Bran

Rice bran is a residue generated from the rice milling process and comprises the brown external layer of the seed, such as pericarp, tegmen, and aleurone, which can be about 10–12% of the total kernel weight (Spaggiari et al., 2021). Many cereal by-products have been used as feedstock, and rice bran is no different. However, as it has a non-negligible amount of lipids (15–20%), a great part of it is used for bran oil extraction (Punia et al., 2021; Spaggiari et al., 2021). Besides this, the material contains relevant content of minerals, such as iron, phosphorus, and magnesium, from 11–15% of crude proteins, up to 50% of carbohydrates, around 11.5% of dietary fibers, mainly β-glucan, pectin, and gum (Spaggiari et al., 2021). The bran (pericarp outer layer) is fibrous, tough, and resilient, and comes out from milling in a form of flakes with low amounts of endosperm attached to it (Hutchinson & Martin, 1970). This layer is a source of hemicellulose (20%), cellulose (30%), lignin (20%), and proteins (up to 17%) (Mohd et al., 2013). Next to the pericarp is the tegmen or seed coat, which contains mainly fatty material, followed by the aleurone layer, which enclosers the endosperm and is rich in proteins (Juliano & Tuaño, 2019). The concentration of target components in rice bran and germ layers, such as proteins, starch, fats, vitamins, and minerals, have been described as depending on the degree of pearling, which have also been used to support the growth of microorganisms (Saman et al., 2019). These fractions represent the majority of those nutrients, while the remaining kernel, which comprises 90% of the endosperm, has lower proportions of them (Saman et al., 2019).

Antioxidants are plentiful in rice bran. Oryzanols, phytosterols, tocotrienols, squalene, polycosanols, phytic acid, and ferulic acid are among the additional bioactive phytochemicals contained in it, which have been classified as antioxidants, free radical quenchers, and chronic disease defenders (Mohd Esa & Ling, 2013; Spaggiari et al., 2021). Rice bran also contains vitamins E and oryzanol, as well as important fatty acids and proteins, cholesterol-lowering lipids, and anti-tumor substances (Pradeep et al., 2014; Saman et al., 2019). Rice bran's high dietary fiber content meets a significant portion of daily fiber needs, allowing feces to pass more easily through the digestive tract (Saman et al., 2019). Due to constant moisture of fecal bulk, reduced dehydration, and hardening of it, which help to prevent colonic stagnation and replenish substrates for bacterial production of butyrate or other beneficial substances to the gastrointestinal system, the increased transit rate can protect against a wide range of bowel disorders (Saman et al., 2019). As a result, distinct heat-stabilization processes should be used for rice bran, which are critical stages given its usage and phytochemicals in various culinary applications (Fărcaş et al., 2021).

10.6.1.1.2 Rice Bran Oil

Rice bran oil (RBO) is a key result of the rice bran extraction process, and it contains a lot of oil, making it a good source of health-promoting chemicals (Punia et al., 2021). It's a by-product of rice bran processing, and this stage yields defatted rice bran, which is normally disposed of in the soil or utilized as a low-cost animal feed ingredient (Punia et al., 2021; Spaggiari et al., 2021). RBO is also high in important lipids and bioactive chemicals, including sterols, tocopherols, and tocotrienols, which are sometimes known as tocols or vitamin E (Punia et al., 2021). RBO's antioxidant and chemopreventive capabilities have been connected to these compounds (Gul et al., 2015). However, refining processes of RBO can significantly decrease the content of tocols and oryzanol in it (Punia et al., 2021). Consequently, less destructive refining techniques can be used to maintain the amount of these beneficial health substances even after processing (Dapčević-Hadnadev et al., 2018).

Many nutraceutical chemicals, including oryzanols, tocopherols, tocotrienols, ferulic acid, phytic acid, lecithin, inositol, and wax, may be discovered in the wastes from RBO refining, which are referred to as waxy sludge, gum sludge, and soap stock (Dapcevic-Hadnadev et al., 2018). Crude RBO typically contains a 4% unsaponifiable fraction, 2–4% free fatty acids, and 88–89% lipids, with 47% monosaturated, 33% polyunsaturated, and close to 20% saturated fatty acids (Gul et al., 2015). Due to high lipid content and the presence of enzymes, such as lipase, rice bran holds a very short shelf-life, which has a relevant influence on the occurrence of rapid hydrolytic rancidity of rice oil, making it inedible (Dapčević-Hadnadev et al., 2018; Mohd Esa & Ling, 2013).

10.6.1.2 Wheat By-Products

10.6.1.2.1 Wheat Bran

The flour milling business produces a significant amount of wheat bran (Duţă et al., 2018). It is made up of several rough outer layers of the wheat grain and can make up to 25% of its weight; however, the amount of bran produced depends on the milling process, which can be wet or dry (Onipe et al., 2015). Wheat bran is made up of several layers, each of which has its own composition and function. Pericarp, testa, and aleurone are the three layers that make up 4–5%, 1%, and 6–9% of the grain, respectively (Prückler et al., 2014). Wheat bran is mainly composed of carbohydrates (56–65%), dietary fibers (~43%), proteins (~13–18%), water (~12%), and lipids (~4%) (Duţă et al., 2018; Prückler et al., 2014). Considering its chemical composition, wheat bran can present different complex structures of either C6-sugars like cellulose or C5-sugars like xylans, which make it possible to differentiate it (Prückler et al., 2014). Besides this, this bran can present different proportions of some phenolic compounds and lignin (Langton & Gutiérrex, 2021; Prückler et al., 2014).

The wheat bran outer layers are formed by dense cellulose cell walls, cuticle materials, high arabinose to xylose ratios (A/X) complex xylans, and polymer chains cross linked by ferulic acid dehydromers (Prückler et al., 2014). Pericarp and testa contain a relevant content of structural polysaccharides embedded with lignin, and the last one also contains the major proportion of alkyl resorcinols of the wheat grain, which can range from 220–400 mg 100 g^{-1} (Prückler et al., 2014). The aleurone layer is made up of cells that contain bioactive chemicals and are surrounded by thick cell walls made of moderately linear arabinoxylans (AX) with a low arabinose-to-xylose ratio (Pavlovich-Abril et al., 2014). Differences in wheat bran composition can be attributed to both internal characteristics such as cultivar and external ones such as growing site and local weather, as well as processing, milling technologies, various analytical methodologies, and other factors (Prückler et al., 2014). Wheat bran is a good source of alkyl resorcinol, ferulic acid, b-glucan, arabinoxylan, lignans, and sterols, in addition to dietary fiber and other beneficial components (Duţă et al., 2018).

Due to an increasing interest among consumers in a natural, functional, and fortified diet, wheat bran has earned a lot of attention as a beneficial food ingredient. Much research has looked at the qualities of bakery items with varied levels of wheat bran, such as pasta, cookies, and extruded cereals, especially because it adds nutritional value and lowers the glycemic index (Duţă et al., 2018). When wheat bran is added to semolina flour, such as in pasta manufacture, it might result in an inhomogeneous mixture and the construction of a gluten network that is impeded by bran particles (Duţă et al., 2018). The production of wheat bran extruded foods is scarce because of the negative aspect, texture, and sensorial properties, which occur due to the increased hardness and decreased expansion volume, as well as the crispness of the final product (Duţă et al., 2018).

The health claims related to the consumption of wheat bran are surpassed by the technological difficulties of incorporating it into foods without altering their sensorial and functional properties. These challenges were reported by Duţă et al. (2018) when wheat bran was incorporated in bakery products, which exhibited poor baking performance and unpleasant aspects, bitter taste, and grittiness. Different pretreatments and alternative bran processing conditions of milling, heating, extraction, extrusion, and fermentation can be used to decrease these negative effects (Duţă et al., 2018; Prückler et al., 2014).

10.6.1.2.2 Wheat Germ

The germ is the embryo structure of the wheat kernel, and it accounts for around 2–3 g 100 g^{-1} of the entire grain weight (Gili et al., 2017). Wheat germ is significantly extracted during the milling process, especially at the beginning of the processing flow after the kernel has cracked (Gili et al., 2017; Onipe et al., 2015). Wheat germ, along with bran, is the most important milling by-product of wheat, with an annual output of roughly 25 million tons globally (Duţă et al., 2018). Other researchers, however, have calculated a far larger figure, nearing 150 million ton of bran every year (Prückler et al., 2014).

Despite its rich compositional qualities, wheat germ is rarely utilized for human consumption, owing to its high lipid content (8–14%), which may compromise the quality of the flour produced and, as a result, the food's stability (Gili et al., 2017). In addition to polyunsaturated fatty acids, wheat germ is a good source of -tocopherol, B-complex vitamins, dietary fiber, minerals, and phytochemicals (Duţă et al., 2018). Due to the high quantity of unsaturated fatty acids and the presence of hydrolytic and oxidative enzymes, this detrimental impact is linked to a significant propensity for rancidity during storage. Several studies have been conducted in this area, with the goal of stabilizing and extending the shelf-life of wheat germ, mostly by inactivation enzymatic activity (Duţă et al., 2018). This process can be achieved by heating, microwave, extrusion, and other cooking treatments, as well as by the extraction of the oily fraction and the incorporation of antioxidant substances in it, although this last one can be expensive and, when using synthetic compounds, be a risk to consumer health (Duţă et al., 2018).

10.6.1.3 Oat By-Products

10.6.1.3.1 Oat Bran

Oat grain processing generates a by-product known as oat bran. This residue is a relevant source of soluble dietary fibers, especially β-glucans, which have been described as possessing health properties linked to cholesterol reduction, prevention of cardiovascular disorders, and reduction of diabetes symptoms and blood pressure (Duţă et al., 2018; Zhang et al., 2011). These bioactive properties are related to their (1→3)/(1→4) chemical bond ratio, as well as their viscosity and molecular weight, being significantly affected by oat processing methods (Zhang et al., 2011). Thus, oat bran with different functional properties can be obtained through the milling processes, and it can be optimized to produce richer β-glucan fractions (Duţă et al., 2018). The content of β-glucans and total dietary fiber can reach up to 20% and 40% of dry matter, respectively (Duţă et al., 2018). Oat bran also has arabinoxylans, oligosaccharides, tocols, and phenolic compounds, as well as niacin, magnesium, iron, copper, and potassium (Duţă et al., 2018). Because oat bran is complexed with lignin, hydrolysable tannins, and organic acids, the phenolics in it, such as phenolic acids, flavonoids, and avenanthramides (found only in oats), can be concentrated by treating it with cell wall polysaccharide-degrading enzymes (Alrahmany & Tsopmo, 2012).

Oat bran is obtained by grinding the cereal groats or rolling them, then sorting them, followed by milling into fractions accounting for less than 50% of the initial grains (Dapčević-Hadnadev et al., 2018). Oat bran is made up of particles from distinct tissue layers, such as aleurone and endosperm, and has a different nutritional content than other cereals treated by dry milling (Heiniö et al., 2016).

Oat bran is frequently used in baked goods, particularly when combined with wheat flour. Due to a high level of dietary fibers, β-glucans, and lipids, as well as a reduction in gluten proteins, this blend increases the nutritional quality of wheat-based bread, but it also affects the overall baking quality of the final product (Duţă et al., 2018). Furthermore, as compared to wheat dough, oat bran promotes increased water absorption and increased mixing needs (Duţă et al., 2018). Researchers have tried to mix oat bran with other ingredients by using an extrusion process and to obtain a ready-to-eat product, though these processes were not so easy to execute (Lobato et al., 2011). Extrusion circumstances that are too harsh, such as high temperature, high shear, and low moisture, can depolymerize and reduce the molecular weight of oat bran glucans, affecting the functional characteristics of food items. Furthermore, these processing issues arise as a result of the high amount of lipids and soluble dietary fibers in the product, which can be reduced by adding additives like monoglycerides, modified starches, modified gelatin, oligofructose, or inulin (Duță et al., 2018; Lobato et al., 2011).

10.6.1.4 Corn By-Products

10.6.1.4.1 Corn Bran

Corn or maize bran is a side stream derived from the corn pericarp and is taken from flaking grits production. The kernel is formed by four structures, bran (pericarp), endosperm, germ, and tip cap, reaching up to 5%, 83%, 11%, and 1%, respectively (Saeed et al., 2021). Dry milling processing of corn produces about 60–70 g of bran per kg of grain (Saeed et al., 2021), while wet milling of corn grains generates major starch from the endosperm fraction and oil from the germ, as well as corn fiber, corn gluten, and corn gluten feed (Dapčević-Hadnadev et al., 2018). These by-products have a high digestible energy value compared with other grains, such as oats, barley, and fodder rice products and treated straw (Dapčević-Hadnadev et al., 2018).

Corn bran is mainly composed of 22% of cellulose and hemicellulose, 20% of starch, 10% of proteins, 4% of phenolics, 5% of lipids, and around 2% of minerals (Saeed et al., 2021). Moreover, it is made mainly of heteroxylans and cellulose and virtually lacks lignin, while 55% of the maize grain is heteroxylans, the bulk being arabinoxylan (Dapčević-Hadnadev et al., 2018; Saeed et al., 2021). Generally, corn bran is mostly derived from the pericarp layer (excluding the endosperm fraction), while corn fiber consists of cellular material from the whole grain (Dapčević-Hadnadev et al., 2018).

Grains of corn are crushed; then the hull, germ, and coarse meal are separated from the other parts. These cakes and meals are higher sources of protein than corn grain and other particular protein-rich foods, such as peas, beans, and sunflower seeds (Vasconcelos et al., 2013). Corn starch is frequently used by paper industries, making up close to 90% of the starch used, due to its properties to physically improve paper and thus reduce production costs. Corn bran is considered a low-price ingredient, a cereal source of food dietary fiber, and an antioxidant supplier, including oil, phytosterol esters, policosinol, and phenolics, which makes it a significant functional ingredient in low-calorie products (Saeed et al., 2021). The incorporation of corn bran into bakery products can improve emulsion and oxidative stability and enhance water absorption capacity, bulking agents (non-caloric), and oil retention agents (Saeed et al., 2021).

10.6.1.4.2 Corn Germ Oil

Corn grains contain up to 5% of lipids (Saeed et al., 2021; Dapčević-Hadnadev et al., 2018). Through starch synthesis, maize oil is extracted from the germ, which contains up to 50% lipids. Corn fiber oil, on the other hand, has gotten a lot of press recently because of its ability to lower plasma cholesterol levels, as established in animal studies (Dapcevic-Hadnadev et al., 2018). Despite the presence of vanillin (2.8 mg/kg), trans-cinnamic acid (0.9 mg/kg), and ferulic acids (0.5 mg/kg) (Dapcevic-Hadnadev et al., 2018), maize oil has low levels of phenolic compounds due to their poor solubility and the refining process that removes them. Maize oil and other by-products, on the other hand, include a substantial amount of carotenoids, which are greater in corn than in other cereals.

10.6.1.5 Barley By-Products

Phenolic compounds, phytate, vitamin E (including tocotrienols), and insoluble dietary fiber are all sources of bioactive components in barley by-products, just as they are in wheat (Dapcevic-Hadnadev et al., 2018). Because chemical contents are not evenly distributed among the tissues that make up the barley kernel, its by-products are a rich source of bioactive compounds such as tocotrienols, tocopherols, vitamin E, and the highest level of phenolics. Due to the unique location of glucans in the cell walls of endosperm, pearled barley has a greater glucan content than the total barley kernel (Dapcevic-Hadnadev et al., 2018).

10.6.1.6 Other Considerations about By-Product Processing

Cereal by-products are produced by all the major milling companies. The improvement of efficient and low-cost processing techniques aiming to potentialize residue transformation into high–value added products is a challenge and an approach to be pursued. This will increase the competition among agri-food corporations and help the environment by the reduction of wastes generated while implementing a circular bioeconomy framework in those industries (Gómez-García et al., 2021; Vasconcelos et al., 2013). Thus, besides the improvement of processing methods, the development of novel and non-destructive evaluation techniques to determine by-product composition and stabilization are issues of great importance (Duţă et al., 2018). Moreover, this could be achieved by a fundamental valorization of food biowastes aiming to maintain their abundance of bioactive molecules, which could be extracted efficiently and used as unique industrial raw materials (Gómez-García et al., 2021). These dietary fibers and different bioactive molecules can be incorporated in food products, giving them functional and health-related properties, especially for the prevention of metabolic diseases (Gómez-García et al., 2021; Martins & Ferreira, 2017).

Some biotechnological approaches that may be used in the recovery and reuse of food by-products include anaerobic digestion, enzymatic aided extraction, solid-state fermentation, aqueous-two phase systems, protein precipitation by polysaccharides, and the encapsulation of food by-products (Comunian et al., 2021; Gómez-Garca et al., 2021). The technical qualities of various food by-products, such as coloring, antioxidants, and antimicrobials, have been studied extensively. Other novel destinations have also been investigated, such as the prebiotic potential of various dietary by-products (Comunian et al., 2021). Other recent research on by-product valorization is contributing to the manufacture of various bioproducts, for example, phytochemicals, enzymes, biofuels, and functional components, among other compounds, that might aid in the enhancement of the social, environmental, and economic sectors (Ravindran & Jaiswal, 2016).

Another significant element to consider is the price. Despite the fact that food by-products are a low-cost resource, the practicality and application of their processing might be costly, affecting the ultimate output. As a result, research into various methods of acquisition and extraction is critical in order to reduce costs and provide a functional meal that is accessible to a wide range of customers. The application of food by-products by industry has a scaling difficulty, which necessitates more study (Comunian et al., 2021).

10.7 RECENT APPLICATIONS IN VARIOUS SEGMENTS OF FOOD SYSTEMS

The agro-industrial processing of cereals produces a high quantity of by-products. These materials can display the greatest content of valuable substances with potential industrial application in diverse food products. Gluten meals (GMs), distillers' dried grains with solubles (DDGS), and brewer's spent grain are by-products from wet-milling and dry-milling of cereal grains (Tapia-Hernández et al., 2019). As previously discussed, after grain milling and processing into flour, valuable fractions remain as by-products, made from cereal bran or germ (Roth et al., 2019). Mainly, bran is a multilayer matter, containing the outer and inner pericarp, testa, and nucellar layer of the grain (Roth et al., 2019; Duţă et al., 2018). Once bran was primarily used for feeding purposes, though many studies have pointed to the potential application of it for use in cereal-based food systems (Roth et al., 2019).

Brewer's spent grain, as the name suggest, is a by-product from the brewing industry. It was estimated that the production of 1.95 billion hL of beer can generate approximately 390 million tons of wet BSG annually (Roth et al., 2019). The composition of BSG can differ depending on the type of cereal used in beer production, although it is mainly composed of the husk, pericarp and seed coat of the grain and contains endosperm and aleurone layer residues (Mussatto, 2014). As with many other industrial residues, BSG is mainly destined to be animal feed, but as this material is rich in sugars and proteins, many biotechnological uses have been proposed, including human intake (Mussatto, 2014). Nevertheless, this material has gained attention since it can be used as a substrate in the production of microbial cell mass and fungal strains (Cooray & Chen, 2018).

Prior to incorporation of BSG into food products, such as in baking, the material needs to be converted in flour, since its original form is too coarse for direct application (Mussatto, 2014). Also, other actions need to be taken before its incorporation in food products, mainly because its brownish color, flavor, and texture characteristics can interfere in sensorial properties and be noted by consumers (Naibaho & Korzeniowska, 2021; Mussatto, 2014). BSG addition into bakery products also increases the amount of dietary fiber, which can lead to a significant increase of water absorption and consequently affect dough formation by disrupting the network linkage and starch gelatinization (Naibaho & Korzeniowska, 2021).

Torbica et al. (2019) looked at the sensory and physico-chemical qualities of a whole-grain wheat bread prepared using chosen food by-products, such as brewers' wasted grain, sugar beet pulp, and apple pomace. When compared to whole grain, the quantity of total dietary fiber increased, while the amount of protein, calories, and nutritional value remained similar. Furthermore, the results of this study demonstrate that consumers have a good attitude toward the use of by-products and a desire to extend their dietary options. Naibaho et al. (2022) investigated the quality of yoghurt made with varying amounts of BSG, up to 20%. The authors discovered that adding BSG to yoghurt reduced fermentation time, increased viscosity, and enhanced shear stress. During 14 days of refrigerated storage, the BSG yogurt maintained flow behavior and stability while also providing nutrients for the development of lactic acid bacteria. Some culinary uses of brewer's waste grain by-products are given in Table 10.3.

Distillers' dried grains with solubles, like BSG, are a byproduct of the dry milling industry. This substance is made up of undigested grains that have been fermented with ethanol (Buenavista et al., 2021). Although, unlike BSG, distillers' dried grains cannot be eaten on their own due to a high quantity of undesirable components and low quantities of soluble carbohydrates, they may be combined with other foods to make a meal (Buenavista et al., 2021). As a result, DDGS are commonly utilized as a feed addition. Several studies were conducted on the composition and sensory qualities of bread and snack products that used distillers' grains (Pourafshar et al., 2018; Roth et al., 2016; Singha et al., 2018).

Table 10.3 Some Food Applications of Rice and Wheat Cereal By-Products

Cereal By-Products	Application	Amount Used	Food Product	Reference
Rice bran	Bakery	10, 15, 20, and 25%	Bread	Doan et al. (2021)
		5, 10, 15, and 20%		Sallam et al. (2019)
		10 and 15%	Corn extruded snacks	Renoldi et al. (2021)
Rice bran wax	Bakery	4.5 and 5.5%	Biscuits	Principato et al. (2021)
Rice bran	Pasta	5–30%	Rice pasta	Nithya et al. (2013)
Wheat bran	Bakery	0, 10, 15, and 20%	Bread	Özkaya et al. (2018)
		6.5–25%		Boita et al. (2016)
		2.5–20%		Gómez et al. (2011)
		—	Cereal food products	Onipe et al. (2015)
	Pasta	20 to 40%	High-fiber pasta	Sobota et al. (2015)

Grain bran and germ are two more by-products derived from cereal milling enterprises. Because of their nutritional and functional qualities, these materials have gotten a lot of attention. Cereal brans are a source of proteins for culinary applications, but their carbohydrate portion has mostly been used by biorefineries to produce other biomolecules such as lactic acid through fermentation (Roth et al., 2019; Tirpanalan et al., 2015). Cereal germ, on the other hand, is high in unsaturated fatty acids, essential amino acids, dietary fiber, vitamins (and useful phytochemicals like flavonoids), and dietary fiber (Wang et al., 2021a).

Due to their high level of antioxidants, bioactive peptides, phytosterols, phytic acid, and polyunsaturated fatty acids, rice bran and rice bran oil are useful by-products that may be used in a variety of culinary applications (Punia et al., 2021). Rice bran is a non-novel food since it is a well-established commodity, particularly in the European market, where rice bran processing firms are well connected with the full rice food chain for processing residue collection (Spaggiari et al., 2021). Rice bran can be used to bread items to increase their nutritional and functional qualities. According to a recent study, cakes with 30–35% rice bran had higher fiber content, phenolic compounds, and antioxidant capacity, as well as lower calorie value (Mendes et al., 2021). Furthermore, researchers highlighted strong consumer acceptance in this study, with 35% of the judges indicating that the product could be consumed at least once a week, highlighting its potential for bakery items.

Researchers employed rice bran to create biocomposite food packaging using low-density polyethylene/low-density polyethylene-grafted acrylic acid in another effort (El-Wakil et al., 2020). RBO may be used as a cooking oil since it has a moderate and pleasant flavor, is resistant to high temperatures, is hypoallergenic, and contains antioxidants that help to reduce oxidative stress (Punia et al., 2021; Spaggiari et al., 2021). Due to its creaming capabilities, RBO is also employed in bakery applications, as are plastic shortenings, which are frequently manufactured by a partial hydrogenation process (Punia et al., 2021).

Grain by-products are important sources of proteins, such as prolamins, which are their endosperm storage proteins (Tapia-Hernández et al., 2019). The incorporation of proteins in food products is important because they are a source of bioactive peptides, which can perform essential biological functions in the body and health, especially in preventing disease (Orona-Tamayo et al., 2019). Prolamins are named differently depending on the cereal type from which they were extracted; for example, glutenin and gliadin are storage proteins from wheat, while zein is from maize, hordein from barley, and avenin from oats (Nuttall et al., 2017; Tapia-Hernández et al., 2019). These bioactive proteins have been studied in biotechnological applications such as development as micro- and nanomaterials for encapsulation of medicines, dyes, and other bioactive compounds (Tapia-Hernández et al., 2019).

In a recent study, gelatin/zein nanofibers were fabricated by hybrid electrospinning for potential application for bioactive delivery in food industry (Deng et al., 2018). The researchers found that zein particles were well dispersed in the gelatin matrix, acting as a hydrophobic plasticizer, maintaining the shape of it, which contributed to preserving their 3D porous structures. Ultrafine zein fibers containing 5% anthocyanins were obtained by electrospinning technique in the work of Prietto et al. (2018) to be used as a natural pH indicator membrane, which could be applied for pharmaceutical, food, and packaging purposes. Protein fractions from brewers' spent grain by alkaline or alcoholic methods were used to prepare composite nanoparticles containing curcumin by Wang et al. (2021b). In this study, BSG proteins extracted with 55% (v/v) ethanol exhibited the greatest encapsulation performance of curcumin, indicating that BSG is a valuable source of functional proteins for encapsulation of water-insoluble bioactive compounds. Figure 10.2 summarizes the functional applications of cereal by-products.

In terms of commodities with a clear economic and environmental effect, pricing grain waste and by-products have sparked a lot of attention (ElMekawy et al., 2013). Despite the fact that valorization processes for certain industrial by-products have been successfully investigated on a laboratory or pilot scale, the process's efficient scaling-up and standardization on an industrial scale remain a challenge (Cassano & Galanakis, 2018).

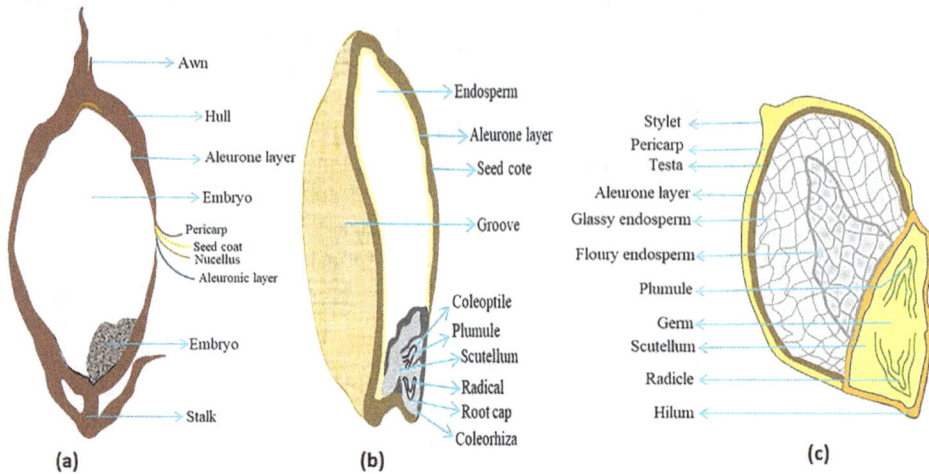

Figure 10.2   Structure of cereals (a) rice, (b) wheat, and (c) millet kernel.

10.8 INNOVATIONS IN POST-HARVEST PROCESSING

10.8.1 Irradiation of Cereal Grains

Ionizing radiation like γ-rays and X-rays are used in the post-harvest processing of cereals to disinfect grains and grain mass with diseases such as *Escherichia coli*, *Campylobacter*, *Salmonella*, and viruses. Food irradiation is now permitted in over 40 countries, with yearly quantities expected to reach 5,000,000 metric ton. Irradiating food using electron beams, X-rays, or γ-rays provides a comparable effect to pasteurization, heating, or other heat treatments with less negative impact on appearance and texture (Bhattacharjee & Sighal, 2016). The primary uses of the irradiation method in the case of cereal grains are sprout suppression, enhancement of grain malting qualities, and grain rehydration features (Diehl, 1995). The major applications of irradiation in cereals are shown in Figure 10.3. When cells are exposed to ionizing radiation, DNA damage occurs that is irreversible. The potential genetic harm from the bigger tissue is removed when these cells undergo apoptosis or develop a nonlethal DNA mutation that is carried down via future cell divisions (Tiwari & Pojic, 2020). Most of the microbial contaminants associated with cereals, including both molds and bacteria, that are located in the outer layers of grain are degraded through permanent DNA damage. According to Hassan et al. (2009), low-dose gamma radiation can reduce phytic acid and tannin levels in maize and sorghum while increasing in vitro protein digestibility. Infrared radiation, in addition to gamma radiation, has been shown to retain the nutritional quality of sorghum, maize, and rice, as well as to minimize anti-nutritional factors in sorghum, maize, and beans (Keya & Sherman, 1997).

10.8.2 Ozone Technology in Post-Harvest Cereal Processing

Ozone is used for (i) the treatment of mycotoxin-contaminated cereals, (ii) microbiological decontamination, (iii) stored grain fumigation, and (iv) the change of the physicochemical properties of main cereal components, including protein and starch, due to its strong oxidant qualities. The color, storage, and germination ability of the grains, as well as the textural and rheological components of the products, may be affected by ozone treatment. Suian José Granella of the State

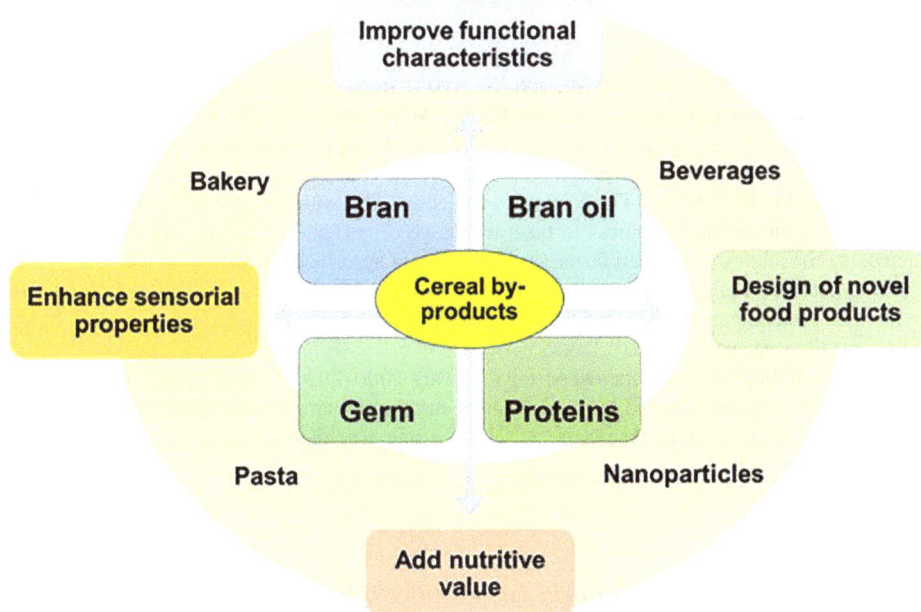

Figure 10.3 Summary of the functional applications of cereal by-products.

University of Western Paraná published a paper in 2018 about the use of ozone technology in combination with drying in the post-harvest processing of naturally infected wheat seeds. The beneficial impact of ozone treatment with drying was indicated by a 92.86% reduction in fungus count, but the wheat grains' physiological qualities—germination, vigor, and electrical conductivity—remained unchanged. Because mycotoxin production is frequently linked to fungal contamination of grains, exposing grains to 60 mM/M ozone for 2 h was found to be an effective treatment for inactivating *Fusarium graminearum* and reducing deoxynivalenol contamination, as well as reducing zearalenone and ochratoxin-A in contaminated corn (Tiwari & Pojic, 2020).

The investigators detected minor changes in color (yellowness decreased, while whiteness increased) and fatty acid content after a 180-minute ozone treatment (Qi et al., 2016). The usefulness of ozone technology for the breakdown of pesticide residues in stored grains has been proven by several authors, with the degradation efficiency being directly related to the exposure period. The use of ozonation to control the enzymatic activity of wheat flour in its fluidized state is also being looked at.

10.8.3 Cold Plasma Technology in Post-Harvest Cereal Processing

Cold plasma technology is employed in the cereal production chain for two reasons: effective bio-decontamination and eco-innovative modification of grain and grain-based product technofunctional qualities. Bio-decontamination of microorganisms, mycotoxins, and pesticides is based on the creation of reactive oxygen and nitrogen species that can react with contaminants on the surface of the grains (Los et al., 2017). The effectiveness of mycotoxin decontamination has been shown to be very variable, depending on a variety of characteristics such as gas composition, plasma exposure time, and byproduct production and toxicity. The nutritional content of cereal matrixes, on the other hand, may be altered by cold plasma treatment (especially those heavy in antioxidants and lipids).

10.9 POLICY AND REGULATIONS

1. National Food Security Mission—The Indian government has initiated a series of policy efforts called the National Food Security Mission (NFSM) to improve rice, wheat, and pulse production. The NFSM has three parts: (i) NFSM-Rice, (ii) NFSM-Wheat, and (iii) NFSM-Pulses, with the goal of accelerating rice output by 10 million ton, wheat production by 8 million ton, and pulse production by 2 million ton, respectively.
2. National Seed Policy (NSP)—The NSP was enacted in 2002 with the goal of creating a conducive environment for the seed business to take advantage of current and future prospects, as well as defending the interests of Indian farmers and preserving agro-biodiversity.
3. New Agricultural Policy (NAP)—The NAP was enacted in 2000 with the goal of providing farmers with price protection and free movement of agricultural goods across the country. It grew at a pace of nearly 4% every year.
4. Agriculture Policy—It was established for the years 2000–2020 with the goals of developing a clear long-term vision in agriculture, establishing sustainable natural resource management, reducing bureaucracy in institutions, enhancing self-help, and articulating a clear vision on agricultural incentive framework.

10.9.1 Regulations

10.9.1.1 Food Safety and Standards Authority of India

In accordance with the guiding principles outlined within the Food Safety and Standards Act of 2011 and India's commitments to the WTO, the Food Safety and Standards Authority of India (FSSAI) developed a food category system (FCS) that's coherent with the categorization system employed in the Codex General Standard for Food Additives (GSFA). Regulation 2.4 of the FSS (Food Product Rules and Food Additives) Regulations, 2011, includes standards for cereal and cereal products. Atta, maida, semolina, besan, barleycorn, food grain, cornflour, corn flakes, custard powder, macaroni products, malted and malt-based goods, oatmeal, and many other items are covered by this regulation.

For sago, *Triticum durum*, wheat semolina and whole *Triticum durum* semolina, *Pennisetum americanum* flour, sorghum, soyabean, soy protein products, ragi, amaranth, and wheat protein products, the FSSAI created new standards in 2017 to supplement the present FSS, 2011 rules.

10.9.1.2 International Organization for Standardization

ISO 24333:2009 defines the standards for dynamic or static sampling of cereals and cereal products, using manual or mechanical methods, in order to check their quality and condition. It is used to sample for the detection of pollutants, unwanted chemicals, and parameters that are typically homogeneously distributed, such as those required to assess quality or conformity with specifications.

10.10 CONCLUSION

The processing of cereals is an important part of the food supply chain. Milling (dry or wet) and malting are the two most common methods used in the grain industry. Despite their potential for use in the enhancement of functional meals or the extraction of bioactive compounds, cereal industry by-products end up as feed, fuel, or waste. Because cereal by-products are being used as food additives, they must meet the same quality and safety criteria as other foods meant for human consumption. In terms of waste, food, and feed legislation, there is still a long way to go before allowing for the reuse of former foodstuffs and the valorization of cereal by-products without jeopardizing food and feed safety. Food waste must also be recovered rather than discarded, which necessitates the use of innovative methods. Conventional solutions, such as animal feed, which is the primary choice for the cereal business, or composting, only utilize a portion of the waste generated by the cereal industry.

REFERENCES

Ačkar, Đ., Jozinović, A., Babić, J., Milićević, B., Panak Balentić, J., & Šubarić, D. 2018. Resolving the problem of poor expansion in corn extrudates enriched with food industry by-products. *Innovative Food Science & Emerging Technologies*, 47:517–524. https://doi.org/10.1016/j.ifset.2018.05.004

Alldrick, A. 2017. Chemical contamination of cereals. In *Chemical contaminants and residues in food*. Woodhead Publishing. pp. 427–449

Alrahmany, R., & Tsopmo, A. 2012. Role of carbohydrases on the release of reducing sugar, total phenolics and on antioxidant properties of oat bran. *Food Chemistry*, *132*(1):413–418. https://doi.org/10.1016/j.foodchem.2011.11.014

Alvarenga, I.C., Ou, Z., Thiele, S., Alavi, S. and Aldrich, C.G. 2018. Effects of milling sorghum into fractions on yield, nutrient composition, and their performance in extrusion of dog food. *Journal of Cereal Science*, 82:121–128.

Awika, J. M. 2011. Major cereal grains production and use around the world. In *Advances in cereal science: implications to food processing and health promotion*. American Chemical Society. 1–13.

Behall, K. M., Scholfield, D. J., Hallfrisch, J. G., & Liljeberg-Elmståhl, H. G. 2006. Consumption of both resistant starch and β-glucan improves postprandial plasma glucose and insulin in women. *Diabetes Care*, 29(5):976–981.

Bhattacharjee, P. and Singhal, R.S. 2016. Effect of irradiation on food texture and rheology. *Novel Food Processing: Effects on Rheological and Functional Properties*. 103–125.

Boita, E. R. F., Oro, T., Bressiani, J., Santetti, G. S., Bertolin, T. E., & Gutkoski, L. C. 2016. Rheological properties of wheat flour dough and pan bread with wheat bran. *Journal of Cereal Science*, 71:177–182. https://doi.org/10.1016/j.jcs.2016.08.015

Buenavista, R. M. E., Siliveru, K., & Zheng, Y. 2021. Utilization of distiller's dried grains with solubles: A review. *Journal of Agriculture and Food Research*, 5:1–9. https://doi.org/10.1016/j.jafr.2021.100195

Bushuk, W. 2004. *Rye. Encyclopedia of grain science* (vol. 3). Academic Press. 85–91.

Cappa, C., & Alamprese, C. 2017. Brewer's spent grain valorization in fiber-enriched fresh egg pasta production: Modelling and optimization study. *LWT—Food Science and Technology*, *82*, 464–470. https://doi.org/10.1016/j.lwt.2017.04.068

Casarotti, S. N., Borgonovi, T. F., Batista, C. L. F. M., & Penna, A. L. B. 2018. Guava, orange and passion fruit by-products: Characterization and its impacts on kinetics of acidification and properties of probiotic fermented products. *LWT—Food Science and Technology*, *98*:69–76. https://doi.org/10.1016/j.lwt.2018.08.010

Cassano, A., & Galanakis, C. M. 2018. Membrane technologies for the fractionation of compounds recovered from cereal processing by-products. In Charis M. Galanakis (Ed.), *Sustainable recovery and reutilization of cereal processing by-products* (1st ed., pp. 159–187). Elsevier. https://doi.org/10.1016/B978-0-08-102162-0.00006-X

Cermeño, M., Dermiki, M., Kleekayai, T., Cope, L., McManus, R., Ryan, C., Felix, M., Flynn, C., & FitzGerald, R. J. 2021. Effect of enzymatically hydrolysed brewers' spent grain supplementation on the rheological, textural and sensory properties of muffins. *Future Foods*, *4*, 100085. https://doi.org/10.1016/j.fufo.2021.100085

Chemat, F., Zill-e-Huma, & Khan, M. K. 2011. Applications of ultrasound in food technology: Processing, preservation and extraction. *Ultrasonics Sonochemistry*, *18*(4): 813–835. https://doi.org/10.1016/j.ultsonch.2010.11.023

Chen, J.J., Lu, S. and Li, C.Y., 1999. Effects of milling on the physicochemical characteristics of waxy rice in Taiwan. *Cereal Chemistry*, 76(5): 796–799.

Chiang, P.Y. and Yeh, A.I., 2002. Effect of soaking on wet milling of rice. *Journal of Cereal Science*, 35(1):85–94.

Comunian, T. A., Silva, M. P., & Souza, C. J. F. 2021. The use of food by-products as a novel for functional foods: Their use as ingredients and for the encapsulation process. *Trends in Food Science and Technology*, *108*:269–280. https://doi.org/10.1016/j.tifs.2021.01.003

Cooray, S. T., & Chen, W. N. 2018. Valorization of brewer's spent grain using fungi solid-state fermentation to enhance nutritional value. *Journal of Functional Foods*, *42*: 85–94. https://doi.org/10.1016/j.jff.2017.12.027

CRA. 2013. From starch to bio-products: the evolution of products of the corn refining industry. Corn Refiners Association 2013 Annual Report, 6–11. http://issuu.com/sweetener into/docs/2013corn_annual?e=6310424/2642287 (accessed October 19, 2015).

Dapčević-Hadnadev, T., Hadnadev, M., & Pojić, M. 2018. The healthy components of cereal by-products and their functional properties. In Charis M. Galanakis (Ed.), *Sustainable recovery and reutilization of cereal processing by-products* (1st ed., pp. 27–61). Elsevier Ltd. https://doi.org/10.1016/B978-0-08-102162-0.00002-2

Deng, L., Zhang, X., Li, Y., Que, F., Kang, X., Liu, Y., Feng, F., & Zhang, H. 2018. Characterization of gelatin/zein nanofibers by hybrid electrospinning. *Food Hydrocolloids*, *75*:72–80. https://doi.org/10.1016/j.foodhyd.2017.09.011

Diehl, J.F. 1995. Potential and Current applications of food irradiation: I. Overview. In *Safety of irradiated foods* (pp. 309–356). CRC Press.

Doan, N. T. T., Lai, Q. D., Vo, H. V., & Nguyen, H. D. 2021. Influence of adding rice bran on physio-chemical and sensory properties of bread. *Journal of Food Measurement and Characterization*, *15*(6):5369–5378. https://doi.org/10.1007/s11694-021-01111-5

Duţă, D. E., Culeţu, A., & Mohan, G. 2018. Reutilization of cereal processing by-products in bread making. In C. M. Galanakis (Ed.), *Sustainable recovery and reutilization of cereal processing by-products* (1st ed., 279–317). Elsevier Ltd. https://doi.org/10.1016/B978-0-08-102162-0.00010-1

Elleuch, M., Bedigian, D., Roiseux, O., Besbes, S., Blecker, C., & Attia, H. 2011. Dietary fibre and fibre-rich by-products of food processing: Characterisation, technological functionality and commercial applications: A review. *Food Chemistry*, *124*(2):411–421. https://doi.org/10.1016/j.foodchem.2010.06.077

ElMekawy, A., Diels, L., De Wever, H., & Pant, D. 2013. Valorization of cereal based biorefinery byproducts: Reality and expectations. *Environmental Science and Technology*, *47*(16):9014–9027. https://doi.org/10.1021/es402395g

El-Wakil, A. E.-A. A., Moustafa, H., & Youssef, A. M. 2020. Antimicrobial low-density polyethylene/low-density polyethylene-grafted acrylic acid biocomposites based on rice bran with tea tree oil for food packaging applications. *Journal of Thermoplastic Composite Materials*, 1–19. https://doi.org/10.1177/0892705720925140

Evers, T. and Millar, S. 2002. Cereal grain structure and development: Some implications for quality. *Journal of Cereal Science*, *36*(3):261–284.

FAO. 1995. Sorghum and millets in human nutrition. *FAO Food and Nutrition Series (FAO)*, (27):16–9.

FAO.2020. FAOSTAT Database: Crop statistics. www.fao.org/faostat/en/#data/QC

Fărcaş, A., Dreţcanu, G., Pop, T. D., Enaru, B., Socaci, S., & Diaconeasa, Z. 2021. Cereal processing by-products as rich sources of phenolic compounds and their potential bioactivities. *Nutrients*, *13*(11):3934. https://doi.org/10.3390/nu13113934

Fernando, B. 2013. Rice as a source of fibre. *Journal of Rice Research*, *1*(2):1–4.

Fierascu, R. C., Fierascu, I., Avramescu, S. M., & Sieniawska, E. 2019. Recovery of natural antioxidants from agro-industrial side streams through advanced extraction techniques. *Molecules*, *24*(23):1–29. https://doi.org/10.3390/molecules24234212

Fratianni, A., Panfili, G., & Cubadda, R. 2015. Carotenoids, tocols, and retinols during the pasta-making process. In Victor Preedy (Ed.), *Processing and impact on active components in food* (pp. 309–317). Elsevier. https://doi.org/10.1016/B978-0-12-404699-3.00037-8

Gili, R. D., Palavecino, P. M., Cecilia Penci, M., Martinez, M. L., & Ribotta, P. D. 2017. Wheat germ stabilization by infrared radiation. *Journal of Food Science and Technology*, *54*(1):71–81. https://doi.org/10.1007/s13197-016-2437-z

Gobbetti, M., De Angelis, M., Di Cagno, R., Polo, A., & Rizzello, C. G. 2020. The sourdough fermentation is the powerful process to exploit the potential of legumes, pseudo-cereals and milling by-products in baking industry. *Critical Reviews in Food Science and Nutrition*, *60*(13):2158–2173. https://doi.org/10.1080/10408398.2019.1631753

Goldberg, G., ed. 2003 *Plants: Diet and Health*. The Report of the British Nutrition Foundation Task Force. Blackwell, Oxford.

Gómez, M., Jiménez, S., Ruiz, E., & Oliete, B. 2011. Effect of extruded wheat bran on dough rheology and bread quality. *LWT—Food Science and Technology*, 44(10):2231–2237. https://doi.org/10.1016/j.lwt.2011.06.006

Gómez-Cruz, I., Cara, C., Romero, I., Castro, E., & Gullón, B. 2020. Valorisation of exhausted olive pomace by an eco-friendly solvent extraction process of natural antioxidants. *Antioxidants*, 9(10):1–18. https://doi.org/10.3390/antiox9101010

Gómez-García, R., Campos, D. A., Aguilar, C. N., Madureira, A. R., & Pintado, M. 2021. Valorisation of food agro-industrial by-products: From the past to the present and perspectives. *Journal of Environmental Management*, 299:1–10 https://doi.org/10.1016/j.jenvman.2021.113571

Gul, K., Yousuf, B., Singh, A. K., Singh, P., & Wani, A. A. 2015. Rice bran: Nutritional values and its emerging potential for development of functional food—A review. *Bioactive Carbohydrates and Dietary Fibre*, 6(1):24–30. https://doi.org/10.1016/j.bcdf.2015.06.002

Gupta, A. K., Jha, A. K., & Singhal, S. (2021). Optimisation of modification parameters for amaranth starch for the development of pudding and study of the quality traits of developed pudding. *Acta Alimentaria*, 50(1):22–32.

Gupta, S., Jaiswal, A. K., & Abu-Ghannam, N. 2013. Optimization of fermentation conditions for the utilization of brewing waste to develop a nutraceutical rich liquid product. *Industrial Crops and Products*, 44:272–282. https://doi.org/10.1016/j.indcrop.2012.11.015

Haque, E., 1991. Application of size reduction theory to roller mill design and operation. *Cereal Foods World*, 36(4):368–375.

Hassan AB, Osman GA, Rushdi MA, Eltayeb MM, Diab EE. 2009. Effect of gamma irradiation on the nutritional quality of maize cultivars (*Zea mays*) and sorghum (*Sorghum bicolor*) grains. *Pakistan Journal of Nutrition*, 8(2):167–71.

Healthy Minor Cereals. 2016. Deliverable D8.1: Report on the market potential of minor cereal crops and consumers perceptions about them in different European regions. https://toolkit.ecn.cz/img_upload/0f728a035bc05faf5df37da80e732579/d-8.1-market-potential-of-mc_final.pdf (accessed 19 October 2021).

Heiniö, R. L., Noort, M. W. J., Katina, K., Alam, S. A., Sozer, N., de Kock, H. L., Hersleth, M., & Poutanen, K. 2016. Sensory characteristics of whole grain and bran-rich cereal foods—A review. *Trends in Food Science & Technology*, 47:25–38. https://doi.org/10.1016/j.tifs.2015.11.002

Heo, S., Lee, S.M., Shim, J.H., Yoo, S.H. and Lee, S. 2013. Effect of dry-and wet-milled rice flours on the quality attributes of gluten-free dough and noodles. *Journal of Food Engineering*, 116(1):213–217.

Heredia-Sandoval, N. G., Granados-Nevárez, M. del C., Calderón de la Barca, A. M., Vásquez-Lara, F., Malunga, L. N., Apea-Bah, F. B., Beta, T., & Islas-Rubio, A. R. 2020. Phenolic acids, antioxidant capacity, and estimated glycemic index of cookies added with brewer's spent grain. *Plant Foods for Human Nutrition*, 75(1):41–47. https://doi.org/10.1007/s11130-019-00783-1

Hoseney, R.C., 1994. Yeast leavened products. *Principles of Cereal Science and Technology*, 229–270.

Humer, E., Schwarz, C. and Schedle, K. 2015. Phytate in pig and poultry nutrition. *Journal of Animal Physiology and Animal Nutrition*, 99(4):605–625.

Hutchinson, J. B., & Martin, H. F. 1970. Nutritive value of wheat bran I—Effects of fine grinding upon bran, and of added bran upon the protein quality of white flour. *Journal of the Science of Food and Agriculture*, 21(3):148–151. https://doi.org/10.1002/jsfa.2740210311

Jahurul, M. H. A., Zaidul, I. S. M., Ghafoor, K., Al-Juhaimi, F. Y., Nyam, K.-L., Norulaini, N. A. N., Sahena, F., & Mohd Omar, A. K. 2015. Mango (*Mangifera indica* L.) by-products and their valuable components: A review. *Food Chemistry*, 183:173–180. https://doi.org/10.1016/j.foodchem.2015.03.046

Jenkins, A. L., Jenkins, D. J. A., Zdravkovic, U., Würsch, P., & Vuksan, V. 2002. Depression of the glycemic index by high levels of β-glucan fiber in two functional foods tested in type 2 diabetes. *European Journal of Clinical Nutrition*, 56(7):622–628.

Jha, A. K., Kumari, S., Gupta, A. K., & Shashank, A. (2021). Improvement in pasting, thermal properties, and in vitro digestibility of isolated Amaranth starch (*Amaranthus cruentus* L.) by addition of almond gum and gum ghatti powder. *Journal of Food Processing and Preservation*, 45(10):e15829.

Juliano, B. O., & Tuaño, A. P. P. 2019. Gross structure and composition of the rice grain. *In* Jinsong Bao (Ed.), *Rice—Chemistry and Technology* (4th Ed., Issue 1, pp. 31–53). Elsevier. https://doi.org/10.1016/B978-0-12-811508-4.00002-2

Keya, E.L. and Sherman, U. 1997. Effects of a brief, intense infrared radiation treatment on the nutritional quality of maize, rice, sorghum, and beans. *Food and Nutrition Bulletin*, *18*(4):1–6.

Koehler, P. and Wieser, H. 2013. Chemistry of cereal grains. In *Handbook on sourdough biotechnology* (pp. 11–45). Springer.

Ktenioudaki, A., Alvarez-Jubete, L., Smyth, T. J., Kilcawley, K., Rai, D. K., & Gallagher, E. 2015. Application of bioprocessing techniques (sourdough fermentation and technological aids) for brewer's spent grain breads. *Food Research International*, *73*:107–116. https://doi.org/10.1016/j.foodres.2015.03.008

Ktenioudaki, A., O'Shea, N., & Gallagher, E. 2013. Rheological properties of wheat dough supplemented with functional by-products of food processing: Brewer's spent grain and apple pomace. *Journal of Food Engineering*, *116*(2):362–368. https://doi.org/10.1016/j.jfoodeng.2012.12.005

Lai, W. T., Khong, N. M. H., Lim, S. S., Hee, Y. Y., Sim, B. I., Lau, K. Y., & Lai, O. M. 2017. A review: Modified agricultural by-products for the development and fortification of food products and nutraceuticals. *Trends in Food Science & Technology*, *59*:148–160. https://doi.org/10.1016/j.tifs.2016.11.014

Langton, M., & Gutiérrez, J. L. V. 2021. The structure of cereal grains and their products. In R. Landberg & N. Scheers (Eds.), *Whole grains and health* (2nd ed., pp. 1–20). Wiley. https://doi.org/10.1002/9781118939420.ch1

Lestienne, I., Buisson, M., Lullien-Pellerin, V., Picq, C. and Trèche, S. 2007. Losses of nutrients and anti-nutritional factors during abrasive decortication of two pearl millet cultivars (*Pennisetum glaucum*). *Food Chemistry*, *100*(4):1316–1323.

Lima, P. M., Rubio, F. T. V., Silva, M. P., Pinho, L. S., Kasemodel, M. G. C., Favaro-Trindade, C. S., & Dacanal, G. C. 2019. Nutritional value and modelling of carotenoids extraction from pumpkin (*Cucurbita moschata*) peel flour by-product. *International Journal of Food Engineering*, *15*(5–6):1–15. https://doi.org/10.1515/ijfe-2018-0381

Lobato, L. P., Anibal, D., Lazaretti, M. M., & Grossmann, M. V. E. 2011. Extruded puffed functional ingredient with oat bran and soy flour. *LWT—Food Science and Technology*, *44*(4):933–939. https://doi.org/10.1016/j.lwt.2010.11.013

Los, A., Ziuzina, D., Boehm, D., Cullen, P.J. and Bourke, P. 2017. The potential of atmospheric air cold plasma for control of bacterial contaminants relevant to cereal grain production. *Innovative Food Science & Emerging Technologies*, *44*:36–45.

Lumdubwong, N. and Seib, P.A., 2000. Rice starch isolation by alkaline protease digestion of wet-milled rice flour. *Journal of Cereal Science*, *31*(1):63–74.

Martins, N., & Ferreira, I. C. F. R. 2017. Wastes and by-products: Upcoming sources of carotenoids for biotechnological purposes and health-related applications. *Trends in Food Science & Technology*, *62*:33–48. https://doi.org/10.1016/j.tifs.2017.01.014

McCarthy, A. L., O'Callaghan, Y. C., Neugart, S., Piggott, C. O., Connolly, A., Jansen, M. A. K., Krumbein, A., Schreiner, M., FitzGerald, R. J., & O'Brien, N. M. 2013. The hydroxycinnamic acid content of barley and brewers' spent grain (BSG) and the potential to incorporate phenolic extracts of BSG as antioxidants into fruit beverages. *Food Chemistry*, *141*(3):2567–2574. https://doi.org/10.1016/j.foodchem.2013.05.048

Mendes, G. da R. L., Souto Rodrigues, P., de las Mercedes Salas-Mellado, M., Fernandes de Medeiros Burkert, J., & Badiale-Furlong, E. 2021. Defatted rice bran as a potential raw material to improve the nutritional and functional quality of cakes. *Plant Foods for Human Nutrition*, *76*(1):46–52. https://doi.org/10.1007/s11130-020-00872-6

Mohd Esa, N., & Ling, T. B. 2013. By-products of rice processing: An overview of health benefits and applications. *Rice Research: Open Access*, *4*(1):1–11. https://doi.org/10.4172/jrr.1000107

Moreira, S. A., Alexandre, E. M. C., Pintado, M., & Saraiva, J. A. 2019. Effect of emergent non-thermal extraction technologies on bioactive individual compounds profile from different plant materials. *Food Research International*, *115*:177–190. https://doi.org/10.1016/j.foodres.2018.08.046

Mussatto, S. I. 2014. Brewer's spent grain: A valuable feedstock for industrial applications. *Journal of the Science of Food and Agriculture*, *94*(7):1264–1275. https://doi.org/10.1002/jsfa.6486

Naibaho, J., & Korzeniowska, M. 2021. Brewers' spent grain in food systems: Processing and final products quality as a function of fiber modification treatment. *Journal of Food Science*, *86*(5):1532–1551. https://doi.org/10.1111/1750-3841.15714

Naibaho, J., Butula, N., Jonuzi, E., Korzeniowska, M., Laaksonen, O., Föste, M., Kütt, M.-L., & Yang, B. 2022. Potential of brewers' spent grain in yogurt fermentation and evaluation of its impact in rheological behaviour, consistency, microstructural properties and acidity profile during the refrigerated storage. *Food Hydrocolloids*, *125*:107412. https://doi.org/10.1016/j.foodhyd.2021.107412

Narwal, S., Gupta, O. P., Pandey, V., Kumar, D., & Ram, S. 2020. Effect of storage and processing conditions on nutrient composition of wheat and barley. In O. P. Gupta, V. Pandey, S. Narwal, P. Sharma, S. Ram, & G. P. Singh (Eds.), *Wheat and barley grain biofortification* (1st ed., pp. 229–256). Elsevier. https://doi. org/10.1016/B978-0-12-818444-8.00009-2

Navarro, S. L., Capellini, M. C., Aracava, K. K., & Rodrigues, C. E. 2016. Corn germ-bran oils extracted with alcoholic solvents: Extraction yield, oil composition and evaluation of protein solubility of defatted meal. *Food and Bioproducts Processing, 100*:185–194.

Neylon, E., Arendt, E. K., Zannini, E., & Sahin, A. W. 2021. Fundamental study of the application of brewers spent grain and fermented brewers spent grain on the quality of pasta. *Food Structure, 30*:100225. https://doi.org/10.1016/j.foostr.2021.100225

Nithya, D. J., Bosco, K. A. S., Jagan Mohan, R., & Alagusundaram, K. 2013. Formulation of rice bran pasta. *Journal of Microbiology, Biotechnology and Food Sciences, 2*(6): 2423–2425.

Nocente, F., Natale, C., Galassi, E., Taddei, F., & Gazza, L. 2021. Using einkorn and tritordeum brewers' spent grain to increase the nutritional potential of durum wheat pasta. *Foods, 10*(3), 502:1–9. https://doi. org/10.3390/foods10030502

Nocente, F., Taddei, F., Galassi, E., & Gazza, L. 2019. Upcycling of brewers' spent grain by production of dry pasta with higher nutritional potential. *LWT-Food Science and Technology, 114*:1–9. https://doi. org/10.1016/j.lwt.2019.108421

Nsogning Dongmo, S., Sacher, B., Kollmannsberger, H., & Becker, T. 2017. Key volatile aroma compounds of lactic acid fermented malt based beverages-impact of lactic acid bacteria strains. *Food Chemistry, 229*:565–573. https://doi.org/10.1016/j.foodchem.2017.02.091

Nuttall, J. G., O'Leary, G. J., Panozzo, J. F., Walker, C. K., Barlow, K. M., & Fitzgerald, G. J. 2017. Models of grain quality in wheat—A review. *Field Crops Research, 202*:136–145. https://doi.org/10.1016/j. fcr.2015.12.011

OECD-FAO.2018. OECD-FAO Agricultural Outlook 2018–2027. Rome: OECD Publishing, Paris/Food and Agriculture Organization of the United Nations https://doi. org/10.1787/agr_outlook-2018-en (accessed 21 October 2021).

Onipe, O. O., Jideani, A. I. O., & Beswa, D. 2015. Composition and functionality of wheat bran and its application in some cereal food products. *International Journal of Food Science and Technology, 50*(12):2509–2518. https://doi.org/10.1111/ijfs.12935

Orona-Tamayo, D., Valverde, M. E., & Paredes-López, O. 2019. Bioactive peptides from selected Latin American food crops—A nutraceutical and molecular approach. *Critical Reviews in Food Science and Nutrition, 59*(12):1949–1975. https://doi.org/10.1080/10408398.2018.1434480

Özkaya, B., Baumgartner, B., & Özkaya, H. 2018. Effects of concentrated and dephytinized wheat bran and rice bran addition on bread properties. *Journal of Texture Studies, 49*(1):84–93. https://doi.org/10.1111/jtxs.12286

Papageorgiou, M. and Skendi, A. 2018. Introduction to cereal processing and by-products. In *Sustainable recovery and reutilization of cereal processing by-products* (pp. 1–25). Woodhead Publishing.

Pavlovich-Abril, A., Rouzaud-Sández, O., Romero-Baranzini, A. L., Vidal-Quintanar, R. L., & Salazar-García, M. G. 2014. Relationships between chemical composition and quality-related characteristics in bread making with wheat flour-fine bran blends. *Journal of Food Quality, 38*(1): 30–39. https://doi. org/10.1111/jfq.12103

Pelembe, L. A. M., Dewar, J., & Taylor, J. R. N. 2003. Food products from malted pearl millet. In *Proceeding of the workshop on the proteins of sorghum and millets: Enhancing nutritional and functional properties for Africa*, Pretoria, South Africa.

Pojić, M.M., Spasojević, N.B. and Atlas, M.Đ. 2014. Chemometric approach to characterization of flour mill streams: chemical and rheological properties. *Food and Bioprocess Technology, 7*(5):1298–1309.

Pomeranz, Y. 1982. Grain structure and end-use properties. *Food Structure, 1*(2):2.

Pradeep, P. M., Jayadeep, A., Guha, M., & Singh, V. 2014. Hydrothermal and biotechnological treatments on nutraceutical content and antioxidant activity of rice bran. *Journal of Cereal Science, 60*(1):187–192. https://doi.org/10.1016/j.jcs.2014.01.025

Prietto, L., Pinto, V. Z., El Halal, S. L. M., de Morais, M. G., Costa, J. A. V., Lim, L.-T., Dias, A. R. G., & Zavareze, E. da R. 2018. Ultrafine fibers of zein and anthocyanins as natural pH indicator. *Journal of the Science of Food and Agriculture, 98*(7):2735–2741. https://doi.org/10.1002/jsfa.8769

Principato, L., Sala, L., Duserm-Garrido, G., & Spigno, G. 2021. Effect of dietary fibre and thermal condition on rice bran wax oleogels for biscuits preparation. *Chemical Engineering Transactions, 87*:49–54. https://doi.org/10.3303/CET2187009

Prückler, M., Siebenhandl-Ehn, S., Apprich, S., Höltinger, S., Haas, C., Schmid, E., & Kneifel, W. 2014. Wheat bran-based biorefinery 1: Composition of wheat bran and strategies of functionalization. *LWT-Food Science and Technology*, *56*(2):211–221. https://doi.org/10.1016/j.lwt.2013.12.004

Punia, S., Kumar, M., Siroha, A. K., & Purewal, S. S. 2021. Rice bran oil: Emerging trends in extraction, health benefit, and its industrial application. *Rice Science*, *28*(3):217–232. https://doi.org/10.1016/j.rsci.2021.04.002

Qi, L., Li, Y., Luo, X., Wang, R., Zheng, R., Wang, L., Li, Y., Yang, D., Fang, W. & Chen, Z. 2016. Detoxification of zearalenone and ochratoxin A by ozone and quality evaluation of ozonised corn. *Food Additives & Contaminants: Part A*, *33*(11):1700–1710.

Randhir, R., Kwon, Y. I., & Shetty, K. 2008. Effect of thermal processing on phenolics, antioxidant activity and health-relevant functionality of select grain sprouts and seedlings. *Innovative Food Science & Emerging Technologies*, *9*(3):355–364.

Ravindran, R., & Jaiswal, A. K. 2016. Exploitation of Food Industry Waste for High-Value Products. *Trends in Biotechnology*, *34*(1):58–69. https://doi.org/10.1016/j.tibtech.2015.10.008

Reis, S. F., & Abu-Ghannam, N. 2014. Antioxidant capacity, arabinoxylans content and in vitro glycaemic index of cereal-based snacks incorporated with brewer's spent grain. *LWT-Food Science and Technology*, *55*(1):269–277. https://doi.org/10.1016/j.lwt.2013.09.004

Renoldi, N., Peighambardoust, S. H., & Peressini, D. 2021. The effect of rice bran on physicochemical, textural and glycaemic properties of ready-to-eat extruded corn snacks. *International Journal of Food Science & Technology*, *56*(7):3235–3244. https://doi.org/10.1111/ijfs.14939

Roth, M., Jekle, M., & Becker, T. 2019. Opportunities for upcycling cereal byproducts with special focus on distiller's grains. *Trends in Food Science & Technology*, *91*:282–293. https://doi.org/10.1016/j.tifs.2019.07.041

Roth, M., Schuster, H., Kollmannsberger, H., Jekle, M., & Becker, T. 2016. Changes in aroma composition and sensory properties provided by distiller's grains addition to bakery products. *Journal of Cereal Science*, *72*:75–83. https://doi.org/10.1016/j.jcs.2016.10.002

Saeed, F., Hussain, M., Arshad, M. S., Afzaal, M., Munir, H., Imran, M., Tufail, T., & Anjum, F. M. 2021. Functional and nutraceutical properties of maize bran cell wall non-starch polysaccharides. *International Journal of Food Properties*, *24*(1):233–248. https://doi.org/10.1080/10942912.2020.1858864

Sallam, A. S., Khalil, A. H., Mostafa, M. M., El Bedawy, A. A., & Atef, A. A. 2019. Quality aspects of pan bread prepared by partial substitution of wheat flour with defatted rice bran. *Menoufia Journal of Food and Dairy Sciences*, *4*(2):89–99. https://doi.org/10.21608/mjfds.2019.174871

Saman, P., Fuciños, P., Vázquez, J. A., & Pandiella, S. S. 2019. By-products of the rice processing obtained by controlled debranning as substrates for the production of probiotic bacteria. *Innovative Food Science & Emerging Technologies*, *51*:167–176. https://doi.org/10.1016/j.ifset.2018.05.009

Sayaslan, A. 2004. Wet milling of wheat flour: industrial processes and small-scale test methods. *LWT-Food Science and Technology*, *37*(5):499–515.

Schettino, R., Verni, M., Acin-albiac, M., Vincentini, O., Krona, A., Knaapila, A., Di Cagno, R., Gobbetti, M., Rizzello, C. G., & Coda, R. 2021. Bioprocessed brewers' spent grain improves nutritional and antioxidant properties of pasta. *Antioxidants*, *10*(5):1–22. https://doi.org/10.3390/antiox10050742

Serna-Saldivar, S. 1995. Structure and chemistry of sorghum and millets. *Sorghum and Millets: Chemistry and Technology*, 69–124.

Shewry, P.R. and Tatham, A.S. 1999. The characteristics, structures and evolutionary relationships of prolamins. In *Seed proteins* (pp. 11–33). Springer, Dordrecht.

Singha, P., Singh, S. K., Muthukumarappan, K., & Krishnan, P. 2018. Physicochemical and nutritional properties of extrudates from food grade distiller's dried grains, garbanzo flour, and corn grits. *Food Science & Nutrition*, *6*(7):1914–1926. https://doi.org/10.1002/fsn3.769

Sobota, A., Rzedzicki, Z., Zarzycki, P., & Kuzawińska, E. 2015. Application of common wheat bran for the industrial production of high-fibre pasta. *International Journal of Food Science & Technology*, *50*(1):111–119. https://doi.org/10.1111/ijfs.12641

Sonar, M. P., & Rathod, V. K. 2020. Microwave assisted extraction (MAE) used as a tool for rapid extraction of marmelosin from *Aegle marmelos* and evaluations of total phenolic and flavonoids content, antioxidant and anti-inflammatory activity. *Chemical Data Collections*, *30*:100545. https://doi.org/10.1016/j.cdc.2020.100545

Spaggiari, M., Dall'Asta, C., Galaverna, G., & del Castillo Bilbao, M. D. 2021. Rice bran by-product: from valorization strategies to nutritional perspectives. *Foods*, *10*(1):85. https://doi.org/10.3390/foods10010085

Steiner, J., Procopio, S., & Becker, T. 2015. Brewer's spent grain: source of value-added polysaccharides for the food industry in reference to the health claims. In *European Food Research and Technology*, 241(3):303–315. https://doi.org/10.1007/s00217-015-2461-7

Tapia-Hernández, J. A., Del-Toro-Sánchez, C. L., Cinco-Moroyoqui, F. J., Juárez-Onofre, J. E., Ruiz-Cruz, S., Carvajal-Millan, E., López-Ahumada, G. A., Castro-Enriquez, D. D., Barreras-Urbina, C. G., & Rodríguez-Felix, F. 2019. Prolamins from cereal by-products: Classification, extraction, characterization and its applications in micro- and nanofabrication. *Trends in Food Science & Technology*, *90*:111–132. https://doi.org/10.1016/j.tifs.2019.06.005

Tirpanalan, Ö., Reisinger, M., Smerilli, M., Huber, F., Neureiter, M., Kneifel, W., & Novalin, S. 2015. Wheat bran biorefinery—An insight into the process chain for the production of lactic acid. *Bioresource Technology*, *180*:242–249. https://doi.org/10.1016/j.biortech.2015.01.021

Tiwari, U., & Cummins, E. 2009. Factors influencing β-glucan levels and molecular weight in cereal-based products. *Cereal Chemistry*, *86*(3):290–301.

Tiwari, U., & Pojić, M. 2020. Introduction to cereal processing: Innovative processing techniques. *Innovative Processing Technologies for Healthy Grains*, 9–35.

Torbica, A., Škrobot, D., Janić Hajnal, E., Belović, M., & Zhang, N. 2019. Sensory and physico-chemical properties of whole grain wheat bread prepared with selected food by-products. *LWT—Food Science and Technology*, *114*:108414. https://doi.org/10.1016/j.lwt.2019.108414

Trigo, J. P., Alexandre, E. M. C., Saraiva, J. A., & Pintado, M. E. 2020. High value-added compounds from fruit and vegetable by-products—Characterization, bioactivities, and application in the development of novel food products. *Critical Reviews in Food Science and Nutrition*, *60*(8):1388–1416. https://doi.org/10.1080/10408398.2019.1572588

Vasconcelos, M. C. B. M. de, Bennett, R., Castro, C., Cardoso, P., Saavedra, M. J., & Rosa, E. A. 2013. Study of composition, stabilization and processing of wheat germ and maize industrial by-products. *Industrial Crops and Products*, *42*(1):292–298. https://doi.org/10.1016/j.indcrop.2012.06.007

Wang, J., Tang, J., Ruan, S., Lv, R., Zhou, J., Tian, J., Cheng, H., Xu, E., & Liu, D. 2021a. A comprehensive review of cereal germ and its lipids: Chemical composition, multi-objective process and functional application. *Food Chemistry*, *362*:130066. https://doi.org/10.1016/j.foodchem.2021.130066

Wang, L., Boussetta, N., Lebovka, N., & Vorobiev, E. 2020. Cell disintegration of apple peels induced by pulsed electric field and efficiency of bio-compound extraction. *Food and Bioproducts Processing*, *122*:13–21. https://doi.org/10.1016/j.fbp.2020.03.004

Wang, L., Ke, L., Rao, P., & Zhang, Y. 2021b. Fabrication and characterization of curcumin-loaded nanoparticles using protein from brewers' spent grain. *LWT-Food Science and Technology*, *150*(149), 111992. https://doi.org/10.1016/j.lwt.2021.111992

Welch, R.W. 2012. *The oat crop: Production and utilization*. Springer Science & Business Media.

Zhang, M., Bai, X., & Zhang, Z. 2011. Extrusion process improves the functionality of soluble dietary fiber in oat bran. *Journal of Cereal Science*, *54*(1):98–103. https://doi.org/10.1016/j.jcs.2011.04.001

CHAPTER 11

Recent Technology in Cereal Science

Bharti Mittu, Zarina Begum, Abida Bhat and Mohammad Javed Ansari

CONTENTS

DOI: 10.1201/9781003252023-11

11.1 INTRODUCTION

Cereal grains and cereal products are rich sources of the total food intake and nutrients that sustain life. Cereals and cereal products are staple foods of the human diet in most parts of the globe (with 60% of calories and protein consumed provided by cereals) and hence are culti-vated in huge amounts, as they provide more energy. In general, cereal grains are composed of three major components: 65–75% carbohydrates (mainly in the form of digestible sugars and starches), 7–13% protein, 2–6.5% lipids, and 11–14% water (Baniwal et al., 2021) and other minor, albeit important, ingredients, such as minerals, fibers, vitamins, and bioactive metabo-lites. Their nutritional contribution consists of 5% of calories and 67% of protein; hence they are the most important source of food all over the world. Cereals provide 10,000–15,000 kJ/Kg of energy, approximately 15–20 times more than fruits and vegetables. In fact, whole grains have more potential to make a good contribution in providing dietary fiber, minerals, vitamins, and antioxidants. This is due to the fact that most of the bioactive components of grain are situ-ated in the bran and germ. Cereal grains are somewhere deficient in vitamins A, D, B-12, and C (Arvola et al., 2007).

Cereals have the capability to grow in adverse environmental conditions even in low fertile soil conditions and give a high yield in comparison with other crops. The conditions for storing har-vested cereal crops influence the nutritional benefits of cereals. The ease of production and storage, together with the relatively low cost and the nutritional contribution of cereal grains, have resulted in their widespread use as foods all over the world. Cereal grains are eaten in many, like whole, pro-cessed, or milled or via other techniques like making flour, with other ingredients like salt, sugar, bran, oil, starch, spices, dried forms, desserts, and soups.

11.2 WHAT ARE CEREALS?

Cereals are mainly composed of carbohydrates, proteins, fat, vitamins, and minerals and ample quantities of soluble and insoluble dietary fiber. The main components present in whole grains with health-enhancing properties (dietary fiber, resistant starch, beta-glucan, inulin, carotenoids, tocotrienols, and tocopherols, phenolics, orzyzenol, vitamins and omega-3 fatty acid, folate, iron, phosphorous, zinc, and magnesium) also play a role in protecting from diseases (hypertension, cardiovascular disorders, type 2 diabetes, obesity, and various types of cancer) in both humans and animals (Baniwal et al., 2021; Macauley, 2015). The major constituents of the principal cereals are listed in Table 11.1.

Table 11.1 Nutritional Composition of Cereal Grains (Sramkova et al., 2009)

Cereal	Energy (kcal)	Carbohydrates (g)	Protein (g)	Fat (g)	Dietary fibers (g)
Wheat	339	71	13.3	2	10.7
Barley	352	77.7	9.9	2.3	15.6
Rice	370	81.68	6.81	1.9	2.8
Brown rice (cooked/raw)	111/357	23	2.6/6.7	1.8/2.8	0.4/1.9
Maize	360	72.2	8.9	3.9	2.2
Oats	180	29	7	3	5
Sorghum	329	72.09	10.62	3.3	6.7
Millet	378	72.9	9.9	2.9	3.2

11.3 RECENT TECHNOLOGY IN CEREAL SCIENCE

There is a need for change in cereals with technology and the latest advancements for enhancing the nutritional characteristics of cereals. The nutritional and functional characteristics of cereal grains are discussed in the following sections.

11.4 GLUTEN INTOLERANCE: NEED FOR ALTERNATIVE FLOURS

Gluten is a type of structural protein found in different cereal grains (barley, wheat, rye). Its unique viscoelastic property provides elasticity to the dough, which is subsequently used to make different products with a chewy texture and the required shape. Some people experience health risks when eating gluten-containing foods if they are intolerant of gluten. Gluten consumption leads to a range of gluten-related disorders, such as celiac disease, dermatitis herpetiformis, gluten ataxia, and non-coeliac gluten sensitivity (Al-Toma et al., 2019).

Hence, a number of flours are gaining ground in the competitive food industry as alternative flours to replace gluten-containing flour for people with gluten allergies and for those who prefer to limit their gluten intake (Barbaro et al., 2018). A diet free from gluten can be effective in treating people who are suffering from gluten intolerance. A gluten-free diet is the talk of the day, and gluten-free products characterized by low fiber content and high starch content, shorter shelf life, or texture issues have flooded the food industry (Demirkesen & Ozkaya, 2020). In general, a gluten-free diet needs the use of gluten-free cereals—oat, sorghum, millets and rice corn, teff—and pseudo-cereals—quinoa, buckwheat, canihua, and amaranth—but other natural gluten-free foods, like potatoes, legumes, oilseeds, tapioca, nut flour, and fruits and vegetables are also available (Gobbetti et al., 2018). However, gluten products are difficult to replace. In addition to nutrient deficiencies, gluten-free dough has less elastic, cohesiveness, and baking quality, so it is difficult to manage (Bender & Schonlechner, 2020; Cappelli et al., 2020). In addition, new naturally gluten-free baking ingredients and new methods of processing traditional ingredients are constantly being sought. Today, the main challenge for food technologists is the manufacture of gluten-free food products like bread, pastries, pasta, and bakery products. Because of the absence of gluten, other additives are required to provide texture, volume, satisfactory crumb, shelf life, and sensory quality to the flour. These include hydrocolloids, enzyme preparations, or sourdough. These additives change the recipe and technology of production (Fernanda and Martins, 2021; Šmídová & Rysová, 2022).

Gluten-free bread and bakery product formulations can be created by combining cereal flour and flour of other crops with starches. A combination of cereals with other legumes has proved nutritionally advantageous. For example, various buckwheat seeds milled in different fractions, rice and corn flour, and fermented buckwheat brans have been used for gluten-free bread preparation. Buckwheat flour, in general, improves the porosity, specific bread volume, and crumb texture. Soya, a traditional gluten-free ingredient, and coconut flour could also be used, as they too bind water very well (Šmídová & Rysová, 2022). Starch, together with flours from gluten-free crops, is one of basic ingredients in gluten-free bread and bakery products. Starch affects gluten-free products by enhancing softness, ensures uniformity in consistency of dough, and also affects starch gelatinization (Šmídová & Rysová, 2022).

Starch is a versatile biomaterial that have a great role in the food industry. It can undergo a number of reactions like hydrolysis, oxidation, esterification, and etherification. These reactions modify the starch, which can be then be used in confectionaries, baked foods, soups, and salad dressings. The functional properties of starch are crucial for food processing and human nutrition, like swelling, pasting, gelatinization retrogradation, and susceptibility of starch to enzymatic digestion (Beta, 2021). Milling techniques produce damaged starch and hence remarkably affect flour quality during dough- and bread-making performance. Recent studies on milling techniques show that production of damaged starch depends on the various types of milling methods. Damaged starch can be defined as the proportion of grain starch granules that are physically broken and fragmented while milling, which is due to the hardness of kernel and intensity at which it is ground. It is a parameter for checking the quality of wheat flour, because if starch is damaged, it will become more prone to enzymatic hydrolysis, which further influences the capacity of the flour for water absorption, stickiness of dough, and color of bread crust and, for leavened products, yeast activity during fermentation (Carcea et al., 2022).

The focus is also on bread products, with rice as a suitable substitute for wheat. Rice has unique nutritional aspects; it is hypoallergenic, colorless, and bland in taste and has a lower amount of prolamin. Brown or unpolished rice containing many bioactive components, including phenolics, amino acids, and fiber minerals, can be an alternative to rice (Yano, 2019).

Sorghum seed flour is substitute source for functional compounds in replacement of whole wheat flour, so it is used as a useful bakery ingredient. Sorghum has a significantly higher content of raw fiber and fat as compared to whole wheat flour. Mineral contents like potassium and magnesium are also higher in sorghum seed flour than whole wheat flour. It has been found to be a useful flour in many food recipes like porridges, breads, and griddle cake (Coulibaly et al., 2020). The use of sorghum in production of bread products still has challenges associated with it, like poor sensory attributes, reduced volume, and hard texture. All these limitations of unconventional flours have been improved with sourdough fermentation, increasing nutritive value while giving attractive flavor and good texture (Montemurro et al., 2019; Nionelli et al., 2018; Ogunsakin et al., 2020). Especially, the utilization of sourdough has shown a positive effect on different bakery products in shelf life, sensory, and nutritional texture (De Vuyst et al., 2009; Gobbetti et al., 2014). Fermentation with sourdough improves the quality and desirability of bread after adding gluten-free matrices. Sourdough fermentation releases acid, which works on polysaccharides and improves swelling just like gluten and produces gluten-free bread with a improved texture and good appearance (Moroni et al., 2009). Quinoa flour fermented with the species *Lactobacillus plantarum* ATCC 8014 has been used in the bakery industry for the production of gluten-free muffins and various bakery products.

Voinea et al. proposed potassium chloride as a substitute for sodium chloride in bread recipes and other bakery products; this is an appropriate measure to meet the requirements of consumers who demand a low-sodium diet. Sodium chloride has a vital role in yeast activity, bread quality, and dough rheology, especially from a sensory point of view. This substitution was only partially (22%) able to obtain the best rheological dough behavior (Voinea, 2020).

11.5 CEREAL-BASED LOW GLYCEMIC INDEX FOODS FOR HEALTHCARE

Jenkins et al. developed the concept of glycemic index (GI) of food products in the early 1980s (1981). This index represents the relative ability of consumed foods containing carbohydrates to increase the level of glucose in the blood (i.e., postprandial glucose concentration). Whole-grain cereals have less potential to raise blood glucose levels compared to refined cereals. Whole grains not only have low GI but also contain various bioactive components, micronutrients, and resistant starch and soluble and insoluble fibers, which play a crucial role in promoting the health of an individual. Various studies have shown that the health benefits of regular consumption of a low-GI diet rich in whole grains have a crucial role in dietary treatment and reduction of developing chronic disorders, such as cardiovascular diseases, type 2 diabetes mellitus, polycystic ovarian syndrome, and different types of cancers. Researchers are continuously developing new multigrain products by keeping the balance between nutrition and sensory parameters and thereby enhancing the health benefits (Slavin, 2004).

There is ample proof that a low glycemic index diet is beneficial to overall health, as it helps in weight control and stimulates a controlled and sustained release of the hormone insulin. There is a slow release of glucose into the bloodstream, causing a slow increase in blood glucose and insulin concentrations over time. In addition to body mass index (BMI) and body weight, there are many other indicators, like body fat, cholesterol percentage, resting energy expenditure, and so on, that indicate the cause of loss of weight and other health effects (Chunxiao et al., 2022).

Recent evidence shows that a low GI diet is far superior to other diets in controlling body mass, fasting blood glucose level, body mass index, and glycosylated hemoglobin percent in patients with one of the four common metabolic diseases (type 2 diabetes, cardiovascular disease, obesity, and metabolic syndrome) (Chunxiao et al., 2022).

11.6 CEREALS AND PSEUDOCEREALS FOR BAKERY INDUSTRY AND HEALTHCARE

Cereals and pseudocereals provide basic nutrition to the human population as well as essential compounds such as fiber, essential amino acids, antioxidants, and vitamins and minerals. Pseudocereals can replace other cereal crops for improving nutrition, health, and people's economic status. Amaranth, quinoa (both *Amaranthaceae*), chia (*Lamiaceae*), and buckwheat (*Polygonaceae*) are the most prevalent pseudocereals used in the food industry (Petrova & Petrov, 2020).

Pseudo-cereals are a good alternative to wheat flour in utilization in various bakery products as they are major sources of proteins, essential amino acids, starch, calories, minerals, vitamins, dietary fiber, and phytochemicals (polyphenols, saponins, phytosterols, phytosteroids, and betalains), with possible future health benefits (Martínez-Villaluenga et al., 2020; Culetu et al., 2021; Stamatovska et al., 2018). Their regular intake is beneficial for preventing malnutrition in children and other immunological disorders in adults (Martínez-Villaluenga et al., 2020).

11.7 BIOFORTIFICATION: STRATEGIES FOR WHEAT AND WHEAT-BASED PRODUCTS

Biofortification is a food-based strategy and a suitable vehicle for micronutrient fortification. It is a promising, sustainable, and cheap means of providing micronutrients and a solution to widespread deficiencies of vitamin A, zinc, and iron, which are common in populations with limited access to nutritional diets. Biofortification can be attained by mineral fertilization or crop breeding. In a few countries, wheat selenium content can be improved by applying fertilizers containing selenium, whereas some studies show the sustainable and cost-effective biofortification of zinc and iron with crop breeding. Crop breeding has proved an effective and sustainable method to biofortify cereal cultivars. Different components like proteins, essential amino acids, amylose, zinc, and vitamin A have been successfully increased in selected cereals (Palacios-Rojas et al., 2020). Nevertheless, this kind of biofortification needs proper identification of genetic resources for determining the interaction of genotype and environment, defining the desired level of increase in micronutrients, and cost effectiveness (Phimolsiripol & Suppakul, 2016). The bioavailability of minerals like iron and zinc could be enhanced by breeding for reducing anti-nutrient compounds like phytates, which provide strength to immune systems and prevents anemia and non-communicable diseases (Poole et al., 2021).

Wheat, rice, and maize are the most popular and accepted foods in people's worldwide diet, and strategies of flour fortification are gaining considerable popularity for meeting the nutritional and health needs of populations (Cardoso et al., 2019). Biofortification of wheat with micronutrients can be a sustainable and cost-effective strategy to tackle zinc, iron, iodine, and selenium deficiencies to prevent malnutrition among target populations. Anthocyanin-biofortified colored wheat has also created interest in industry due to its antioxidant and anti-inflammatory activity (Cardoso et al., 2019).

Fortification of breads with soyabean or oat flour has substantially improved their protein quality and fiber content. Oat protein has higher lysine content, so it is superior to wheat protein. The amount of dietary fiber in breads can also be increased by using barley, flaxseed, and rye flours. Dietary fiber protects from cardiovascular diseases, cancer, and diabetes (Yano, 2019).

Another important public health intervention to date is the fortification of flour with folic acid. Flours of wheat and maize have been fortified with folic acid to meet the daily requirement of folic acid, which also reduces the risk of neural tube defects (Sayed et al., 2008; Bower et al., 2009; Santos et al., 2015, to treat people with spina bifida (Grosse et al., 2016) and other adverse health outcomes. In countries where folate intake is not sufficient, population groups such as women and children are at a high risk of deficiency disorders. Additionally, folic acid fortification has been instrumental in reducing the risk of other health issues, including cardiovascular disease, cancer, and depression.

11.8 ENHANCED BIOACTIVE COMPOUNDS IN CEREALS

Cereals are a very rich source of many bioactive components, such as dietary fiber, polysaccharides (starch), protein, peptides, saponins, flavonoids, and polyphenols. Cereals contain essential amino acids, and the lipid fraction of cereals has essential fatty acids like linoleic acids; palmitic, fat-soluble vitamins; and phytosterols. They also have a remarkable quantity of vitamin B, thiamine, riboflavin, pyridoxine, and niacin; some folic acid and biotin; and tocol derivative precursors of vitamin E. They have ample amounts of calcium, zinc, selenium, phosphorous, magnesium, manganese, potassium, and copper. The polar lipid fraction of cereals reduces cholesterol absorption and improves the gastrointestinal microbiome. Antioxidant and phytochemical components provide protection against impaired immune systems, cardiovascular diseases, neurological damage, cataracts, and cancers (Poole et al., 2021).

These components of whole-grain cereals have been demonstrated to exert various biological functions. Epidemiological studies have further suggested that the health benefits of coarse cereals are explained by the combined impacts of phytochemicals, micronutrients, and dietary fiber. These bioactive substances are covalently linked to the cellular matrix of the grain and hence hinder bioavailability. There have been enhancements in strategies and processing techniques (such as germination, fermentation, and extrusion) that have resulted in functional cereal-based foods carrying increased content of bioactive compounds and dietary fiber. Cooking techniques like baking and roasting and the use of microwaves also greatly influence the amounts of bioactive components in cereals.

Dietary fiber acts as a substrate for gut microbiota, which influences undigested polysaccharides and oligosaccharides, glycoprotein, proteins, and peptides (Machate et al., 2020). Dietary fibers could be extracted from a number of sources by different physical, mechanical, chemical, and enzymatic processes (Tejada-Ortigoza et al., 2016). These fiber-rich by-products have well-established functional and physiological properties and hence could be used as sources of functional ingredients for improving hydration properties, oil-holding capacity, and rheological characteristics of different food products, with a key impact on nutritional health benefits. Recently, dietary fibers of fruits and vegetables have been of interest to food scientists (Machate et al., 2020).

Cereal brans (a by-product generated during milling) are often discarded or reserved for animal feeding. These brans are a rich source of many bioactive substances like flavonoids, phenolic compounds, glucans, and pigments, and their addition to flours influences the physico-chemical and nutritional characteristics of bread and bakery products. The addition of bran in bread recipes causes a reduction in the volume of the bread loaf, denser crumb, and darker crust texture as compared to traditional products (Blandino et al., 2013), but the water, fiber, protein, ash, lipid, and phytate content is enhanced noticeably (Pauline et al., 2020). Therefore, they have numerous health benefits and have been mentioned in studies on coronary heart diseases, colon and several other forms of cancer, obesity, regulation of blood glucose level, and gastrointestinal disease prevention (Ibidapo et al., 2015; Previtali et al., 2016; Rasha et al., 2016; Dhingra et al., 2012; Gil et al., 2011). Research on the use of brans to enrich bread and other food products reveals that phytate contents of bread samples also varied significantly ($p < 0.05$) with bran type, and maize bran–enriched bread recorded the highest value (0.53 g/100 g DM), though lower than the standard (2.5 g/100 g DM). Moreover, breads enriched with brans were lower in calories compared to control bread. The addition of maize, sorghum, and rice brans also significantly ($p < 0.05$) improved the iron, zinc, potassium, magnesium, and manganese contents when compared to control bread, but there was no significant change in calcium and copper contents. But there was a drastic decrease in levels of manganese and iron in rice-bran–enriched bread, and a decrease in calcium content was also observed in maize bran–enriched breads (Pauline et al., 2020).

11.9 CEREALS FOR PREVENTION OF DISEASE

Gluten-intolerant individuals may encounter three types of medical conditions associated with gluten: celiac disease, allergy to wheat, and non-coeliac gluten sensitivity. Prolamines are components of gluten that are responsible for stimulating immune response. Celiac individuals have many nutrient deficiencies, so there is a challenge in the development of products free from gluten (Saturni et al., 2010; Vici et al., 2016). Management through a gluten-free diet using alternative sources is the only solution for gluten-intolerant patients. Various formulations of gluten-free bread, pastries, biscuits, cookies, cakes, spaghetti, and pasta have been standardized by food scientists. Consumer requirements have provided an impetus for the food industry to constantly look for improved formulations and processing methodologies for gluten-free foods.

Celiac disease, wheat allergy, and non-celiac gluten sensitivity may invite severe health issues for people ingesting even a small amount of gluten (Conte et al., 2019). In recent times, non-celiac gluten sensitivity has been increasingly recognized as a clinical condition of patients and both wheat allergy and celiac disease have been diagnosed. This disease is diagnosed by intestinal and extra-intestinal symptoms that are triggered by the ingestion of gluten-containing foods (Barbaro et al., 2018).

Sorghum bicolor has been reported to have various applications in African and Indian traditional medicine. These applications are attributed to the presence of different phenolic compounds and dietary fibers in sorghum and its antioxidant activity (Coulibaly et al., 2020). In India, a decoction prepared from the seeds of sorghum has been in use as a diuretic for treating urinary tract infections and kidney-related disorders. Work by Lim reported that biocomponents responsible for the red color of sorghum also have antifungal, anti-anemic, and antimicrobial properties. On the same ground, Ben Slima et al. extracted a soluble polysaccharide from *Sorghum bicolor* (L.) seeds and reported its ability in wound healing to treat burns caused by fractional CO_2 laser. The extracted polysaccharide was efficient in closing the wound and hence can be a proven agent in wound healing (Ben Slima et al., 2022).

11.10 FLOUR IMPROVERS TO IMPROVE TECHNICAL QUALITY OF FLOUR

Flour, mainly wheat flour, is a major raw ingredient for food and bakery industries. Every type of bakery product needs a specific texture and composition of flour, either alone or in combination with other improvers to achieve the required quality of the end product. So for improving the quality of bakery products, many flour improvers are added in very low quantities to provide desired properties to flour during processing.

These flour improvers are chemical and biological compounds and could be categorized into the following types: antimicrobial agents, maturing agents, bleaching agents, enzymes, emulsifiers, minerals, and vitamins. Examples of some of these are given in the following (Skendi et al., 2021).

11.10.1 Bleaching Agents

The principal agents used for bleaching of wheat flour are chemicals nitrogen peroxide, nitrogen trichloride, chlorine dioxide, chlorine, benzoyl peroxide, and acetone peroxide. These chemical bleaching agents accelerate the bleaching property of the natural pigment (yellowish pigment-xanthophyll) when the flour is stored in large amounts (Martins et al., 2018).

11.10.2 Maturing Agents

Maturing agents improves bread quality by accelerating the maturing process (or ageing) of the flour, which otherwise would take storage of 1–2 months. Chemical maturing agents work by causing modification in the physical properties of gluten during fermentation in which dough becomes less sticky. The improvement in handling properties increases tolerance of dough to different conditions of fermentation and finally results in the production of larger-volume loaves with fine-textured crumb in the bread. Potassium bromate, ascorbic acid, acid calcium phosphate, azo-dicarbonamide, ammonium/potassium persulphate, stearoyll actylates, and sodium pyrophosphate are most frequently used as maturing agents in wheat flour. Acetone peroxide, chlorine, and chlorine dioxide function as both bleaching and maturing agents.

The action of the maturing agent is supposed to oxidize sulfhydryl or thiol (–SH) groups of wheat gluten. The stress in the dough releases during the reaction and hence dough is tightened.

11.10.3 Enzymes

Enzymes like oxidases, amylases, and proteases are frequently used in the food industry to improve fermentation, storage properties, texture, and flavor development of baked products. Starch molecules break down into glucose due to amylases, which in turn are used by yeast for liberating carbon dioxide gas. Proteases are used in biscuit dough for decreasing mixing time. Glucose oxidase enzyme is a replacement for chemical maturing agents like potassium bromate, which has been banned in some countries. Lipoxygenase can be used as a bleaching and maturing agent for bread dough (Zhygunov, 2018).

11.10.4 Emulsifiers

Emulsifiers are widely used in bakery industries in the production of icings, cakes, and biscuit dough for improving the texture by reducing the surface tension and thereby facilitating the mixing of fat and aqueous phases. Mono/diglycerides, polyglycerol esters, polysorbates, sodium/calcium, stearoyll actylate, and lecithin are commonly used surfactants in baked food product manufacturing. Emulsifiers with a low hydrophilic–lipophilic balance value (such as mono/diglycerides and lecithin) are recommended for high-fat products, while emulsifiers with greater hydrophilic–lipophilic balance values (like sodium stearoyll actylate) are suitable for low-fat and high-moisture food products (ML Khaleel et al., 2018).

11.10.5 Antimicrobial Agents

Bakery products are very susceptible to microbial contamination because of high moisture and nutrient-rich content. Bakery products may get contaminated in any of the steps in processing like cooling, slicing, and wrapping. So flours are added with antimicrobial agents or inhibitors like lactic acid, acetic acid, acid calcium phosphate, sodium diacetate, sorbic acid or its calcium/potassium salts, and calcium/sodium propionate.

Microbial fermentation, particularly lactic acid fermentation, has become the most popular way to improve cereal taste, texture, safety, and appearance. Lactic acid further improves the organoleptic characteristics of the food (new flavor, aroma, and texture) and also increases the nutritional value of various starch-containing raw materials, It has become a primary tool for preservation of food and a method for decreasing contamination by pathogenic bacteria. Lactic acid fermentation delivers enrichment to the human diet by producing a number of positive changes in cereal food composition like enriching cereal substrates with protein, fatty acid, and essential amino acid content; detoxification; and partial hydrolysis of gluten and starch (Petrova & Petrov, 2020). Lactic acid fermentation of cereals containing starch produces an ample amount of vitamins B and K, folate, amino acids like lysine, and other micronutrients in different fermented foods (Tangyu et al., 2019). Amylolytic lactic acid bacteria plays an important role in the fermentation of probiotic starch-containing and hypoallergenic foods, especially for infants (Terpou et al., 2019; Nguyen et al., 2007). Lately, their active role in type 2 diabetes prevention has been explored (Cabello-Olmo et al., 2019).

Lactic acid bacteria has the potential to ferment gluten-free flours. *Lactobacillus plantarum* was chosen for synthesis of γ-aminobutyric acid (GABA) from fermenting wheat sourdough, rice, soyabeans, rye, millet, oats, buckwheat, amaranth, and sorghum (Apostol et al., 2020).

11.10.6 Vitamins and Minerals

Whole grains are a rich source of minerals and vitamins, but milling of the grain removes the bran, germ, and aleurone layers of the grain. In fact, these layers are mainly rich in vitamin and mineral content. Therefore, there is a need to fortify staple cereal grain flour with vitamins and minerals. Milling and baking processes substantially remove the essential amino acid lysine, so flours are enriched with lysine.

Maize flour is enriched with several micronutrients, like iron, folic acid, zinc, calcium, vitamin B complex, and vitamin A. In a study by Das et al. (2013), cereal grains fortified with iron, vitamin A, and multiple micronutrients led to significant improvements in serum concentrations of micronutrients, serum ferritin, and hemoglobin concentrations (WHO Guidelines, 2016).

11.11 CEREALS AS INFANT FOOD DURING WEANING

Cereals are used as a choice of food for infants during weaning for the following reasons (Klerks et al., 2019).

1. Cereals are an excellent source of food for energy; they also give a substantial amount of carbohydrate and dietary fiber to the baby, along with vitamins, proteins, minerals, and other bioactive compounds. Breastfeeding alone cannot meet the nutritional requirements of a 6-month-old baby.
2. Cereals as infant food could be effectively fortified at the start of complementary feeding to meet nutritional requirements.
3. Non-digestible carbohydrates in cereals lead to the development of a microbiota, with increased *Bacteroides* population in the gut of the infant.
4. Cereals with a mild taste and a semi-solid texture help in the transition of infant food habits, from breast feeding to solid food intake at the beginning of complementary feeding.

11.12 CEREALS FOR BEVERAGES

The quality and composition of different raw materials are crucial for the beer industry. The chemical composition of cereal grains used as raw materials is according to the soil and climatic conditions, crop management, seed maturity, species, date of harvest, storage conditions, and drying conditions after harvest. Starch is the most important raw material required in the beer industry.

Starch is the main component of cereals that is converted in to fermentable sugars during beer production. The brewing qualities of a cereal are determined by the starch content. The highest amount of starch is found in maize (60–80%, followed by sorghum (55–76%) and then by barley (52.00–69.0%) (Adiamo et al., 2018; Dabija et al., 2021).

Barley is a suitable cereal grain for the brewing industry (Dabija et al., 2021). Many countries also successfully use unconventional malted grains. A few examples are rice in Asia, millet and sorghum in Africa, and maize in America for beer production (Zhang et al., 2019; Evera et al., 2019; Chaves-López et al., 2020). This substitution is around 10–30% in European countries, 40–50% or more in America and Australia, and 50–70% in Africa (Hernández-Becerra et al., 2020).

The limited replacement of barley malt with additional sources is also increasing and are considered supplementary sources of carbohydrates. (Bogdan & Kordialik-Bogacka, 2017; Hernández-Becerra et al., 2020). Adjuvants in varying proportions (10–50%) have regularly found use in the beer industry, sometimes to give adjuvant sources of fermentable yeast sugars and at other times to improve the stability of foam and adjust the color and flavor of beer (Dabija, 2019). Very recently rice has become the most prevalent additive for beer, and as the demand for this cereal is ever increasing, a substitute for rice is sought (He et al., 2018). Bioactive compounds like vitamins and minerals, and phenolic compounds like phytoestrogens, could be featured compounds (Mellor et al., 2020).

Dabija et al. (2021) reported the use of maize and sorghum as raw materials in the process of brewing, either as simple additives or as complete substitutes in the brewing of beers made from 100% sorghum or maize malt. They showed that sorghum has potential as a raw material in brewing and is also a valuable ingredient in the baking industry (Dabija et al., 2021). The addition of sorghum seed flour improves the nutritional value of bread in various percentages to wheat flour and increases content of fibers, minerals, and fats (Apostol et al., 2020).

An interesting approach of brewing industry is the production of gluten-free beer. Gluten-free or non-gluten beer can be defined as beer with less than 20 mg/kg of gluten defined by Codex Alimentarius and EU Regulation 41/2009 (Kerpes et al., 2017; Ciocan et al., 2020).

11.13 CEREAL BY-PRODUCTS AND THEIR APPLICATIONS

The milling process of the cereal grain removes the germ, bran, and aleurone layers of endospermic tissue, which then constitute cereal by-products. All these grain fractions are a rich source of a number of bio-functional molecules, dietary fiber, lignans, phytoestrogens, phenolic compounds, and vitamins and minerals. The milling process determines the resulting nutritional value of the by-products, and they are suitably applied in food and non-food products apart from use only as feed. Furthermore, processes like brewing, malting, and distilling of cereals also release different by-products that are becoming very alluring materials in other industries (Oluwajuyitan et al., 2021).

1. **Corn by-products**

Corn by-products obtained after the milling process are commonly used as animal feed. The milling process provides a number of products diverse in dry matter, fiber, protein, fat, and energy. Moreover, dry milling of corn gives distillers' grains and distillers' soluble grains, whereas wet milling gives corn gluten meal, gluten feed, liquefied corn product, germ meal, starch molasses, condensed fermented corn extractives, and hydrolyzed corn protein. Dry milling fractions such as standard meal germ cake, broken kernels, and pericarp are usually mixed together and hammer-milled to produce hominy feed, which is an exceptional animal feed. Recently, hominy feed has also been used as raw material in ethanol production due to its rich starch content.

The germ fraction of the milling process is used for oil extraction. The recovery of oil (varying from 15–25%) from the germ fraction depends on the oil content of the corn, fraction of the germ obtained from the milling process, and the efficiency of the oil extraction process. The resulting cake left after the oil extraction process has residual oil and proteins that can further be added to gluten feed.

These by-products are well used by pharmaceutical companies as a growth media for microorganisms or the production of antibiotics and other related compounds. Corn bran has been in use as a human food supplement. Distillers' grains and distillers' soluble grains in dried form are also used for production of distilled alcohol, and when fermented, they become a valuable raw material for bread and the bakery industry.

2. Rice by-products

Rice by-products consist of husk, bran, and germ amounting to 20, 8, and 2%. These milling by-products can be used alone or with combinations of different feeds used as animal feeds. In the present era, the food and pharmaceutical industries are also using rice by-products due to their unique functional properties and lower allergens. However, there are many challenges; the lipids in rice grain are mostly found in the bran layer, which is rapidly destroyed during the milling process. Moreover, milling of rice grain produces phospholipase, which hydrolyses and causes oxidation of lipids (Price & Welch, 2012; Esa et al., 2013). So stabilization of rice bran immediately after milling is important. Further, rice bran contains micronutrients like oryzanol, tocopherols, and phytosterols, which makes it an attractive candidate as a nutraceutical.

Another important rice by-product is rice bran oil, which has appreciably higher bioactive minor components (i.e., g-oryzanol, phytosterols, and tocotrienols) in comparison to other common vegetable oils. These compounds have the potential to decrease cholesterol absorption, lower the cholesterol level of blood, and prevent coronary heart diseases and some cancer types, and they display antioxidant activities.

Rice bran is a important source of insoluble dietary fiber. It contains polysaccharides like cellulose, hemicellulose, and pentosans. Its protein is hypoallergic and has good digestibility (90%). In general, all the by-products produced while rice milling are gluten free and hence could be used in the production of functional gluten-free products. The bran could also be used for improving and managing disease conditions like oxidative stress, hypertension, and type 2 diabetes mellitus.

Rice husk is mainly used as fuel and as biomass for the production of energy by using various biochemical and thermo-chemical processes.

3. Wheat by-products

Wheat milling by-products include wheat bran and wheat shorts, wheat middlings, wheat red dog, and wheat feed flour. The by-products are commonly used as livestock feed. The quality and composition of the by-products differs because of soil and climatic conditions, variety/species of wheat, and processing techniques.

Dry milling of the wheat grain generates bran and shorts as the principal by-products. Wheat bran is a excellent source of dietary fiber, but it has low lipid content. It contains non-starch polysaccharides, starch, protein, and lignin at 38, 19, 18, 6% and also phytochemicals and vitamins. The main phytochemical is a phenolic compound, ferulic acid attached to arabinoxylans, and other phenolics like p-hydroxybenzoic acid, vanillin, p-hydroxybenzaldehyde, vanillic acid, and trans-cumaric acid are present in the wheat bran.

Dietary fiber–rich bran is utilized in the bakery industry for producing bread, muffins, and cookies (Allen et al., 2014). Wheat milling also involves a debranning process where each individual bran layer is removed. Debranning by-products are used for producing novel breads of different types and cereal-based fermented food and also act as a source of arabinoxylans. Allen et al. observed that debranning of wheat produces bran fractions suitable for functional food or nutraceutical purposes; they are highly enriched in dietary fiber and have antioxidant activity (2014).

Wheat bran contains starch and hemicellulose/cellulose, which are valuable for producing bio-ethanol. An array of physical or chemical pre-treatments are needed since enzymatic treatments of wheat bran are not sufficient for production of simple bio-compounds. Wheat germ also has bioactive compounds and could be used in the food industry.

4. Millet by-products

Millet by-products are generally obtained by dehulling of the millet grain. Millet grain, being smaller in size, is severely treated before consuming to increase its edible character and other sensory attributes. By-products are mostly used for preparing different types of meals, alcoholic beverages, and distilled liquors (Liu et al., 2012; Shobana & Malleshi, 2007).

5. Barley by-products

Pearlings are produced in the pearling process of barley. Pearlings are an important and nutrition-rich by-product of barley. They are a excellent source of many bioactive compounds, like phytate, phenols, dietary fiber, and vitamin E. The pearling process also separates the pericarp, aleurone, and sub-aleurone layers present in oat kernels. These β-glucan–rich fractions are associated with many health benefits. Barley middling obtained from milling of the barley grain is a valuable source of dietary fiber value, particularly a high β-glucan value. The addition of barley middling to bread provides various health and economic benefits to the consumer and food industry. Wet-milling barley by-products could be used for food and non-food applications after starch isolation. Hulls, fiber such as β-glucan and arabinoxylan, and protein by-products can be used as livestock feed. These could also find application in the production of ethanol (Menrad, 2003; Capettini et al., 2010).

6. Sorghum by-products

Sorghum by-products include sorghum germ meal, sorghum gluten meal, sorghum bran, sorghum brewers' grains, sorghum wine, sorghum distillers' dried grains and solubles, and malted sorghum sprouts. Sorghum can be consumed in different fermented forms and sorghum wine. By-products produced from the fermentation of indigenous sorghum liquor are used in many parts of the world (Lazaro & Favier, 2000; Adiamo et al., 2018).

11.14 MICROSTRUCTURE AND TEXTURAL PROPERTIES OF CEREALS

Maize, famous as the queen of cereals; barley; and sorghum, all belonging to the *Poaceae* family, are a type of monocotyledonous plants with roots and stems interconnected by nodes and internodes. Grains of barley with protective coating have average size of (2–4.5 mm thick, 3–5 mm wide, and 8–12 mm long) which compared to sorghum and barley grains without coating have average size of (4–8 mm thick, 3–8 mm wide, and 2.5–22 mm long). The chemical composition of these cereals depends on soil nature, grain maturity, climate, and crop conditions. Maize, barley, and sorghum cereals are beneficial in the brewing industry, and with new technologies and modified sensory characteristics, research is towards developing gluten-free beer (Dabija et al., 2021). Cereal grains (wheat, rice, and oats) are composed of three structural parts, endosperm, germ, and bran, which are potential sources of minerals, starch, and fiber. Wheat starch microstructure consists both small spherical (2–10 um) and large lenticular (20–40 um) granules compared to oats granules (20–100 um). Scanning electron microscopy (SEM) reveals a Maltese cross microstructural pattern of wheat and oat grain kernels (Yiu, 1993). Chemical constituent distribution in these cereals is mainly starch in the endosperm and minerals reserved in the germ and bran portions of the grains. Wheat, rye, oats, and other cereal grains are utilized in the bakery industry for making bread and

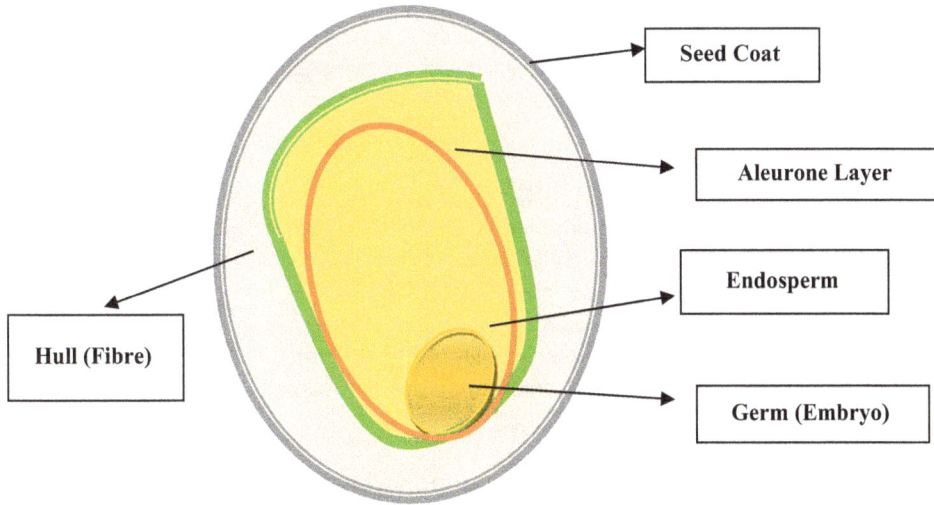

Figure 11.1 Microstructure of cereal grains.

(*Source:* Langton et al., 2021)

different bakery items. With the advancement of new techniques, gluten-free food items are in more demand. The microstructure of different cereal grains has distinct characteristics that are greatly changed by process such as mixing, proofing, milling, and baking. SEM and laser scanning confocal fluorescence microscopy (LSCFM) techniques revealed the microstructure of dough, which shows a high proportion of strands with starch granules embedded and gluten in lower proportion through-out the dough (Jekle & Becker, 2015). Another class of cereals are pseudo-cereals such as qui-noa, amaranth, teff, chia, broomcorn millet, and buckwheat, which receive great attention for their advantages in making gluten-free products. Fermentation of quinoa sourdough with *Lactobacillus plantarum* improves its rheological features such as viscosity and nutritional content and lowers pH. It is used in making muffins and gluten-free products. The chemical composition (proteins, miner-als, fatty acids [linolenic acid], flavonoids, and dietary fiber) of pseudo-cereals compared to other cereals is more efficient. The germination or triticale process affects properties such as stability of starch, protein weakening, and starch gelatinization, which cause reduced dough stability (Banu et al., 2014). Protein distribution and structure making up the microstructure of different cereals and pseudo-cereals differ in the kernel part of grains. Gel electrophoresis and SEM methods revealed different protein microstructure patterns between cereals (maize) and pseudo-cereals (amaranth) sharing a similar pattern with soybean (Gorinstein et al., 2001). Canary grass (*Phalaris canariensis* L.), belonging to the *Poaceae* family, is a protein-rich cereal compared to pseudo-cereals or other cereals of the same plant family. Using SEM technology, the microstructure of canary grass reveals starch granules in the endosperm with crack-like margins; germ and aleurone in a single layer containing oil and protein; and proteins present in the endosperm, germ, bran, and aleurone layers.

11.15 SENSORIAL ANALYSIS OF CEREALS

11.15.1 Fluorescence Microscopy

The morphological and chemical composition of cereals is revealed, with fluorescence micros-copy being the most sensitive method. Fluorescent markers (8-anilino-1-naphthalene sulfonic acid) for proteins and starch (orange G and fluoescamine) are available to reveal the different cereal components in the microstructure. The microstructure of all cereals is composed of three compartments (germ, bran, and endosperm) containing proteins, fats, and starch packed as granules in them. This method provides fast, simple, selective specificity and improved sensitivity over other conventional microscopy techniques (Fulcher et al., 1982).

11.15.2 Karl Fischer Titration

Determination of water composition and migratory behavior of starch and gluten in cereals affects their textural properties. Karl Fischer titration (KFT) or other titration methods depend on the chemical properties of the water in a sample. The KFT method has emerged as a useful method in water content determination of both soluble and insoluble samples. Advantages of the KFT method include single- or two-component volumetric analysis, temperature analysis, free or bound water determination, increased sensitivity, and coupling with other methods such as oven or distillation (Popescu et al., 2020).

11.15.3 R5 Sandwich Enzyme-Linked Immunosorbent Assay

Gluten components present in cereals trigger many immune disorders such as allergy, celiac disease, and wheat sensitivity. Gluten proteins (prolamins) are stored in seeds called gliadins (in wheat), glutenins (in barley), secalins (in rye), and hordeins (in barley). R5 enzyme linked immunosorbent assay (ELISA) is a type1 method which is devised by Codex Alimentarius standard. The R5 antibody detects prolamin content in the sample. The recognition epitope (QQPFP) of the R5 antibody shows a range of reactivity towards wheat cultivators with varying reactivity using a 10% C-hordein or gliadin calibration standard (Huang et al., 2020).

11.15.4 Scanning Electron Microscopy

SEM is used to determine cereal grain microstructure and anatomy. The structural image generated is a result of the interaction of an electron beam with sample constituents. Different physical characteristics of cereals can be determined using this microscopy technique (Machado Pereira et al., 2021).

11.16 CONCLUSION

Cereal grains are a stable food with a number of nutritional and bioactive compounds that are consumed all over the world. They can be processed into a diverse range of food products, alone or as multigrain food products in order to increase the bioavailability of their bioactive compounds or nutrients. Innovations in cereal and the bakery industry and the increased concern of consumers about food safety have enhanced the development of functional foods with an emphasis on healthcare (Rossana et al., 2019). This provides a promising future with greater opportunities for researchers and entrepreneurs to develop cereal-based novel products. Further, biofortification of grains for reviving nutritional composition could help in preventing deficiencies due to micronutrients having public health significance in many countries.

Cereal-based industry has generated a need for the development of whole grain–based multigrain food products that can cater to the nutritional needs of consumers due to several health benefits. Gluten-free products; foods fortified with vitamins, minerals, and other nutritional additives; and high-fiber foods are some of the health-benefitting characteristics of cereal grains that have been explored.

Novel whole-grain foods with a primary focus on healthcare and nutrition are currently needed. An increase in gluten intolerance and a rise in the awareness of health-conscious consumers are factors that attracted food manufacturers to develop food product formulations that meet the definitions of whole-grain foods and dietary requirements of consumers.

Future growth will depend on increases in yield, and this will need attention and a continued focus on the spread of technologies requiring yield-enhancing, up-skilled natural resource management, research in the field of technical efficiency, and application of technology in cereal science.

REFERENCES

Adiamo, O.Q.; Fawale, O.S.; Olawoye, B. Recent trends in the formulation of gluten-free sorghum products. J. Culin. Sci. Technol. 2018, 16, 311–325.

Allen, K.J.; Turner, P.J.; Pawankar, R.; Taylor, S., Sicherer, S., Lack, G., . . . Mills, E.C. Precautionary labelling of foods for allergen content: Are we ready for a global framework? World Allergy Org. J. 2014, 7(1), 1–14.

Al-Toma, A.; Volta, U.; Auricchio, R.; Castillejo, G.; Sanders, D.S.; Cellier, C.; Mulder, C.J.; Lundin, K. European society for the study of coeliac disease (ESsCD) guideline for coeliac disease and other gluten-related disorders. United Eur. Gastroenterol. J. 2019, 7, 583–613.

Apostol, L.; Belc, N.; Gaceu, L.; Oprea, O.B.; Popa, M.E. Sorghum flour: A Valuable Ingredient for Bakery Industry? Appl. Sci. 2020, 10, 8597.

Arvola, A.; Lahteenmaki, L.; Dean, M.; Vassallo, M.; Winkelmann, M.; Claupein, E.; Saba, A.; Shepherd, R. Consumers' beliefs about whole and refined grain products in the UK, Italy, and Finland. J. Cereal Sci. 2007, 46, 197–206.

Baniwal, P.; Mehra, R.; Kumar, N.; Sma, S.; Kumar, S. Cereals: Functional constituents and its health benefits. Pharm. Innov. 2021, 10(2), 343–349.

Banu, I.; Dragoi, L.; Aprodu, I. From wheat to sourdough bread: A laboratory scale study on the fate of deoxynivalenol content. Qual. Assur. Saf. Crops Food. 2014, 6, 53–60.

Barbaro, M.R.; Cremon, C.; Stanghellini, V.; Barbara, G. Recent advances in understanding non-celiac gluten sensitivity; F1000 Res. 2018, 7: F1000 Faculty Rev-1631.

Bender, D.; Schonlechner, R. Innovative approaches towards improved gluten-free bread properties. J. Cereal Sci. 2020, 91, 102904.

Ben Slima, S.; Ktari, N.; Chouikhi, A.; Trabelsi, I.; Hzami, A.; Taktak, M.A.; Msaddak, L.; Ben Salah, R. Antioxidant activities, functional properties, and application of a novel Lepidium sativum polysaccharide in the formulation of cake. Food Sci. Nutr. 2022, 10(3), 822–832.

Beta, T. Improving the nutritional and nutraceutical properties of wheat and other cereals. Burleigh Dodds Science Publishing; 2021 May 20.

Blandino, M.; Sovrani, V.; Marinaccio, F.; Reyneri, A.; Rolle, L.; Giacosa, S.; Locatelli, M.; Bordiga, M.; Travaglia, F.; Coïsson, J.D. Nutritional and technological quality of bread enriched with an intermediated pearled wheat fraction. Food Chem. 2013, 141, 2549–2557.

Bogdan, P.; Kordialik-Bogacka, E. Alternatives to malt in brewing. Trends Food Sci. Technol. 2017, 65, 1–9.

Bower, C.; D'Antoine, H.; Stanley, F.J. Neural tube defects in Australia: Trends in encephaloceles and other neural tube defects before and after promotion of folic acid supplementation and voluntary food fortification. Birth Defects Res. Part A Clin. Mol. Teratol. 2009, 85(4), 269–273.

Cabello-Olmo, M.; Oneca, M.; Torre, P.; Sainz, N.; Moreno-Aliaga, M.J.; Guruceaga, E.; Díaz, J.V.; Encio, I.J.; Barajas, M.; Araña, M. A fermented food product containing lactic acid bacteria protects ZDF rats from the development of type 2 diabetes. Nutrients. 2019, 11, 2530.

Capettini, F.; Ceccarelli, S.; Grando, S. Barley production, improvement, and uses (pp. 210–220). Blackwell; 2010.

Cappelli, A.; Oliva, N.; Cini, E. A systematic review of gluten-free dough and bread: Dough rheology, bread characteristics, and improvement strategies. Appl. Sci. 2020, 10, 6559.

Carcea, M.; Narducci, V.; Turfani, V.; Finotti, E. Stone milling versus roller milling in soft wheat (part 2), influence on nutritional and technological quality of products. Foods. 2022, 11, 339.

Chaves-López, C.; Rossi, C.; Maggio, F.; Paparella, A.; Serio, A. Changes occurring in spontaneous maize fermentation: An overview. Ferment. 2020, 6, 36.

Chunxiao Ni; Qingqing Jia; Gangqiang Ding; Xifeng Wu; Min Yang. Low-glycemic index diets as an inter-vention in metabolic diseases: A systematic review and meta-analysis. Nutrients. 2022 Jan; 14(2), 307.

Ciocan, M.; Dabija, A.; Codină, G.G. Effect of some unconventional ingredients on the production of black beer. Ukr. Food J. 2020, 9, 322–331.

Conte, P.; Fadda, C.; Drabi´nska, N.; Krupa-Kozak, U. Technological and nutritional challenges, and novelty in gluten-free breadmaking: A review. Pol. J. Food Nutr. Sci. 2019, 69, 5–21.

Coulibaly, W.H.; Bouatenin, K.M.J.-P.; Boli, Z.B.I.A.; Alfred, K.K.; Bi, Y.C.T.; N'Sa, K.M.C.; Cot, M.; Djameh, C.; Djè, K.M. Influence of yeasts on bioactive compounds content of traditional sorghum beer (tchapalo) produced in Côte d'Ivoire. Curr. Res. Food Sci. 2020, 3, 195–200.

Culetu, A.; Elena Susman, I.; Eglantina Duta, D.; Belc, N. Nutritional and functional properties of gluten-free flours. Appl. Sci. 2021, 11, 6283.

Dabija, A. Biotehnologies in the food industries. Performantica Press; 2019.

Dabija, A.; Ciocan, M.E.; Chetrariu, A.; Codină, G.G. Maize and sorghum as raw materials for brewing, a review. Appl. Sci. 2021, 11, 3139.

Das, J.K.; Salam, R.A.; Kumar, R., Bhutta, Z.A. Micronutrient fortification of food and its impact on woman and child health: A systematic review. Syst Rev. 2013, 2, 67–93.

Demirkesen, I.; Ozkaya, B. Recent strategies for tackling the problems in gluten-free diet and products. Crit. Rev. Food Sci. Nutr. 2020, 1–27.

De Vuyst, L.; Vrancken, G.; Ravyts, F.; Rimaux, T.; Weckx, S. Biodiversity, ecological determinants, and metabolic exploitation of sourdough microbiota. Food Microbiol. 2009, 26, 666–675.

Dhingra, D.; Michael, M.; Rajput, H., Patil, R.T. Dietary fiber in foods: A review; J. Food Sci. Technol. 2012, 49, 255–266.

Esa, N.M.; Ling, T.B.; Peng, L.S. By-products of rice processing: An overview of health benefits andapplica-tions. Rice Res. Open Access. 2013.

Evera, E.; AbedinAbdallah, S.H.; Shuang, Z.; Sainan, W.; Yu, H. Shelf life and nutritional quality of sorghum beer: Potentials of phytogenic-based extracts. J. Agric. Food. Technol. 2019, 9, 1–14.

Fernanda, C.O.L. Martins et al. Categories of food additives and analytical techniques for their determination. Innovative Food Anal. 2021.

Fulcher, R.G. Fluorescence microscopy of cereals. Food Struct. 1982, 1(2), 167–175.

Gil, A.; Ortega, R.M.; Maldonado, J. Whole grain cereal and bread: A duet of the Mediterranean diet for the prevention of chronic diseases. Public Health Nutr. 2011, 14, 2316–2322.

Gobbetti, M.; Pontonio, E.; Filanninob, P.; Rizzellob, C.G.; De Angelisb, M.; Di Cagnoa, R. How to improve the gluten-free diet: The state of the art from a food science perspective. Food Res. Int. 2018, 110, 22–32.

Gobbetti, M.; Rizzello, C.G.; Di Cagno, R.; De Angelis, M. How the sourdough may affect the functional features of leavened baked goods. Food Microbiol. 2014, 37, 30–40.

Gorinstein, S.; Delgado-Licon, E.; Pawelzik, E.; Permady, H. H.; Weisz, M.; Trakhtenberg, S. Characterisation of soluble amaranth and soybean proteins based on fluorescence, hydrophobicity, electrophoresis, amino acid analysis, circular dichroism, and differential scanning calorimetry measurements. J. Agric. Food Chem. 2001, 49, 5595–5601.

Grosse, S.D. Retrospective assessment of cost savings from prevention: Folic acid fortification and spina bifida in the U.S. Am. J. Prev. Med. 2016, 5(Suppl 1), S74–S80.

He, Y.; Cao, Y.; Chen, S.; Ma, C.; Zhang, D.; Li, H. Analysis of flavour compounds in beer with extruded corn starch as an adjunct. J. Inst. Brew. 2018, 124, 9–15.

Hernández-Becerra, E.; Contreras-Jiménez, B.; Vuelvas-Solorzano, A.; Millan-Malo, B.; Muñoz-Torres, C.; Oseguera-Toledo, M.E.; Rodriguez-Garcia, M.E. Physicochemical and morphological changes in corn grains and starch during the malting for Palomero and Puma varieties. Cereal Chem. 2020, 97, 404–415.

Huang, X.; Ma, K.; Leinonen, S.; Sontag-Strohm, T. Barley C-Hordein as the calibrant for wheat gluten quan-tification. Foods. 2020, 9(11), 1637.

Jekle, M.; Becker, T. Wheat dough microstructure: The relation between visual structure and mechanical behavior. Crit. Rev. Food Sci, Nutri. 2015, 55(3), 369–382.

Jenkins, D.J.; Wolever, T.M.; Taylor, R.H.; Barker, H.; Fielden, H.; Baldwin, J.M.; Bowling, A.C.; Newman, H.C.; Jenkins, A.L.; Goff, D.V. Glycemic index of foods: A physiological basis for carbohydrate exchange. Am. J. Clin. Nutr. 1981, 34, 362–366.

Kerpes, R.; Fischer, S.; Becker, T. The production of gluten-free beer: Degradation of hordeins during malting and brewing and the application of modern process technology focusing on endogenous malt peptidases. Trends Food Sci. Technol. 2017, 67, 129–138.

Khaleel, M.L.; Sharoba, A.M.; El-Desouky, A.I., Mohamed, M.H. Use of some emulsifiers to improve the quality of pan bread product. Tikrit Jj. Agri. Sci. 2018, 18, 150–161.

Klerks, M.; Bernal, M.J.; Roman, S.; Bodenstab, S.; Gil, A.; Sanchez-Siles, L.M. Infant cereals: current status, challenges, and future opportunities for whole grains. Nutrients. 2019, 11(2), 473.

Langton, M.; Gutiérrrex, J.L. The structure of cereal grains and their products. Whole Grains Health. 2021, 1–20.

Lazaro, E.; Favier, J. Alkali debranning of sorghum and millet. Cereal Chem. 2000, 77(6), 717–720.

Liu, J.; Tang, X.; Zhang, Y.; Zhao, W. Determination of the volatile composition in brown millet, milled millet and millet bran by gas chromatography/mass spectrometry. Molecules. 2012, 17(3), 2271–2282.

Macauley, H.; Ramadjita, T. Cereal Crops: Rice, Maize, Millet, Sorghum, Wheat: Background Paper, Feeding Africa, 21–23 October 2015, Dakar, Senegal; The African Development Bank Group and the African Union: Abidjan, Ivory Coast, 2015; pp. 1–31: Publisher university of cape coast.

Machate, D.J.; Figueiredo, P.S.; Marcelino, G.; Guimarães, R.C.A.; Hiane, P.A.; Bogo, D.; Pinheiro, V.A.Z.; Oliveira, L.C.S.; Pott, A. Fatty acid diets: Regulation of gut microbiota composition and obesity and its related metabolic dysbiosis. Int. J. Mol. Sci. 2020, 21(11), 4093.

Martínez-Villaluenga, C.; Peñas, E.; Hernández-Ledesma, B. Pseudocereal grains: Nutritional value, health benefits and current applications for the development of gluten-free foods. Food Chem. Toxicol. 2020, 137, 111–178.

Martins, Z.E.; Pinho, O.; Ferreira, I.M. Impact of new ingredients obtained from Brewer's spent yeast on bread characteristics. J. Sci. Techno. 2018, 55(5), 1966–1971.

Mellor, D.D.; Hanna-Khalil, B.; Carson, R. A review of the potential health benefits of low alcohol and alcohol-free beer: Effects of ingredients and craft brewing processes on potentially bioactive metabolites. Beverages. 2020, 6, 25.

Menrad, K. Market and marketing of functional food in Europe. J. Food Eng. 2003, 56(2), 181–188.

Montemurro, M.; Coda, R.; Rizzello, C.G. Recent advances in the use of sourdough biotechnology in pasta making. Foods. 2019, 8, 129.

Moroni, A.V.; Bello, F.D.; Arendt, E.K. Sourdough in gluten-free bread-making: An ancient technology to solve a novel issue? Food Microbiol. 2009, 26, 676–684.

Nguyen, T.T.; Loiseau, G.; Icard-Verniere, C.; Rochette, I.; Treche, S.; Guyot, J.-P. Effect of fermentation by amylolytic lactic acid bacteria; in process combinations; on characteristics of rice/soybean slurries: A new method for preparing high energy density complementary foods for young children. Food Chem. 2007, 100, 623–631.

Nionelli, L.; Montemurro, M.; Pontonio, E.; Verni, M.; Gobbetti, M.; Rizzello, C.G. Pro-technological and functional characterization of lactic acid bacteria to be used as starters for hemp (Cannabis sativa L.) sourdough fermentation and wheat bread fortification. Int. J. Food Microbiol. 2018, 279, 14–25.

Ogunsakin, A.; Sanni, A.; Banwo, K. Effect of legume addition on the physiochemical and sensorial attributes of sorghum-based sourdough bread. LWT. 2020, 118, 108769.

Olubunmi, I.P.; Babatunde, K.S.; Bolanle, O.O.; Seyioba, S.O.; Taiwo, L.T.; Olukayode, O.A.; Nwankego, E.G. Quality evaluation of fibre-enriched bread. Int. J. Nutr. Food Sci. 2015, 4, 503–508.

Oluwajuyitan, T.D.; Ijarotimi, O.S.; Fagbemi, T.N.; Oboh, G. Blood glucose lowering, glycaemic index, carbo-hydrate-hydrolysing enzyme inhibitory activities of potential functional food from plantain, soy-cake, rice-bran and oat-bran flour blends. J. Food Measure. Charact. 2021, 15(4), 3761–3769.

Palacios-Rojas, N.; McCulley, L.; Kaeppler, M.; Titcomb, T.J.; Gunaratna, N.S.; Lopez-Ridaura, S.; Tanumihardjo, S.A. Mining maize diversity and improving its nutritional aspects within agro-food systems. Compr. Rev. Food Sci. Food Saf. 2020, 19(4), 1809–1834.

Pauline, M.; Roger, P.; Nina, N.E.; Arielle, T.; Eugene, E.E.; Robert, N. Physico-chemical and nutritional characterization of cereals brans enriched breads. Sci. Afr. 2020, e00251.

Petrova, P.; Petrov, K. Lactic acid fermentation of cereals and pseudocereals: Ancient nutritional biotechnologies with modern applications. Nutrients. 2020, 12(4), 1118.

Poole, N.; Donovan, J.; Erenstein, O. Continuing cereals research for sustainable health and well-being. Int. J. Agric. Sustain. 2021, 16, 1–2.

Popescu, G.; Radulov, I.; Iordănescu, O.A.; Orboi, M.D.; Rădulescu, L.; Drugă, M.; Bujancă, G.S., David, I.; Hădărugă, D.I.; Hădărugă, N.G.; Riviş, M. Karl Fischer water titration—principal component analysis approach on bread products. Appl. Sci. 2020 Jan, 10(18), 6518.

Previtali, M.A.; Mastromatteo, M.; Conte, A.; De Vita, P.; Ficco, D.B.; Del Nobile, M.A. Optimization of durum wheat bread from a selenium-rich cultivar fortified with bran. J. Food Sci. Technol. 2016, 53, 1319–1327.

Price, R.K.; Welch, R.W. *Cereal grains* (pp. 307–316). 2013, Encyclopedia of Human Nutrition, Academic Press, Cambridge.

Rasha, M.O.E.; Tajul, A.Y.; Abdel, H.R.; Khogali, E.A., Hassan, A.M.; Saifeldin, M.K. Chemical composition and functional properties of wheat bread containing wheat and legumes bran. Int. J. Food Sci. Nutr. 2016, 5, 10–15.

Rossana, V.C.C.; Fernandes, Â.; Gonzaléz-Paramás, A.M.; Barros, L.; Ferreira, I.C.F.R. Flour fortification for nutritional and health improvement: A review. Int. Food Res. J. Nov. 2019, 125, 108576.

Santos, L.M., Lecca, R.C., Cortez-Escalante, J.J., Sanchez, M.N., Rodrigues, H.G. Prevention of neural tube defects by the fortification of flour with folic acid: A population-based retrospective study in Brazil. Bull. World Health Organ. 27 Oct 2015, 94(1), 22–29.

Saturni, L.; Ferretti, G.; Bacchetti, T. The gluten-free diet: Safety and nutritional quality. Nutrients. 2010, 2, 16–34.

Sayed, A.R.; Bourne, D.; Pattinson, R.; Nixon, J.; Henderson, B. Decline in the prevalence of neural tube defects following folic acid fortification and its cost-benefit in South Africa. Birth Defects Res. A, Clin. Mol. Teratol. 2008, 82(4), 211–216.

Shobana, S.; Malleshi, N. Preparation and functional properties of decorticated finger millet (*Eleusine coracana*). J. Food Eng. 79(2), 2007, 529–538.

Slavin, J. Whole grains and human health. Nutr. Res. Rev. 2004, 17(1), 99–110.

Šmídová, Z.; Rysová, J. Gluten-free bread and bakery products technology. Foods. 2022, 11(3), 480.

Sramkova, Z.; Gregova, E.; Sturdik, E. Chemical composition and nutritional quality of wheat grain. Acta Chim Slovaca. 2009, 2(1), 115–138.

Stamatovska, V.; Nakov, G.; Uzunoska, Z.; Kalevska, T.; Menkinoska, M. Potential use of some pseudocereals in the food industry. ARTTE. 2018, 6, 54–61.

Tangyu, M.; Muller, J.; Bolten, C.J.; Wittmann, C. Fermentation of plant-based milk alternatives for improved flavour and nutritional value. Appl. Microbiol. Biotechnol. 2019, 103, 9263–9275.

Tejada-Ortigoza, V.; Garcia-Amezquita, L.E.; Serna-Saldívar, S.O.; Welti-Chanes, J. Advances in the functional characterization and extraction processes of dietary fiber. Food Eng. Rev. 2016, 8(3), 251–271.

Terpou, A.; Papadaki, A.; Lappa, I.K.; Kachrimanidou, V.; Bosnea, L.A.; Kopsahelis, N. Probiotics in food systems: Significance and emerging strategies towards improved viability and delivery of enhanced beneficial value. Nutrients. 2019, 11, 1591.

Vici, G.; Belli, L.; Biondi, M.; Polzonetti, V. Gluten free diet and nutrient deficiencies: A review. Clin. Nutr. 2016, 35, 1236–1241.

Voinea, A.; Stroe, S.-G.; Codină, G.G. Use of response surface methodology to investigate the effects of sodium chloride substitution with potassium chloride on Dough's rheological properties. Appl. Sci. 2020, 10, 4039.

WHO guideline: fortification of maize flour and corn meal with vitamins and minerals. Geneva: World Health Organization; 2016. Licence: CC BY-NC-SA 3.0 IGO.

Yano, H. Recent practical researches in the development of gluten-free breads. NPJ Sci. Food. 2019, 3(1), 1–8.

Yiu, S.H. Food microscopy and the nutritional quality of cereal foods. Food Struct. 1993, 12(1), 13.

Zhang, T.; Zhang, H.; Yang, Z.; Wang, Y.; Li, H. Black rice addition prompted the beer quality by the extrusion as pretreatment. Food Sci. Nutr. 2019, 7, 3664–3674.

Zhygunov, D.; Mardar, M.; Kovalyova, V. Use of enzyme preparations for improvement of the flour baking properties. Food Sci. Appl. Biotechnol. 2018, 1(1), 26–32.

CHAPTER 12

Allergens in Cereal Grains

Muhammad Afzaal, Farhan Saeed, Bushra Niaz, Muhammad Ahtisham Raza, Muzzamal Hussain, Misbah Aslam, Tabussam Tufail, Amir Sasan Mozaffari Nejad and Mohammad Javed Ansari

CONTENTS

12.1 INTRODUCTION

Food allergens are most commonly caused by foods derived from plants, especially in adults. These allergens are the leading factors in widespread food allergies and life-threatening anaphylaxis, according to recent clinical investigations. The most common allergens are naturally present in nuts, cereals, and seeds, notably peanuts, although fresh fruits and vegetables have also been linked to allergic responses of varying severity (Costa et al., 2021; Skypala, 2019). Clinically, essential allergenic features of plant-based diets have yet to be established. All plant-based foods are not equally allergenic, making in vitro and in vivo cross-reactivity difficult to differentiate (Cau et al., 2021; Hari, 2019). The majority of plant allergens are either inflammation-associated proteins or seed-retention proteins, and their activity in plants can vary depending on development, maturation, disease infection, or climate change. Processing can eliminate allergy epitopes or create new allergens not seen in natural foods, resulting in considerable allergenic content changes.

Wheat (*Triticum* sp.), rice (*Oryza sativa*), and corn (*Zea mays*) are the three most important cereal crops in the world, each yielding around 600 million tons annually. Wheat has a huge variety of cultivars, with approximately 25,000 varieties generated by breeding programs across the world. Moreover, due to its accessibility and the functional qualities of wheat starch and gluten proteins, wheat flour is commonly used as a food processing ingredient (Bresciani & Mart, 2019). Due to its commercial influence, wheat flour is used and processed all over the world, and many individuals and employees in the grinding, baking, and commercial food sectors are subjected to it in the environment, increasing the chance of asthmatic allergies (Jeebhay et al., 2019). Cereals are the most common food source on the planet, and they might trigger respiratory and food allergies when consumed (Sharma et al., 2020)

Recent studies have revealed novel food allergens in grains, such as gliadin, a wheat allergenic protein connected to gluten intolerance and also recognized as the key allergen in wheat-dependent exercise-induced sensitivity (Juhasz et al., 2020). Buckwheat (*Fagopyrum esculentum*), a grain that belongs to the *Polygonaceae* family that is commonly used as a replacement for widely accepted grains, particularly in Asia, where it has been described as a cause of severe respiratory and food allergies, was the first allergenic 2S albumin ever found in grains. Various allergens with molecular weights ranging from 9–67 kDa were found in an epidemiological investigation regarding either allergy (displaying symptoms after consumption of buckwheat) or sensitized (showing no symptoms after consumption of buckwheat but positive RAST) individuals. Three of them have been identified and arranged: the 9 kDa allergen, which is similar to the buckwheat proteolytic enzymes; the 16 kDa allergen, which is analogous to millet α-amylase/trypsin inhibitor; and the 19 kDa allergen, which is analogous to rice's globulin, a 2 S albumin. α-globulin and 2 S albumins are specifically recognized as buckwheat allergens. These allergens and the two synthetic 2 S albumins showed no immune cross-reactivity.

Another recently discovered cereal allergy is a 9 kDa protein belonging to the lipid transfer protein (LTP) family in maize. These proteins are abundant in plants, and they contribute significantly to pathogenic conditions and are involved in stress defense (Wang et al., 2019). It's worth noting that maize LTP has roughly 63% similarity with peach LTP and rice LTP (79%) but has less similarity with other grain LTPs like barley (57%) and wheat (59%). This could describe maize-hypersensitive people who have recurrent allergic responses to rice and peaches as well as sensitivity to ingested wheat (bread and pasta).

Many studies on wheat allergies have focused on breathing allergies (bakers' asthma), which is one of the most common allergies in several countries, such as the United Kingdom, and celiac, a type of gluten allergy that affects about 1% of the population in North Africa, North and South America, Europe, and the Indian subcontinent (Sabenca et al., 2021). Wheat allergy is considerably less frequent in the general population, while it affects about 1% of children and can cause hypersensitivity and death in its most severe form. The proteins that cause wheat food allergies are much better understood than those that cause bakers' asthma, but new research suggests that there are fascinating parallels and distinctions between both disorders (Di Francesco et al., 2021; Sabença et al., 2021).

Globulins (salt soluble) and albumins (water-soluble) are two types of food allergens that are soluble in water or saline solutions. Even though the exact amounts of a particular protein necessary to sensitize a person are unknown, people with IgE-mediated allergies can react to very small amounts of the irritating food (Zhao et al., 2021). The physicochemical or immunochemical features of food allergies that account for their specific allergenicity are poorly known.

12.2 FOOD ALLERGIES

Food allergies can cause several symptoms in the skin, lungs, and gastrointestinal tract, with anaphylaxis being the most severe and potentially fatal symptom. Major advances in fundamental, cognitive, and medical trials have resulted in improved knowledge of the actual immunological processes that contribute to the breaking of therapeutic and immunological resistance to food allergens that can contribute to IgE-mediated or non-IgE-mediated responses. Food allergies are induced by lifestyle, dietary habits, social interactions, and stress, as well as the microbiota's descriptive and analytical characteristics (Gomaa, 2020). These factors are considered to have the biggest impact during childhood, leading to the creation of theories to explain the gluten intolerance outbreak, such as the dual-allergen theory. Such theories have fueled investigation into standard precautions aimed at establishing desensitization and acceptance of allergens in people who are allergic to them. Many epidemiological studies have investigated allergen-nonspecific therapy techniques, which will presumably enhance the therapeutic possibilities for food intolerance patients (Renz et al., 2018).

Wheat is the third most prevalent food in Japan that causes allergies in children (Jiang et al., 2021). Wheat-dependent exercise-induced anaphylaxis is a well-known anaphylactic condition linked to a specific kind of grain protein. Anaphylaxis, atopic dermatitis (AD), and urticaria, among other allergic reactions, relate to wheat proteins. Although less well known than Wheat-dependent exercise induced anaphylaxis (WDEIA), it is clear that the bulk of the proteins connected to wheat food allergies have storage or defensive functions, but some also lead to asthmatic allergies (Sabença et al., 2021).

It is important to provide adequate and well-balanced nutrition to maintain energy. However, food-related adverse reactions are relatively common nowadays, leading to allergic reactions in the body (Crinò et al., 2018). A food allergy is defined as an immunological reaction to proteins in nature (Crowe, 2019). Some major food allergens cause nasal obstruction, bronchitis, sinusitis, intestinal cramp leading to bloody stools, asthma, diarrhea, eczema, hives, and migraines. Food allergies to a variety of foods are becoming more common, although wheat allergy (celiac disease), egg, peanut, soy, and fish are among the most common. About 90% of all food-allergic responses are caused by these foods. However, the incidence of apparent food hypersensitivity varies between populations (1.4–19.1%) (Dragland, 2020). Food allergies affect roughly 1.5% of adults and 5–8% of children under the age of 8.

Genetics, social and dietary practices, and exposure to allergic foods play a role in food allergies. At the outset, food intolerance must be distinguished from food allergies, which might be classified as IgE, Non-IgE, or mixed responses. IgE is linked to mast cells, epithelial cells in the mucous and skin, and blood eosinophils with a high affinity for particular Fc receptors. These cells have a variety of granulocytes, the most prominent of which is histamine, which stores prepared mediators. When allergenic proteins bind to certain IgEs related to eosinophils in the gut, they cause allergic reactions. When IgEs are conjugated with allergens, these cells degranulate, exposing immuno-mediators to the immediate ecosystem. Various mediators, including prostanoids, leukotrienes, and cytokines, are produced as a result. The release of histamine causes an immediate response several minutes after coming into contact with the allergen. Vasodilation, mucous production, smooth muscle contraction, and tissue fluid exudation are at the root of the response. The immediate reaction, which starts 4 to 6 hrs after interaction with the allergen and lasts for many days, is followed by an early reaction. Chemokine mediators generated at the same time as the acute response stimulate preferential recruitment of inflammatory cells, mostly basophils and macrophages, resulting in this reaction. When these cells penetrate the tissues, they cause inflammation that lasts 4 to 5 days. However, two clinical factors are necessary to identify IgE mediated allergy: the existence of IgE-specific antibodies and a confirmed link between food consumption and the onset of disease.

Non-IgE-mediated food allergy, on the other hand, is defined as immunological responses triggered by antibodies besides IgE (e.g., IgA, IgG, and IgM) and immunological components (food, food antibodies, and cell-mediated immunity). It's vital to note that there is no evidence linking an allergenic diet to an allergic response to non-IgE antibodies in any allergic condition. Normal people, on the other hand, have a rise in IgG food antibodies after eating a meal. Similarly, there is no substantial evidence that cell-mediated resistance has a role.

The use of foodstuffs with relatively less allergenicity is proposed as a strategy for allergy prevention. Structural proteins are life-threatening factors of response severity; occurrence and cross-reactivity are typically predicted by them. The clinical history must guide the diagnostic workup.

Standard procedures for skin testing and food-specific IgE are highly useful, and surgery may be prescribed in some cases. Utilizing foods with the minimum allergens is the best diagnostic criteria until the whole pathophysiology is determined. To determine allergens, the skin prick test and prick to prick test are recommended for the diagnosis of allergies.

12.3 SYMPTOMS OF GRAIN ALLERGY

Food allergies are a major condition that affect people all over the world. In Japan, for example, allergies to cow's milk, eggs, shrimp, peanuts, wheat, and buckwheat have been reported regularly. Except for eggs and cow's milk, animal foods include a limited number of allergens, while plant allergens are significantly more diverse.

Cereals are the world's most dominant food source, and they can cause both respiratory and food allergy responses when consumed. Until recent times, the -amylase/trypsin inhibitor was found to be the most common cereal allergen, and it can cause hypersensitivity through both inhalation and gastrointestinal routes: it has been characterized as the main allergen in baker's asthma, as well as in eczemic children with positive double-blind placebo control food challenge to wheat as well as in rice allergic patients (Raole, 2020; Ei Hassouni et al., 2021).

Most allergens found in plants may now be classified into three categories: structural proteins, defense-related proteins, and seed storage proteins. Symptoms of different cereal allergens are shown in Figure 12.1. The current study is based on major allergens in important crops such as wheat, maize, barley, rice, and oats.

12.3.1 Wheat

Wheat is one of the world's most frequently grown, packaged, and eaten crops, and it is associated with allergies and intolerances (Asrani et al., 2021). Two types of allergies are particularly clearly defined in many studies. One is bakers' asthma, which is induced by flour and dust aspiration during

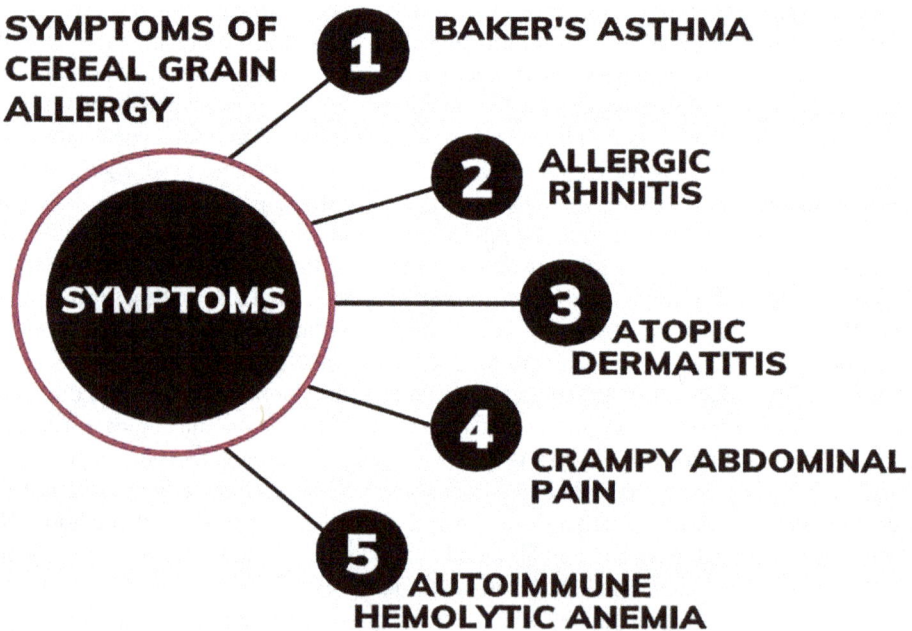

Figure 12.1   Symptoms of cereal grain allergens.

grain processing. Despite the discovery of a variety of wheat binding proteins IgE from bakers' asthmatics, there is little question that a well-defined collection of alpha-amylase inhibitors remains the fundamental culprit (Raulf, 2021).

Allergic reactions to wheat as well as other cereal grains are most frequent in children, and they usually decrease after a few years. Cereal allergy mediated by IgE can result in a minor local infection on the skin and sometimes becomes more adverse due to severe anaphylactic reaction leading to death. Celiac disease is an autoimmune reaction to gluten (a protein component present in wheat and related grains) including barley, rye, and oats (Kosová et al., 2020; Cabanillas, 2020). The gastrointestinal tract of susceptible persons is particularly affected by the loss of their capacity to absorb nutrients. Because there is no treatment, gluten-free items must be avoided for the rest of the life. Avoidance is a big difficulty for the consumer because of the widespread use of gluten-containing flours in processed foods (Demirkesen & Ozkaya, 2020). Furthermore, gluten can be found in dietary supplements, cosmetics, and pharmaceuticals.

Celiac disease is caused by gluten in wheat or gluten-like proteins in highly associated cereals such as rye and barley (Tanner et al., 2019). The stomach reacts to gluten by becoming smooth and reducing its capacity to absorb nutrients, resulting in symptoms such as gastroenteritis and vitamin deficiencies (Sharma et al., 2020). It can appear in infancy, impacting children's growth and development, or in adulthood, affecting adults' growth and development. Individuals with celiac disease must avoid gluten for the rest of their lives because there is no treatment. However, if gluten is unintentionally consumed, celiac disease does not induce anaphylaxis, which can be severe (Bucchini et al., 2019).

Gluten-containing pre-packaged food cereals should be labeled as such, according to the Codex Alimentarius Commission's proposal. There are also specific "gluten-free" eating suggestions (Meijer et al., 2021).

12.3.2 Maize

Maize is a grass that belongs to the *Poaceae* family and is farmed all over the world as a major cereal crop. Maize is the most industrialized cereal grain, with several byproducts such as bran husk or hull that are consumed not only by humans but also for animal feed (Saeed et al., 2021). It is widely distributed around the world, due to its excellent mechanical properties, efficient digestion, and lower price than other grains. It is an adaptable commodity that may be produced in a variety of agro-ecological regions.

Along with its numerous health benefits, many people are allergic to it. Corn allergy is a kind of food allergy that is mediated by the IgE protein, which is activated by the proteins that occur in its kernels (Dange & Patil, 2021). Lipid transfer protein (LTP) is a potent corn allergen that was first recognized as a fruit allergen but is now found in a variety of cereals, nuts, and vegetables. LTP is a very resilient protein that is resistant to food processing and can withstand excessive heating application as well as against gastrointestinal digestion (Wong et al., 2019). Because of these characteristics, LTP is a strong food allergen that can trigger severe immune responses (Scheurer & Schülke, 2018) Corn is linked to wheat, barley, oat, rye, and rice, among other cereals. Corn LTP allergy is the only well-studied category among other corn allergies because they contain homologous lipid transfer proteins. Such similar reactions are termed cross-reactions.

In the past, there was limited literature that looked at how often people are affected by maize-associated allergies. So far, IgE antibodies to LTP have been identified as the only known risk factor for maize allergy. However, epidemiological studies show that this infection is more frequent in young patients. Peach is another foodstuff that possesses lipid transfer protein, and it has been observed in several studies that humans allergic to maize seem to be frequently allergic to peach. As a result, lipid transfer protein syndrome is the name given to this infection. This is especially common in adults who are allergic to fruits. The peach is the most commonly implicated fruit. Almost every person who had a maize allergy connected to LTP had a peach allergy as well. While this should be investigated further, allergy to LTP appears to exist. It's not been determined how little maize is required to cause an allergic response.

Another probable allergy in maize is storage proteins, which have previously been found as allergens in other grains such as wheat. Many of the allergens found in corn pollen can also be found in the kernel. Such proteins are likely to perform a function as allergens in maize food intolerances. Corn allergy can cause skin problems such as atopic dermatitis or eczema, as well as hypersensitivity (Lovell et al., 2020). It can also cause inflammation of the skin, itching, gut sensations including diarrhea, nausea, constipation, vomiting, runny nose, wheezing, and septic shock in exceptional instances. The allergic reaction provoked by corn is greatly dependent on the presence of proteins and the person's susceptibility.

Maize-allergic people should avoid it, as avoidance is the only treatment so far. Corn and its derivatives are found in a variety of food products, and it can be difficult to identify the presence of corn, so consumers should be vigilant. Maize and corn derivatives are not required to be labeled under EU regulations; thus consumers should take the necessary steps to avoid any incidents.

12.3.3 Barley

Wheat-related harmful effects have been observed more commonly than with other grains of food. Celiac disease and IgE-dependent wheat allergy are the most well known of these, which are also caused by immunological failure. Other grain allergies, such as corn, millet, and sorghum, are uncommon and unrelated to rye, wheat, oats, and barley responses. As a result, those who have a wheat allergy, such as celiac disease, may generally eat corn meals like polenta. However, allergic responses to corn can occur as a consequence of fruit allergies, which are common in the southern part of Europe and commonly begin with reactions to peaches.

On the other hand, IgE-mediated grain allergies are induced by the immunoglobulin protein IgE's binding to a wide range of proteins, not simply gluten. In a condition known as exercise-induced anaphylaxis, the responses (usually acute) arise only if a person exercises for a few hours after eating wheat or comparable grains. Wheat allergy sufferers are more likely to respond to highly associated grains like barley and rye than to their more distant relatives, oats. The fact that certain blood tests for grain food allergies might mistakenly identify an allergy to grass pollen can make diagnosing grain allergies more difficult.

As a result of these negative responses, gluten-containing grains (oats, maize, rye, wheat, barley, or their hybridized forms) and derivative goods have been added to Annex IIIa of the food labeling requirement. Wheat-based glucose syrups, such as dextrose, wheat-based maltodextrins, barley-based glucose syrups, and grains used in refined products for alcohol have all been allowed limited exclusions to the labeling regulation.

12.3.4 Rice

Rice allergy symptoms are caused by 9-, 14-, 31-, and 52 kDa protein bands, which can be present in foodstuffs such as oil, wheat, and milk. In polished rice, RAG2 and 19 kDa globulin were found, while bran of rice had 52 kDa globulin. 52-kDa globulin has been recognized as the most likely allergen responsible for rice bran allergy (Satoh et al., 2019).

Rice allergy is still uncommon, but it can affect anybody at any time owing to people's changing lifestyles and eating habits. The utilization of processed food alters the metabolic cycles, which make your immune system suspicious to be allergic to the specific diet or specific dietary components (Rinninella et al., 2019). People who have problems with rice may have a sensitivity rather than an allergy, similar to wheat or gluten difficulties. The major effects related to rice allergy reported include rice allergy rashes (urticaria), asthma, itchy skin, gastrointestinal disorders, and anaphylactic shock, rarely.

Haneda et al. (2021) observed that direct skin contact with rice bran caused sensitivity because the patient had rashes when showering while consuming food since he was 5 years old.

12.4 MAJOR ALLERGENS

Food allergies, which are characterized as an abnormal immunological reaction to dietary proteins, affect approximately 6% of children and 3–4% of adults. Food-induced allergy responses can produce a wide range of effects in the respiratory and gastrointestinal tract, and skin can be triggered by both IgE-mediated and non-IgE-mediated (cellular) processes. Our comprehension of how food allergies cause a loss of normal oral tolerance is changing. Elimination of food, proper laboratory medication, and proper history is crucial for the treatment of patients who are allergic to food. Food allergens are classified into two classes, gastrointestinal tract and respiratory tract provoke mild to severe allergic reactions in the body, and in extreme reactions, many episodes of anaphylactic shock are also observed. Different cereal allergens cause many diseases, which are described in Table 12.1. Moreover, major cereal allergens are discussed in the following.

12.4.1 Wheat Germ Agglutinin

Wheat germ agglutinin is one of the most-studied lectins and is important in biomedicine. The lectin is a copolymer that breaks down into monomers when exposed to an acidic pH (Sivaji et al., 2019; Gautam et al., 2018). It is a carbohydrate-free, non-metallic protein that's made up of four iso-lectins with varied electrophoretic mobility. Because earlier lignins such as ricin and abrin are poisonous, lectins have been considered toxic since their identification. Although few lectins are very hazardous at low concentrations, cytotoxicity is generally dose dependent (Ribeiro et al., 2018). Lectin-mediated cell toxicity and rapid cellular toxicity are two types of lectin cytotoxicity. Lectins bind to immune cells and serve as antigens, resulting in cytotoxicity and an inflammatory reaction (Balčiūnaitė & Dzikaras, 2021; Elumalai et al., 2019).

Table 12.1 Major Cereal Allergens and Their Health Problems

Allergens	Sources	Diseases	References
Wheat germ agglutinin (WGA)	Grains, pseudo-grains, wheat kernels, and vegetables	Bakers' asthma, autoimmune hemolytic anemia, and rhinitis	(Mumolo et al., 2020)
Profilins	Celery, peanut, soybeans, barley, wheat, rye, grass pollens, peach, pumpkin, and tomato	Serological allergy, gastrointestinal and respiratory hypotension	(Nilsson et al., 2018; Male & Rhinoconjunctivitis, 2021)
Non-prolamins	Wheat, corn, sorghum, barley	Celiac disease, systemic autoimmune disease, and villous atrophy	(Gell et al., 2017)
2S albumin	Peanuts, tree nuts, eggs, cereal grains	Asthma, atopic dermatitis Hypoalbuminemia, vomiting, and crampy abdominal pain	(Blazowski et al., 2019)
Glycoproteins	Nuts, peach pollens, rice, and soybean	Hypersensitivity, atopic dermatitis, rhinitis and asthma	(Trcka et al., 2012: Haneda et al., 2021)
ω-5 gliadins	Wheat, oat, rye, barley	Celiac disease, Crohn's disease, and wheat-dependent exercise-induced anaphylaxis	(Altenbach et al., 2019)

12.4.2 Profilins

Profilins make up an enormous fraction of class 2 allergens, which typically show a link between pollen and food. They also play a crucial function in regulating the polymerization of actin filaments. Many people are allergic to these pollens, their immune system can't respond to these allergens, and the patient faces episodes of mild to severe anaphylactic shock. The bulk of these class 2 allergens appear to be made up of structural epitopes, making them thermally tortuous, prone to hydrolytic enzymes, and difficult to separate, with inadequate standardized extracts for diagnostic purposes.

Profilin allergy causes extensive damage to the epithelial lining of the mucosal surface, allowing the protein to infiltrate the mucosa and induce an inflammatory reaction. Furthermore, highly allergic individuals had an 81% increase in effector cell sensitivity (Rosace et al., 2019).

12.4.3 2S Albumin

2S storage proteins are a major group of proteins that are well known in many mono- and di-cotyledon seeds that support the seed from early (germination) to late (seedling) developmental stages due to the presence of methionine and cysteine (sulfur-containing amino acids) and provide an acceptable level of S to the plant (Souza, 2020). These proteins can cross the gut mucosal barrier and stimulate an allergic response, which is supported by cysteine. They are immunologically dominant and have the power to bond with IgE from allergic patients' plasma due to flexible and solvent-exposed hypervariable regions present in their structures. 2S albumins have disulfide bridges that provide a compact structure that is resistant to heat and enzymatic treatments (Souza, 2020).

2S albumin is considered a major allergen found in seeds of many plants due to which there is a special area of interest in its epidemiological trails. Moreno and Clemente (2008) reported that the interaction between dietary allergens and the immune system, as well as the formulation of particular allergy immunotherapy, requires immune-dominant areas containing IgE-binding epitopes.

12.4.4 Glycoproteins

Glycosylation is the process wherein glycoproteins are formed by the regulated assembly of sugar moieties to form glycans attached covalently to proteins after translation occurs in the ribosome. It is the most prevalent of the post-translational modifications (PTMs) that proteins undergo (Alla & Stine, 2022). Glycoproteins are polymers of proteins and carbohydrates, covalently attached, through a process of glycosylation. The extracellular segments of proteins have segments that extend outside the cell membrane that are often glycosylated (Reithmeier et al., 2016). The most common food allergies known as class 1 allergens are water-soluble globular proteins with a particle size of 10 to 70 kDa that are generally resistant to heat, acid, and proteases (Ekezie et al., 2018).

12.4.5 Prolamin

The major allergenic and antigenic components of cereal that drive illness have been well defined and are primarily found within the proline-rich cereal storage prolamin proteins, gliadin and glutenin, although other non-gluten proteins have been implicated in some allergic responses as well. Water-soluble albumins and salt-soluble globulins are made up of non-prolamin. In comparison to glutenins and gliadins, there has been little research on non-prolamins. Glutenins are made up of high- and low-molecular-weight components. Non-prolamins play a variety of roles in wheat development and growth. For example, globulins and albumins contain proteins and inhibitory enzymes that govern growth at various stages. Essential amino acids such as lysine, tryptophan, aspartate, and threonine are more plentiful in globulins and albumins than in storage proteins but insufficient in globulins and albumins.

12.4.6 ω-5 Gliadins

The role of gliadins and their generated peptides in lowering celiac disease's adverse responses has yet to be fully understood. Gliadins are further split by subunits, which include α-, β-, γ-, and ω-gliadins. On the base of sequence and motility, ω-gliadins are also split into ω1, ω2, and ω5 elements. α-, β-, and γ-gliadins are hazardous to people with celiac disease, whereas ω-gliadins are not.

The primary allergen in wheat-dependent, exercise-induced anaphylaxis has been recognized as wheat ω-5 gliadin. IgE to ω-5 gliadin has recently been found to be a strong predictor of acute allergy to consumed wheat in infants, with higher levels correlated with favorable oral wheat challenge outcomes. IgE antibodies to ω-5 gliadin were found in more than 80% of youngsters with acute wheat symptoms. Individuals with acute cutaneous responses to digested wheat proteins and patients with wheat-dependent, exercise-induced anaphylaxis had unique characteristics of sensitivity. Similarly, Ricci et al. (2019) have shown that patients with IgE-mediated wheat allergy and celiac disease are familiar with various wheat epitope profiles. ω5-gliadins, for example, caused the release of histamine in the eosinophils of wheat-dependent, exercise-induced anaphylaxis patients but not in controls.

Fast-gliadin (which also contains ω5-gliadin) is a significant allergen in WDEIA and IgE, against which it interacts with other gliadins. In adults with wheat-dependent, exercise-induced anaphylaxis, ω5-gliadin has been found as a significant allergen (Ito et al., 2008).

12.5 DIAGNOSIS

The presence and concentration of key allergens in cereals will determine the diagnosis of cereal allergy extracts that are employed in diagnostic testing. It is therefore critical to identify and characterize wheat allergens (Kampen et al., 2013).

Celiac disease and IgE-mediated wheat allergy are not well understood in the scientific community, and the likelihood of a combination of T helper 1 (Th1) and Th2-type illnesses is currently being debated. Wheat allergy has a 0.2 to 0.7% occurrence, and it is largely mediated by Th2-type cells since it is an IgE-mediated reaction. In many epidemiological studies, celiac patients have digestive effects similar to those of inflammatory bowel disease, making identification difficult (Borghini et al., 2018).

An oral food challenge (OFC) is commonly employed to establish if a medical history, skin prick test, and specific IgE findings are appropriate. The skin prick test plays a critical role in the diagnosis of IgE-mediated food allergies, including cereal allergies. Nonetheless, the consistency, strength, and standardization of allergenic extraction, which are sometimes ill defined in commercialized grain extracts, are all important factors in the effectiveness of skin testing. A food challenge or exclusion diet is the most effective technique to identify a rice allergy. A food challenge is eating in a controlled, clinical environment while an allergist monitors for a response. An eating plan involves eliminating rice from your diet for a few weeks to see whether your health improves. An eating plan is significantly safer, but the findings will not be accurate. Previously, blood allergy testing was performed by the allergists at NY Allergy & Sinus Centers to detect which proteins will cause an allergic reaction. This form of allergy test does not require any preparation, making it a practical alternative. In addition, unlike other allergy tests, it does not cause an allergic reaction. While you wait for the results of your blood test, stay away from rice-based meals and goods (Kumar, 2020).

Patients with signs of malfunctioning, inflammatory, and/or fibrotic esophagus may be diagnosed with eosinophilic esophagitis, which is a clinical and pathological diagnosis. Eosinophilic esophagitis symptoms, including swallowing and food obstruction, are caused by esophageal fibrosis, which is more common in older children and adults (Cianferoni, 2016).

12.6 TREATMENTS FOR GRAIN ALLERGIES

LTP, ω-5 gliadin, and ATIs are the most dangerous grain allergens. These allergens can play a role in a variety of allergies. Testing IgE reaction to salt-soluble and salt-insoluble protein content from wheat flour in patients with different clinical profiles of wheat allergy (food, WDEIA, bakers' asthma) indicated a wide variability among recognized allergens in people with varying clinical profiles, including within every group (Gilissen et al., 2014).

Wheat gluten proteins are important triggers of inflammatory immunological responses. High hydrostatic pressure (HHP) technology primarily breaks intramolecular and intermolecular noncovalent interactions within both protein complexes, changing secondary and tertiary structures and allergenicity (Yao et al., 2022). The current prevalence of bakers' asthma among younger bakers is between 0.3 and 2.4 instances per 1000 person per year, and it is rising among commercial bakery employees.

Three types of treatment can be used to avoid the onset of allergy symptoms:

- The primary treatment, the elimination of powerful dietary allergens (e.g., wheat bread) during the first 4 months of life, can minimize sensitivity. Antibiotics should be used with care in children since infectious diseases strengthen the stability of the growing immune system and reduce the development of allergies. Foods that boost immune system strength, such as oats, which contain particular β-glucans, may aid in prevention.
- Secondary treatment usually concentrates on socially sensitized individuals. Avoiding irritating allergens in the surroundings and diet can help to stop the onset of symptoms. In Europe, allergen labeling on packaged goods is now required by law. "May include" labeling, on the other hand, may protect the maker rather than the customer. The marketplace for "free from" food goods, particularly the "free from gluten" market, is quickly expanding due to increased knowledge in the food business and customer demand.
- In tertiary treatment, non-dietary therapeutic (medical) treatment of (chronic) symptoms in newly diagnosed cases is included. In the case of CD, the only established therapy is a strictly gluten-free diet for the rest of one's life.

12.7 CONCLUSION

Foods allergens are the leading factors in widespread food allergies and life-threatening anaphylaxis. The majority of plant allergens are either inflammation-associated proteins or seed-retention proteins, or their activity in plants can vary depending on development, maturation, disease infection, or climate change. Wheat is increasing the chance of asthmatic allergies. Globulins and albumins are two types of food allergens that are soluble in water or saline solutions. People with IgE-mediated allergies can react to very small amounts of irritating food. Corn allergy is a kind of food allergy that is mediated by the IgE protein, which is activated by the proteins that occur in its kernels. Food-induced allergy responses can produce a wide range of effects in the respiratory and gastrointestinal tract, and skin can be triggered by both IgE-mediated and non-IgE-mediated (cellular) processes. It is important to provide adequate and well-balanced nutrition to maintain energy. However, food-related adverse reactions are relatively common nowadays, leading to allergic reactions in the body. Food allergens cause nasal obstruction, bronchitis, sinusitis, intestinal cramp leads to bloody stools, asthma, diarrhea, eczema, hives, and migraines. Elimination of food, proper laboratory medication, and proper history are crucial for the treatment of patients who are allergic to food.

REFERENCES

Alla, A. J., & Stine, K. J. 2022. Recent strategies for using monolithic materials in glycoprotein and glycopeptide analysis. *Separations, 9*(2), 44.

Altenbach, S. B., Chang, H. C., Yu, X. B., Seabourn, B. W., Green, P. H., & Alaedini, A. 2019. Elimination of omega-1, 2 gliadins from bread wheat (Triticum aestivum) flour: Effects on immunogenic potential and end-use quality. *Frontiers in Plant Science, 10*, 580.

Asrani, P., Ali, A., & Tiwari, K. 2021. Millets as an alternative diet for gluten-sensitive individuals: A critical review on nutritional components, sensitivities and popularity of wheat and millets among consumers. *Food Reviews International*, 1–30.

Balčiūnaitė-Murzienė, G., & Dzikaras, M. 2021. Wheat germ agglutinin—From toxicity to biomedical applications. *Applied Sciences, 11*(2), 884.

Blazowski, L., Majak, P., Kurzawa, R., Kuna, P., & Jerzynska, J. 2019. Food allergy endotype with high risk of severe anaphylaxis in children—Monosensitization to cashew 2S albumin Ana o 3. *Allergy, 74*(10), 1945–1955.

Borghini, R., Donato, G., Marino, M., Casale, R., Di Tola, M., & Picarelli, A. 2018. In extremis diagnosis of celiac disease and concomitant wheat allergy. *The Turkish Journal of Gastroenterology, 29*(4), 515.

Bresciani, A., & Marti, A. 2019. Using pulses in baked products: Lights, shadows, and potential solutions. *Foods, 8*(10), 451.

Bucchini, L., Daly, M., & Mills, E. C. 2019. Food allergen labelling regulation. *Health Claims and Food Labelling, 22*, 107.

Cabanillas, B. 2020. Gluten-related disorders: Celiac disease, wheat allergy, and nonceliac gluten sensitivity. *Critical Reviews in Food Science and Nutrition, 60*(15), 2606–2621.

Cau, S., Tilocca, M. G., Spanu, C., Soro, B., Tedde, T., Salza, S., . . . & Mudadu, A. G. 2021. Detection of celery (*Apium graveolens*) allergen in foods of animal and plant origin by droplet digital PCR assay. *Food Control, 130*, 108407.

Cianferoni, A. 2016. Wheat allergy: diagnosis and management. *Journal of Asthma and Allergy, 9*, 13.

Costa, J., Villa, C., Verhoeckx, K., Cirkovic-Velickovic, T., Schrama, D., Roncada, P., . . . & Holzhauser, T. 2021. Are physicochemical properties shaping the allergenic potency of animal allergens? *Clinical Reviews in Allergy & Immunology*, 1–36.

Crinò, A., Fintini, D., Bocchini, S., & Grugni, G. 2018. Obesity management in Prader–Willi syndrome: current perspectives. *Diabetes, Metabolic Syndrome and Obesity: Targets and Therapy, 11*, 579.

Crowe, S. E. 2019. Food allergy vs food intolerance in patients with irritable bowel syndrome. *Gastroenterology & Hepatology, 15*(1), 38.

Dange, H. V., & Patil, R. A. 2021. Involvement of some allergens of plant and animal origin in allergic reactions related to respiratory system. *Journal of Medical Pharmaceutical and Allied Sciences, 1*(1947), 67–74.

Demirkesen, I., & Ozkaya, B. 2020. Recent strategies for tackling the problems in gluten-free diet and products. *Critical Reviews in Food Science and Nutrition*, 1–27.

Di Francesco, A., Cunsolo, V., Saletti, R., Svensson, B., Muccilli, V., De Vita, P., & Foti, S. 2021. Quantitative label-free comparison of the metabolic protein fraction in old and modern Italian wheat genotypes by a shotgun approach. *Molecules, 26*(9), 2596.

Dragland, V. 2020. *Self-reported food hypersensitivity in relation to biomarkers: The Fit Futures Study* (Master's thesis, UiT Norges arktiske universitet).

Ekezie, F. G. C., Cheng, J. H., & Sun, D. W. 2018. Effects of nonthermal food processing technologies on food allergens: A review of recent research advances. *Trends in Food Science & Technology, 74*, 12–25.

El Hassouni, K., Sielaff, M., Curella, V., Neerukonda, M., Leiser, W., Würschum, T., . . . & Longin, C. F. H. 2021. Genetic architecture underlying the expression of eight α-amylase trypsin inhibitors. *Theoretical and Applied Genetics, 134*(10), 3427–3441.

Elumalai, P., Rubeena, A. S., Arockiaraj, J., Wongpanya, R., Cammarata, M., Ringø, E., & Vaseeharan, B. 2019. The role of lectins in finfish: a review. *Reviews in Fisheries Science & Aquaculture, 27*(2), 152–169.

Fernandez-Feo, M., Wei, G., Blumenkranz, G., Dewhirst, F. E., Schuppan, D., Oppenheim, F. G., & Helmerhorst, E. J. 2013. The cultivable human oral gluten-degrading microbiome and its potential implications in coeliac disease and gluten sensitivity. *Clinical Microbiology and Infection, 19*(9), E386-E394.

Gautam, A. K., Srivastava, N., Nagar, D. P., & Bhagyawant, S. S. 2018. Biochemical and functional properties of a lectin purified from the seeds of *Cicer arietinum* L. *3 Biotech, 8*(6), 1–11.

Gell, G., Kovács, K., Veres, G., Korponay-Szabó, I. R., & Juhász, A. 2017. Characterization of globulin storage proteins of a low prolamin cereal species in relation to celiac disease. *Scientific Reports, 7*(1), 1–10.

Gilissen, L. J., van der Meer, I. M., & Smulders, M. J. 2014. Reducing the incidence of allergy and intolerance to cereals. *Journal of Cereal Science, 59*(3), 337–353.

Gomaa, E. Z. (2020). Human gut microbiota/microbiome in health and diseases: a review. *Antonie Van Leeuwenhoek*, 1–22.

Haneda, Y., Kadowaki, S., Furui, M., & Taketani, T. 2021. A pediatric case of food-dependent exercise-induced anaphylaxis due to rice bran. *Asia Pacific Allergy, 11*(1).

Hari, V. 2019. *Plant foods for nutritional good health.* Notion Press.

Ito, K., Futamura, M., Borres, M. P., Takaoka, Y., Dahlstrom, J., Sakamoto, T., . . . & Morita, E. 2008. IgE antibodies to ω-5 gliadin associate with immediate symptoms on oral wheat challenge in Japanese children. *Allergy, 63*(11), 1536–1542.

Jeebhay, M. F., Moscato, G., Bang, B. E., Folletti, I., Lipińska-Ojrzanowska, A., Lopata, A. L., . . . & Siracusa, A. 2019. Food processing and occupational respiratory allergy-An EAACI position paper. *Allergy, 74*(10), 1852–1871.

Jiang, N., Xu, W., & Xiang, L. 2021. Age-related differences in characteristics of anaphylaxis in Chinese children from infancy to adolescence. *World Allergy Organization Journal, 14*(11), 100605.

Juhász, A., Colgrave, M. L., & Howitt, C. A. 2020. Developing gluten-free cereals and the role of proteomics in product safety. *Journal of Cereal Science, 93*, 102932.

Kosová, K., Leišová-Svobodová, L., & Dvořáček, V. 2020. Oats as a safe alternative to triticeae cereals for people suffering from celiac disease? A review. *Plant Foods for Human Nutrition, 75*(2), 131–141.

Kumar, A. 2020. Food allergy: Symptoms, diagnosis and treatment. *SunText Review of Biotechnology, 1*(1), 101.

Lovell, C., Paulsen, E., Lepoittevin, J. P. 2020. Adverse Skin Reactions to Plants and Plant Products. In: Johansen, J., Mahler, V., Lepoittevin, J. P., Frosch, P. (eds) Contact Dermatitis. Springer, Cham. https://doi.org/10.1007/978-3-319-72451-5_88-2

Male, S. M., & Rhinoconjunctivitis, C. R. 2021. Profilins and food-dependent exercise-induced anaphylaxis. *Journal of Investigational Allergology & Clinical Immunology, 31*(4), 332–359.

Meijer, G. W., Detzel, P., Grunert, K. G., Robert, M. C., & Stancu, V. 2021. Towards effective labelling of foods. An international perspective on safety and nutrition. *Trends in Food Science & Technology, 118*, 45–56.

Moreno, F. J., & Clemente, A. 2008. 2S albumin storage proteins: what makes them food allergens? *The Open Biochemistry Journal, 2*, 16.

Mumolo, M. G., Rettura, F., Melissari, S., Costa, F., Ricchiuti, A., Ceccarelli, L., . . . & Bellini, M. (2020). Is gluten the only culprit for non-celiac gluten/wheat sensitivity?. *Nutrients, 12*(12), 3785.

Nilsson, N., Nilsson, C., Ekoff, H., Wieser-Pahr, S., Borres, M. P., Valenta, R., . . . & Sjölander, S. 2018. Grass-allergic children frequently show asymptomatic low-level IgE co-sensitization and cross-reactivity to wheat. *International Archives of Allergy and Immunology, 177*(2), 135–144.

Raole, V. M. 2020. Be aware of pollen, fungal and food allergens! *Environment at Crossroads Challenges and Green Solutions*, 147.

Raulf, M. 2021. Immediate-type hypersensitivity by occupational materials. *Contact Dermatitis*, 499–512.

Reithmeier, R. A., Casey, J. R., Kalli, A. C., Sansom, M. S., Alguel, Y., & Iwata, S. 2016. Band 3, the human red cell chloride/bicarbonate anion exchanger (AE1, SLC4A1), in a structural context. *Biochimica et Biophysica acta (BBA)-Biomembranes, 1858*(7), 1507–1532.

Renz, H., Allen, K. J., Sicherer, S. H., Sampson, H. A., Lack, G., Beyer, K., & Oettgen, H. C. 2018. Food allergy. *Nature Reviews Disease Primers, 4*(1), 1–20.

Ribeiro, A. C., Ferreira, R., & Freitas, R. 2018. Plant lectins: Bioactivities and bioapplications. *Studies in Natural Products Chemistry, 58*, 1–42.

Ricci, G., Andreozzi, L., Cipriani, F., Giannetti, A., Gallucci, M., & Caffarelli, C. 2019. Wheat allergy in children: a comprehensive update. *Medicina, 55(7)*, 400.

Rinninella, E., Cintoni, M., Raoul, P., Lopetuso, L. R., Scaldaferri, F., Pulcini, G., . . . & Mele, M. C. 2019. Food components and dietary habits: Keys for a healthy gut microbiota composition. *Nutrients, 11*(10), 2393.

Rosace, D., Gomez-Casado, C., Fernandez, P., Perez-Gordo, M., del Carmen Dominguez, M., Vega, A., . . . & Barber, D. 2019. Profilin-mediated food-induced allergic reactions are associated with oral epithelial remodeling. *Journal of Allergy and Clinical Immunology, 143*(2), 681–690.

Sabença, C., Ribeiro, M., Sousa, T. D., Poeta, P., Bagulho, A. S., & Igrejas, G. 2021. Wheat/gluten-related disorders and gluten-free diet misconceptions: A review. *Foods, 10*(8), 1765.

Saeed, F., Hussain, M., Arshad, M. S., Afzaal, M., Munir, H., Imran, M., . . . & Anjum, F. M. 2021. Functional and nutraceutical properties of maize bran cell wall non-starch polysaccharides. *International Journal of Food Properties, 24*(1), 233–248.

Satoh, R., Tsuge, I., Tokuda, R., & Teshima, R. 2019. Analysis of the distribution of rice allergens in brown rice grains and of the allergenicity of products containing rice bran. *Food Chemistry, 276*, 761–767.

Scheurer, S., & Schülke, S. 2018. Interaction of non-specific lipid-transfer proteins with plant-derived lipids and its impact on allergic sensitization. *Frontiers in Immunology, 9*, 1389.

Sharma, N., Bhatia, S., Chunduri, V., Kaur, S., Sharma, S., Kapoor, P., . . . & Garg, M. 2020. Pathogenesis of celiac disease and other gluten related disorders in wheat and strategies for mitigating them. *Frontiers in Nutrition, 7*, 6.

Sivaji, N., Suguna, K., Surolia, A., & Vijayan, M. 2019. Structural biology of plant lectins and macromolecular crystallography in India. *Current Science, 116*(9).

Skypala, I. J. 2019. Food-induced anaphylaxis: role of hidden allergens and cofactors. *Frontiers in Immunology, 10*, 673.

Souza, P. F. 2020. The forgotten 2S albumin proteins: Importance, structure, and biotechnological application in agriculture and human health. *International Journal of Biological Macromolecules, 164*, 4638–4649.

Tanner, G., Juhász, A., Florides, C. G., Nye-Wood, M., Békés, F., Colgrave, M. L., . . . & Tye-Din, J. A. 2019. Preparation and characterization of avenin-enriched oat protein by chill precipitation for feeding trials in celiac disease. *Frontiers in Nutrition, 6*, 162.

Trcka, J., Schäd, S. G., Scheurer, S., Conti, A., Vieths, S., Gross, G., & Trautmann, A. 2012. Rice-induced anaphylaxis: IgE-mediated allergy against a 56-kDa glycoprotein. *International Archives of Allergy and Immunology, 158*(1), 9–17.

van Kampen, V., De Blay, F., Folletti, I., Kobierski, P., Moscato, G., Olivieri, M., . . . & Raulf-Heimsoth, M. 2013. Evaluation of commercial skin prick test solutions for selected occupational allergens. *Allergy, 68*(5), 651–658.

Wang, J., Vanga, S. K., & Raghavan, V. 2019. Effect of pre-harvest and post-harvest conditions on the fruit allergenicity: A review. *Critical Reviews in Food Science and Nutrition, 59*(7), 1027–1043.

Wong, L. H., Gatta, A. T., & Levine, T. P. 2019. Lipid transfer proteins: the lipid commute via shuttles, bridges and tubes. *Nature Reviews Molecular Cell Biology, 20*(2), 85–101.

Yao, Y., Jia, Y., Lu, X., & Li, H. 2022. Release and conformational changes in allergenic proteins from wheat gluten induced by high hydrostatic pressure. *Food Chemistry, 368*, 130805.

Zhao, J., Li, Z., Khan, M. U., Gao, X., Yu, M., Gao, H., . . . & Lin, H. 2021. Extraction of total wheat (Triticum aestivum) protein fractions and cross-reactivity of wheat allergens with other cereals. *Food Chemistry, 347*, 129064.

CHAPTER 13

Reception of Grains and Their Global Standards

Amara Rasheed, Farhan Saeed, Muhammad Afzaal, Ali Ikram, Muhammad Ahtisham Raza, Muzzamal Hussain, Tabussam Tufail, Gulzar Ahmad Nayik and Mohammad Javed Ansari

CONTENTS

13.1 INTRODUCTION

Cereals are edible seeds or grains from the *Gramineae* grass family (Baladhiya et al., 2018: Sarwar & Biswas., 2021), which are grown almost in every region of the world due to their consumption as a staple diet (Sarrocco et al., 2019). Wheat, rice, and maize are considered major cereals due to their consumption and processing as compared to other cereals (Li et al., 2020);

however, oat, rye, barley, triticale, sorghum, and millet are categorized as minor cereals due to their limited consumption and poor baking properties (Torbica et al., 2021), but the continuous efforts of researchers have opened the way to characterize these grains and reported the presence of various bioactive moieties (e.g., polyphenols, vitamins, and antioxidants), along with their health-endorsing properties (Baniwal et al., 2021).

Cereals are consumed as staple foods in different countries; are major sources of complex nutrients, especially carbohydrates; and represent 60% of the world's cultivated areas (Hossain et al., 2020). In different studies, cereals and cereal-based products are reported as good sources of macro- and micronutrients such as vitamins and minerals. Cereals are processed in several ways to produce a variety of non-food items, which are used in the cosmetic industry (Skendi et al., 2020). Cereal grains have been esteemed as a food source since prehistoric times because of their long shelf life, but some microbiological and physicochemical threats are being faced in grains due to deterioration from rodents and insects. In light of recent statistics, about 60 million tons of land is cultivated specifically for cereal grains, and 2.7 billion tons of cereal were produced worldwide in 2018 (FAO STAT, 2019).

Grain quality depends on growing procedures, harvesting time, post-harvest processing, storage management, and transportation practices (Manandhar et al., 2018: Zhu et al., 2020). Beyond local farms, the critical stage is evaluating several aspects of quality that will decide the rate paid to the farmer. This quality assessment also dictates how grains are managed throughout the whole supply chain, ensuring that grains should not be mixed with lower-grade or defective seeds and any other foreign matter that can jeopardize grain quality.

Beyond this point, the value of the collected grains that have been rated as being of a given quality must be retained by storage and transferring them with that quality type intact. It is critical for cereal processing enterprises to identify the exact quality standards required for optimal processing quickly (Kong et al., 2021). Grain quality can be affected by two factors:

- Intrinsic factors
- Extrinsic factors

Grain color, size, shape, bulk density, odor, and aroma are all fundamental predictors of grain quality. A near-infrared radiation (NIR) analyzer is useful as a quick and non-destructive determinant of grain quality and is employed in industries for the specific qualitative needs of consumers (Ndlovu, 2021). Damaged/immature grains, foreign matter, and moisture content are all external factors that can deteriorate the grain quality (Guru & Mridula, 2021). In studies, it was reported that rodents/insects directly contaminate grains through their fecal matter, and microbes directly deteriorate the germ and endosperm of the grain, which lowers grain quality, and the seeds become unfit for consumption (Bhargava & Kumawat, 2010).

To achieve customer satisfaction with the selection of grains, the single kernel characterization method slows down the rapid detection of external deteriorating factors such as color/shape of the grains, which directly affects the yield during milling.

Grain quality is assessed by both qualitative and quantitative methods. Physical evaluation (PE) of all grain constituents is required for this procedure. PE is a simple qualitative factor in the quality assessment of the cereal grains and is greatly dependent on the expertise of the analysts. To recognize changes like raw materials/feeds, one must be skilled. Following harvest, effective planning ensures that grains are delivered in the shortest time possible before being infected and deteriorating. The silo facility must be thoroughly cleaned; all materials and equipment must be serviced; all mechanical shifting parts must be lubricated; all weighing appliances must be recertified; and all electrical vehicles, machinery, and the rest of the system must be tested long before the cereals arrive in vehicles. Within the silo, the elevator cells aeration duct/lids, which protect the routes, should be open, and empty containers must be cleaned properly. The documentation required for the exercise is also crucial. Staff members responsible for analysis, inspection, and grading, as well as weighing and administrative workers, plant operation, maintenance, and security, must be identified and adequately informed and advised of their allocated specialized roles and tasks. Before shipping samples for chemical or biological analyses, a physical evaluation can be performed immediately after receiving the grain sample.

In cereal reception in milling industries, different quality analyses are performed to ensure the quality of the grains, which will be discussed briefly in this chapter.

13.2 PHYSICAL EVALUATION

Physical evaluation involves the color, aroma, particle size, shape, damaged/degraded grain, bulk density, and foreign matter. The study of grain characteristics is necessary for the thorough evaluation of cereal grains (Ponce-García et al., 2017). Hazards linked with cereals, like other food systems, might be chemical, physical, or microbiological in character and can be introduced either by natural processes or by human actions (Adams et al., 2000). To preserve the integrity of grain, post-harvest handling is crucial throughout the supply chain and even the potential for events at any point in the supply chain to affect consumer health.

"Grain quality" in this context means that the grains being purchased must be appropriate for the flour miller to make flour with the characteristics required for the precise baking operations or other types of processing (Wrigley & Batey, 2012). The best recommended time for grain quality assessment is immediately after harvesting, when the grain is received and separated into distinct storage rooms, each holding grain of a different quality. Processing is another component that occurs farther down the grain chain; the grain buyer's purpose is to examine the quality of grain so that the processor only receives grain of acceptable and consistent quality (Mobolade et al., 2019).

13.3 GRAIN RECEPTION

Grain-reception criteria in wheat-growing countries establish a range of qualities, based on specific grades assigned to the grains (Acosta-Navarrete et al., 2014). This is the crucial stage of evaluating the numerous characteristics of grain quality that will determine the price given to the grower (Wrigley, 2017). This quality assessment also dictates how the grain is segregated throughout delivery, confirming that it is mixed with the same quality (Walker et al., 2018). Grain value and standards are assessed differently depending on the organization, establishment, country, market, and client, all of which are influenced by the intended use. Such criteria include size, color, shape, density, insect damage, moisture content, hectoliter weight, foreign materials, and broken grain (Awulachew, 2020). If appropriate analytical data are available in a timely manner, harvest time provides a substantial opportunity to monitor grain quality. Despite the time constraints of assessing grain quality prior to merging deliveries in the same cell, smaller amounts of all deliveries contributing to each storage cell can be kept and evaluated in detail later, providing details about the quality of grain (Barai et al., 2019). If this information is accessible, it can be used to market precise lots, each with a known quality, in order to meet the processing requirements of the feed or food maker at the next phase of the grain chain. Physical appearance is critical, followed by bulk density testing and sieving to establish how much grain is undersized and to check for impurities (Jewiarz et al., 2020). The first step in establishing grain soundness is to evaluate it visually, looking for evidence of sprouting, frost or heat damage, and fungal or insect damage (Awulachew, 2020). Several grain testing facilities cut the samples into pieces, mix them properly, and analyze one of the samples. Cups, often known as "pelicans," are manual sampling devices that are put in the running grain stream (Delwiche & Miskelly, 2017). This corresponds to at least two sets for trucks, with each set representing 500 bushels. Aside from moisture content, bulk density, and hectoliter weight, all other measurements or constituents are expressed as a percentage of the total (Ramli et al., 2019).

The vehicles transporting the consignment are directed to the weigh station for weighing if the grain meets these standards following the assessment; otherwise, the shipment is rejected.

13.4 RAW MATERIAL INSPECTION AND STANDARDS

A sample is collected from each batch during the transportation of grains towards flour mill (Simmons et al., 2005: Salarikia et al., 2021). The grain is usually given to small-scale millers in bags, and a sample should be obtained from the sacks through the application of thief sampler (Vijayalakshmi et al., 2015). The following quality factors should be carefully assessed during the sample assessment:

- Grain variety
- Grain quantity
- Contaminants
- Damaged grains
- Moisture content
- Grain color
- Grain maturity
- Microbial count
- Presence of flour mites

13.4.1 Moisture Content

Moisture content (MC) is a predictor of grain quality and milling profitability (Tzatzani et al., 2020: Liu et al., 2019); it should not exceed 14.5% (Iskenderov et al., 2019) because it attracts mold, insects, and bacteria (Atungulu et al., 2019), all of which lower grain quality (Bisht et al., 2021). The moisture content of grain traded on a weight basis has a significant effect on the sale price. Low-moisture wheat or flour is more stable throughout storage, necessitating wheat moisture content management during receipt and storage (Kumar et al., 2021).

13.4.2 Test Weight

Test weight is the most appropriate method of determining grain plumpness in situations that often limit the chance for extensive testing at the point of grain reception (Wrigley & Batey, 2003). Grain weight is expressed in pounds per bushel (lb/bu) per unit volume or kilograms per hectoliter (kg/hl). Grain test weight is important to millers because it is linked to higher grinding flour production, whereas lower test weights are linked to shriveled but less sound kernels, resulting in lower flour yields. As a consequence, a chondrometer is frequently employed in laboratory mills to help millers forecast wheat behavior and extraction rates (Howarth et al., 2021: Kumar et al., 2022).

13.4.3 Defects and Contaminants

Moisture above 14.5% is susceptible to the hazardous condition of mycotoxins, due to the presence of molds like *Fusarium* and *Alternaria* (Pascari et al., 2022). They aren't usually found in moldy grain, but it's evident that keeping grain dry and free of molds is the best approach to avoid them. Immunoassay kits are now available for quickly determining various mycotoxins in field circumstances, allowing checking for risk on the spot, as opposed to relying on laboratory-based analysis that took several days to return results a few years ago. The enzyme-immuno-assay employed for quantitative detection of various mycotoxins or some pesticide residues could replace traditional High Performance Liquid Chromatography (HPLC) or Gas Chromatography-Mass Spectrometry

(GC-MS) studies with routine screenings once field test validation for mold deterioration and insect identification has been conclusive.

13.4.4 Grain Hardness

Grain hardness is a basic quality parameter that differentiates grades in international trade because it is important in determining the appropriateness of the generated flour for bread-making (hard grain) versus a need for soft grain for a range of special products, such as cakes, biscuits (cookies), and grocery and pastry flours. Grain hardness is a varietal trait; thus, if the varietal identity is known, this quality can be determined (Fradgley et al., 2022). Wheat endosperm texture and hardness are important factors in evaluating wheat's compatibility for various end products (Greffeuille et al., 2006).

13.4.5 Variety Identification

Outer bran coating, embryo/germ, and endosperm are the three main components of a wheat kernel (Shewry et al., 2020). Because the breeder has purposefully selected each variety to fit a specific range of processing needs, knowledge of variety is crucial in order to determine the grain quality. Many wheat-producing countries have implemented a system of variety declaration for delivered wheat, as well as the capacity to verify that the variety declaration is accurate (Joshi & Braun, 2022). Laboratory testing, most commonly by extracting grain proteins and assessing protein composition, provides a more precise technique of identification. Grain protein gel electrophoresis has been the standard method for variety evaluation in recent decades; however, HPLC and capillary electrophoresis are also prescribed in many studies due to improved and accurate testing (Salgotra & Bhat, 2022).

13.4.6 Protein

Protein is a major determinant in the evaluation of grain quality, as gluten is the main constituent in the baking industry (Zhang et al., 2022). Protein concentration is a critical criterion for millers throughout the procuring process since it is linked to water absorption, gluten content, and strength (Djordjević et al., 2021). Crispy or tender food, such as snacks or cakes, should have a low protein content, while chewy products, such as pan bread and hearth bread, should have a high protein content. As a result, millers must determine the protein content of their wheat in order to classify it and use each variety to produce the appropriate flour. It's usually determined indirectly by using procedures like Kjeldahl (AACC Method 46–11A) and combustion to measure nitrogen concentration (AACC Method 46–30). The protein content is then calculated using a correction factor ($\times 5.7$) that accounts for amino acid composition and nonprotein nitrogen.

13.4.7 Damaged Starch

Starch granules are destroyed to varying degrees during the grinding process, depending on the hardness of the grain and milling process (Wang et al., 2020). Due to damaged starch networks, the ability of starch to absorb water is increased (An et al., 2022) up to tenfold (Roman et al., 2021), due to which it is a critical component in determining the mixing properties. As a result, damaged starch has a direct impact on dough behavior during fermentation. Its actions will have an impact not only on the completed product's volume but also on its color. Failure to regulate the quantities of damaged starch during the transformation of flour into cooked items might result in a variety of problems.

13.4.8 Wet Gluten

In the milling sector, this test is routinely performed in any laboratory. Glutenin and gliadin, which compose wet gluten, are responsible for the dough's elasticity and extensibility, respectively (Singh et al., 2022). It is one of the most essential properties of wheat flour and must be assessed using the Glutomatic system, which determines the gluten content and the gluten index, which explains the quantity and quality of wet gluten (Barros et al., 2022).

13.4.9 Rheological Analysis

The role of dough rheology in predicting and controlling baked product quality is widely understood. In bakery applications, variations in the elastic and viscous behavior of different wheat flour doughs are regarded important quality considerations (Hoehnel et al., 2019). The major goal of rheological measures is to distinguish between wheat varieties based on their baking performance without the baking. Time, material, and labor cost can be precluded if rheological tests could accurately forecast the baking performance of any wheat variety.

The mechanical properties of wheat flour doughs, as assessed empirically by the Extensograph, Farinograph, Alveograph, and Mixograph instruments, have a major impact on dough-handling properties during processing and finished good quality attributes (Aaliya et al., 2021). The tests performed by these devices are useful for providing practical information to the baking sector, but they are unsuitable for analyzing basic baking quality behavior and dough processing.

13.4.10 Baking Test

Baking tests offer information on flour quality attributes for end-users. Furthermore, the findings of these tests can be used to optimize processing conditions before baking on a commercial scale. Using the following criteria, one can achieve their baking goals:

- Extensibility and gluten strength of dough
- Machinability of dough
- Fermentation process in various situations
- Effect of additives as well as flour corrector solutions
- Sensory attributes of the baked good

13.5 GRAIN ASSESSMENT AND GLOBAL STANDARDS

Grain assessment and the development of standard techniques for quality assessment is a time-consuming process that entails collaboration between a diverse group of scientists from a variety of facilities participating in the types of investigations being assessed. Implementations of these procedures as a standard requires a thorough discussion with the relevant scientific community.

13.5.1 Transportation Equipment

The transportation vehicles allocated to food items should not be engaged in any non-food product transportation that can contaminate the food products (Okpala & Korzeniowska, 2021). Before the loading of every single container, the vehicle should be inspected for infestations, spills, rodents, insects/pests, and any vehicle defects or spillages.

13.5.2 Sanitation Facility

Grain storage facility sanitation is a critical factor of integrated grain management. This includes removing pest-infested grasses/bushes, cleaning the various pit/tunnel segments, and preparing the silo cells. Others are washing grain from dust particles at the entrance and cleaning all the equipment, pipelines, including elevators, chutes, conveyors, and discharge auger lines. During the cleaning process, any debris, dust, or remaining grains are burned, and the tunnels are then left open for ventilation. The silo cells are prepared in a specific way for reception, and physical cleaning of the interior silo body and floor for old sticking grains is required, as well as scraping-off of the difficult ones (Vishwakarma et al., 2022).

All ventilation ducts must be cleaned and opened, with the covers inspected for openings that could allow seeds to enter the duct. All silo edges and structural gaps can be sprayed with liquid insecticides, and the silo must be fumigated. The surface ventilation doors are closed, and the silo is locked until it is needed. The surrounding and neighboring areas of the silo facility must be cleaned, disinfected, and pesticide-treated or burned. Throughout the facility, including the workhouse, tunnels, pits, and grain storage, insecticide must be sprayed.

13.5.3 Lubrication/Servicing of Equipment

To ensure that grain analysis laboratory equipment is performing properly, it must be calibrated and checked. Good-quality grains will be obtained for storage thanks to good equipment. The quality of the grains can be harmed by inaccuracy in the equipment. Every silo apparatus, including reception pits, grain handling systems, blowers, dust collectors, cleaners, storage bins, and heat monitoring, should be monitored, cleansed, and repaired as needed to ensure a trouble-free operation. Spare parts, fuel, oil, and lubricants should all be found or simple replacement techniques developed. All equipment, including sweep augers, conveyors, cleaners, elevators, intake pits, cyclones, and ventilation fans, should be inspected for damage. All of their revolving parts and bearings should be checked and greased. This should be done thoroughly, with any weak parts or links being replaced before the action begins. All electric motors must be inspected and checked from the control panel to ensure that everything is in working order.

Examine and clean the cleaners, as well as the exhaust system. All transmission components, such as belts, gears, chains, and sprockets, should be inspected and, if necessary, replaced. If electric motors employ speed reducers with an oil-filled box or gear, the gear oil should be checked and replenished if it falls below the recommended level. Open the elevator top and inspect and clean the overall chain junction for effective movement. Loose belts, broken cups, and nuts are also checked for indicators of wear on the elevator cups and belts.

Any parts that are broken or loose should be replaced or tightened. The sweep/discharged augers, as well as their electric motors, must be inspected and tested. If the weighbridge is analog, it must be oiled, the knife edge/test point must function properly, and all other lubrication parts must be completed. A test weight must be used to ensure accuracy. To avoid damage to the wires, which could result in erroneous output if the gauges are digital, the wires and connections must be removed. The load cells must be inspected as well as cleaned. It's critical to examine electronic display units and other gadgets twice. Before beginning operations, the entire electrical system should be tracked from the generator house to the control panel to the equipment for rats chewing wires, bridges, and broken wires. Generators must be serviced at least once a year. The transformer should be inspected for optimal performance. The storage compartment must be closed for final treatment.

13.5.4 Arrival of Trucks

Prior to the entry of the first vehicle, a certification system for record keeping and truck queuing must be in place. In many silo sites, the security department is in charge of keeping such records, while other supporting documents are stored at the weigh station and testing labs. When a truck bringing grains arrives at the silo, the silo management will conduct preliminary inspections to ensure the legality of the grain supply.

13.5.5 Grain Sampling/Grading

Grain testing is the process of gathering a sample of grains for analysis that is supposed to be representative of the whole grain. To gather samples, grain probes, samplers, and sampling spears are employed. Primary samples are gathered and carefully mixed from as many bags as possible. The samples are collected in sample bags that are neatly labelled with identification information such as dates, names, agents, products, suppliers, farmers, and truck numbers. The size of the first sample drawn is governed by the size of the consignment, as described in the following: Dimensions of the shipment. The total number of bags sampled was less than ten. Each and every sack had been empty. From a minimum of 10 bags, 10–100 bags are chosen. There are over a hundred bags in all. The total number of bags is divided by the square root of the total number of bags. The purpose of evaluating grain quality is to determine whether the grain and its homogeneity meet the requirement in terms of intrinsic accurate attributes. These findings can be used to create a quality analysis report that can be used to analyze the grain's suitability for storage, use, or other purposes, as well as to evaluate it; estimate its price; and estimate its appropriateness for storage, use, or other purposes.

13.5.6 Quality Standard of Received Grains

Grain quality standards for animal feed manufacturers will be separate from grain quality standards for storage. However, most grain assessments are based on physical/biological traits because quality characteristics and biochemical processes might take several days to determine, especially when destructive methods of inspection are applied. If non-destructive methods (ultrasound, radiography, magnetic particle) are used, it may only take hours. Such criteria include size, color, shape, density, insect damage, moisture content, hectoliter weight, foreign materials, and broken grain. Several grain testing facilities cut the samples into pieces, mix them properly, and analyze one of the samples. Aside from moisture content, bulk density, and hectoliter weight, all other measurements or constituents are expressed as a percentage of the total.

13.6 NEW APPROACHES FOR GRAIN ASSESSMENT

Grain inspection at the point of receipt is still mainly reliant on human visual examination. To reduce the time and labor needs of inspection, researchers have been working on developing objective techniques that can help or replace traditional inspection procedures. Ruggedness, reliability, ease of use, speed, and precision will be required for these techniques to be adopted successfully at the country elevator. NIR spectroscopy, which was first used in the grain sector in the 1970s, is possibly the best example of a technology that revolutionized the industry. NIR whole-kernel transmittance is currently widely employed in national inspection programs and in commerce to determine the protein content of wheat. New technologies that have the same potential for adoption at reception locations as NIR spectroscopy are being sought.

- Digital imaging
- Hyperspectral imaging

- ELISA test kits for insect activity
- PCR-based detection methods
- Electronic noses
- X-ray imaging for internal insects

NIR technologies have become widely used at mill or elevator receipt for determining moisture and protein content in cereal grains in general, both for single samples and for continuous monitoring online, in recent decades. This is the standard method for grading grains, particularly wheat and barley. NIR is also used to determine wheat grain toughness. Originally, NIR technology was only used on milled grain, but advancements have allowed this technology to be used on whole grain. As a result, no sample preparation is required—a grain sample can simply be tipped into the top of the NIR machine, and results are automatically generated. NIR works in the same way that any other spectroscopic approach does: distinct chemical groups in the sample absorb radiation at specific wavelengths, and the extent of absorption is proportional to the analyte concentration. While NIR has proven the most successful method for measuring key grain components such as moisture, protein, and oil content, determining smaller components may be more difficult. Nonetheless, NIR technology offers a lot of room for expansion into a variety of analytical applications.

13.7 GLOBAL STANDARDS FOR GRAIN RECEPTION

Despite all of these advancements in grain quality characterization, there are still two major issues: low rejection rate in the quality parameters (e.g., insect density, sprouted kernels, *Fusarium* spotted kernels, insect-damaged kernels, etc.), and if affected kernels have a clumped distribution within the grain mass, it is difficult to obtain a representative sample in bulk grain.

All efforts to develop rapid, precise quality characterization methods are geared toward predicting end-use industrial (technical) quality. However, in many grain trade scenarios, there may be a weak link between the broker's qualitative needs and the standard quality parameters necessary for the grain's planned or stated end-use by the end-user.

In grain quality testing, there is a tendency toward measuring a wider range of characteristics at lower and lower concentrations (e.g. ELISA for mycotoxins and pesticide residues). We must, however, be cautious that these new tools are not utilized to erect non-tariff trade obstacles. Nonetheless, the whole grain storage, handling, transporting, and processing industries have made developing methods for assessment of end-use processing quality of delivered grain a priority. The global grain trade should increasingly shift to a quality-based market.

To attain this purpose, a more widespread application of new technology for quick examination of quality qualities should be combined with an attempt to standardize internationally. The inverse logic appears to be incorrect or senseless (regulate first and then find analytical technologies consistent with the regulations). The ICC, AACC, and other international organizations that promote the standardization of novel analytical or sampling procedures are engaged in this type of international coordination endeavor. However, further information on the factors that influence grain quality and how it changes throughout storage is required.

REFERENCES

Aaliya, B., Navaf, M., & Sunooj, K. V. 2021. Dough handling properties of gluten-free breads. In *Gluten-free Bread Technology* (pp. 49–70). Springer, Cham.

Acosta-Navarrete, M. S., Padilla-Medina, J. A., Botello-Alvarez, J. E., Prado-Olivarez, J., Perez-Rios M, M., Díaz-Carmona, J. J., . . . & Fernandez-Jaramillo, A. A. 2014. Instrumentation and control to improve the crop yield. In *Biosystems Engineering: Biofactories for Food Production in the Century XXI* (pp. 363–400). Springer, Cham.

Adams, M. R., Moss, M. O., & Moss, M. O. 2000. *Food microbiology*. Royal society of chemistry.

An, D., Li, H., Li, D., Zhang, D., Huang, Y., Obadi, M., & Xu, B. 2022. The relation between wheat starch properties and noodle springiness: From the view of microstructure quantitative analysis of gluten-based network. *Food Chemistry*, 133396.

Atungulu, G. G., Kolb, R. E., Karcher, J., & Shad, Z. M. 2019. Postharvest technology: Rice storage and cooling conservation. In *Rice* (pp. 517–555). AACC International Press.

Awulachew, M. T. 2020. Understanding basics of wheat grain and flour quality. *Journal of Health and Environmental Research*, 6(1), 10–26.

Baladhiya, H. C., Sisodiya, D. B., & Pathan, N. P. 2018. A review on pink stem borer, *Sesamia inferens* Walker: A threat to cereals. *Journal of Entomology and Zoology Studies*, 6(3), 1235–1239.

Baniwal, P., Mehra, R., Kumar, N., Sharma, S., & Kumar, S. 2021. Cereals: Functional constituents and its health benefits. *The Pharma Innovation*, 10(2), 343–349.

Barai, A., Uddin, K., Dubarry, M., Somerville, L., McGordon, A., Jennings, P., & Bloom, I. 2019. A comparison of methodologies for the non-invasive characterisation of commercial Li-ion cells. *Progress in Energy and Combustion Science*, 72, 1–31.

Barros, J. H., Montenegro, F. M., & Steel, C. J. 2022. Characterization and regeneration potential of vital wheat gluten treated with non-thermal plasma. *Journal of Cereal Science*, 104, 103402.

Bhargava, M. C., & Kumawat, K. C. 2010. *Pests of stored grains and their management*. New India Publishing.

Bisht, A., Kamble, M. P., Choudhary, P., Chaturvedi, K., Kohli, G., Juneja, V. K., . . . & Taneja, N. K. 2021. A surveillance of food borne disease outbreaks in India: 2009–2018. *Food Control*, 121, 107630.

Delwiche, S., & Miskelly, D. 2017. Analysis of grain quality at receival. In *Cereal Grains* (pp. 513–570). Woodhead Publishing.

Djordjević, M., Djordjević, M., Šoronja-Simović, D., Nikolić, I., & Šereš, Z. 2021. Delving into the role of dietary fiber in gluten-free bread formulations: Integrating fundamental rheological, technological, sensory, and nutritional aspects. *Polysaccharides*, 3(1), 59–82.

Fradgley, N. S., Gardner, K., Kerton, M., Swarbreck, S. M., & Bentley, A. R. 2022. Trade-offs in the genetic control of functional and nutritional quality traits in UK winter wheat. *Heredity*, 1–14.

Greffeuille, V., Abecassis, J., Rousset, M., Oury, F. X., Faye, A., L'Helgouac'h, C. B., & Lullien-Pellerin, V. 2006. Grain characterization and milling behaviour of near-isogenic lines differing by hardness. *Theoretical and Applied Genetics*, 114(1), 1–12.

Guru, P. N., & Mridula, D. 2021. Safe storage of food grains. ICAR-Central Institute of Post-Harvest Engineering and Technology, Ludhiana (Punjab). *Technical Bulletin No.: ICAR-CIPHET/Pub./2021–22/01*, 32.

Hoehnel, A., Axel, C., Bez, J., Arendt, E. K., & Zannini, E. 2019. Comparative analysis of plant-based high-protein ingredients and their impact on quality of high-protein bread. *Journal of Cereal Science*, 89, 102816.

Hossain, A., Krupnik, T. J., Timsina, J., Mahboob, M. G., Chaki, A. K., Farooq, M., . . . & Hasanuzzaman, M. 2020. Agricultural land degradation: Processes and problems undermining future food security. In *Environment, Climate, Plant and Vegetation Growth* (pp. 17–61). Springer, Cham.

Howarth, C. J., Martinez-Martin, P. M., Cowan, A. A., Griffiths, I. M., Sanderson, R., Lister, S. J., . . . & Marshall, A. H. 2021. Genotype and environment affect the grain quality and yield of winter oats (Avena sativa L.). *Foods*, 10(10), 2356.

Iskenderov, R., Lebedev, A., Zacharin, A., Lebedev, P., & Marjin, N. 2019, December. Constructive and regime parameters of horizontal impact crusher of grain materials. In *IOP Conference Series: Earth and Environmental Science* (Vol. 403, No. 1, p. 012057). IOP Publishing.

Jewiarz, M., Mudryk, K., Wróbel, M., Frączek, J., & Dziedzic, K. 2020. Parameters affecting RDF-based pellet quality. *Energies*, 13(4), 910.

Joshi, A. K., & Braun, H. J. 2022. Seed Systems to Support Rapid Adoption of Improved Varieties in Wheat. In *Wheat Improvement* (pp. 237–256). Springer, Cham.

Kong, J., Yang, C., Wang, J., Wang, X., Zuo, M., Jin, X., & Lin, S. 2021. Deep-stacking network approach by multisource data mining for hazardous risk identification in IoT-based intelligent food management systems. *Computational Intelligence and Neuroscience*, 2021.

Kumar, D., Narwal, S., Verma, R. P. S., & Singh, G. P. 2022. Advances in Malt and Food Quality Research of Barley. In *New Horizons in Wheat and Barley Research* (pp. 697–728). Springer, Singapore.

Kumar, P. K., Parhi, A., & Sablani, S. S. 2021. Development of high-fiber and sugar-free frozen pancakes: Influence of state and phase transitions on the instrumental textural quality of pancakes during storage. *LWT*, *146*, 111454.

Li, M., Xu, J., Gao, Z., Tian, H., Gao, Y., & Kariman, K. 2020. Genetically modified crops are superior in their nitrogen use efficiency—A meta-analysis of three major cereals. *Scientific reports*, *10*(1), 1–9.

Liu, J., Fu, Y., Wang, H., Peng, Z., Fan, X., Weng, Y., & Yao, W. 2019. Simulation of Winter Durum Wheat Flour Yield Based on Structural Equation Modeling.

Manandhar, A., Milindi, P., & Shah, A. 2018. An overview of the post-harvest grain storage practices of smallholder farmers in developing countries. *Agriculture*, *8*(4), 57.

Mobolade, A. J., Bunindro, N., Sahoo, D., & Rajashekar, Y. 2019. Traditional methods of food grains preservation and storage in Nigeria and India. *Annals of Agricultural Sciences*, *64*(2), 196–205.

Ndlovu, P. F. 2021. *Rapid monitoring and quantification of unripe banana flour adulteration using visible-near infrared spectroscopy* (Doctoral dissertation).

Okpala, C. O. R., & Korzeniowska, M. 2021. Understanding the relevance of quality management in agrofood product industry: From ethical considerations to assuring food hygiene quality safety standards and its associated processes. *Food Reviews International*, 1–74.

Pascari, X., Marin, S., Ramos, A. J., & Sanchis, V. 2022. Relevant *Fusarium* mycotoxins in malt and beer. *Foods*, *11*(2), 246.

Ponce-García, N., Ramírez-Wong, B., Escalante-Aburto, A., Torres-Chávez, P. I., & Serna-Saldivar, S. O. 2017. Grading factors of wheat kernels based on their physical properties. *Wheat Improvement, Management and Utilization*, *275*.

Ramli, N. A. M., Rahiman, M. H. F., Kamarudin, L. M., Zakaria, A., & Mohamed, L. 2019, November. A review on frequency selection in grain moisture content detection. *IOP Conference Series: Materials Science and Engineering*, *705*(1), 012002.

Roman, L., Gomez, M., & Martinez, M. M. 2021. Mesoscale structuring of gluten-free bread with starch. *Current Opinion in Food Science*, *38*, 189–195.

Salarikia, A., Jian, F., Jayas, D. S., & Zhang, Q. 2021. Segregation of dockage and foreign materials in wheat during loading into a 10-m diameter corrugated steel bin. *Journal of Stored Products Research*, *93*, 101837.

Salgotra, R. K., & Bhat, J. A. 2022. Efficient high-throughput techniques for the analysis of disease-resistant plant varieties and detection of food adulteration. *Current Protein and Peptide Science*, *23*(1), 20–32.

Sarrocco, S., Mauro, A., & Battilani, P. 2019. Use of competitive filamentous fungi as an alternative approach for mycotoxin risk reduction in staple cereals: State of art and future perspectives. *Toxins*, *11*(12), 701.

Sarwar, A. K. M. G., & Biswas, J. K. 2021. Cereal grains of Bangladesh—Present status, constraints and prospects. *Cereal Grains, 1*, 19.

Shewry, P. R., Wan, Y., Hawkesford, M. J., & Tosi, P. 2020. Spatial distribution of functional components in the starchy endosperm of wheat grains. *Journal of Cereal Science*, *91*, 102869.

Simmons, R. W., Pongsakul, P., Saiyasitpanich, D., & Klinphoklap, S. 2005. Elevated levels of cadmium and zinc in paddy soils and elevated levels of cadmium in rice grain downstream of a zinc mineralized area in Thailand: implications for public health. *Environmental geochemistry and health*, *27*(5), 501–511.

Singh, A., Gupta, O. P., Pandey, V., Ram, S., Kumar, S., & Singh, G. P. 2022. Physicochemical components of wheat grain quality and advances in their testing methods. In *New Horizons in Wheat and Barley Research* (pp. 741–757). Springer, Singapore.

Skendi, A., Zinoviadou, K. G., Papageorgiou, M., & Rocha, J. M. 2020. Advances on the valorisation and functionalization of by-products and wastes from cereal-based processing industry. *Foods*, *9*(9), 1243.

Torbica, A., Belović, M., Popović, L., Čakarević, J., Jovičić, M., & Pavličević, J. 2021. Comparative study of nutritional and technological quality aspects of minor cereals. *Journal of Food Science and Technology*, *58*(1), 311–322.

Tzatzani, T. T., Kavroulakis, N., Doupis, G., Psarras, G., & Papadakis, I. E. 2020. Nutritional status of 'Hass' and 'Fuerte' avocado (*Persea americana* Mill.) plants subjected to high soil moisture. *Journal of Plant Nutrition*, *43*(3), 327–334.

Vijayalakshmi, D., Srividhya, S., Vivitha, P., & Raveendran, M. 2015. Temperature induction response (TIR) as a rapid screening protocol to dissect the genetic variability in acquired thermotolerance in rice and to identify novel donors for high temperature stress tolerance. *Indian Journal of Plant Physiology*, *20*(4), 368–374.

Vishwakarma, R. K., Kumar, N., Sharma, K., Kumar, Y., & Kumar, C. 2022. Storage. In *Agro-Processing and Food Engineering* (pp. 353–413). Springer, Singapore.

Walker, S., Jaime, R., Kagot, V., & Probst, C. 2018. Comparative effects of hermetic and traditional storage devices on maize grain: Mycotoxin development, insect infestation and grain quality. *Journal of Stored Products Research*, *77*, 34–44.

Wang, Q., Li, L., & Zheng, X. 2020. A review of milling damaged starch: Generation, measurement, functionality and its effect on starch-based food systems. *Food Chemistry*, *315*, 126267.

Wrigley, C. 2017. Assessing and managing quality at all stages of the grain chain. In *Cereal Grains* (pp. 3–25). Woodhead Publishing.

Wrigley, C. W., & Batey, I. L. 2003. Assessing grain quality. *Bread Making: Improving Quality*, 71–96.

Wrigley, C. W., & Batey, I. L. 2012. Assessing grain quality. In *Breadmaking* (pp. 149–187). Woodhead Publishing.

Zhang, M., Jia, R., Ma, M., Yang, T., Sun, Q., & Li, M. 2022. Versatile wheat gluten: functional properties and application in the food-related industry. *Critical Reviews in Food Science and Nutrition*, 1–17.

Zhu, G., Liu, H., Xie, Y., Liao, Q., Lin, Y., Liu, Y., . . . & Hu, S. 2020. Postharvest processing and storage methods for *Camellia oleifera* seeds. *Food Reviews International*, *36*(4), 319–339.

CHAPTER 14

Grain Storage and Transportation Management

Sakshi Sharma, Anil Dutt Semwal, M Pal Murugan, Mohammed Ayub Khan
and Dadasaheb Wadikar

CONTENTS

DOI: 10.1201/9781003252023-14

14.1 INTRODUCTION

Continuous and timely availability of food, water, clothes, and shelter are basic requirements for the survival of human beings on this earth. Humankind discovered the science of cultivation of useful and beneficial crops to ensure the availability of food to them. Our main concern is to provide enough food for the ever-increasing human population. The idea of food security was described by Food and Agriculture Organization (FAO) in 1983, which means "all the people at all times have the physical and economical access to the basic food they need." The World Bank subsequently in 1986 stated, "food security is access by all the people at all the times to enough food for an active, healthy life." This implies that for successful achievement of food security goals, there are two major requirements: (i) production of a sufficient quantity of food and (ii) safe storage of food and timely transportation to ensure nationwide physical supply at all periods, which include years with inadequate food production (World Bank, 1986).

Food grains are the most essential diet components to provide calories globally for the majority of human beings (Rajashekar et al., 2016). The large population of rural farmers involved in food grain activities not only produce food for their own consumption, but it is also a source for their employment and income. With advancements in production techniques in agriculture, improving crop productivity, genetically coupled with improved practices for crop cultivation, food grain production has also been increasing steadily both at the global and country level, thus ensuring sufficient production. As a staple food, there is regular need for food grains across the whole year, while the production of food grains is confined to some months only. Therefore, there is demand to store the available excess food during the harvesting period safely to meet average food requirements evenly throughout the year.

Food grains are an essential consumable commodity used for human consumption at a large scale. Food grains mainly include cereal (rice, wheat, sorghum, maize, millets, etc.) and pulse (gram, pigeon pea, lentil, green gram, black gram, field pea, kidney bean, etc.) crops. However, a huge quantity of available food grains produced through hard labor by the farmers spoils due to inadequate and improper storage conditions. An estimated total of 12–16 million metric tons (MMT) of food grains spoil due to the unavailability of proper post-harvest handling every year, which is sufficient to feed 33% of the population of India. A total of yearly grain losses has been estimated as nearly INR 50,000 crores (World Bank Report). Phillips and Throne (2010) reported that food grains losses in storage because of poor storage capacity and infestation by insects and pests accounted for 10–20% of overall production. By improving proper storage conditions through adopting scientific methods, these huge losses can be minimized, which will lead to better accessibility of food grains for human use. Per Sheahan and Barrett (2017), realizing the vital consequences of safe food storage with negligible losses, the United Nations ambitiously set a target to reduce global food wastage by 50% by 2030 as part of its Sustainable Development Goals. Development of production systems which can meet the goals of UN Sustainable Development Goals needs extensive research on crop rotation and management tools as well as farming systems for achieving sustainable agriculture. One example of sustainable agriculture practice can be the utilization of leguminous plants, such as grass pea plants because of their ability to fix and use atmospheric nitrogen, thus, making it an excellent green manure (Sharma et al., 2022).

14.2 OBJECTIVES OF FOOD GRAIN STORAGE

The principal aim of storage is the avoidance of decline in the quality of food grains for a specified time period so that quality food is available to humans when needed. It is realized through regulation of moisture and movement of air and avoiding food spoilage caused due to microorganisms, insects, and rodents. The major purpose during storage is to substantially reduce the metabolic activities of stored grains so that there is minimal deterioration.

In the complete cycle from production to consumption, food grains have to pass through a sequence of operations before they reach consumers for consumption. These operations include harvesting, threshing, winnowing, bagging, transportation, storage, and processing. At all these stages, there are variable losses in the crop produce. For that reason, researchers across the countries are currently paying attention to work out improved grain storage techniques and facilities to reduce food wastage. This reduction of food wastage in storage is equivalent to additional grain production.

During 2018, India produced ~344.8 million tons of food grains (cereals and pulses) against the total global production of 3055.24 million tons. India emerged as the third-largest producer of food grains, after China and the United States, sharing about 11.29% of total global production. Commodity-wise, India represents 10.74 and 27.63% of the total global production of cereals and pulses and is ranked third and first, respectively (GOI, 2020).

According to the estimate of the GOI (2020), 157.8 million ha of farming land is owned by 146.5 million farming families in India, with an average landholding size of 1.08 ha during 2015–2016. The report also revealed that 29.23% of total agricultural land is owned by 5.7% of farmer families, which have ≥4 ha land, but ~94.3% farmers in the country with small landholdings accounted for ~70.77% of agricultural land. Farmers with smallholdings retain approximately 60–70% of their produce for their own consumption and store it in locally available storage structures (Kanwar & Sharma, 2003). The remaining surplus grains may be sold by them within a couple of months after harvest to government agencies or private traders. On the other hand, big farmers sell a larger proportion of their produce and hold it in storage installations run by government authorities such as the Food Corporation of India (FCI). Therefore, grain storage systems operate at different levels in the country, farmer, trader, and commercial or government levels, based on the purpose and period of storage. Storage at the farmer level is normally inter-seasonal for household consumption, as a seed for next season, or for cash or barter exchange. Traders store food grains for a shorter period ranging from few days to months. They buy and sell quickly, with a major objective of earning more profit. Millers and cooperatives store the food grain for shorter durations to meet their clients' demand. Government involvement in grain storage often has the purpose to create a food reserve for urban population, to have buffer stock to maintain food security reserves, to provide a stimulus to production, to stabilize food grain prices, and so on. Usually the government has its own agencies or marketing boards involved in this specific activity.

14.3 BASIC REQUIREMENTS OF FOOD GRAIN STORAGE SYSTEMS

Hall (1970) highlighted the fundamental necessities of a grain storage system to save grains from storage pests during storage and to avoid grain spoilage caused by microorganism activities. After harvesting, grains being the living entity, continue to carry out all biological activities and respire during the storage period and are consequently attacked by insects, rodents, microorganisms, and so on. Hence, there is a critical requirement for a scientific modern approach to store them safely. Intensive studies of destructive insects and pests (insects, microorganisms, rodents, birds and squirrels, etc.) and factors (moisture, relative humidity [RH], temperature and location of storage, etc.) that provide favorable conditions for the growth of these destructive agents are necessary. The effect of microorganisms and insects/pests can be controlled up to certain extent with the development of adequate conditions within storage structures. The most suitable temperature considered was 21°C, with a moisture level of nearly 9%, within a bin (Mehrotra et al., 1987).

Storage structures vary greatly in design based on their size (small or big), location (indoor or outdoor), duration (shorter or longer), nature (temporary or permanent), and level of user (individual or community). These structures may have an option for a closed, semi-open, or open system of storage (Gwinner et al., 1990). The availability of natural resources in a particular environment and prevailing customs play a significant part in the selection of storage systems, which is often relevant to specific climatic conditions (Hall, 1970). Traditional storage structures/systems have been used by farmers for a long time successfully due to the utilization of scientific values, however inadvertently.

Knowledge of traditional methods of storage has evolved in communities over long years of experience. This traditional knowledge continues to pass on from one generation to the next (Natarajan & Govind, 2006). Some traditional methods of grain storage are unique to a specific society and vary considerably among villages, locales, communities, and nations. Based on cultural connections and close awareness of specific environmental conditions, these native storage practices evolved.

A good storage system must be able to prevent or minimize any kind of losses during storage and be easy to operate and cost effective. Thus, a storage system must (i) have sufficient safety from rodents, birds, insects, mites, and so on; (ii) allow need-based aeration and fumigation facilities; (iii) be able to avoid losses due to moisture and temperature; (iv) have appropriate facilities for easy handling, cleaning, inspections, transport, and so on; and (v) be economical on unit storage cost basis.

14.4 LOSSES DURING STORAGE OF FOOD GRAINS

Food grains are a non-perishable or shelf-stable type of food at room temperature. Storage losses are of two types: (i) physical loss to commodities, termed direct loss or loss, and (ii) loss in quality and nutrition of the commodities, which is termed indirect loss or damage. Both losses and damage caused by storage insect pests during storage are equally important. The term "damage" is an indication of decay or deterioration, one of the example being holes found in the grains, which chiefly impacts the quality, while term "loss" refers to the total disappearance or non-availability of the food that can be measured by quantitative methods (Boxall, 2002). A loss in quality or damage leads to lowering of the product's value. Depending upon the economic status of the consumers and the degree of severity of the damage, such a product may be used or completely rejected. A person with low income may eat food that is damaged to some degree, while wealthy customers might refuse a food product with slight damage. However, in the case of loss, the commodity cannot be utilized. Some reduction in quantity of storage grain also occurs in the manner of spillage due to leakage in bags that could be recognized when the storage is empty: grains can be found present on the floor. An FAO report on wastage of food grains at various stages revealed that the aggregate waste of food grains in the less developed countries may range from 30 to 50%. The extent of losses to various causes is estimated as (i) field losses (25%), (ii) storage losses (15%), (iii) rodents (2.5%), (iv) birds (0.8%), (v) moisture (0.7%), (vi) handling and processing (7.0%), and (vii) miscellaneous (6.0%).

14.4.1 Factors Influencing Storage Losses

Factors that influence storage losses can be categorized into three types: (i) biotic factors, which include insects/pests, microorganisms, rodents, birds, and so on; (ii) abiotic factors, which include moisture, temperature, and so on; and (iii) other factors, which include quality of grains at the time of storage and storage structure (Abedin et al., 2012). Information as to how these factors contribute to storage losses follows.

14.4.1.1 Insect Pests

Storage pests are categorized as: (i) primary storage pests refers to those insects that damage sound and healthy grains, and (ii) secondary storage pests refers to those insects that damage broken or already damaged grains. Various types of losses like quantitative and qualitative losses, seed viability losses, and any damage to storage structures are caused by insects during storage.

14.4.1.1.1 Quantitative Loss

This refers to losses caused in quantity of stored grains due to direct feeding by insects that results in reduction in weight or quantity among stored grains. For example, *Sitophilus oryzae* L. (rice weevil) can consume 14 mg out of 20 mg of a rice grain during its developmental period; however, from commercial point of view, there is complete loss of the grain. A female *Sitophilus oryzae* L. is known to be able to produce 1,500,000 progeny in three generations a year that will infest and eat 1,500,000 rice grains, equivalent to 30 kg of rice.

14.4.1.1.2 Qualitative Loss

This refers to quality losses in stored grains. Due to insect infestation, chemical changes occur in grains. Grains also become contaminated with molted skin and body parts, which enhances the possibility of pathogenic microorganism development.

14.4.1.1.3 Seed Viability Loss

Insect pests can cause a reduction in seed viability to an extent of 3.6 to 41% in rough rice as they eat or damage the embryo portion of grains.

14.4.1.1.4 Damage Caused to Storage Structures

Rhyzopertha dominica (lesser grain borer) has the potential to damage structures made of wood, containers, polythene lined bags, and so on, thus causing damage to storage structures.

14.4.1.1.5 Major Insect Pests of Stored Food Grains

Major insect pests offer a serious threat to grains. Some of these are known as primary insects or internal feeders because they destroy the whole seed by feeding inside the grain. Some injure the embryo or feed on the germ part, affecting its viability. They are known as external feeders and are equally destructive. The important ones are:

1. **Rice Weevil** (*Sitophilus oryzae* L): A crucial pest of rice, sorghum, wheat, and maize. It prefers a temperate but humid climate and breeds wherever it finds undisturbed conditions of storage. Both adults and larva attack the whole grains. It is a tiny weevil, generally reddish brown or dark brown in color, with a snout-like head. They feed voraciously, rendering seed unfit for human consumption.
2. **Lesser Grain Borer** (*Rhyzopertha dominica*): A prominent pest of all the cereal grains. A badly infested seed is hollowed out until a thin shell remains. The newly hatched larva is quite active and campodeiform in shape. It burrows at once into the seed or crawls actively, feeding on loose starch material. The adult *Rhyzotherpa dominica* is a dark brown, small, and cylindrical-shaped beetle, which measures around 3×1 mm. It has a deflexed head, which is enclosed with a crenulated hood-shaped pronotum. The adult lesser grain borer is a powerful flier and moves around from one godown to the other, thus initiating fresh infestation. There is substantial quantity of frass generated by adults, thus spoiling more as compared to the amount consumed by them.
3. **Khapra Beetle** (*Trogoderma granarium*): A serious pest to whole cereal grain, especially to wheat. Generally, its infestation occurs on the top layer of seed because the insect fails to penetrate beyond a certain depth. As a rule, it first attacks the embryo point and then the whole seed. The adults are incapable of flying. The adult beetle is small, about 2–3 mm in length and dark-brown colored, with a retractile head, and its antennae are clubbed. Its body is completely clothed with fine hairs.
4. **Pulse Beetle or Bruchids** [*Callosobruchus maculatus* (Fabricius), *Callosobruchus chinensis* (L)]: It is the most damaging insect to pulses both in the field as well as in storage. It attacks many stored pulse crops, like mungbean, cowpea, pigeonpea, chickpea, and so on. Infestation starts from the field when the crop is maturing. They also infest seed godowns. The insect remains active throughout the year; however, maximum infestation is found during July to September, causing 40–50% losses.
5. **Grain Moth** (*Sitotroga cerealella* Oliv): Like the pulse beetle, the grain moth also infests seed crops in the field. Only the larva stage of grain moth is injurious to seed, as adults only lay eggs on crops in the field or on grains in grain bags. After emerging from the egg, the larva enters the grain through a minute hole, feeds, and completely destroys it.

6. **Rice Moth** (*Corcyra cephalonica* Staint) attacks rice, maize, millets, and whole wheat. Only the larva stage of this moth feed on silken webs. When it attacks the whole grain, kernels are bound into bumps.

There are also some insect pests that attack many cereal crops during storage, but of minor importance are the granary weevil (*Sitophillus granaries*), saw-toothed grain beetle (*Oryzaephilus surinamensis* L.), cadella beetle (*Tenebroides masritanicus* L.), and flat grain beetle (*Cryptrolestres mauritanicus* L.).

Penicillium and *Aspergillus* fungi in association with stored insect pests contribute to faster deterioration. Pulses are generally consumed as *dal* or in dehusked split form by consumers. Due to the lack of seed coat cover, dehusked split pulses or *dal* have a tendency to gain moisture and lead to fungal infestation. *R. dominica*, *T. granarium*, *T. castaneum*, and *Cadra cautella* attack split pulses during storage. Due to insect infestation, not only physical loss, reduction in carbohydrates and proteins, and nutritional losses but product contamination with uric acid, fragments, and fecal matter were also observed.

Insect growth in different storage structures depends upon favorable conditions like moisture, temperature, air, and so on. Hence, there is a need to regulate the moisture, air, and temperature properly within storage to minimize losses. Insects/pests grow very rapidly if storage structures in which food grain are stored unable to discourage insects/pests. Insect multiplication takes place because of insect damage to stored grains and laying of eggs inside. They spread odor, reduces weight, affect quality aspects like nutritional value and viability, and raise the temperature.

14.4.1.2 Microorganisms

Bacteria, yeast, and fungi are microorganisms that cause significant damage of stored food grains because they cause poor odor and flavor, hot spots, grain clogging, moisture enhancement due to high respiration, and so on. Fungi behave like parasites in stored grain and can cause severe disease in consumers. They also help to increase the respiration rate of food grain within storage gradually and create hot spots, which hamper the good milling characteristics of stored grains. Highly toxic mycotoxins have also been produced due to multiplication of fungi (Minjinyawa, 2010). Spores of fungus are very light and small and can spread from one place to another with the help of wind and insects. It is extremely challenging to eradicate them completely from food grain storage. Fungus spores are able to reproduce themselves, cause spoilage of stored food grains, and lower the quality. Humid and hot atmospheric conditions are favorable for fungus growth; sometimes mold growth also takes place at lower temperatures and high air relative humidity. Dry weather retards the growth of fungi but is not able to destroy spores due to their high resisting capacity in dry conditions (Groot, 2004).

14.4.1.3 Rodents

Rodents are disastrous, as they damage both storage structures and stored grains. They make holes in storage structure and facilitate easy stored grain infestation. They also consume a significant amount of stored food grains. Consumption of rodent-contaminated food also spreads hazardous disease. Rodents also transport several diseases by means of decay of dead animals and excreta; therefore, proper care has to be taken during the construction of storage structures and proper hygienic conditions observed surrounding it (Minjinyawa, 2010).

14.4.1.4 Birds

Birds mainly carry food grains for consumption, during which a minor part of uncovered grains can be infected and facilitate microorganism growth, but quantity losses are significant. Generally infestation of insects, birds, and rodents affects quantity losses because of changes in moisture level and temperature as well as qualitative losses affected by a reduction in nutritional value and sensory characteristics during storage.

14.4.2 Abiotic Factors

The main decisive factors affecting storage life are moisture content and temperature. Also, atmospheric conditions and their interaction with stored product also play a significant role in storage losses.

14.4.2.1 Grain Moisture and Relative Humidity

The moisture factor is characteristic of water activity. There is a limit for optimum values of relative humidity below or beyond which, at a given temperature, the causes of spoilage are initiated. Most grains in tropical environments deteriorate quickly due to high humidity coupled with temperature.

For secure storage, the grain moisture level in grains at storage time is of crucial importance. A moisture level below 13% for grains is desirable to check the growth of most microorganisms and mites, while below 10% limits development of most stored grain pests. Hence, moisture content below 10% is considered good for storage purposes. Moisture content within bulk storage is not uniformly distributed. It is changeable from season to season and one climatic zone to another. A low moisture level maintains the relative humidity at a level lower than 70% and restricts the growth of mold successfully. The moisture content requirements of different storage pests are given in the following.

Storage insects	Moisture content (%)	
	Minimum	Maximum
Sitophilus oryzae	9.5–11.0	14.0–14.7
Rhyzopertha dominica	9.0–10.0	11.0–14.0
Trogoderma granarium	0.0–0.19	11.5
T. castancum	10.0	11.5–16.0
C. cephalonica	9.0	15.0–20.0
C. cautella	10.0	16.0

Source: Sinha: Safe Storage of Seeds: Insect Pest Management, 1999.

14.4.2.2 Temperature

Temperature is one of the various significant factors that influence the metabolic activities of all biological organisms. All organisms remain alive and flourish at specific range of temperature. The air temperature, grain temperature, and inter-granular air temperature are all considered crucial for safe and prolonged storage of grains. At higher temperatures (>20°C), most microbes, mites, and insects multiply and infest seeds, thus affecting its health. The growth rate of these organisms decrease with a decrease in temperature. The thermal requirements of different groups of pests are as follows.

Pest or bio-agent	Temperature (°C)			Minimum Moisture Content (%)
	Minimum	Optimum	Maximum	
Insects	15–20	27–37	33–38	9.0–11.0
Mites	5–12	19–31	32–36	–
Fungi	2–5	19–31	34–40	15.4–16.8
Microbes	5–6	28–35	50–58	–

Source: Sinha: Safe Storage of Seeds: Insect Pest Management, 1999

The level of temperature experienced during storage in the tropics and sub-tropics is therefore favorable for insect pests to attack grains. Generally, insects do not develop below 15°C, mites below 5°C, and storage fungi below 0°C; this information is important from a storage point of view. Temperature effects on an organism could easily be correlated with level of moisture present because a rise in temperature corresponds with a decrease in the relative amount of moisture in the atmosphere.

Respiration rate of grains, growth of microorganisms, and chemical reactions during storage also accelerate up to a certain temperature. Respiration results in oxidative breakdown or deficiency of O_2 and generation of CO_2, moisture as well as heat energy. This energy is utilized by cells for metabolic processes, which result in heat. The metabolic heat produced is in range of 1.3×10^{-5} by dry grains and 1×10^{-7} cl/sec/cm^3 by wet or damp grains. However, the magnitude of heat generated by biotic factors like fungi, insects, and other organisms infesting the grains is much higher, and when the grain temperature rises to around 20°C, it is easier for insect pests and microorganisms to infest it. At the same time, the respiration rate also increases.

14.4.2.3 Temperature and Moisture Migration

Fluctuation in temperature due to changes in weather conditions results in accumulation of moisture in traditional storage structures. The moisture accumulates either at the top or bottom, subject to the direction of air convection. As movement of warm air takes place from warmer to cooler areas in the stored grains, the air gives both moisture and heat. Due to condensation, there is a buildup of moisture in the storage bin, either at the top or bottom, according to the flow of usual convection of the air within it. During winter, moisture condenses on the top, while it condenses at bottom during summer. Therefore, spoilage may occur in the absence of preventive measures. Moisture migration is generally slow in small as compared to large storage bins.

The upper surface and center of the grain remain warmer than the remaining parts of bin, as during winter convective air currents flow upwards through the center of the storage bin. Contrary to this, during summer, air currents move upward along the warm walls and downward through the center of the storage bin, and the center of bin near the bottom remains cold. Hence, temperatures changes occurring due to changes in season also affect storage.

14.4.2.4 Grain Conditions at Time of Harvesting

Quality of harvested grains just prior to storage is an important parameter for losses during storage. The climatic conditions prevailing at the time of maturity of the crops also determine the storage capacity of the grains. Unfavorable weather conditions such as rain, hail storms, high humidity, sharp changes in the environment, and disease and insect pests prevailing at the time of maturity of crops influence grains and their longevity. During harvesting and threshing operations, mechanical damage may occur and can result in injured spots on grains. These injured or damaged spots may act as centers for infection, resulting in quality deterioration (Shah, 2013).

14.4.2.5 Storage Conditions and Storage Structures

Structures and conditions of storage are components of crucial importance. In the developing countries of Africa and South Asia, grains are generally stored in traditional storage structures, which are made of local materials (straw, bamboo, mud, bricks) that are easily available to farmers. Baloch (2010) reported bins and pots made of mud, *bokharies* and *kothis*, are commonly used storage structures in Asia. For short-duration storage, bags (gunny or polythene) are used, while for long-term storage *dole, berh, gola,* and *motka*, steel/plastic drums, are used. In African countries, different types of granaries are used. For example, *ebli-va* and *kedelin*, in-house smoked storage, is used for storage of maize in Togo (Pantenius, 1988). Using jute or polypropylene bags, raised platforms, conical structures, and baskets, food grains are stored in West Africa (Abass et al., 2014). Farmers of East and South Africa use bags, wood cribs, pits, iron drums, and metal bins for storing grains (Abass et al., 2014; Wambugu et al., 2009). Locally available materials and traditional skills are used to construct most of these structures, and they lack scientific design. Grains stored in such storage structures are prone to losses due to different biotic, abiotic, and other factors.

Improved and advanced storage structures efficiently minimize food grain losses and lengthen the shelf life of food grain without quality deterioration (Groot, 2004). Storage structures should incorporate controlled, safe, and hygienic storage conditions for maintaining the high quality of grains required to meet the market demand of a healthy food grain supply (Rezende, 2002). Adequate atmosphere should be provided within storage structures, preventing stored food grain degradation. Proper cleaning around storage structures is necessary for protecting the grain from several insects/pests.

14.5 GRAIN STORAGE SYSTEMS

Food grains are generally stored in two ways, (i) bag storage and (ii) loose bulk storage. Both systems have advantages and limitations. The choice of system depends upon local factors, like availability of storage type and duration, climatic conditions, labor cost, and transport system.

14.5.1 Bag Storage

In this system, food grains are stored in bags of varying capacity, like 35, 50, 75, and 100 kg. These bags may have or may not have plastic linings. Bags are generally made of jute, canvas, or low-density polyethylene sheets. The advantages of bag storage are: (i) each bag holds a fixed quantity of grains that can be purchased, sold, and dispatched easily; (ii) loading and unloading operations can be performed with ease; (iii) easy removal and treatment of infested bags; (iv) due to exposure of bags to the atmosphere, there is no sweating of grains; and (v) they require low capital cost. Some limitations of bag storage systems are: (i) slow handling, (ii) considerable level of spillage, (iii) high operating cost, (iv) high probability of rodent loss, and (v) chances of occurrence of re-infestation.

14.5.2 Loose Bulk Storage

In this system, grains are stored in different types of storage structures in loose form. The advantages with this system are: (i) large quantities of grains are stored, (ii) rapid handling, (iii) little spillage, (iv) low operating cost, and (v) less chance of rodent loss. The limitations with this system are (i) high capital cost and (ii) little protection against re-infestation.

14.6 CLASSIFICATION OF STORAGE STRUCTURES

Food grains have been stored in different traditional, improved, and modern storage structures to prevent quality deterioration of food grains. Depending upon various criteria, these storage structures can be grouped into different categories. Generally, grain storage structures are classified into the following types based on different criteria.

- Short-term (temporary) or long-term (permanent structures)
- Indoor and outdoor storage structures
- Above-ground and underground
- Rooms/bins/pots constructed with mud
- Wood or bamboo storage structures
- Metallic drums, bins, or containers

- Structures made with straw of paddy and wheat, and so on
- Traditional, improved, and modern structures

Storage structures are termed temporary when they store food grains for a shorter period and permanent when they serve the purpose of long-term storage of food grains. Storage structures kept inside the house are generally smaller in size, known as indoor storage structures such as *kothi*, *kanaja*, and *sanduka*, and earthen pots of different sizes have been used for a small quantity of grain storage, while outdoor storage structures are relatively big and hence kept outside of the house. Some structures like *khatties* or *banda* are built underground, but the majority of structures, like *kothi, kuthar, kuthla, kanaja, thekka, puri*, and *morai* storage structures, are examples of storage structures built above ground level. These above-ground structures facilitate easy inspection of food grain and turning for avoiding hot spots. Based on the construction material used, they are termed mud, wooden, or bamboo structures; metallic bins or containers; or paddy and wheat straw structures. The available information on storage structures in the literature is grouped into traditional, improved, and modern storage structures in the following.

14.6.1 Traditional Storage Structures

Different traditional storage structures developed by farming communities of different regions are based on their indigenous skills, ethics, and available materials (Srivastava et al., 1988). Traditional storage structures are economical, require fewer construction and maintenance skills, have small storage capacity, and are used for holding food grain for shorter durations (Minjinyawa, 2010). Understanding of traditional storage methods has been generated through experience, and the same has been continuously passed to succeeding generations with or without need-based modifications (Natarajan & Govind, 2006). Some methods are exclusive to the culture of a society. Considerable variation occurs in these methods in different villages and communities. About 60 to 70% of the total food grains produced in developing nations is stored in threshed or unthreshed form using traditional storage structures. Some details about commonly used traditional storage structures are given in the following.

14.6.1.1 Morai-Type Storage Structures

This type of storage structure with variable storage capacity (3.5–18 tons) is used to store paddy, maize, and sorghum (jowar) in eastern and southern parts of India. Structurally, it is an inverted cone-shaped structure raised on a supported platform on pillars (masonry or wooden). The improved version of Morai structure types has a circular wooden plank floor supported on pillars by means of timber joints. A metal cylinder of about 90 cm height is nailed around the wooden floor. A 7.5-cm-thick rope made from paddy straw is placed inside the cylinder starting from the floor up to 90 cm height, followed by bamboo splits in vertical positions along the inner space with no gap between them. Food grains are poured into the metal cylinder, maintaining the vertical position of bamboo splits. Pouring of grains and winding of the rope are done at the same time. The pouring-in practice continues until the required height is attained. A mud plaster layer (1 cm thick) is applied over the rope to create a smooth surface. A conical roof with enough overhangs all around is provided on top of the structure.

14.6.1.2 Bukhari

These structures may be cylindrical or square in shape and are utilized for storing maize, paddy, wheat, bengal gram, sorghum, and so on, with a varied storage capacity (3.5 to 18 tons). The materials used to construct these structures are mud, bricks, and cement. A space is provided in the bottom portion for taking food grains out easily. The storage structures are raised above ground level by a wooden or masonry platform. Timber planks or bamboo splits are used to construct the floor and are plastered with mud mixed with dung and paddy straw. A timber or bamboo framework and bamboo matting are used to construct the wall of the structure, and plasters is applied on the both sides. Polythene sheets provide moisture protection, and mud is used for proper sealing. Some intensive studies revealed that some farmers provide a layer of sand before storing the wheat grains, which acts as primary barrier to protect the food grains from insects/pests to some extent. The cylindrical structure is provided with a cone-type overhanging roof made up of bamboo and straw. To check rat entry, rat-proofing cones may also be positioned on each pillar.

A more durable and safe improved bukhari structure is used today. It has a double layer of bamboo strips and a gap between two bamboo strip layers. This gap is filled with mud to provide better strength to the structure, and the conical roof is plastered with a 4–5-cm-thick mud layer to prevent leakage during the rainy season (Sahay & Singh, 2009).

14.6.1.3 Kanaja

This is a traditional storage structure with a round base constructed using bamboo. A mixture of soil and cow dung is used to plaster it. Paddy straw is used to cover the structure. For grain filling, it has a large inlet at the top, which used to be sealed with mixture of cow dung and soil after grain filling. A small opening is provided in the lid for taking out grain without opening the sealed lid. The height and storage capacity of *kanaja* structures are variable, and they can store 3–12 quintals of food grain. Some k*anaja* structures have two or more partitions in which more than one type of grain can be stored. In order to prevent soil moisture in the storage structure, a foundation made with wooden planks, stones, and bricks, which has height of 12 inches from the ground, is provided (Naik & Kaushik, 2011). It is also called *thombai* in Tamil Nadu, where both its outer and inner sides are plastered with a mixture of cow dung and clay and left to dry in the sun. After drying, the top portion of the structure is covered with crop residue, long grasses, straw and leaves of locally available trees in a conical shape to prevent the entrance of rain water into the structure. Generally ginger grasses are used to protect it from rain (Kiruba et al., 2006). *Kanja* are very commonly used structures in Karnataka and Maharashtra in India.

14.6.1.4 Kothi

A widely used traditional structure with low storage capacity, a *kothi* is constructed in a cylindrical shape of different dimensions for storing food grains. However, rectangular-shaped *kothi* are also common in some regions. It is made up of unbaked clay, straw, and cow dung (Dhaliwal & Singh, 2010) and consists of a wide door for pouring food grains into the structure, with provision of a small outlet for drawing out grains. It has a storage capacity of 1 to 50 tons to store sorghum (jowar) and paddy (Naik & Kaushik, 2011).

14.6.1.5 Kuthla

Kuthla is an indoor type of storage structure that is widely used by farmers in rural areas of Bihar and Uttar Pradesh to store food grains. Made up of baked mud, these structures are usually kept inside the house and used for small amounts of food grains.

14.6.1.6 Palmyra Leaf Bin (Vattappetti)

In southern India, storage structures made with Palmyra leaves are widely used traditionally for short-term grain storage. As the name suggests, it is made by stitching Palmyra leaves into a cylindrical shape. Its dimensions can vary depending upon the requirements of farmers; however, the commonly used bins are 2.5 meters in height, 1 meter in width, and 2 meters deep. It can store up to 5 tons of grain (Kiruba et al., 2006).

14.6.1.7 Earthen Pots

Rural areas of South Asian countries like India, Bangladesh, Pakistan, and Nepal widely use earthen pots to store food grains for shorter durations in small quantities. These earthen pots are made with baked clay in various shapes and sizes according to the amount of grain they will store. These earthen pots kept inside houses. To avoid moisture or air migration, the mouths of these structures are sealed with mud, cow dung, or a mixture. More earthen pots are used to keep one above the other vertically from ground level; this arrangement of storage structures is also called *dokal* (Channel et al., 2004). It is also called *mataka* in Uttar Pradesh, Bihar, Uttarakhand, Jharkhand, and Madhya Pradesh. In the southern part of India, it is also known as *paanai* or *addukkupaanai* and is used as a single pot for storing small amounts, and piles of earthen pots arranged vertically one above another with placement of bigger pots on the bottom are used (Karthikeyen et al., 2009b; Kiruba et al., 2006).

14.6.1.8 Gummi/Kuthar

A *gummi/kuthar* is a circular or hexagonal outdoor-type storage structure made of bamboo strips or reeds with mud plaster applied. The base is constructed using reeds or stone slabs. In order to prevent moisture movement from the ground, these structures are placed approximately 1 meter above the ground to store food grains.

14.6.1.9 Hagevu

A short-term, round food grain storage container, *hagevu*, is made of paddy straw ropes by arranging the straw ropes one above the other in a round shape. After putting the food grains inside this structure, it is covered with thick layer of straw and plastered using mud to provide proper air sealing. An inlet opening is provided at some height from the bottom (Channel et al., 2004). To ensure safe storage of food grains inside, regular mud plastering is required (Naik & Kaushik, 2011).

14.6.1.10 Bharola

Bharola is an easily transportable lightweight traditional storage structure that can store a small quantity (40–80 kg) of food grains for daily consumption. It is made with mud in a round shape with one round opening at the top of the structure that can be used for both pouring and removing the stored grain (Dhaliwal & Singh, 2010). It is widely used in Punjab, Haryana, and other northern parts of India and the Sindh and Punjab regions of Pakistan.

14.6.1.11 Kupp

This is an indigenous temporary storage structure used for storing food grains for short periods on agricultural farms. It is conical with wide a circular base for better stability. It is made of bamboo splits or wooden sticks, rope, and paddy straw. Bamboo splits or thin wooden sticks are attached with the help of rope made with jute or paddy straw to form a round base and conical top. It has one opening at the top, which should be covered with thick layer of paddy straw, crop residues, bamboo splits, and cotton sticks to prevent the entry of rain, direct sunlight, and air (Dhaliwal & Singh, 2010).

14.6.1.12 Kodambae

This is made of locally available soil, clay, and mud by mixing with cow dung. A round wall up to 3 feet in height on a nearly 0.5–0.7-foot-high platform from ground is made. Kodambae is a traditional storage structure used in South Asian regions, mainly in south India (Tamil Nadu, the southern part of Andhra Pradesh, and Telangana), Bangladesh, and Pakistan for storage of food grains and seeds of other fodder crops. It can store nearly 1 ton of food grains, but this may vary per requirements. For loading and unloading grains, one small door is included.

14.6.1.13 Bamboo Baskets

Bamboo baskets have been used for grain storage for a long time in rural areas. They are used throughout the southern part of Asia, and their size varies according to the quantity of grain to store. They are generally classified as indoor, temporary, small-capacity, and short-duration storage structures. Bamboo splits are stitched in the shape of basket. One circular opening is provided at the top for storing and taking out the grain during storage. It is provided with a flat cover made from bamboo or wooden sticks. After filling the grain, this structure is completely layered with a mixture of cow dung and soil to seal small holes, which makes it hermetic and protects from rodent, insect, and pest attack. It is known as *urai* in South India (Karthikeyen et al., 2009b). Similar structures are also used in the Himachal Pradesh region with the name of *peru*, constructed by stitching bamboo strips (width 2 cm). In Himachal Pradesh, it is kept on a platform made with locally available wooden planks called *tarein* to avoid microorganism and insect attack (Kanwar & Sharma, 2003, 2006).

14.6.1.14 Wooden Boxes (Sanduk/Arisi Petti)

Wooden boxes are a very popular traditional, small-capacity indoor storage structure used in different parts of southern Asia, especially in different Indian states. It is generally made with wood of strong and locally available trees and is rectangular or square in shape. Its dimensions depend on the quantity of grain to store. It is also called by different names in different regions, like *sanduk* or *peti* in North India, Pakistan, and Bangladesh and *arisi petti* in the southern region of India. It has a lock system to protect grain from thieves. This storage structure is suitable to prevent attack by weevils, insects, rodents, and moisture migration (Karthikeyen et al., 2009b). It is called *peti*, particularly in Himachal Pradesh, where it is made with preferable *tuni* (*Cedrela toona* Roxb.) and *akhrot* (*Juglans regia* Linn.) wood, which discourages termites and other insects (Kanwar & Sharma, 2006).

14.6.1.15 Mud Houses

This is a traditional storage structure long used in rural areas of India and other parts of south Asia for storage of food grains. It is cost effective and categorized as an indoor, temporary, and small-capacity storage structure. In this structure, the walls are constructed with earthen bricks and plastered with cow dung. It is covered with boards prepared with wood or bamboo and called *mankattai* in Tamil Nadu (Kiruba et al., 2006).

14.6.1.16 Earthen Bins

The traditional earthen bin is used for grain storage for up to 2 quintal and kept inside the house. It has a cylindrical shape with a narrow bottom and wider upper portion with a narrow neck-like opening for pouring food grains and a small opening at bottom to take the grains out. After pouring grains, the top and bottom openings should be closed with the help of mud, clothes, and straw. It is successfully used for storing paddy, black gram, and millets and also known as *kulukkai* in Tamil

Nadu (Kiruba et al., 2006). The size of this structure varies according to requirements. It should be painted with good-quality paint or primer to increase the longevity of the storage structure and prevent the attack of termites, insects, and pests.

14.6.1.17 Solarization

Solarization, or heating food grains in sunlight, aiming to destroy insect pests and reduce grain moisture is an age-old practice used by farmers in areas where outside temperatures reach 20°C or higher (Chua & Chou, 2003). Based on the product, the duration required for solarization process varies. Farmers also sun-dry stored grains by spreading the grains on the bare ground, spread polythene, tarpaulins, bamboo mats, roadsides, or rooftops. Kiruba et al. (2008) found that when green gram stored in bags of different colors was exposed to sunlight for 24 hours, there was complete killing of *Callosobruchus chinensis* eggs and grubs located in infested green gram. Also, complete grub mortality was documented in black bags.

14.6.1.18 Open Fireplaces

Sarangi et al. (2009) reported that a large number of farmers store their food grains in close proximity to the kitchen, bearing in mind the fact that the heat and smoke due to burning firewood in the kitchen enter food grains and protect them from insect infestation. In particular, raised barns are constructed to store more food grains in which a slow-burning flame is lit and controlled hot air renders the grains dry (De Lima, 1982; Sarangi et al., 2009). Farmers usually store a small quantity of grains over the kitchen fire either in the farm hut or in the open. In the open, the high temperature because of direct solar radiation helps to destroy the growing larvae of insect pests in food grains.

14.6.1.19 Gourds

Gourds are normally economical indoor storage structures used to store small amounts (5–30 kg) of grain for 6–12 months for home consumption in areas where they are safe from insect infestation (Bodholt & Diop, 1987; Proctor, 1994). The hard dried outer skin of fruits belonging to the *Cucurbitaceae* family present in the tropics and the subtropics is used for gourds (Wehner & Maynard, 2003). After putting grains in these, varnish, paint, or linseed oil is applied. To make it airtight, mud or cow dung is used to cover the lid. To protect grains from moisture absorption, they are placed on a platform. Farmers can easily check for insect infestation (Makalle, 2012).

14.6.1.20 Cribs

Cribs are traditionally used for storage of unthreshed maize cobs, but are currently also used to store other crops. A crib is a raised (0.5 to 1 m above ground) structure with support columns that is rectangular in shape. It is made of straw, palm leaves, bamboo, or wire netting with ventilated sides. Wood, metal, bamboo, or wire mesh is used to construct it. The roof can be made of thatched straw or an iron sheet and placed such that the prevailing winds blow perpendicular to the length. Rat-proof devices are placed on the legs to avoid rodent attack. The design of the crib allows continuous drying due to the free flow of air over the stored produce. Maize cobs with husk are protected for 3–6 months without insecticide (Mijinyawa, 2002).

14.6.1.21 Rhombus

Traditionally used rhombus storage structures in Africa may be the mud or thatched rhombus type based on the material it is made of. Mud rhombuses usually have a bin-like structure placed on large stones and covered with a thatched roof and are used to store unthreshed millet, maize, and sorghum for 2–7 years. The shape of mud rhombuses is variable; it may be cylindrical, spherical, or circular in shape. It is made of a mix of clay and dry grass with a storage capacity of 1–8 tons of food grains. Mud rhombuses are generally not moisture proof and are therefore mostly suited to dry climatic conditions, where the moisture level of the harvested grain can be reduced simply by sun-drying. Unloading of the stored food grains is labor intensive. In rhombus bins of higher storage capacity, an opening for easy offloading of stored grains is provided by breaking the shell, which can subsequently be sealed after unloading.

Unthreshed food grains are usually stored in cylindrical or circular thatched rhombuses with varying storage capacities. The woven grass walls of the thatched rhombus normally have two layers and are further strengthened by having two to three tension rings. Some farmers fill in cow/animal dung among the layers to stop animals from eating the walls of the rhombus (Adejumo & Raji, 2007). However, these structures are associated with some disadvantages, like gain of moisture; attack by rodents; alteration in taste, color, and smell of stored material; and pest infestation.

14.6.1.22 Kihenge, Kichenga, Reli, and Dari

Traditionally food grains are stored in storage structures known as *kihenge, kichenga, reli*, and *dari* in Tanzania. A *kihenge* is a 1.5-m raised hut with poles fitted with rodent guards. It is grass thatched and plastered with mud and has a door on one side. Bags of maize are stored inside the structure. A *kihenge* is usually constructed outdoors, within a residential compound, and therefore does not offer any protection against rainfall, insect and rodent attack, and theft (Rugumamu, 2003).

A *kichanga* is a table-like platform, usually erected inside a residential house. It is made with pieces of wood tied together with ropes and is raised about a meter above the ground. A *reli* is made with racks of big pieces of wood laid on the floor inside a residential house, while a *dari* is made with racks of pieces of wood laid on top of the fireplace inside a house. These storage techniques are very prone to insect infestation.

14.6.1.23 Straw Bins

To construct this, paddy straw is dried completely while keeping it straight. A rope is woven using dried straw arranged in concentric manner over a large area, and bark of *Erythrina indica* and *E. variegate* are kept along with the paddy straw. Food grains are mixed with sifted ash before being placed in the straw bin. After putting food grains in the bin, straw ropes are folded over it. Straw bins are kept suspended from roof rafters (Jain et al., 2004). Straw bins are inexpensive, and the low temperature keep the grains cool.

14.6.1.24 Nahu

Traditional storage structures like a *nahu* are commonly used by poor farmers to preserve food grains in the West Siang district in Arunachal Pradesh, India. The life of a *nahu* structure is about 15–20 years, and storage capacity varies from 5.0 to 8.0 tons. *Nahu* are constructed in clusters near a residential region in the village to prevent fire outbreak. *Nahu* look similar to cribs in appearance. It has three compartments: the lowest, middle, and uppermost compartments. The bottom one is used for firewood, the middle one is empty, and the uppermost compartment is used for storage of grains. Grains to be stored are thoroughly dried before storage. Thinly woven bamboo mats are used to make an airtight compartment for storing of grains. Subsequently grains are kept for storage, a bamboo mat is used to cover them tightly, and a stone is kept above it to prevent rodent entry. *L. jenkensiana* is utilized for roofing that is changed every 5 years (Sarangi et al., 2009).

14.6.1.25 Storage Bags

Storage bags or sacks are extensively used in farms, villages, and commercial storage centers. Made from woven jute, sisal, local grass, cotton, and other locally available materials, these bags are largely used in both Nigeria and India. Although polypropylene bags are also used, farmers still use jute or sisal bags. The size and storage capacity of these bags varies, and generally bags with a storage capacity of 25 to 100 kg/bag are more in use. To create moisture-resistant and airtight storage conditions, a layer of polythene bags is placed inside the storage bags (Mutungi et al., 2015; Ng'ang'a et al., 2016). The storage of grains in sacks has many benefits: bags of grains may be piled under any suitable kind of shelter, they can be transferred and handled without the use of any special equipment; they permit gaseous exchange and have insect control through the use of fumigants inside a closed storehouse or beneath a plastic sheet (Hall, 1980). Farmers stack the sacks on pallets or raised platforms. The raised platform is superior to using a plastic sheet underneath, as it permits movement of air beneath the sacked grains (David, 1998).

14.6.1.26 Storage with Table Salt

The technique of mixing table salt with food grains is used to prevent insect infestation by a few farmers in developing countries for shorter durations. Approximately 200 g of table salt is mixed well with 1 kg of pigeonpea (*Cajanus cajan*) and *P. vulgaris* and sealed in jute bags for storage over a 6–8-month period. The abrasive action of the salt checks the movement of insects within the container, which results in suppression of population build-up of insects (Jeeva et al., 2006).

14.6.1.27 Camphor

Considering the pungent odor of camphor acts as repellent or antifeedant to storage insect pests, it is used in short-term storage of seeds required for planting during the next season. Camphor is put into bags or containers consisting of sun-dried shelled grains or paddy (Karthikeyan et al., 2009a).

14.6.1.28 Obeh

This is a traditional storage structure widely used for the storage of unthreshed rice by the resource-poor farmers of the Senapati district of Manipur state in India. *Obeh* have a storage capacity ranging from 5.0–10.0 t. Tightly interwoven bamboo sticks are used to make an oval storage platform with a square bottom and tapering top with an airtight compartment. Loading and offloading operations are performed through the removable roof.

14.6.1.29 Storage of Grains with Natural Products/Botanicals

Considering various reports of synthetic insecticides and their harmful effects both on human health and the surrounding environment, there is increasing interest in using natural products or bio-pesticides to control storage insects (Isman, 2008). The use of natural products or botanicals for safe storage of grains has long been practiced in Africa and Asia. Because they are economically cheaper, with prompt availability, repellency, anti-feeding, and ovipositional deterrence properties, farmers use natural products to control infestation of insects (Rajashekar et al., 2012). Pulverized plant components are mixed with grains in the storage container, and the obnoxious smell from these products prevents insects from infesting the stored grains. Plant parts of *Acorus calamus, A. indica, Pongamia glabra, Artemisia* species, *Citrus aurantium, Curcuma longa, Khaya senegalensis, Capsicum* species, and others are commonly used for this purpose.

For the development of biopesticide/green pesticide, China has employed computational chemistry techniques. These techniques include docking (Sharma et al., 2020, 2021) and Quantitative Structure Activity Relationship (QSAR) studies (Sharma et al., 2016), which are widely utilized in the field of pharmaceuticals. China conducted a Green Pesticide Research Program (Qian et al., 2010) to discover new green crop protection chemicals with novel mechanism of action (MoA), high selectivity to pest species, and low environmental, animal, and human risk. In addition to this, another study by Sparks et al. (2001) employed Structure Activity Relationship (SAR) and QSAR techniques for optimizing biopesticide activity. Thus, the emerging interest in sustainable pesticides predicts great potential in moving forward in the development of new tools to improve agricultural practices and combat new resistances using computational techniques.

14.6.1.30 Storage with Cow Dung

Some farming communities consider cow dung to have pesticidal properties, so by using it, stored grain can be saved from insect infestation. Farmers collect fresh cow dung and make it into a round plate. Seeds are embedded in it. These plates are then subjected to sun drying for 2–3 days. This dried cow dung containing seeds is stored in the open or placed in a box made of wood. This way, the seed required for next season's planting is stored (Karthikeyen et al., 2009a).

14.6.2 Improved Village-Level Small-Scale Storage Structure

Keeping in mind the disadvantages and limitations associated with traditionally used storage, focused efforts have been made to improve traditionally used grain storage structures by removing some of their disadvantages. These improved storage structures provide improved security for stored grains. These improved storage structures, like Pusa bins, Punjab Agricultural University (PAU) bins, and *hapur tekka*, are used for small-quantity storage in India. Bins are used for food grain storage, generally placed inside houses.

14.6.2.1 Pusa Bins

Pusa bins were designed by the Indian Agricultural Research Institute (IARI) (popularly known as the Pusa Institute), New Delhi (Acharya & Agarwal, 2009). The Pusa bin is very popular on the Indian subcontinent and in a few African countries. It gives better results when dried grains are stored in it. Mud, bricks, and polythene sheets are used to construct a Pusa bin. It is constructed on a mud brick platform covered with a low-density poly ethylene sheet of 700 gauge to avoid rat burrowing. The outer walls are constructed using baked bricks up to 45 cm height. Using appropriate care, stored grains can be kept safely inside the bin for more than 1 year.

14.6.2.2 PAU Bins

The PAU bin was designed by PAU, Ludhiana. It is constructed using a galvanized iron sheet. Its storage capacity ranges from 1.5 to 15 quintals, which is dependent on the designed size (Acharya & Agarwal, 2009).

14.6.2.3 Hapur Tekka

Hapur tekka is cylindrical in shape, locally constructed on a base of a metal tube with the help of bamboo and expandable clothes. In the improved version, bamboo is replaced with galvanized iron or aluminum sheets for a more durable structure. Grain is taken out of a small circular or rectangular outlet provided at the bottom of the structure. The Indian Grain Storage Institute located at Hapur, UP, designed *hapur tekka* to store food grains safely (Acharya & Agarwal, 2009). Generally, it can store 2 to 10 tons of grain (Said & Pradhan, 2014).

14.6.2.4 Coal Tar Drums

The Central Institute of Agricultural Engineering (CIAE) of Indian Council of Agricultural Research (ICAR) at Bhopal has developed coal-tar drums of 200 kg capacity for storage of food grains. It is a low cost, durable storage bin used for small storage capacity. It is easily available at the domestic level.

14.6.3 Improved Larger-Scale Storage Systems

Improved storage structures are improved versions of traditional structures, with a higher storage capacity of 1.5 to 150 tons and long-term storage of food grains. Several agencies like FCI, warehousing corporations, and grain marketing co-operatives provide storage facilities to farmers on a rental basis. Large-scale grain storage uses storage structures like cover and plinth (CAP) and silos. Some improved storage structures are described in the following.

14.6.3.1 Pusa Cubicle

This is a modification of the Pusa bin that is able to store food grains up to 24 tons. The Pusa cubicle looks like a room (3.95 × 3.15 × 2.60 m) and provides large storage capacity. It is raised on a platform with dimensions of 3.73 × 2.93 × 0.07 m and is made of unbaked bricks on a concrete floor (except 22 cm of outer sides with baked bricks). A sheet made of polyethylene is kept over the platform, and another platform of corresponding magnitude is constructed of bricks. The inner wall is approximately 22 cm thick, while the height is 2.6 m. A door with a wooden framework (1.89 × 1.06 m) is secured at the fore side of the 3.95-m wall. The roof is generally constructed of wood beams spaced at about 15 cm space and covered with unbaked bricks.

14.6.3.2 CAP Storage Structure

The CAP storage structure has plinths at the lower part and a cover at the upper part. Brick pillars (height 14') are constructed on the ground with grooves. Crates of wood are secured in these grooves for piling bags containing food grains. Low-density polyethylene sheets of 250 microns are used to cover the pile of bags. Food grains are generally stored in CAP storage for a period of 6–12 months. It is a widely utilized storage structure by the FCI because of its economic feasibility. This storage structure can be constructed in less than three weeks. Cost-wise, it is an economical storage structure to store food grains on a larger scale (India Agronet, 2009).

14.6.3.3 Silo Type of Storage Structures

Silos or storage towers can be cement, concrete, or metal. Metal silos are costlier than concrete silos. Conveyor belts are used for unloading grains in bulk through mechanical operations Galvanized silo storage systems are a confirmed scientific system used in Europe and America for storage of food grains with almost zero wastage. This system has been adapted to some extent by the private sector since 1990 in India. The storage capacity of each of these silos is about 25,000 tons. The system has many advantages, like: (i) lower running costs, (ii) lower labor requirements, (iii) rapid handling, (iv) low losses due to spillage and rodents, (v) easy and effective fumigation operation, (vi) requires less land area, (vii) complete control of aeration, (viii) storage of grain for longer durations, (ix) it is possible to mechanize all operations, and (x) it is possible to store moist grain for brief periods.

14.6.4 Advanced Storage Methods

The following advanced storage methods are in use to resolve problems faced in storage in bulk quantity.

14.6.4.1 Grain Aeration

Aeration is broadly utilized for the protection of grains during storage. To improve the storing ability of the grains, the ambient air is chilled and then sent to the stored grain mass across the aeration system to maintain uniformity in moisture and temperature all through the storage. It is a satisfactory process to lower the temperature of the product by the use of mechanized aeration systems such as fans. It is a suitable system for a low-humidity environment. Using aeration, safe storage can be provided for grains with 1–2% moisture above normal. The measure of air utilized for aeration varies from 0.021–0.014 m^3/min ton of grains.

14.6.4.2 Modified Atmosphere Technology

Many studies have assessed various atmospheric compositions for the safety of stored grains (Adler et al., 2000; Navarro, 2006). Modified atmospheres (MAs) and controlled atmospheres (CAs) are a secondary approach to prevent infestation by insects and pests that attack food grains during storage as compared to the usage of typical chemical fumigants, which are known to produce residues. Controlled atmosphere systems are known to hinder the growth of fungi, maintaining the quality of the product.

14.6.4.3 Hermetic Storage

Hermetic storage (HS) is also called sealed, airtight, or sacrificial sealed storage. Based on its effectiveness and advantages in terms of avoiding the use of chemicals, it is becoming popular in developing countries. The HS approach allows insects and other aerobic organisms for production of the modified atmosphere on their own by decreasing the concentration of O_2 and increasing the concentration of CO_2 through the respiration process. The process of respiration in live beings forms an atmospheric state with approximately 20% of CO_2 and 1 to 2% of oxygen (White & Jayas, 2003). A reduction in O_2 and increase in CO_2 not only destroy insects and mite pests but also checks the growth of aerobic fungi (Weinberg et al., 2008). Additionally, HS has been found successful in eliminating losses (losses during storage lower than 1%) for global merchandise (Villers et al., 2010). HS has many advantages like easy installation, avoiding use of pesticides, affordable costs, and moderate infrastructure necessities that make it an attractive option (Global Agricultural Productivity Report, 2014).

Concentration of CO_2 at the inner of the bags serves as an indicator of the biological activities of the food grains. High level of initial moisture and increased temperature leads in increment of CO_2 content due to increasing respiration (Bartosik et al., 2008).

HS is very popular in several countries like Ghana, Philippines, Rwanda, Sri Lanka, and Sudan. The construction of HS structures depends upon the amount of grains to store. According to the capacity range of the structure, HS structures are called super grain bags (between 0.59 and 1 ton) or bunkers and cocoons (5000–30,000 tons) (Villers et al., 2008). Navarro et al. (1984) stated that hermetic bunkers are suitable for storing wheat of 12.5% moisture content for 4 years without any qualitative degradation. HS structures provide easy portability, avoid insect and pest infestation, and retain qualitative characteristics during the storage period.

14.6.4.4 Refrigerated Storage

In subtropical climates with higher atmospheric temperatures, refrigerated aeration is utilized for chilling of dry grains to control insects efficiently. Although refrigerated storage systems require high initial financial funding, coupled with dehumidifiers, they hold the potential to become useful for secure commercial storage in tropical climates (Navarro & Noyes, 2002). By using dehumidifiers, the air is chilled and then flows over the grain mass through the aeration system.

14.6.4.5 Storage Structures Used by the FCI and Other Agencies

The FCI uses the CAP system for storing important food crops such as wheat and paddy. For the CAP system, appropriate care is needed in the selection of place and construction. The storage site is constructed relatively high as compared to the adjoining ground. The storage site should be away from drainage, canals, and flood-prone areas to prevent flooding of the area. Usually bricks and mortar are used to make the plinth, which is at least 0.45 meters above ground level. Anti-termite treatment is needed during construction to prevent termite attacks in the future. The bags are stacked on dunnage material that is generally wooden crates (frames), which are placed on a raised platform (plinth), and the stack of grain bags is protected with 800–1000-gauge-thick polyethylene sheets.

Dunnage can be made of wooden planks or polyethylene sheets either exclusively or pressed between two mat sheets, bamboo mats, or ballies. Recently, plastic crates have been utilized instead of crates of wood. The dunnage is important for maintaining the appropriate aeration at the lower layers in the food grain bags, thus keeping them from any damage. The CAP type of storage system is generally utilized for shorter-duration (typically <1 year) storage. The utility of this storage type is in storing a higher amount of food grains, especially right after harvesting operations.

The FCI prefers a stack size of 9.3 × 9.3 × 6.2 m. A maximum of 15 bags could be put over each other. The upper side of the stack is shaped to form an inverted V, which allows easier motion of rainwater after covering. For minimizing possible harm caused because of birds, rain, and so on, a bed of bags containing straw is typically utilized. For regions with high wind velocity, it is necessary to lash the stack with ropes after covering it with polythene. Conveniently, a plastic net-type cover on the polyethylene sheets for tying the stacks is used. Knowledge of proper stack formation and maintenance of the standard distance between two consecutive stacks is important for easy handling operations. A 3-m-wide gallery is reserved next to the warehouse to allow easy passage of bags inside the godown. A gangway of 2 m is kept along the length and in the center of the warehouse. For inspection of stacks, 1 m space around the entire stacking area should be kept. Adequate protection like proper sealing of the doors, maintenance of inspection doors, and ventilation control are necessary.

To maintain the proper temperature and moisture aeration, operation of the stack is very important. A generally utilized approach is lifting the plastic cover to provide aeration in the CAP system. Aeration is generally carried out once weekly and should be done when the sky is clear. In addition, remedial options like fumigation are implemented in several areas.

14.6.5 Bag Storage Godown/Warehouse used by FCI

Warehouses are mainly used by the FCI to store bagged food items for their protection from various components present in the environment. Cereals like wheat and rough rice can be kept using the CAP storage system; however, milled or white rice grains needs a covered godown for storage. Many developing countries also use covered storage structures for storing grains. The FCI constructs warehouses using its own set of standards and specifications.

Warehouses for keeping bagged food grains are a labor-intensive operation and involve high operational cost. Large numbers of laborers are required for stacking at initial storage and also for braking stacks at the time of liquidating food grain bags. In the case of application of standard protocols, grain loss because of attack by insects and pests and any spillage during the material handling operation are limited. Poorly constructed godowns may face the problem of water seepage.

14.6.5.1 Stacking of Seed Bags

Before issuing food grains to consumers, they have to be stored and preserved in godowns based on scientific recommendations. The bags containing food grains should be properly stacked. If they are just dumped in the godown in a haphazard manner, it will create problems in handling, loading, disinfection work, and checking. The filled bags should always be stacked on pallets (platforms) made of wooden or plastic. Use of pallets prevents stored grains picking up the moisture from the floor. Appropriate stacking methods, such as simple, crossed, and so on, may be used. Proper stacking of food grain bags helps not only in inspection and effective disinfestation but also provide hassle-free and easy accessibility of stock present in the godown.

14.7 NECESSARY MEASURES FOR GOOD STORAGE PRACTICES WITH RESPECT TO FOOD GRAINS

Scientific and improved grain storage conditions usually reduce insect infestation effectively. Storage of food grains at low temperature and low relative humidity is recommended for good and safe storage, but this is generally neglected by people involved in storage. Careful selection of the storage site is very important. The site should be sufficiently away from rivers or canals and should be high enough to avoid possible damage due to rain water/floods. Timely cleansing and application of fumigants, facilities to provide appropriate aeration to food grains, and systematic checking of stored stocks are also needed to safely store grains. Infestation by insects and pests at the time of storage is influenced by the level of moisture present in grains, temperature, relative humidity, types of structures used for storage, duration of storage, various operating processes, sanitation around the storage, and frequency of application of fumigants. Pests predominantly present in food grains during storage include weevils, beetles, moths, and rodents. Prophylactic and curative types of treatments are normally in practice. The former include application of pesticides such as Malathion, Dichlorvos (2,2-dichlorovinyl dimethyl phosphate or DDVP), and Deltamethrin (2.5% WP). The latter utilizes application of fumigants such as aluminum phosphide for prevention of infestation in airtight conditions. To prevent rodent attacks, cages to catch rats, poison baits, and using rat burrow fumigation are generally preferred.

14.7.1 Measures Required before Storage

The following measures are generally preferred to ascertain a secure grain storage area prior to storing of grains.

14.7.1.1 Grain Godowns

In the case of grain stored in godowns, it is important that seed stores be made free from leaks, loose plaster, cracks, and crevices. The floor and walls should be smooth, and there should be provision of good and controlled ventilation. Godowns should be made bird and rodent proof by providing bird screens to ventilators. The site should be at high elevation to avoid entry of rain water and also should be away from canals to avoid the chance of floods.

14.7.1.2 Storage Bags/Containers

Generally grains are stored either in bags made of jute or canvas or low-density polyethylene sheet or containers made of metal, wood, or mud. These bags or containers are either kept in large sheds or godowns or any other structure under natural conditions. If necessary, these containers should be repaired before storage to plug any holes and smooth sharp corners. The use of new bags reduces the chances of insect infestation. In case already used or old bags have to be used, they should be properly cleaned and treated with suitable insecticides.

14.7.1.3 Preparation of Seed Store

Before the arrival of a new harvest, all storage structures should be properly cleaned and thoroughly disinfested by spray of appropriate insecticides at least 3–4 weeks before. Arrangements should be made to keep the new harvest separate and not to mix it with carried-over stock.

14.7.2 After Receipt of Food Grains

The grain lot should be checked for soundness of quality, infestation, and moisture level. Proper care should be taken for these aspects. If a grain remains damaged or moist, it has to be segregated from the others.

14.7.3 Care Needed at the Time of Storage

Maintain cleanliness if stored in warehouse or godowns. After properly stacking, treat the leftover space (alleyways), stack, and doorways with appropriate insecticides. Ensure necessary and timely aeration. Leakage checking is required after rains. Biweekly inspection for pests, insects, mites, and rats is required. If live insects or other evidence of infestation are detected, fumigate the grain lot under airtight conditions. The arrangements necessary to separate, recover, and process at places where there has been damage because of issues like water leaking should be made.

14.8 TRANSPORT MANAGEMENT OF FOOD GRAINS

As essential components of staple foods for people, food grain production also has significant bearing on the economy and food security of nations. Realizing the vital importance of food grains, the government of India formed the national food security mission to remove the bottlenecks in the system. Transport management is required at various steps for transportation of grains from farmers (producers) to end users (consumers) to ensure timely food availability. These steps include transporting food grains from farmers' fields to local processing centers, from local processing centers to storage centers, and from storage centers to consumers through local distributors or traders.

In India, *kharif* and *rabi* are the two major seasons for growing and harvesting food grains (Mahapatra & Mahanty, 2018). Wheat, barley, lentil, chickpea, and field pea are grown during October to April, and their produce is procured from April to June. Similarly, rice, maize, sorghum, pearl millet, green gram, and black gram are cultivated mainly in *kharif* (June to November), and subsequently the harvest is procured from October to February (Sharma et al., 2022; Mogale et al., 2017). In this way, in our country, two distinct procurement periods and crop production periods are intermixed together, and all related activities continue the whole year. Likewise, two types of market systems prevail in the country: (i) a public market system, which is regulated by the government, and (ii) an open–private market system, and the shares of both systems are almost equal. About half of food grain production is marketed through the government-controlled system and the other half by the private market. Nodal organization of the central government, like the FCI, together with other state agencies, maintains food supply chain (FSC) operations in the country. Food grains procured by these government agencies are utilized for maintenance of buffer stock as national food reserve and for meeting requirements of the public distribution system (PDS) and several other government-operated welfare programs. The FCI is responsible not only for procuring food grains but also properly storing them and giving them timely transport to the distribution system. The PDS system of the country is the biggest retail system of its kind globally. In general, on average, more than 2 million bags, which is approximately 50 kg per bag of food grains, are transported daily using different available means of transportation like railways, roadways, and inland waterways covering around 1500 km (www.fci.gov.in/movements/view/5). This transport costs around 47.2737 billion INR annually as per report by Comptroller and Auditor General of India 2013. The huge transport cost is one of the major concerns, and a challenging task is to bring it down to manageable levels.

The share of the private sector is represented by a large number of traders involved in the business, and there is sufficient competition among them. These trading companies or traders involved in short-term storage of food grains subsequently make it available to consumers by selling and transporting to retail traders of local markets at appropriate times, aiming to get better profits.

The existing supply chain in India is more complex and creates difficulty in management because of the unorganized nature and many intermediaries in comparison to that of developed countries. The range of intermediaries involves suppliers of inputs, farmers involved in selling and producing food grains, the business community, commission agents, processors, and distributors. Due to poor basic infrastructure, there is huge wastage of agriculture produce annually, resulting in low income to farmers and instability in prices. Also, insufficiency of logistical infrastructure, like roads, railways, airports, and seaports, also contributes to the inefficiency of the supply chain (Sahay & Mohan, 2003). However, there has been a lot of improvement in these conditions.

In developed countries, transportation loses are relatively very low because of improved infrastructure of roads and engineered installations in the field as well as better loading and unloading services, which result in minute damage or no damage at all. On the other hand, poor transportation infrastructure and an absence of mechanized loading and unloading facilities in developing or not-so-developed nations results in damage/losses of food grains through bruising and spillage. At the field level, the majority of field crop produce is shipped in bullock carts or open trolleys in south Asian countries. Food grains for self-usage are generally transferred in bags from field storage to processing facilities in bullock carts, tractor trolleys, small motor vehicles, or open trucks.

Baloch (2010) reported that in south Asian countries like India and Pakistan, loading and unloading of stored wheat grain in bags from vehicles is carried out frequently prior to milling. Some portion of grain is wasted due to spillage each time. The loading and unloading operations of food grains from trucks, wagons, and rails are undertaken by manual means in developing countries, which results in higher spillage. However, in developed countries, there are bulk handling systems, resulting in negligible losses. The poor quality of jute bags used in transportation and storage processes is also an important cause of higher spillage. Hooks used to lift larger quantities (usually 100 kg of grains) lead to tearing of the jute bags, resulting in higher spillage (Baloch, 2010). Similarly, the trucks involved in transport are not in a perfect state for shipping cereal grains in developing countries. Alavi et al. (2012) reported approximately 2 to 10% losses at the time of handling and transporting rice in Southeast Asia.

An overall analysis of the present status of production of food grain, procurement, and distribution systems revealed that the FSC network in India has four major points:

1. Procurement of grains.
2. Transportation of grains within the state.

3. Transport (inter-state) of grains from states that produce surplus amounts to states where grain is deficient.
4. Transport of food grains from large-scale storage locations to the end fair price shops, the ultimate delivery point to consumers. The FCI manages to transport food grains from storage centers to consumers mainly via railways in the country. Wherever railway links are not available, especially in hilly regions, the movement of food grains is undertaken by road transport through trucks.

14.9 CONCLUSION

To achieve the goal of food and nutritional security, sufficient production of food grains, safe storage, and ensuring the timely availability of a sufficient quantity of food grains through an effective transport system are required, as food grains are a staple food and main source of dietary requirements of humans. There is a considerable magnitude of losses during post-harvest handling of food grains. Among these, 10–15% losses are due to unavailability of science-based storage structures. Focused attention is needed to minimize losses during storage, which are caused mainly due to biotic and abiotic factors, their interaction with each other, and the structure and conditions of storage. Traditional food grain storage structures and practices needed to be strengthened, improved upon, or modified based on scientific information. The use of improved and advanced storage structures and practices will efficiently minimize food grain losses and lengthen the shelf life of food grain without quality deterioration and thus provide a safe and economical means of grain storage for long durations. An efficient and effective transport system is of vital importance for the procurement of food grains from producers, to bring food grains to storage locations, and to subsequently transport them to end users or consumers throughout the year within the country in time.

REFERENCES

Abass, A.B.; Ndunguru, G.; Mamiro, P.; Alenkhe, B.; Mlingi, N.; Bekunda, M. 2014. Post-harvest food losses in a maize-based farming system of semi-arid savannah area of Tanzania. *Journal of Stored Products Research*, 57: 49–57.

Abedin, M.; Rahman, M.; Mia, M.; Rahman, K. 2012. In-store losses of rice and ways of reducing such losses at farmers' level: An assessment in selected regions of Bangladesh. *Journal of Bangladesh Agricultural University*, 10: 133–144.

Acharya, S.S.; Agrawaal, N.L. 2009. *Agriculture Marketing in India*. New Delhi, India: Oxford and IBH Publishing Company Pvt Limited, 111–116.

Adejumo, B.A.; Raji, A.O. 2007. Technical appraisal of grain storage systems in the Nigerian Sudan savannah. *Agricultural Engineering International, International Commission of Agricultural Engineering (CIGR) E-Journal*, IX (11).

Adler, C.; Corinth, H.G.; Reichmuth, C. 2000. Modified atmospheres. In: Subramanyam, B.H.; Hagstrum, D.W. (Eds.), *Alternatives to Pesticides in Stored Product IPM* (vol. 5). Norwell, MA: Kluwer Academic Publishing, 105–146.

Alavi, H.R.; Htenas, A.; Kopicki, R.; Shepherd, A.W.; Clarete, R. 2011. *Trusting Trade and the Private Sector for Food Security in Southeast Asia*. Washington, DC: World Bank Publications.

Baloch, U.K. 2010. Wheat: Post-harvest operations. In *Postharvest Compendium*, Lewis, B.; Mejia, D. (Eds.). Islamabad, Pakistan: Pakistan Agricultural Research Council, 1–21.

Bartosik, R.; Rodríguez, J.; Cardoso, L.; Malinarich, H. D. 2008. Storage of corn, wheat soybean and sunflower in hermetic plastic bags. In *International Grain Quality and Technology Congress*. Proceedings Chicago, IL. sp.

Bodholt, O.; Diop, A. 1987. *Construction and Operation of Small Solidwall Bins*. FAO Agricultural Services Bulletin No. 69. Rome: Food and Agriculture Org, 20.

Boxall, R. 2002. Damage and loss caused by the larger grain borer. *Prostephanus Truncatus: Integrated Pest Management Reviews*, 7: 105–121.

Channal, G.; Nagnur, S.; Nanjayyanamath, C. 2004. Indigenous grain storage structures. *Leisa India*, 6(3): 10.

Chua, K.J.; Chou, S.K. 2003. Low-cost drying methods for developing countries. *Trends in Food Science and Technology*, 14: 519–528.

David, D. 1998. *Manual on Improved Farm and Village Level Grain Storage Methods*. Geneva, Switzerland: GTZ, 9–177.

De Lima, C.P.F. 1982. *Strengthening the Food Conservation and Crop Storage Section*. Field Documents and Final Technical Report, Project PFL/SWA/002. Rome: FAO.

Dhaliwal, R.K.; Singh, G. 2010. Traditional food grain storage practices of Punjab. *Indian Journal of Traditional Knowledge*, 9(3): 526–530.

Global Harvest Initiative. 2014. *Global Agricultural Productivity Report—Global Revolutions in Agriculture: The Challenges and Promise of 2050*. Washington, DC: GHI.

GOI. 2020. *Agricultural Statistics at a Glance 2020*. New Delhi: Government of India. Ministry of Agriculture and Farmers Welfare, Department of Agriculture, Cooperation and Farmers Welfare. Directorate of Economics and Statistics.

Groot, I.D. 2004. *Protection of Stored Grains and Pulses*. Wageningen, The Netherlands: Agrodok 18, Agromisa Foundation, 8–20.

Gwinner, J.; Harnish, R.; Muck, O. 1990. *Manual on the Preservation of Post-Harvest Grain Losses*. Hamburg: GTZ, Post-Harvest Proect, 11.

Hall, C.W. 1980. *Drying and Storage of Agricultural Crops*. Pullman, WA: Washington State University, 99164, 381.

Hall, D.W. 1970. Handling and storage of food grains in tropical and subtropical areas. *Food and Agriculture Organization*, 90–350.

India Agronet. 2009. *Storage and Warehousing*. www.indiaagronet.com/indiaagronet/Agri_marketing/contents/Storage%20and%20Warehousing.htm

Isman, M.B. 2008. Botanical insecticides: For richer, for poorer. *Pest Management Science*, 64: 8–11.

Jain, D.; Satapathy, K.K.; Wahlang, E.L. 2004. *Traditional Postharvest Technology of North East Hill Region: ICAR Research Complex for NEH Region, India*.

Jeeva, S.; Laloo, R.C.; Mishra, B.P. 2006. Traditional agricultural practices in Meghalaya, North East India. *Indian Journal of Traditional Knowledge*, 5: 7–18. Online publishing at NIScPR.

Kanwar, P.; Sharma, N. 2003. An insight of indigenous crop storage practices for food security in Himachal Pradesh. In: Kanwar, S.S.; Sardana, P.K.; Satyavir, K. (Eds.), *Food and Nutritional Security, Agrotechnology and Socio-Economic Aspects*. India: SAARM, 175–179.

Kanwar, P.; Sharma, N. 2006. Traditional storage structures prevalent in Himachali homes. *Indian Journal of Traditional Knowledge*, 5(1): 98–103.

Karthikeyen, C.; Veeraragavathatham, D.; Karpagam, D.; Firdouse, S.A. 2009a. Traditional storage practices. *Indian Journal of Traditional Knowledge*, 8: 564–568.

Karthikeyen, C.; Veeraragavathatham, D.; Karpagam, D.; Ayisha Firdouse, S. 2009b. Indigenous storage structures. *Indian Journal of Traditional Knowledge*, 8(2): 225–229.

Kiruba, S.; Jeeva, S.; Kanagappan, M.; Stalin, I.S.; Das, S.S.M. 2008. Ethnic storage strategies adopted by farmers of Tirunelveli district of Tamil Nadu, Southern Peninsular India. *Journal of Agricultural Technology*, 4: 1–10.

Kiruba, S.; Manohar Das, S.A.; Papadopoulou, S. 2006. Prospects of traditional seed storage strategies against insect infestation adopted by two ethnic communities of Tamil Nadu, Southern Peninsular India. *Bulletin of Insectology*, 58(2): 129–134.

Mahapatra, M.S.; Mahanty, B. 2018. India's national food security programme: A strategic insight. *Sadhana— Academy Proceedings in Engineering Sciences,* 43(12): 1–13.

Makalle, M. 2012. Postharvest storage as a rural household food security strategy in Tanzania. ARPN. *Journal of Science and Technology*, 12(9): 814–821.

Mehrotra, S.N.; Verma, N.; Datta, A.; Batra, Y.K. 1987. *Building Research Note, Central Building Research Institute, India, Small Capacity Grain Storage Bins for Rural Areas*. Roorkee, India: Central Building Research Institute, 120.

Minjinyawa, Y. 2010. *Food and Crop Storage Technology, Farm Structures* (2nd ed.). Ibadan, Nigeria: Ibadan University Press, 110.

Mogale, D.G.; Kumar, S.K.; Márquez, F.P.G.; Tiwari, M.K. 2017. Bulk wheat transportation and storage problem of public distribution system. *Computers & Industrial Engineering,* 104: 80–97.

Mutungi, C.; Affognon, H.D.; Njoroge, A.W.; Manono, J.; Baributsa, D.; Murdock, L.L. 2015. Triple-layer plastic bags protect dry common beans (*Phaseolus vulgaris*) against damage by *Acanthoscelides obtectus* (Coleoptera: Chrysomelidae) during storage. *Journal of Economic Entomology*, 108: 2479–2488.

Naik, S.N.; Kaushik, G. 2011. *Grain Storage in India: An Overview.* New Delhi, India: Centre for Rural Development and Technology, IIT, 119.

Natarajan, M.; Govind, S. 2006. Indigenous agricultural practices among tribal women. *Indian Journal of Traditional Knowledge,* 5.

Navarro, S. 2006. Modified atmospheres for the control of stored-product insects and mites. In: Heaps, J.W. (Ed.), 105–146.

Navarro, S.; Donahaye, E.; Kashanchi, Y.; Pisarev, V.; Bulbul, O. 1984. Airtight storage of wheat in a PVC covered bunker. In Ripp, B.E. et al. (Eds.), *Controlled Atmosphere and Fumigation in Grain Storages.* Amsterdam: Elsevier, 601–614.

Navarro, S.; Noyes, R. (Eds.). 2002. *The Mechanics and Physics of Modern Grain Aeration Management.* Boca Raton, FL: CRC Press, 647.

Ng'ang'a, J.; Mutungi, C.; Imathiu, S.M.; Affognon, H. 2016. Low permeability triple-layer plastic bags prevent losses of maize caused by insects in rural on-farm stores. *Food Security,* 8(3): 621–633.

Pantenius, C.U. 1988. Storage losses in traditional maize granaries in Togo. *International Journal of Tropical Insect Science,* 9(6): 725–735.

Phillips, T.W.; Throne, J.E. 2010. Biorational approaches to managing stored-product insects. *Annual Review of Entomology,* 55: 375–397.

Proctor, D.L. 1994. *Grain Storage Techniques: Evolution and Trends in Developing Countries.* FAO Agricultural Service Bulletin no: 109. Rome, Italy: FAO, 277.

Qian, X.H.; Lee, P.W.; Cao, S. 2010. China: Forward to the green pesticides via a basic research program. *Journal of Agricultural and Food Chemistry,* 58(5): 2613–2623.

Rajashekar, Y.; Bakthayatsalam, N.; Shivanandappa, T. 2012. Botanicals as grain protectants. *A Journal of Entomology,* 2012, 1–13.

Rajashekar, Y.; Tonsing, N.; Shantibala, T.; Manjunath, J.R. 2016. 2, 3-Dimethylmaleic anhydride (3, 4-Dimethyl-2, 5-furandione): A plant derived insecticidal molecule from *Colocasia esculenta* var. esculenta (L.) Schott. *Scientific Reports,* 6(1): 1–7.

Report of the Comptroller and Auditor General of India. 2013. *Storage Management and Movement of Food Grains in Food Corporation of India.* Union Government Ministry of Consumer Affairs, Food and Public Distribution.

Rezende, A.C. 2002. Good storage practices: Hazard analysis and critical control points. In: Campinas, S.P. (Ed.), *Grain Storage.* Portugal: Bio Genezis, 1000.

Rugumamu, C.P. 2003. Insect infestations and losses of maize Z. *mays* (L.) in Indigenous storage structures in Morogoro region Tanzania. *Tanzania Journal of Science,* 29(2): 1–9.

Sahay, B.S.; Mohan, R. 2003. Supply chain management practices in Indian industry. *International Journal of Physical Distribution and Logistics Management,* 33(7): 582–606.

Sahay, M.K.; Singh, K.K. 2009. *Unit Operations of Agricultural Processing* (2nd ed.). Noida, India: Vikas Publishing House Private Limited, 169–173.

Said, P.P.; Pradhan, R.C. 2014. Food grain storage practices—A review. *Journal of Grain Processing and Storage,* 1(1): 1–5.

Sarangi, S.K.; Singh, R.; Singh, K.A. 2009. Indigenous method of rat proof grain storage by Adi tribe of Arunachal Pradesh. *Indian Journal of Traditional Knowledge,* 8: 230–233.

Shah, D. 2013. *Assessment of Pre and Post-Harvest Losses in Tur and Soybean Crops in Maharashtra, AERC Report.* Pune, India: Agro-Economic Research Centre, Gokhale Institute of Politics and Economics, 119.

Sharma, S.; Katyal, M.; Singh, N.; Singh, A.M.; Ahlawat, A.K. 2022. Comparison of effect of using hard and soft wheat on the high molecular weight-glutenin subunits profile and the quality of produced cookie. *Journal of Food Science and Technology,* 59(7): 2545–2561.

Sharma, S.; Paliwal, S.; Singh, S.; Akhter, M. 2016. Molecular modeling of viral nucleocapsid protein Zn fingers modulators. *Indian Journal of Biochemistry and Biophysics,* 53: 24–38.

Sharma, S.; Semwal, A.D.; Murugan, M.P.; Wadikar, D.D.; Sharma, R.K. 2022. BOAA: A neurotoxin. In *Handbook of Plant and Animal Toxins in Food.* Chennai: CRC Press, 251–274.

Sharma, S.; Sharma, S.; Pathak, V.; Kaur, P.; Singh, R.K. 2021a. Drug repurposing using similarity-based target prediction, docking studies and scaffold hopping of lefamulin. *Letters in Drug Design and Discovery,* 18(7): 733–743.

Sharma, S.; Srivastav, S.; Singh, G.; Singh, S.; Malik, R.; Alam, M.M.; Shaqiquzamman, M.; Ali, S.; Akhter, M. 2021b. In silico strategies for probing novel DPP-IV inhibitors as anti-diabetic agents. *Journal of Biomolecular Structure and Dynamics,* 39(6): 2118–2132.

Sharma, S.; Srivastav, S.; Shrivastava, A.; Malik, R.; Almalki, F.; Saifullah, K.; Alam, M.M.; Shaqiquzamman, M.; Ali, S.; Akhter, M. 2020. Mining of potential dipeptidyl peptidase-IV nhibitors as anti-diabetic agents using integrated in silico approaches. *Journal of Biomolecular Structure and Dynamics*, 38(18): 5349–5361.

Sheahan, M.; Barrett, C.B. 2017. Food loss and waste in Sub-Saharan Africa: A critical review. *Food Policy*, 70(1): 1–12.

Sparks, T.C.; Crouse, G.D.; Durst, G. 2001. Natural products as insecticides: The biology, biochemistry, and quantitative structure-activity relationships of spinosyns and spinosoids. *Pest Management Science*, 57(10): 896–905.

Srivastava, S.; Gupta, K.C.; Agrawal, A. 1988. Effect of plant product on *Calloso bruchuschinensis* L. infection on red gram. *Seed Research*, 16(1): 98–101.

Villers, P.; Navarro, S.; De Bruin, T. 2008. *Development of Hermetic Storage Technology in Sealed Flexible Storage Structures Controlled Atmosphere and Fumigation (CAF)*. Conference Paper, Grain Pro, Inc. Organic Storage Systems, Accessed on 16 May 2016, URL: http://grainpro.com/gpi/images/PDF/Commodity/CAF_Presentation_Development_of_Hermetic_Storage_Technology PU2015PV0708.pdf

Villers, P., Navarro, S., De Bruin, T. 2010. New applications of hermetic storage for grain storage and transport. *Proceedings 10th International Working Conference on Stored Product Protection*, 446–452.

Wambugu, P.; Mathenge, P.; Auma, E.; van Rheenen, H. 2009. Efficacy of traditional maize (*Zea mays* L.) seed storage methods in Western Kenya. *African Journal of Food, Agriculture, Nutrition and Development*, 9: 110–1128.

Wehner, T.C.; Maynard, D.N. 2003. Cucumbers, melons and other cucurbits. In: Katz, S.H. (Ed.), *Encyclopedia of Food and Culture*. New York: Scribner & Sons, 474–479.

Weinberg, Z.G.; Yan, Y.; Chen, Y.; Finkelman, S.; Ashbell, G.; Navarro, S. 2008. The effect of moisture level on high moisture maize (*Zea mays* L.) under hermetic storage conditions-*in vitro* studies. *Journal of Stored Products Research*, 44: 136–144.

White, N.D.G.; Jayas, D.S. 2003. Quality changes in grain under controlled atmosphere storage. In: Navarro, S.; Donahaye, E. (Eds.), *Proceedings of the International Conference on Controlled Atmosphere and Fumigation in Grain Storages*. Kansas: Caspit Pr.

World Bank. 1986. Poverty and hunger: Issues and options for food security in developing countries. In: *World Bank Policy Study*. Washington, DC: World Bank.

CHAPTER 15

Quality and Safety Aspects of Cereal Grains

Farhan Saeed, Muhammad Afzaal, Bushra Niaz, Amara Rasheed, Maryam Umar,
Muzzamal Hussain, Gulzar Ahmad Nayik and Mohammad Javed Ansari

CONTENTS

15.1 OVERVIEW

Cereals are the most common staple foods and a major energy source. Cereals are classified as *Triticum aestivum* L. (wheat), *Secale cereale* L. (rye), *Hordeum vulgare* L. (barley), *Avena sativa* L. (oat), *Oryza sativa* L (rice), *Pennisetum typhoides* L. (millet), *Zea mays* L. (corn), *Sorghum sudanensis* L. (sorghum), and *Triticosecale* (triticale). Cereals and cereal-based products provide essential nutrients to the body in the form of minerals, vitamins, fiber, essential fatty acids, and proteins for good health. Wheat, maize, rice, and barley are widely cultivated cereals also called major cereals, but rye, millet, oat, and sorghum are less cultivated and consumed (Su et al., 2017). In 2015–2016 the world production and consumption of cereals was 2529.6 million tons and 2527.8 million tons,

DOI: 10.1201/9781003252023-15

respectively, while world trade production of cereal was 14.4% (Papageorgiou et al., 2018). Cereals and other grains provide a significant amount of macronutrients and micronutrients (protein, lipids, carbohydrates, vitamins, minerals, and phytochemicals). Insoluble dietary fibers and phenolics have effects on intestinal health, and cereals are rich in dietary fibers (Monk et al., 2016). Whole grain consumption has been related to reducing the risk of chronic illnesses. Cereals can be contaminated at any level along the supply chain, from farm to fork, compromising their safety and quality. Cereals and their products are essential for the human diet and livestock feed all over the world.

Cereals include rye, maize bran, germ, oat flours, semolina, corn meal, corn grits, bread, pasta, snack foods, dry mixes, and tortillas. Cereals are used in batters, coatings, thickeners, sweeteners, processed meats, infant meals, confectionery, and beer. Because of their use in human diets and as livestock feed, microbiological safety and quality of cereals are a concern. Food safety and quality have been a hot topic in food policy, industry, and research for the past ten years (Saeed et al., 2021). Fungal infections cause a variety of plant diseases, which reduce production, induce discoloration, and cause grain shriveling, reducing grain size and product quality. During cultivation, harvesting, storage, and transportation, cereal grains are contaminated from different sources. Cleaning, grinding, grading, processing, and packaging material contaminate grains due to residues in containers, equipment, and screw conveyors.

15.2 GRAIN QUALITY

Depending on your location, "grain chain" refers to the process of growing and managing crops, harvesting, sorting, storing, and transporting them, as well as processing and end-use. For optimal grain production and processing management, grain quality assessment is essential. Grain quality analysis provides critical information about the safety and nutritional properties of the grain, allowing those involved in the value-added chain to better manage processing for the financial benefit of the specific stage of the overall process (Wrigley et al., 2017). The type and quality of grain determine how it is used in the end. It possesses physical and chemical characteristics that represent overall quality. Chemical parameters are fat, fiber, starch, protein, minerals, and phenolic acids, whereas physical factors are hardness, moisture content, and kernel size (Ratnavathi, 2019). Grain quality is affected by physico-chemical factors like moisture content, bulk density, kernel size, kernel hardness, vitreousness, kernel density, and damaged kernels, as well as safety parameters like fungal infection, mycotoxins, insects and mites and their fragments, foreign material odor and dust, and compositional factors like milling yield, oil content, protein content, starch content, and viability (Barr, 2019). Mold degrades the physical quality of grain, although it does not always impair the chemical quality. As a result, quality testing may be primarily concerned with assessing if a grain type qualifies for a specific use to improve the final product quality. However, farmers, traders, home economists, and consumers can determine the chemical composition, digestibility, and other quantitative attributes of grains.

15.3 GRAIN SAFETY

In most countries, food safety regulations are being designed and implemented, and public awareness of food safety issues is increasing. The main risk for grain safety is contamination of cereals. For both health and economic reasons, food safety is important to consumers and food business. Pathogens are adapting to the environment and develop emerging mycotoxins that create fungal disorder. Infection in cereal crops is common with fungal disorder in the food chain. Grain contamination occurs in different ways, but the most relevant source is the environment in which grains are cultivated, handled, and processed. Microorganisms that infect cereal grains are found in insects, animals, rodents, birds, human, storage and shipping containers, and handling and processing

Table 15.1 Types of Food Safety Hazards in Cereals and Their Sources

Source	Type	References
Inherent	Phytotoxins	Bazaz et al., 2016
Bioaccumulation	Heavy metals	dos Santos et al., 2017
Agronomic	Mycotoxins	Kamala et al., 2018
Processing	Microbiological	Di Cola Bucciarelli et al., 2021
	Pesticides	Alldrick, 2017
	Toxicants	Alldrick, 2016
Criminal acts	Sabotage	British Standards Institute, 2014

equipment. Physical, chemical, and biological hazards are the three basic forms of food safety issues (Negash, 2018). Before being consumed, the majority of grains are processed into food products, and some are consumed with only modest processing such as cleaning. To ensure that the end food product prepared from grains is safe and healthy, grain safety must be improved from production to processing. Therefore, grain safety should be ensured at all stages, including harvesting, storage, transportation, processing, and storage of processed foods (Bonciu, 2018). Cereal grain decontamination is an important food safety concern due to the global significance of cereals and their products. Cereal microorganisms are a critical control point since their growth can impact the grain's safety and quality characteristics (Los et al., 2018). Different food safety hazards in cereals and their sources are also discussed in Table 15.1.

A contaminant has been defined by European Food Safety Agency (EFSA) and the Codex Alimentarius Commission (CODEX). A contaminant is defined as "any substance present in food as a result of the manufacturing, processing, packaging, shipping, or holding of food, or as a result of environmental contamination." Microorganisms that can contaminate cereal grains include yeast, mold, bacteria, bacterial pathogens, coliforms, and enterococci. The most significant pollutants are mycotoxins, heavy metals, and acrylamide (Fatima, 2019). *Clostridium botulinum, Escherichia coli, Salmonella, Bacillus cereu, Clostridium perfringens*, and *Staphylococcus aureus* are bacterial pathogens that cause contamination in wheat grains.

15.4 FACTORS AFFECTING GRAIN QUALITY AND SAFETY

Improper post-harvest management and storage conditions deteriorate grain quality. Different factors affecting grain quality and safety are discussed in the following.

15.4.1 Physical Contamination

Physical contamination can be organic (chaff or various types of grain) and inorganic (dust, stones, soil). Physical hazards are organic materials such as rodent hairs, skin, and dead insects (Gibbs et al., 2018). Physical contamination occurs as a result of poor harvesting, drying, and threshing practices (Befikadu, 2018).

15.4.2 Mechanical Losses during Handling

Grain breakage occurs due to improper handling throughout post-harvest processing and storage processes, but it is also done during threshing. Corn is threshed by farmers by the sack and a process of hitting with rods. As a result, many grains crack (Yasin et al., 2019). All cereal quality is degraded by broken grains, and these broken grains are more vulnerable to quality damage (mold and insect attack).

15.4.3 Insufficient Drying

During insufficient drying, microbial activity increases and causes contamination and deterioration. Fungi and mold cause a decline in grain quality due to sufficient moisture content after harvesting. Mold growth increases on grain when moisture content is higher than the permissible level, which is 14%. More moisture content is ideal for insect invasion and grain discoloration. The word "biodeterioration" refers to the combination of mold and insect attack (Nayak et al., 2018).

15.4.4 Poor Conditions during Storage

Water, insects, and rodents enter through poor storage conditions, causing chemical browning and grain discoloration (Befikadu, 2018). These external factors affect storage stability of cereal grains and lower the safety and quality of different cereals.

15.4.5 Biodeterioration

Chemical changes and pesticides cause biodeterioration in the grains. Pesticides and chemical changes are accelerated by high temperature and relative humidity. The chemical changes accelerate with every 10°C increase in temperature. Changes in pH are an example of chemical changes that occur with time (Schmidt, 2018). Insect infestation, as well as high temperature and humidity level, cause rancidity in milled rice and cause decolorized corn grains, yellow grains in milled rice, and non-viable seed grains. With efficient post-harvest management and storage processes, all of these quality changes in cereal grains are delayed (Srivastava et al., 2021). The most common microorganisms that lower cereal grain quality and safety during storage and post-harvest processing are:

- Insects
- Birds
- Molds
- Rodents

15.4.6 Fungal Infestation

Crops are susceptible to plant pathogenic disorders. Virus, bacteria, fungus, yeast, molds, and protozoa constitute the indigenous microbiota in wheat grains. Grains have roughly 10 million germs per gram and around 150 mold species. *Pseudomonadaceae, Micrococcaceae, Lactobacillaceae*, and *Bacillaceae* are the most common bacteria discovered in grains, while *Alternaria, Fusarium, Helminthosporium*, and *Cladosporium* are the most common molds. Molds begin by growing on the surface of stored grain before slowly entering and killing it. On wheat grains, there are 150 species of yeasts and molds. Molds are the most important of them. Filamentous fungi are a major health risk because they produce mycotoxins, which build in grains before and after harvest and are linked to serious health problems. Carcinogenic, mutagenic, genotoxic, teratogenic, neurotoxin, and estrogenic toxicity, as well as reproductive and developmental toxicity, can be caused by mycotoxins. Field and storage fungi are two different types of fungi in cereal grains related to moisture content. *Alternaria cladosporium, Fusarium*, and *Helminthosporium* are field fungi. Blights, discoloration, and mycotoxin production caused by field fungi affect seed quality and cause disorder in crops. Principal storage fungi include *Eurotium, Aspergillus*, and *Penicillium* species and cause quality loss in cereals during storage. Storage fungi attack grain when moisture content, water activity, and relative humidity are less. Relative humidity of about 70% causes mold spoilage at the grain surface. Moisture content in grains shows relative humidity at the surface of grains. When the relative humidity is 70%, most cereals have the same moisture content (Neme et al., 2017). Germination

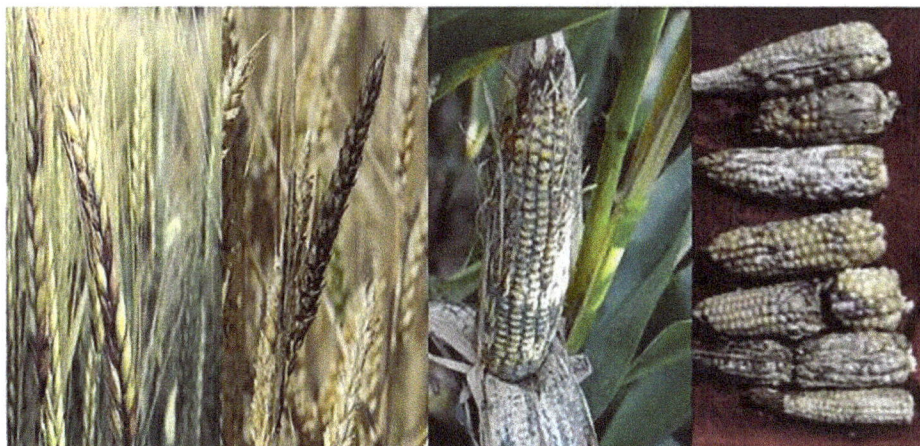

Figure 15.1 Different species of mold and fungus damage cereal crops.

loss, discoloration, warmth and mustiness, dry matter loss, mycotoxin production, and nutritional alterations are all prominent effects of storage fungus on grain. The significance of these impacts is determined by the grain's final application. Fungi infestations can affect grain quality and even entirely destroy grain's usability, depending on its utilization. Grains must be stored at or below a safe storage moisture level to avoid mold growth. Because the plant's natural defense against mold growth weakens as it ages, physiologically ripe (Kumari et al., 2021). *Blumeriagraminis, Puccinarecondita, Puccinia graminis, Puccinastriiformis, Septoria tritici, Septoria nodorum,* and *Fusarium graminis* are all economically significant grain diseases. (Różewicz et al., 2021). Fusarium head blight is the most frequent disease in wheat, caused by *Fusarium culmorum* and *Fusarium graminearum*, two fungi belonging to the genus *Fusarium*. The fungi multiply, resulting in a major loss and decline in quality of crops (Figure 15.1).

15.4.7 Mold Causes Grain Damage

- During post-harvest processing, grains that are not properly dried cause mold growth with the passage of time.
- Due to water loss, production of hot spots, and insect activity during storage, grains in a big grain bulk are damp (Tola & Kebede, 2016). Improve storage conditions to avoid leakage of water, and ensure pest control procedures on large grain bulks are completed on time.
- Growing crops are stressed by dryness or when damp grain is exposed due to insect infestation in the field. Carefully detect and remove damaged grain during post-harvest treatment to avoid this (Hodges, 2007).

15.4.8 Mycotoxins

Mycotoxin is formed from the words *mykes* and *toxicum*, which respectively indicate fungus or mold and poison or toxin. The term "mycotoxin" refers to fungus poison or mold toxin in general. Certain fungi create mycotoxins that are major toxic compounds in cereal. Fungi are airborne or soil borne and can infect plants in the field and during the manufacturing process (Alshannaq et al., 2017). According to the FAO, fungi kill nearly a quarter of all food crops. Mycotoxins cause a 1-billion-ton loss of food each year, resulting in billions of dollars in losses. Biological, environmental, harvesting, storage, handling, and processing factors affect mycotoxin contamination of grain. Different mycotoxins in different cereal grains are shown in Table 15.2. Some fungal genera can produce

Table 15.2   Mycotoxins in Cereal Grains

Grains	Fungal Sources
Wheat, maize, barley	*Fusarium graminearum, Fusarium crookwellense, Fusarium culmorum*
Corn, wheat	*F. graminearum, F. culmorum, F. crookwellense*
Barley, wheat	*Aspergillus ochraceus, Penicillium verrucosum*
Maize	*Fusarium moniliforme*
Corn, rice	*Aspergillus flavus*
	Aspergillus parasiticus

Table 15.3   Conditions for Mycotoxin Formation

Mycotoxin	Temperature (°C)	Water Activity
Aflatoxin	33	0.99
Ochratoxin	25–30	0.98
Fumonisin	15–30	0.9–0.995
Zearalenone	25	0.96
Deoxynivalenol	26–30	0.995
Citrinin	26–30	0.75–0.85

mycotoxins in response to various variables such as rain, crop type compatibility with agricultural environment and activity, dryness stress, and insect-borne sickness. The temperature and moisture level of the grains are crucial parameters that promote mycotoxin production. Another factor that influences the moisture level of stored grain is relative humidity (conditions for mycotoxin formation in cereal grains are presented in Table 15.3). Molds thrive in temperatures ranging from 10°C to 40.5°C, with a relative humidity of 70% and a pH of 4–8 (Channaiah & Maier, 2014). Aflatoxins, zearalenone, ochratoxin, fumonisins, and deoxynivalenol/nivalenol (DON) are mycotoxins that are particularly harmful to human health when found in grains. Aflatoxins have a 55% and 1642 g/kg incidence rate, ochratoxin A has a 29% and 1164 g/kg incidence rate, fumonisins have a 61% and 71,121 g/kg incidence rate, deoxynivalenol has a 58% and 41,157 g/kg incidence rate, and zearalenone has a 46% and 3049 g/kg incidence (Lee et al., 2017). Mycotoxins found in cereals are frequently classified as field and storage, depending on when they are produced either before or after harvest. Field mycotoxins (deoxynivalenol) are produced during the growth of the crop, whereas storage mycotoxins (ochratoxin A) are produced after it has been harvested. Prevention is now commonly acknowledged as being better than cure (Pleadin et al., 2019). The majority of strategies necessary to carry out this plan are based on Hazard Analysis and Critical Control Point (HACCP) principles. Field mycotoxins are mostly controlled before harvest, whereas storage mycotoxins are controlled after harvest.

To avoid mycotoxin contamination of stored cereals, consider the following factors:

1. The current state of sanitary grain harvesting and storage.
2. Transport time from the field to the dryer for high-moisture grain.
3. Thermodynamics of drying to reduce moisture content
4. In the long run, the lower moisture content is preserved.

15.4.9 Aflatoxins

Aflatoxins are harmful secondary metabolites produced by *Aspergillus* spp. (*Aspergillus flavus* and *Aspergillus parasiticus*) (Mousavi khaneghah et al., 2017). Food safety is jeopardized by aflatoxins, which also cause significant infections in humans and livestock. Humans become infected

Figure 15.2 Aflatoxin effect on corn grains.

with aflatoxins (AFs) by regularly eating contaminated food, which leads to nutritional deficiencies, immunological suppression, and hepatocellular carcinoma. Acute aflatoxicosis is caused by ingesting extremely high quantity of AFs, which causes serious liver damage, bleeding, edema, and even death. The time of harvest and method of drying and storage have an impact on aflatoxin contamination. AFs B1, B2, G1, and G2, secondary metabolites of *Aspergillus flavus, Aspergillus parasiticus*, and *Aspergillus nomius*, produce a wide range of food and feed products during the growing, harvesting, and shipping processes (Somorin et al., 2012). AFB1 has high prevalence in cereal products. A total AF (4 g/g) is allowed in cereals and its products (Silva et al., 2019). Maize, sorghum, wheat, and rice are among the cereals that are commonly infected by *Aspergillus* spp. (Figure 15.2).

15.4.10 Heavy Metals

"Heavy metals" refers to a group of metallic elements that are known to be harmful. Arsenic, cadmium, and lead are particularly important in grains. Inhalation, nutrition, and manual handling are sources of entry in plants, animals, and human tissues. Mercury, lead, cadmium, and arsenic not only compete for absorption with minerals (calcium, magnesium, and iron) but can also attach to key biological components (structural proteins, enzymes, and nucleic acids), preventing them from fulfilling their tasks (Andrade et al., 2017). Heavy metal exposure over time can cause a variety of health issues. Arsenic and cadmium affects the body parts (skin, lungs, brain, kidneys, liver, metabolic system, cardiovascular system, immunological system, and endocrine system) (Kosek-Hoehne et al., 2017).

15.5 DIFFERENT TECHNIQUES IMPROVE QUALITY AND SAFETY

Traditionally, gas chromatography (GC), enzyme-linked immunosorbent assay (ELISA), polymerase chain reaction (PCR), high-performance liquid chromatography (HPLC), gas chromatography mass spectrometry (GC-MS), and liquid chromatography mass spectrometry (LCMS MS) have been used to assess cereal and other food safety and quality (Rammanee & Hongpattarakere, 2011). HPLC and GC are extensively used for grain inspection (both qualitatively and quantitatively); however, the ELISA approach is typically employed for harmful chemicals in cereal products. The main disadvantages of these techniques are their inherent destructiveness, time commitment, and laboriousness. Furthermore, they usually demand lengthy sample preparation procedures involving the use of chemicals that could contaminate the environment (Bahadoran et al., 2016). So the industry places a premium on developing precise, dependable, quick, and nondestructive methods for detecting cereal items. Several spectral imaging techniques for cereal quality and safety testing have recently been developed (Ma et al., 2016). Hyperspectral imaging (HSI), Raman imaging (RI), fluorescence imaging (FI), laser light backscatter (LLB), soft X-ray imaging, and magnetic resonance imaging are all examples of spectral imaging techniques. These were designed to give high selectivity and quick, nondestructive agricultural product analysis, even at trace levels (Hussain et al., 2018; Yaseen et al., 2017). Online applications can benefit greatly from spectral and imaging approaches. Fungal contamination, quality discrimination, and adulteration have all been detected using estimation methods based on FS, NIRS, FTIRS, and HIS (Su et al., 2017; Tao et al., 2018).

15.6 HAZARD ANALYSIS AND CRITICAL CONTROL POINTS

Hazard analysis and critical control points is a food safety management system that involves the systematic identification and assessment of food hazards, as well as the execution of control measures (Gehring & Kirkpatrick, 2020). A hazard analysis and critical control point is a scientific approach for food quality and safety that is used to avoid food poisoning. Rather than reacting to microbiological, chemical, and physical threats, food management systems are in charge of securing the food supply. It was developed 30 years ago as a result of a collaboration between National Aeronautics and Space Administration (NASA), Pillsbury, and US army scientists to produce "totally safe food" for US astronauts. It was so successful that the whole food processing sector immediately adopted it, and it is now regarded as a global standard for safe practices. This is largely due to the fact that it allows for a proactive approach through process and material management (Panghal et al., 2018).

Following are its seven clearly defined principles (Figure 15.3).

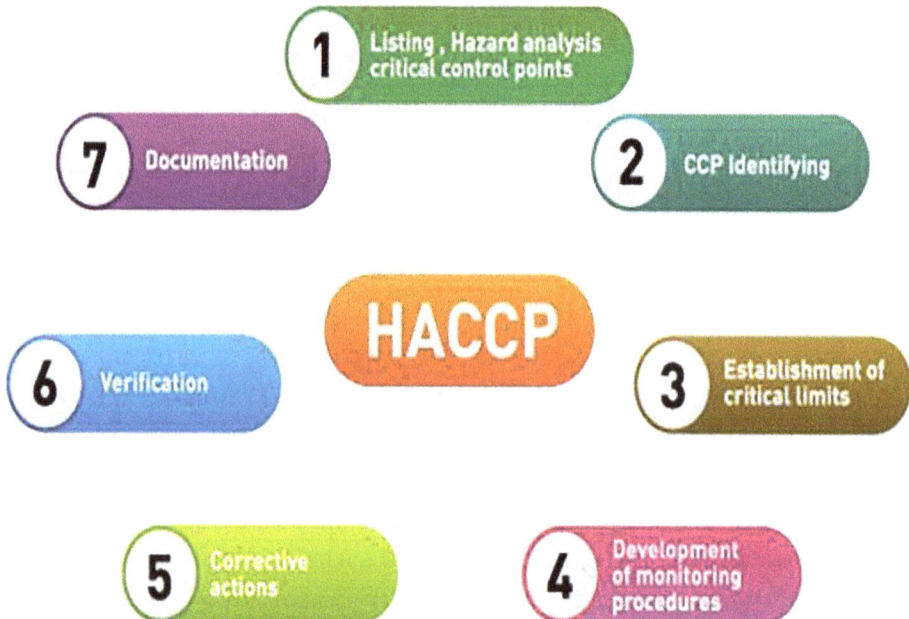

Figure 15.3 HACCP principles.

15.6.1 The Seven Principles of HACCP

1. Hazard identification, risk assessment, and control method description
2. The critical control points must be identified (CCP)
3. Critical limits are established
4. Procedures for CCP monitoring and evaluation are being established
5. Actions to be taken to correct the situation
6. Establishment of processes for verification
7. Record-keeping and documentation

HACCP development and execution are complex undertakings, and not all countries have the technical expertise and experience needed to build effective HACCP-based integrated mycotoxin management systems in cereals. Given the importance of HACCP in food safety programs, FAO has made it a top priority to provide training on the HACCP concept and its application to experts in developing countries (De Oliveira et al., 2016). Hazards must be taken into account in HACCP-based integrated mycotoxin management systems at all stages of manufacturing, handling, and processing. Good manufacturing practices (GMPs) and good agricultural practices (GAPs) are also followed in HACCP programs (Okpala & Korzeniowska, 2021). Food industry and certified food control agencies have employed the HACCP system to prevent and reduce hazards associated with pathogenic microorganisms and chemical toxicants in food products. HACCP was developed to ensure food safety from harvest to consumption by removing potentially dangerous biological, chemical, and physical components. HACCP has developed quick testing for rapid detection of toxins.

Mycotoxins can be discovered in large levels in cereal crops. Drought, pest infestation, primary inoculum, and delayed harvesting are external factors that cause an impact. Some of these features are influenced by the environment, and humans have limited control over them. To reduce mycotoxin contamination, the hazard analysis and critical control point system is broadly used in food and feed manufacturing. During food processing, the main goal is to avoid the formation of mycotoxins. Storage and processing are two critical areas where contamination can be avoided in the post-harvest period (Agriopoulou et al., 2020).

15.7 GOOD MANUFACTURING PRACTICES

In good manufacturing practices, the recommended procedures are based on the hygienic principles and sanitary practices integrated in all processes. Good manufacturing practices and hazard analysis and critical control points are widely utilized in the food industry for assurance of quality and safety of food, as well as loss prevention (De Oliveira et al., 2016).

Good manufacturing practices entail the following procedures:

- Installations and facilities
- Personnel
- Operations and processes in production
- Tools and equipment
- Cleaning and sanitation
- Code
- Storage and distribution
- Pest control
- Quality assurance and control
- Training and supervision

These strategies not only help the company to compete in an increasingly competitive and demanding market but also allow for high quality food production, fewer losses, reduced accident risks, and cost savings (Sameen et al., 2022). These benefits could be clearly obtained by extrapolating these benefits into a grain storage facility plan, primarily highlighting the grain storage segment's true participation in the "productive chain" of any food, and thus, the safety of stored products will produce qualitative products. The issue of workplace safety and health has become increasingly important for global society not only from a humanistic point but also from an economic point.

15.8 CONCLUSION

Cereals are contaminated at any level along the supply chain, from farm to fork, compromising their safety and quality. Novel techniques should be used to improve the nutritional, microbiological, and physicochemical properties of cereal grains. At the industrial scale, cereal grain safety and quality can be accomplished through the implementation of quality assurance principles such as good manufacturing practice and hazard analysis and critical control point techniques based on understanding of the origin and occurrence of the hazards concerned.

REFERENCES

Agriopoulou, S., Stamatelopoulou, E., & Varzakas, T. 2020. Advances in occurrence, importance, and mycotoxin control strategies: Prevention and detoxification in foods. *Foods*, *9*(2), 137.

Alldrick, A. J. 2016. Chemical Hazards. In *Handbook of Hygiene Control in the Food Industry* (pp. 89–102). Woodhead Publishing.

Alldrick, A. J. 2017. Food safety aspects of grain and cereal product quality. In *Cereal grains* (pp. 393–424). Woodhead Publishing.

Alshannaq, A., & Yu, J. H. 2017. Occurrence, toxicity, and analysis of major mycotoxins in food. *International Journal of Environmental Research and Public Health*, *14*(6), 632.

Andrade, V. M., Aschner, M., & Marreilha Dos Santos, A. P. 2017. Neurotoxicity of metal mixtures. *Neurotoxicity of Metals*, 227–265.

Bahadoran, Z., Mirmiran, P., Jeddi, S., Azizi, F., Ghasemi, A., & Hadaegh, F. 2016. Nitrate and nitrite content of vegetables, fruits, grains, legumes, dairy products, meats and processed meats. *Journal of Food Composition and Analysis*, *51*, 93–105.

Barr, S. 2019. *Technology of cereals, pulses and oilseeds*. Scientific e-Resources

Bazaz, R., Baba, W. N., & Masoodi, F. A. 2016. Development and quality evaluation of hypoallergic complementary foods from rice incorporated with sprouted green gram flour. *Cogent Food & Agriculture*, *2*(1), 1154714.

Befikadu, D. 2018. Postharvest losses in Ethiopia and opportunities for reduction: A review. *International Journal of Sciences: Basic and Applied Research (IJSBAR)*, *38*, 249–262.

Bonciu, E. 2018. Food processing, a necessity for the modern world in the context of food safety: A review. *Annals of the University of Craiova-Agriculture, Montanology, Cadastre Series*, *47*(1), 391–398.

British Standards Institute, 2014. *PAS 96:2014—Guide to Protecting and Defending Food and Drink From Deliberate Attack*. BSI Group.

Channaiah, L. H., & Maier, D. E. 2014. Best stored maize management practices for the prevention of mycotoxin contamination. *Mycotoxin Reduction in Grain Chains*, 78.

De Oliveira, C. A. F., Da Cruz, A. G., Tavolaro, P., & Corassin, C. H. 2016. Food Safety: Good manufacturing practices (GMP), sanitation standard operating procedures (SSOP), hazard analysis and critical control point (HACCP). In *Antimicrobial food packaging* (pp. 129–139). Academic Press.

Di Cola, G., Fantilli, A. C., Pisano, M. B., & Ré, V. E. 2021. Foodborne transmission of hepatitis A and hepatitis E viruses: A literature review. *International journal of food microbiology*, *338*, 108986.

dos Santos, G. M., Pozebon, D., Cerveira, C., & de Moraes, D. P. 2017. Inorganic arsenic speciation in rice products using selective hydride generation and atomic absorption spectrometry (AAS). *Microchemical Journal*, *133*, 265–271.

EFSA Panel on Biological Hazards (BIOHAZ), Ricci, A., Chemaly, M., Davies, R., Fernández Escámez, P. S., Girones, R., . . . & Bolton, D. 2017. Hazard analysis approaches for certain small retail establishments in view of the application of their food safety management systems. *EFSA Journal, 15*(3), e04697.

Fatima, F. 2019. *Microbiological and chemical risk assessments of the addition of selected cereal grains as non-dairy ingredients to dairy products: a thesis presented in partial fulfilment of the requirements for the degree of Master in Food Technology at Massey University, Manawatū, New Zealand* (Doctoral dissertation, Massey University).

Gehring, K. B., & Kirkpatrick, R. 2020. Hazard analysis and critical control points (HACCP). In *Food safety engineering* (pp. 191–204). Springer

Gibbs, M., & Steele, P. 2018. *Post harvest technology of horticultural crops.* Scientific e-Resources.

Hodges, R. J. 2007. *Session 6: Grain quality and food safety.* Springer.

Hussain, A., Pu, H., & Sun, D. W. 2018. Innovative nondestructive imaging techniques for ripening and maturity of fruits—a review of recent applications. *Trends in Food Science & Technology, 72*, 144–152.

Kamala, A., Shirima, C., Jani, B., Bakari, M., Sillo, H., Rusibamayila, N., . . . & investigation team. 2018. Outbreak of an acute aflatoxicosis in Tanzania during 2016. *World Mycotoxin Journal, 11*(3), 311–320.

Kosek-Hoehne, K., Panocha, B., & Śliwa, A. 2017. Heavy metals—a silent threat to health. *Journal of Education, Health and Sport, 7*(1), 121–132.

Kumari, P., Kumar, V., Kumar, R., & Pahuja, S. K. 2021. RETRACTED ARTICLE: Sorghum polyphenols: plant stress, human health benefits, and industrial applications. *Planta, 254*(3), 1–14.

Lee, H. J., & Ryu, D. 2017. Worldwide occurrence of mycotoxins in cereals and cereal-derived food products: Public health perspectives of their co-occurrence. *Journal of Agricultural and Food Chemistry, 65*(33), 7034–7051.

Los, A., Ziuzina, D., & Bourke, P. 2018. Current and future technologies for microbiological decontamination of cereal grains. *Journal of food science, 83*(6), 1484–1493.

Ma, J., Sun, D. W., Qu, J. H., Liu, D., Pu, H., Gao, W. H., & Zeng, X. A. 2016. Applications of computer vision for assessing quality of agri-food products: a review of recent research advances. *Critical reviews in food science and nutrition, 56*(1), 113–127.

Monk, J. M., Lepp, D., Zhang, C. P., Wu, W., Zarepoor, L., Lu, J. T., . . . & Power, K. A. 2016. Diets enriched with cranberry beans alter the microbiota and mitigate colitis severity and associated inflammation. *The Journal of nutritional biochemistry, 28*, 129–139.

Mousavi Khaneghah, A., D Chaves, R., & Akbarirad, H. 2017. Detoxification of aflatoxin M1 (AFM1) in dairy base beverages (acidophilus milk) by using different types of lactic acid bacteria-mini review. *Current Nutrition & Food Science, 13*(2), 78–81.

Nayak, S., Mukherjee, A. K., Sengupta, C., & Samanta, S. 2018. Association of microbial diversity with post harvest crops and bioprospecting of endophytic microorganisms for management. ICAR-NRRI.

Negash, D. 2018. A review of aflatoxin: occurrence, prevention, and gaps in both food and feed safety. *Journal of Applied Microbiological Research, 1*(1), 35–43.

Neme, K., & Mohammed, A. 2017. Mycotoxin occurrence in grains and the role of postharvest management as a mitigation strategies. A review. *Food Control, 78*, 412–425.

Okpala, C. O. R., & Korzeniowska, M. 2021. Understanding the relevance of quality management in agro-food product industry: From ethical considerations to assuring food hygiene quality safety standards and its associated processes. *Food Reviews International*, 1–74.

Panghal, A., Chhikara, N., Sindhu, N., & Jaglan, S. 2018. Role of Food Safety Management Systems in safe food production: A review. *Journal of food safety, 38*(4), e12464.

Papageorgiou, M., & Skendi, A. 2018. Introduction to cereal processing and by-products. In *Sustainable Recovery and Reutilization of Cereal Processing By-Products* (pp. 1–25). Woodhead Publishing.

Pleadin, J., Frece, J., & Markov, K. 2019. Mycotoxins in food and feed. *Advances in food and nutrition research, 89*, 297–345.

Rammanee, K., & Hongpattarakere, T. 2011. Effects of tropical citrus essential oils on growth, aflatoxin production, and ultrastructure alterations of *Aspergillus flavus* and *Aspergillus parasiticus*. *Food and Bioprocess Technology, 4*(6), 1050–1059.

Ratnavathi, C. V. 2019. Grain structure, quality, and nutrition. In *Breeding sorghum for diverse end uses* (pp. 193–207). Woodhead Publishing.

Różewicz, M., Wyzińska, M., & Grabiński, J. 2021. The most important fungal diseases of cereals—problems and possible solutions. *Agronomy, 11*(4), 714.

Saeed, F., Afzaal, M., Hussain, M., & Tufail, T. 2021. Advances in assessing product quality. In *Food losses, sustainable postharvest and food technologies* (pp. 191–218). Academic Press.

Sameen, A., Sahar, A., Tariq, F., Khan, U. M., Tariq, T., & Ishfaq, B. 2022. Food Microbial Hazards, Safety, and Quality Control: A Strategic Approach. In *Food safety practices in the restaurant industry* (pp. 67–98). IGI Global.

Schmidt, M. 2018. *Fundamental study on perspectives for post-harvest bioprotection of cereal crops* (Doctoral dissertation, University College Cork).

Silva, A. S., Brites, C., Pouca, A. V., Barbosa, J., & Freitas, A. 2019. UHPLC-ToF-MS method for determination of multi-mycotoxins in maize: Development and validation. *Current Research in Food Science, 1*, 1–7.

Somorin, Y. M., Bertuzzi, T., Battilani, P., & Pietri, A. 2012. Aflatoxin and fumonisin contamination of yam flour from markets in Nigeria. *Food Control, 25*(1), 53–58.

Srivastava, S., & Mishra, H. N. 2021. Ecofriendly nonchemical/nonthermal methods for disinfestation and control of pest/fungal infestation during storage of major important cereal grains: A review. *Food Frontiers, 2*(1), 93–105.

Su, W. H., He, H. J., & Sun, D. W. 2017. Non-destructive and rapid evaluation of staple foods quality by using spectroscopic techniques: A review. *Critical Reviews in Food Science and Nutrition, 57*(5), 1039–1051.

Tao, F., Yao, H., Hruska, Z., Burger, L. W., Rajasekaran, K., & Bhatnagar, D. 2018. Recent development of optical methods in rapid and non-destructive detection of aflatoxin and fungal contamination in agricultural products. *TrAC Trends in Analytical Chemistry, 100*, 65–81.

Tola, M., & Kebede, B. 2016. Occurrence, importance and control of mycotoxins: A review. *Cogent Food & Agriculture, 2*(1), 1191103.

Wrigley, C., Batey, I., & Miskelly, D. 2017. Grain Quality: The Future Is with the Consumer, the Scientist and the Technologist. In *Cereal grains* (pp. 691–725). Woodhead Publishing.

Yaseen, T., Sun, D. W., & Cheng, J. H. 2017. Raman imaging for food quality and safety evaluation: Fundamentals and applications. *Trends in Food Science & Technology, 62*, 177–189.

Yasin, M., Wakil, W., Ali, K., Ijaz, M., Hanif, S., Ali, L., . . . & Ahmad, S. 2019. Postharvest Technologies for Major Agronomic Crops. In *Agronomic crops* (pp. 679–710). Springer.

CHAPTER 16

Future Aspects of Grain Quality and Role of Technologists in Its Management

Tridip Boruah, Barsha Devi, Twinkle Chetia, Kamini Choubey, Karabi Talukdar, Mohammad Javed Ansari and Gulzar Ahmad Nayik

CONTENTS

16.1 INTRODUCTION

A rapidly increasing global population and increasing demand of food supply are emerging as the greatest challenges to current civilization. The total population of the world is estimated to reach approximately 9.1 billion by 2050, which will demand about 70% extra food production (Godfray et al., 2010). Global changes in climate, urbanization, and use of fertile land for the production of non-food crops deepen this distress related to increasing food demand. Global food security is highly compromised by the threat of climate change. Agriculture is primarily challenged by the continuously changing climate, and the increase in natural calamities due to changing climates such as drought, flood, storms, new pests, and disease incidence is the most relevant driver of food insecurity. Moreover, increasing water scarcity, soil salinization due to heavy irrigation, invasion of invasive weed species, and the resistance of pests and diseases against a growing number of agrochemicals are other challenges for agricultural production. The average global temperature is predicted to be 1.5 to 4.5°C higher, and atmospheric CO_2 concentration is estimated to reach 550 $\mu mol\ mol^{-1}$ towards the middle of this century, with many recurrent occurrences of intense climatic proceedings (Carter, 2011). These changes in environmental variables will disturb the growth of plants, quality, and grain yield either directly or indirectly (Ainsworth et al., 2008). With technological advancements, new technologies to combat biotic and abiotic stresses and improve soil fertility and water availability can potentially increase the amount of crop productivity.

Cereal grains have been the most extensively consumed agricultural food globally since prehistoric times. About 60% of cultivated land worldwide is occupied by cereal grains (Koehler & Wieser, 2013). The yield and production of cereal grains have been largely augmented in the last five decades to cope with the requirements of the increasing global population. Cereal grains and cereal-based foods are the chief constituents of the human diet, particularly in developing counties. Cereal grain–associated health benefits could be credited to the coactive effects of phytochemicals, micronutrients, and dietary fiber (Kaur et al., 2014). Phytochemicals present in cereal grains have the capacity to prevent oxidative damage through antioxidant properties (Yu et al., 2020). Dietary fiber helps in disease prevention, along with preventing immunity related health problems and enhancing the qualities of both the cereal based products and cereals (Foschia et al., 2013). Cereals contribute to a significant part of global food production and constitute most harvested crops. The nutritional quality of cereal has a substantial impact on human health. Current and near-future environmental conditions will severely affect cereal yield and nutritional value of grains. Therefore, the major challenge for plant breeders is to increase cereal grain production while considering a satisfactory grain nutrient content. Various authors have documented numerous effects of elevated atmospheric CO_2 on cereals. According to Loladze (2002), rising CO_2 is likely to lead to "globally imbalanced plant stoichiometry," which in turn would intensify the acute problem of micronutrient malnutrition. Grain mineral and protein content are quality characteristics that can change due to elevated temperature. A decrease in mineral and protein content in rice, wheat, and barley grains and an increase in the ratio of non-structural carbohydrate to proteins in grains due to elevated CO_2 have been reported (Loladze, 2002).

Microbial contamination of grains during storage is a chief worry for the grain processing industry that affects grain quality and quantity. With the wide consumption of cereal grain–related products and cereals as a whole globally, cereal decontamination and preservation have become the most vital food security issues. Harvested grains are either transported or stored for further use. Therefore, conserving valuable grains in order to sustain their quality is a necessary element of grain management. Stored grains are an appealing object to many predators such as insects, microorganisms, birds, and animals. Thus, systematic evaluation of cleanliness and soundness of grain is fundamental for storage management. Therefore, proper management of microbiological safety before storing grains is of the utmost importance.

Weeds are another big challenge to the production of cereal grains in both developed and developing countries. They compete with crops for nutrients, sunlight, space, and water and cause maximum loss to crops in terms. Furthermore, they harbor pathogens and insects that damage the crop. The intensity of financial damage caused by weeds can be predicted by the fact that the grain growers in Australia saw losses of 3.3 billion AUD annually and weeds accounted for 2.7 million tons of grain in terms of yield losses (Chauhan, 2020). In India, these costs were much higher. According to Gharde et al. (2018), India loses more than 11 billion USD per year to the diverse effects of weeds, while in The United States, the amount can be as high as USD 33 billion annually (Chauhan, 2020). These global studies specify the extensive economic losses and loss of yield caused by weeds. Even though biological management of weeds is not regarded as the universal solution, it has great prospects if the conditions are favorable. In the meantime, with technological advancement, vast opportunities have been created for management of weeds by means of thermal techniques such as solarization, electrocution, microwave, and flaming technology (Bajwa et al., 2015). Modern weed management techniques come with techniques like precision weed management that include the application of remote sensing, robotics, and modeling in a refined manner.

In many cereal-growing countries, the major inadequacy is always is water, which can be either in the form of irrigation or rain. As the major consumer of freshwater, the global irrigation sector is under a tremendous amount of pressure to maximize its efficiency. The application of micro-irrigation technology has been thriving in various climatic conditions ranging from arid to humid and semi-arid regions. The advancement in dripper and emitter technology, the development of low-cost sand and screen filters, and the introduction of inexpensive drip tape are certainly responsible for the expansion of the micro-irrigation technique. Apart from that, the application of emerging techniques such as soil moisture sensors is making these systems more automatic and easy to use. This chapter discusses various aspects of grain quality and technological advancements for its management in the near future.

16.2 GLOBAL CONCERNS FOR CEREAL GRAIN QUALITY

For thousands of years, we could not look beyond cereals for our staple diets, and they are still the largest food source for people all over the world. These are plants producing grains that offer more than 50% of the protein needs. Among total cultivated land, cereals are grown on more than two-thirds of it (Shahbaz & Ashraf, 2013). Due to their specific characteristics such as easy cultivation technique, storage, and transportation, they are the primary food of most human communities and are also regarded as the most primitive cultivated crops. They are cosmopolitan in distribution and composed of various varieties and species adapted to even extreme climatic conditions (Shahbaz & Ashraf, 2013). Adaptations by crops are either structural or physiological. Cereals contribute more than 50% to the total daily caloric intake of the world. Compared with industrialized countries, developing countries rely more on cereal grains to meet their nutritional needs. Developing nations require a huge amount of cereals to derive almost 60% of the required calories, and the figure is close to 30% for developed countries (Honfoga, 2003).

Climate is the most important factor affecting productivity in agriculture. The global change in climate is expected to affect agriculture and animal production, water balance, resource supply, and many other constituents of the agricultural system. Cultivated plants are instantly exposed to changes in climatic factors such as precipitation and temperature, as well as the severity and frequency of unpredictable events such as floods, droughts, and cyclones. The agriculture sectors are severely affected by extreme changing environments. Both anthropogenic and natural actions result in magnified stress, which disturbs crop productivity. Many reviews point out that due to rising temperature, water, and salinity, the production of different crops in various parts of the globe has declined (Wang & Frei, 2011). The average global temperature is mounting at a disturbing rate of 0.3°C and is expected to rise to 13°C by 2025. Excessive accumulation of metals in plants will reduce crop yields and adversely affect the development of plants and physiological parameters such as growth, photosynthesis, and germination (Ali et al., 2015). Since carbon dioxide in the atmosphere is the only carbon source for plants, the enrichment of carbon dioxide will act as a carbon fertilizer, which will have a profound effect on crop productivity, crop yield, and plant vitality, especially food quality. The growing levels of CO_2 have a direct impact on photosynthesis and have been revealed to improve plant productivity and growth under appropriate conditions (Nosberger et al., 2006). In addition to the increase in biomass production, it is often observed that the chemical composition of plants grown under carbon dioxide enrichment has undergone significant changes. According to Loladze (2002), an increase in carbon dioxide changes the overall stoichiometry of the plant and causes the concentration of most elements to decrease. These alterations will have a harmful impact on the nutritional properties of food and feed or the cycle of ecosystem elements. A notable example of the effect of carbon dioxide is the reduction of the concentration of N_2 in vegetation plant components as well as grains and seeds, in which the protein concentration is reduced (Hogy & Fangmeier, 2008). There is a strong interaction between the nitrogen supply of plants and the increased influence of carbon dioxide, which affects grain quality and plant metabolism. It has additionally been revealed that multiplied carbon dioxide will lower grain nitrogen and as a result protein concentration regardless of the nitrogen accessibility, that is, the reduced grain quality might not simply be surmounted by further nitrogen fertilization. Kimball et al. (2001) provided information that the decrease in the concentration of protein caused by carbon dioxide can be partially offset by nitrogen fertilizer. The current knowledge about changes in the quality of grains caused by carbon dioxide is almost entirely derived from the study of some form of controlled environment or field chamber. The possible negative effects of carbon dioxide on Fe and Zn concentrations are particularly of concern, as most people in developing countries do not get these elements in their diets. The important point is whether the various experimental methods used to control atmospheric CO_2 affect crops in the same way. Regrettably, direct comparisons of the impact of CO_2 on major crop variables such as yields due to different CO_2 application methods are limited. The only research on composition of grain elements with the use of Free-Air CO_2 Enrichment FACE technology is a couple of experiments on rice-growing conditions in Japan and China and two experiments on wheat in the United States and Germany (Kimball et al., 2001).

In a recent study, Ortiz-Bobea et al. (2021) found that since 1961, anthropogenic climate change has had a deleterious effect on the overall productivity of global agricultural factors by about 21%, affecting warmer regions such as Africa (–34%) more than colder regions such as Europe and Central Asia (–7.1%). Due to the increase in CO_2 and the complex effects of changes in temperature and rainfall, change in climate is predicted to impact cereal grains worldwide, impacting their quantity and quality in the coming decades. Numerous results of accelerated atmospheric CO_2 on plants have been reported by a photosynthesis-mediated CO_2 fertilization effect, together with growing carbon assimilation, growth, yield, and carbon content (Ainsworth et al., 2008). Therefore, increasing CO_2 can increase the concentration of photosynthetic-derived carbohydrates in grain, of which starch is the main component. Since grains are mainly composed of carbohydrates, it has been suggested that increasing the concentration of starch will liquefy other nutrients, including proteins, lipids, vitamins, and minerals. In addition, the regulation of photosynthetic apparatus and the redistribution of senescent leaves to grains are key mechanisms (Chernyad'ev, 2005). Therefore, one of the biggest challenges that plant bleeders are facing today is to increase grain production while keeping in mind the proper nutrient content of the grain.

16.3 MICROBIAL CONTAMINATION IN GRAINS AND ITS IMPACT ON GRAIN QUALITY

Cereals are among the foremost vital agricultural commodities on the planet, both as human foods and as the major component of animal nutrition. In prehistoric times, the development of agriculture was greatly allied with the domestication of cereals, and from the moment of their first cultivation, most civilizations began to depend on cereals for the bulk of their food supply. A key problem of the grain industry is the contamination of stored grains by microorganisms and insects, as it affects the grain both qualitatively and quantitatively. Due to the presence of microorganisms and insects, up to 50–60% of the cereals can be lost during processing, making the grain unsuitable for both human and animal consumption (Kumar & Kalita, 2017). Microorganisms found in cereals are a key control point, as their growth can additionally have an effect on the quality, properties, and safety of the cereals.

Certain types of mold can produce mycotoxins that can be harmful and pose a severe risk to the physical wellbeing to consumers. During storage, the loss of cereals by molds and mycotoxins is predicted to be between 5 to 30%, with 5% due to insects and close to 2% for rodents, with an average loss of yield up to 1% for developed countries, and this figure ranges from 10–30% for developing nations. Aflatoxins, produced by some fungi, are generally considered the most dangerous group of mycotoxins (Sawicka & Egbuna, 2020). Microbial contamination of cereal grains happens in the course of cultivation, harvesting, drying, and post-harvest storage, and it comes from numerous sources, which include water, soil, air, birds, insects, dust, contaminated devices, and unhygienic handling. The sort of microbial infection varies according to the developing area and is highly dependent on environmental situations such as temperature, rainfall, sunlight, and drought, along with improper harvesting, handling, storage, and processing equipment. The field microbiome includes microorganisms found in or on grain prior to harvest and relies upon the situation in which the cereals were grown (Reid et al., 2021). Cereals are dominated by bacteria, and yeast is the next most ample component. During the last stage of maturation, the number of filiform fungi increases. When the grains are high in moisture (18–30%), fungi, together with species of *Cladosporium*, *Helminthosporium*, *Alternaria*, and *Fusarium*, infect grain in fields with high relative humidity (90–100%). For preharvest fungal contamination, *Phyllosphere* fungi are responsible (Los et al., 2018).

Lactic acid bacteria present in immature grains can be transmitted during processing and spoil flour and cornflour doughs. Deterioration of grains by filamentous fungi at some stage in storage is often due to ineffective drying, which promotes growth of microbes and can increase mycotoxin levels. Storage temperature closely impacts the rates and types of microbial contamination (Los et al., 2018). Aflatoxin, which is produced by *Aspergillus*, has a major influence on proteins, RNA, and DNA. It causes severe effects in humans, acquiring carcinogenic and mutagenic activity. Aflatoxins act and modify DNA, causing improper regulation of cells, which ultimately leads to the death of the cells (Zegura, 2016). Fumonisins are chiefly formed by the actions of *Fusarium verticilliodes* and *Fusarium oxysporium*. The intake of this mycotoxin can cause a critical impact on humans, including equine leukoencephalomalacia, porcine lung edema, and brain necrosis. Weanimix, a popular baby food in Ghana made from maize, groundnuts, and beans, has been reported to be contaminated by aflatoxins and fumonisins. The legumes and seeds consumed had a chance of being contaminated with both toxins, which were transmitted to the prepared food (Temba et al., 2017). Pathogens colonize and grow on equipment employed in the food industry, which leads to the penetration of pathogens into food production. Food contamination by microbes has emerged as the biggest problem concerning most of the food industry. Contamination affects not only food processing but also food packaging. Although the industry has properly followed hygiene strategies, it becomes very tough to overcome contamination. Various diseases such as food poisoning, botulism, and other intestinal infections are caused by pathogenic microbes, which produce harmful toxic metabolites (Chatterjee & Abraham, 2018).

The sources of cereal grain contamination through microbes are plenty: crop production, preharvest and post-harvest stages, transportation, processing, and storage. Humans, animals, water, air, and contaminated equipment are several mediums by which contaminant microbes travel (Los et al., 2018). Some molds can produce harmful mycotoxins, which are among the most significant food contaminants that affect the final grain product. Mycotoxins are mainly produced by *Aspergillus*, *Penicillium*, and *Fusarium* genera, which may contaminate food during storage or in the field. Molds and mycotoxins cause 5–30% losses of cereal grains during storage. Some countries have reported up to 50% losses of cereal grain during storage. Current agricultural production of cereal grains relies heavily on chemical pesticides, which assist cereal growers to produce more yield on fewer acres of land by shielding crops from diseases and pests. However, the utilization of pesticides has undesirable effects on human and animal health. Therefore, there is an urgent need to develop environment-friendly technologies for cereal grain decontamination, sustainable management of pests, and agricultural methods.

16.4 CURRENT AND FUTURE TECHNOLOGIES FOR MICROBIAL DECONTAMINATION OF CEREAL GRAINS

Current technologies for microbial decontamination productively decrease the microbial load of grains; however, there are some limitations in the current techniques, like changes in the technological characteristics and quality of cereals and generation of harmful environmental impacts. Some of the current decontaminating technologies are discussed in the following.

16.4.1 Pesticides

Globally, production of cereal grains relies heavily on pesticides. Generally, the term "pesticide" covers a extensive range of combinations including herbicides, fungicides, insecticides, nematicides, molluscicides, rodenticides, and many more. The main aim of pesticide application is to protect crops from pests and plant disease vectors that will ultimately result in higher yield and better-quality grains. But there are a huge number of environmental concerns connected with the use of pesticides like the killing of beneficial and non-target species; contamination of soil, air, and non-target vegetation; groundwater contamination; and the decline of beneficial soil microorganisms (Aktar et al., 2009). Frequent use of pesticides can inherently entail the risk of resistance developing in the pest population. Therefore, to reduce pesticide input and the associated risk to the environment and human health, the idea of integrated pest management (IPM) has been introduced that emphasizes long-standing avoidance of pests or their destruction through a blending of skills such as use of resistant varieties, alteration of cultural practices, biological control, and habitat manipulation.

16.4.2 Drying

Drying is among the most common and low-cost techniques to preserve cereal grains. Usually, cereal grains retain a high level of moisture after harvest. So, drying is post-harvest processing when grains are dried in order to bring the moisture content to an optimum level, ensuring safe storage. Properly dried grains contain a moisture level of 10–14%. This value allows minimal qualitative and quantitative losses and ensures safe storage. The most cost-effective method for drying is to spread the grains out under the sun and dry them. In humid climates, there may be a requirement for artificial dryers. One of the major limitations of grain drying is the inconsistency of the process. Under- or over-drying results in grains with dissimilar moisture content. Also, if the drying of grain is not carried out under appropriate conditions it intensifies the threat of growth of mycotoxin-generating molds (Los et al., 2018). Excessive temperature may also cause loss of viability of grains.

16.4.3 Debranning

Debranning is the process of gradual removal of the bran layers of cereal grains. This technique is used in the flour milling industry to improve the quality of flour and to produce good-quality milled products from poor-quality grains. Debranning of grains has been reported to reduce the microbial loads from the exterior of the grain (Laca et al., 2006).

16.4.4 Chlorine and Hypochlorite

Chlorine gas, because of its high oxidizing capacity, is used for decontamination purposes. However, chlorination is ineffective for heavily contaminated foodstuffs. Positive results were obtained by treating corn grains with sodium hypochlorite solution. They observed that with increasing the duration of treatment, the percentage of infected seeds was found to be greatly reduced. However, later researchers have also obtained negative results by treating corn and wheat grains with sodium hypochlorite solutions (Sun et al., 2017). Chlorine-based methods generate toxic by-products after treatments. They also change the taste and odor of the treated foodstuff.

16.4.5 Ionizing Radiation

Irradiation in food processing refers to the use of ionizing radiation on edible products. The use of this method as an operative decontamination process for providing biosafety of food products has emerged in recent years (Khaneghah et al., 2020). The aim of this technology is to prolong the shelf-life of food products during storage under various conditions and to eliminate pathogenic microorganisms that cause crop loss. Three key methods of food irradiation technologies are X-ray irradiation, gamma irradiation, and electron beam irradiation (EBI). The electron beam and X-ray methods are considered better implementations of ionizing radiation technology than gamma irradiation, as both of them use sources other than radioactive materials, as opposed to gamma irradiation, which involves materials with radioactive properties. So, the use of X-ray and EBI technologies limits the danger of radioactive infection compared with the risk of handling isotopic sources, which may cause an added threat of radiological mishaps.

With the help of technological advancement, new strategies for microbial decontamination have emerged to avoid the limitations of conventional methods. An ideal method can decontaminate all the treated cereal grains efficiently with nominal changes in sensory and nutritional properties and without any form of toxic residue. Some of the potential techniques for microbial decontamination are discussed in the following.

16.4.6 Microwave

Microwaves (MWs) are electromagnetic waves that travel at the speed of light, mirrored by metals, transmitted through neutral material and absorbed by electrically charged material. Causes that affect microwave heat transfer are thickness, dielectric properties, and geometry of food. There is a vast application of microwaves in food processing for microbial decontamination. MW irradiation has the ability to destroy fungal pathogens, bacteria, and even bacteriophages. The heat generated through MWs causes the destruction of bacterial cells. Researchers have effectively reduced *F. graminearum* contamination of wheat seeds to below 7% while upholding seed germinability. MW can also be used for insect disinfestation of stored grains (Vadivambal et al., 2008). The main benefit of MW over other conventional heating methods is its fast and selective heating ability. Second, it leaves no chemical residues in food and is hence safe for human consumption. MW treatment has been proposed as an alternative to chemical fumigation. One of the major limitations of MW decontamination is non-uniform temperature distribution and limited penetration into the grain (Horikoshi & Serpone, 2019). If not applied properly, microwave drying may result in poor-quality grains.

16.4.7 Non-Thermal Cold Plasma

Non-thermal cold plasma (NTCP) is a auspicious and efficient technology to increase the quality and safety of agricultural produce. Being a non-thermal process, there is very little risk of thermal destruction of treated food products. Cold plasma (CP) treatment has been confirmed to promote growth, development, and seed germination (Tamošiūnė et al., 2020). Studies indicated that plasma possesses germicidal properties and therefore can be used as an effective sterilization and decontaminating agent. The stabilization of microbes using CP is primarily associated with the production of reactive species. The possible mode of action for microbial inactivation may include rupturing of bacterial cell walls through bombardment of free radicals. Generated free radicals affect macromolecules such as proteins and DNA and result in oxidation of various cellular components, accumulation of charged particles on microbial surfaces, and membrane breakdown. Kordas et al. (2015) applied plasma technology for decontaminating fungal pathogen of winter wheat grain where they observed a decrease in the quantity of fungal colonies subsequent to handling. In a similar study, Los et al. (2018) initiated this reduction by two orders. This technology was found to be effective against two fungal species with pathogenic characters, *Aspergillus* spp. and *Penicillium* spp., artificially injected on seed surfaces of numerous cereal grains. The fungal contamination from the seed surface was found to reduce by 1% of the initially inoculated load in 15 min of handling. Zahoranova et al. (2016) achieved a decrease of 1 log CFU/g for bacteria and complete inactivation for yeast and fungi when wheat seeds were treated with CP for 120 seconds.

16.4.8 Pulsed Ultraviolet Light

This is a developing non-thermal technology to stabilize surface microorganisms employing short-duration, high power pulses of light in the spectral band between 200–280 nm. The intensity of each blaze of light, which lasts only for a few hundred millionths or thousandths of seconds, is 20,000 times more than that of sunlight and is reported to contain some ultraviolet (UV) light. This UV light is absorbed by DNA, resulting in the formation of DNA photoproducts, which interrupt transcription and translation processes and result in cell death (Gayan et al., 2014). In the United States, pulsed UV light technology has been made functional in the food processing industry for decontaminating food products. This technology is not believed to be beneficial for cereal grains because of their irregular surface. The microbial decontaminating efficiency of this technology on stored wheat grains was considered by Aron Maftei et al. (2014). They achieved a reduction in microbial growth of about 4 logs CFU/g. Pulsed UV light can upset food texture and color, depending on the distance from the lamp, dose of energy, and surface of the foodstuff. Also, because of the shadow effects produced by uneven surfaces of cereal grains, the effectiveness of the treatment dose is reduced.

16.4.9 Organic Acids

Organic acids are used as preservatives and food additives aiming at avoiding food worsening and prolonging the shelf life of unpreserved food ingredients. Organic acids such as lactic, acetic, citric, benzoic, formic, and propinoic acid are regarded as potent chemical preservative agents exhibiting broad-spectrum antimicrobial activity. Organic acids have been reported to prevent mold spoilage of bakery products. The application of organic acids for grain preservation has been evaluated. Sabillón et al. (2016) reported that the accumulation of organic acids alone or in combination with NaCl in the tempering water reduces the microbial load in wheat grains.

16.5 EXISTING IRRIGATION AND WEED MANAGEMENT SYSTEMS AND THEIR ROLE IN GRAIN QUALITY

Irrigation systems play a vital role in cereals and show adaptation with response to climate change. Irrigated lands have a higher efficiency in crop production. In 1989, according to the FAO, in worldwide total irrigated area, approximately 15–20% was for cultivation. There are several methods of irrigation that may differ from place to place, crop to crop, and farmer to farmer such as surface, sprinkling, subirrigation, intrasoil, and mist irrigation. Since the success of the green revolution, international water-related debate was focused on irrigation systems in order to increase food yield. Grains like rice and wheat are the main source of carbohydrates provided all over the world in households (Shiferaw et al., 2013). But due to climate change and variations in weather, yield of grains, especially wheat, have been reduced and influence the livelihood of farmers. This can be overcome by a system of irrigation, which helps in photosynthesis and even in the process of translocation to increase grain yield and size. Various differences were found within the full irrigation treatment in the case of leaf area index, straw, and grain yield. Normally after applying irrigation, the protein content of grain was found to be 11.20 to 13.78%, whereas in H_2O stress conditions, the protein content was high, ranging from 12.47 to 13.92% (Noorka & Silva, 2012). But when irrigation was applied at the milk stage, the protein ratios were found to be increased, as well as the gluten content. Irrigation systems help crops tolerate soil salinity. Plant biotechnology has a great importance in salt-tolerant crops and has worked on *Hordeum vulgare* L. (barley). Various researchers have applied a triple-line source sprinkler system in soil salinity to show response in barley crops for high yields. In this system, the salt present in water is absorbed through leaves, and the absorption of salt could hamper salt tolerance activity. The drip-injection irrigation system was also found to be accurate and low cost, and it was suitable for soil tolerance of crops.

Population overgrowth in the last few decades puts tremendous pressure on the productivity of crops. Weeds are the major biological constraints in the crop production sector that seriously cause reduction in crop germination, growth, and yield through crop rotation and allelopathic interaction. Climatic change, intensive weed management practices, and ecological shifts cause rapid changes in weed behavior and weed infestations (Chauhan, 2020). In technologically progressive countries, chemical weed management is prevalent, but in developing nations, mechanical or manual weed management strategies are still being practiced. In India, weeds are the major restrictions to crop productivity. It is likely that weeds in India decrease crop productions by 31.5% every year. Other scientists have estimated yield loss in some crops as 10–100%. Other studies reported 33% of entire losses in revenue in addition to weakening quality of production (Pradhan et al., 2022). The existing weed management practices have various limitations, like need of manpower, health hazards, risk of eco-degradation in chemical control options, and development of herbicide resistance in weeds. So present-day agriculture requires an improved weed management system to manage problems associated with traditional practices.

16.6 IMPROVEMENT IN IRRIGATION SYSTEMS AND WEED MANAGEMENT PRACTICES

Irrigation is one of the key inputs for efficient and sustainable agricultural production. Water resources are becoming scarce day by day. According to predictions made by the National Commission for Integrated Water Resource Development Plan, in India, there will be a 50% increase in water requirement for irrigation in the next 50 years. With reduced water availability, little water productivity, unpredictable monsoon behavior, growing drought incidence, and uncertainties associated with changing climate, there is a need for India's irrigation system to accept technologies with greater water productivity (Praharaj et al., 2017). As the agriculture sector alone consumes a maximum (80%) percentage of water, its management has become crucial for food security. Precision irrigation or site-specific irrigation can be a possible answer to increase productivity and decrease the environmental effect of irrigated agriculture. The demand for efficient technologies for higher water use is increasing due to the increasing scarcity of water supply under a changing climate scenario. Micro-irrigation (MI) systems can play a dynamic part in the improvement of water use efficiency. MI systems use small devices that sprinkle, drip, or spray water directly on the crop root zone at required intervals and quantities. The irrigation system has been confirmed to increase crop yield under augmented water use proficiency over conventional systems. These systems suggest a great level of regulation over water applications using a small volume of water at low pressure, causing minimal energy costs (Varma et al., 2006). MI mainly includes drip and sprinkler irrigation methods. Both methods differ in terms of pressure requirement, flow rate, mobility, and wetted area. The drip irrigation method (DIM) is the most common in which water is brought directly to the root zone by a web of pipes with the help of emitters, thereby minimizing water losses through evaporation, while the sprinkler irrigation method (SIM) sprinkles water through nozzles that then produce small water droplets and tumble on the plant surface. It is a kind of artificial rain that fulfils the normal requirements of the plant and results in uniform distribution of water. Implementation of MI systems has not only increased water use effectiveness in agriculture but has also brought many social and economic profits to society. The gain in productivity due to the use of MI is projected to be 20–90% for various crops. An increase in productivity subsequently reduces problems of soil erosion, weeds, cost of cultivation, and energy use in terms of electricity, which is essential to lift water from irrigation boreholes (Narayanamoorthy, 2003). Development of MI systems combined with fertilizer injectors led to adoption of the system, thereby improving the water and nutrient use efficiency. Such innovations are helpful in applying water and fertilizer in judicious amounts throughout the growing season to meet crop requirements. Precise irrigation scheduling enables presentation of water at a rate and time that are based on crops' accurate water requirements. Through precision irrigation, there is a potential to increase crop yield by irrigating the crops with a accurate amount of water well matched to the growth stage of crops (Adeyemi et al., 2017). Precision irrigation reduces runoff and energy consumption while saving water and money. Irrigation scheduling is the process of determining the accurate duration and frequency of watering. The objective of irrigation scheduling is to put on enough water to completely wet the root zone of the plants while reducing overwatering and then allowing the soil to dehydrate out between waterings for gaseous exchange and proper root development. For scheduling irrigation for the greatest effective use of water and optimizing crop productivity, it is required to regularly monitor the soil water conditions in the root zone of the crop. Several methods for irrigation scheduling have been developed with varying degrees of success. Automation of irrigation scheduling has been made possible today with the advancement of frequency and time-domain reflectometry sensors that offer information on soil moisture levels on real-time basis. Earlier irrigation was based on farmer's instinct, weather, and physical conditions of soil and plant. These new sensors consider several parameters such as crop type, soil type, growth stage, and soil moisture status. Modern equipment provides real-time data on soil moisture right in the field, allowing farmers to make judicious irrigation conclusions. However, at present, sensors can provide soil moisture status only up to a particular point and depth. In such conditions, appropriate placement of sensors is crucial to offer a characteristic measurement for making an inclusive irrigation decision. Application of geographic information system (GISs), satellite positioning systems, remote sensing, and automated machine guidance technologies is an emerging area of research for precision irrigation scheduling. Drones, unmanned airborne vehicles, or satellites can define the water status of plants or soil at great spatial irrigation and accordingly prompt irrigation accurately in a specific location (Awais et al., 2021). In the future, there is the potential to apply nanotechnology and biotechnology in micro-irrigation for improved water quality, less emitter clogging, and filtration techniques.

Improvement in tillage, weed-seed predation, growth in the ground of allelopathy, proper crop nutrient management, herbicide tolerant crops, bioherbicide, and thermal techniques are some probable weed-managing strategies in modern agriculture.

16.6.1 Tillage and Weed Management

Tillage has been an indispensable part of traditional agricultural systems since ancient times. In a broad sense, tillage is mechanical management of plant and soil residues to formulate a seedbed before planting crops. The sole purpose of performing this technique is to loosen the soil for better nutrient and gaseous exchange and improved regulation of water and air circulation within the soil

(Reicosky & Allmaras, 2003). However, the tillage technique also plays a significant role in weed control. Tillage prevents weed emergence by uprooting, disorienting, and concealing them deep in the soil. It also changes the soil's environment inhibiting or sometimes promoting weed growth. Tillage system affect the weed dynamics by interacting with various features such as soil type, cropping system, and weed flora. According to Blackshaw et al. (2001), weed infestation in conservation tillage is a problem of concern, as it reduces crop yield. The effects of tillage on weeds are different from species to species and respond differently under different tillage regimes.

16.6.2 Crop Nutrient Management

Though proper plant nutrition is a chief contributor to a high-yielding crop, weed growth is also multiplied by nutrients. Weeds take up a substantial amount of nutrients from the soil, like crops. Moreover, studies show that weeds can uptake nutrients more efficiently and quickly than crops. In this way, they accumulate higher nutrients than crops do, thus depleting soil nutrient concentration and reducing crop productivity (Blackshaw et al., 2001). It has been established that fertilizers intensify weed growth, dynamics, distribution, emergence, competitiveness, and persistence. Therefore, there is a need for the development of strategies that will make nutrients available to soil and not to weeds. The role of fertilizer placement in weed management is vital. Fertilizer placement refers to the specific application of fertilizer formulations in the close vicinity of seeds and plant roots to guarantee higher nutrient availability (Nkebiwe et al., 2016). Weed seeds most commonly reside adjoining to the soil surface, and use of fertilizers in that zone may encourage their better growth and emergence. Fertilizer placement can increase nutrient availability to crop and not to weeds, enhance crop competitiveness, and reduce weed interference. It has been observed that surface banding of nitrogen and phosphorus decreased weed emergence more than broadcasting due to lack of availability of nutrients to weeds. A noteworthy decrease in weed biomass was observed by application of subsurface fertilizer in dehydrated direct-seeded rice. Different weeds respond differently to nutrient management. Studies showed that nutrient availability may alter crop-weed competition duration. Sweeney et al. (2008) reported that use of nitrogen fertilizer altered density, competitive ability, and emergence patterns of various weeds. The variability in weed response to fertility indicates that weeds can be organized through proper fertilizer management. Moreover, variation in fertilizer application timing, doses, and techniques can alter weed-crop interaction (Blackshaw et al., 2001).

16.6.3 Weed Seed Predation

Seed predation through mammals and insects is considered an important way to eradicate weed from agricultural fields. Predators that feed on weed seeds can lead to suppression of annual weeds from fields. Seed predation is of two types: pre-dispersal and post-dispersal. Pre-dispersal seed predation is considered more effective than post-dispersal, as seeds of weeds are more exposed at this phase. Pre-dispersal insect predators mostly belong to the orders Diptera, Lepidoptera, Hymenoptera, and Coleoptera and are specific to plant family, genus, or species, whereas post-dispersal seed predation occurs after the seeds have been shed on the surface of the soil. Predators for post-dispersal predation include a wide range of vertebrates and invertebrates. In North America, several species of field cricket and carabid beetles have been detected as potential predators of several weeds such as large crabgrass, redroot pigweed, giant foxtail, and velvet leaf seeds (Carmona et al., 1999). In Canada it was observed that rats consumed up to 21% of seeds of common lambsquarters and barnyard grass in soybean [*Glycine max* (L.) Merr.] and no-till corn (*Zea mays* L.), leading to significant weed suppression. A seed removal rate of up to 91% for southern crabgrass [*Digitaria ciliaris* (Retz.) Koel.], goosegrass [*Eleusine indica* (L.) Gaertn.], and jungle rice [*Echinochloa colona* (L.) Link] over a 14-d period were reported in the Philippines (Chauhan & Johnson, 2010). Seed predation is a probable biological weed control method that can be employed either alone or in combination.

16.6.4 Allelopathy

Existing research on allelopathy shows that it can be a significant organic weed management practice. Allelochemicals released by plants to the soil suppress physiological functioning and growth of other plants growing in the vicinity. Allelopathy for weed management can be presented in two different ways: cultural means and allelopathic extract application. Allelopathic plants exhibit self-poisoning. Farmers observed a decrease in crop productivity on subsequent cultivation in traditional farming. The reduction in crop productivity is due to the accumulation of harmful poisonous chemicals released by one crop that is planted recurrently. Therefore, crops with allelopathic properties are incorporated in a planned rotation to evade gathering of self-toxic allelochemicals in an exclusive single-cropping system, and also their residual effect suppresses weed flora and offers an environment free of weeds for the succeeding crop. Allelopathic residues-based mulches are decent weed resisting options. There are several reports on weed control in the field through allelopathic mulching. Sorghum releases some natural allelochemicals to the soil. Crops growing in the field after sorghum take advantage of this natural herbicide for weed suppressing. Cheema and Khaliq (2000) reported that integration of vegetative parts of sorghum minimized weed density in wheat field. But traditional weed flora are quite tough to eliminate through allelopathic mulching. Further, surface mulching of sorghum, cone marigold, and brassica controlled weeds in mung bean, cotton, and rice. Mulching of sorghum residues either alone or in combination with rice, sunflower, and *Brassica* spp. residues provided control against various weeds present in wheat, corn, and cotton fields. Allelopathic crops can be employed as cover crops directed at particular weeds and crops. For example, barley is planted as a cover crop for weed regulator in soyabean, and legumes are planted as cover crop for weed regulator in maize (Farooq et al., 2013). Allelochemicals are the group of secondary metabolites with complete solubility in water. Application of allelopathic water extracts is a more eco-friendly method for weed eradication than using synthetic chemicals. Water extracts prepared with two or more allelopathic compounds provide a better weed control effect than using a single allelopathic compound (Farooq et al., 2013). Concentrated sorghum water extracts controlled weed biomass and density in wheat crops. Likewise, sunflower aqueous extracts controlled wild oat, broadleaf dock (*Rumex obtusifolius* L.), and yellow sweet clover [*Melilotus officinalis* (L.) Lam.] in wheat fields. Combined extracts of sorghum, sunflower, and rice along with butachlor controlled weed biomass and density in rice fields (Rehman et al., 2010). The influence of allelopathy on future weed management practices can be understood by highlighting the salient points presented in Table 16.1.

16.6.5 Thermal Weed Management

Thermal weed management strategies are centered on susceptibility of plants to higher temperatures. Increased temperature causes disruption of various physiological functions in plants like enzyme inactivation, membrane rupture, and protein denaturation. According to Zimdahl (2013), a large portion of plants do not survive after introduction to temperatures ranging between 45 and 55°C. The main advantage of thermal weed control is that they do not have any residual effect on soil like herbicides, nor do they disturb the soil. Various ways in which thermal weed management is carried out are discussed in the following.

16.6.6 Microwaves and Radiation

This technique allows efficient and targeted killing of weeds and reduces the risk of non-targeted loss. Microwaves have been reported to successfully eradicate various weeds in Denmark. The only disadvantage of using microwave is the high cost of energy production. However, through

Table 16.1 Weed Management through Allelopathic Approach

Application Method	Allelopathic Source	Weed Controlled	Cereal Crops Benefitted	References
Soil incorporation	Sorghum	Common lambsquarters, fumitory, toothed dock, littleseed canarygrass	Wheat	Cheema and Khaliq (2000)
Soil incorporation	Sorghum, sunflower, and rice in combination	Purple nutsedge, horse purslane	Wheat, corn, cotton	Khaliq et al. (2010)
Aqueous extract	Sorghum	Rice flatsedge (*Cyperus iria* L.), purple nutsedge, jungle rice	Rice	Wazir et al. (2011)
Aqueous extract	Sorghum	Fumitory, toothed dock, common lambsquarters, littleseed canarygrass,	Wheat	Cheema and Khaliq (2000)
Aqueous extract	Sunflower, sorghum, rice and butachlor combination	Barnyardgrass, flatsedge, crowfootgrass [*Dactyloctenium aegyptium* (L.) Willd.], rice	Rice	Rehman et al. (2010)
Aqueous extract	Combination of sorghum and sunflower/tobacco/ brassica	Littleseed canarygrass, wild oat	Wheat	Jamil et al. (2009)
Intercropping	Wheat and chickpea	Purple nutsedge, barnyardgrass, wild oat, lesser swinecress [*Coronopus didymus* (L.) Sm.], scarlet pimpernel (*Anagallis arvensis* L.),	Wheat, chickpea	Banik et al. (2006)
Cover crops	Velvet bean	Barnyard grass	Soyabean and rice grown after wheat	Peters et al. (2003)

stimulation of thermal runaway in weed plants and flux configuration, its energy budget may be reduced, making it affordable like other weed control tools (Brodie et al., 2012). Likewise, laser beams can also be used effectively to eradicate weeds. Laser radiation was effectively used to destroy noxious water hyacinth plants in the United States. Mathiassen et al. (2006) were able to control scentless chamomile and common chickweed under varying levels of laser beam exposure.

16.6.7 Bioherbicides

Biological weed control mechanisms utilize natural substances, natural enemies, or biotic agents to defeat the growth and germination of weed populations. Bioherbicides are prepared from pathogens and extra microbes and toxins derived from insects, microbes, or plant extracts that are applied exogenously and repeatedly for weed control. The first ever bioherbicide was developed in the mid-1970s, and since then, numerous bioherbicides have been registered worldwide (Hasan et al., 2021). The chief advantage of bioherbicides is that they have a less adverse effect on non-targeted weeds. In the international market, the majority of the registered bioherbicides were derived from microorganisms. There were seven bioherbicides registered in the United States, six in Canada, and one in both Japan and Ukraine by 2012. There were nine bioherbicides derived from fungi, three from bacteria, and one from plant extracts by 2016. By 2020, six bioherbicides were extracted from essential oils, all of which are commercially available (Verdeguer et al., 2020).

16.6.8 Weed Modeling

Computer stimulation modeling is a significant tool for understanding and predicting weed-crop interaction. The three primary purposes for biological weed modeling are control, prediction, and understanding, while secondary objectives include synthesizing and summarizing information, providing a intangible framework, identifying areas of unawareness, and providing insight. Modeling crop-weed competition is the most obvious way to predict the future. With proper sophisticated modeling, it is possible to interpret the likely production loss coupled with diverse weed densities in crops. A crop-weed competition model also helps in short- and long-term strategic preparation decisions in reaction to existing conditions (Renton et al., 2015). Successful modeling is centered on data compiled through sensing technologies that provide a comprehensible picture regarding weed emergence patterns, canopy architecture, competitiveness, seed bank dynamics, possible loss of productivity, and replacement trends. A high-quality management assessment support model is one that predicts the results of diverse management alternatives where solitary cost-effective future plans may be considered. According to Christensen (2009), modeling for weed dynamics in corn, barley, sugar beet, and wheat provided excellent results in terms of decision support and precision. Weed modeling can be a notable approach for the functioning of precision weed management.

16.6.9 Robotics

The application of robotics for weed control is a ground-breaking development in the field of agriculture. Robotic weed management involves the following processes: mapping, identification, guidance, and precision robotic elimination of weed species. Weed detection is done based on visual textures, spectral features, and morphological features. According to Brown and Noble (2005), morphological features are regarded as probable tools for detection through machine vision, particularly for making the distinction between weeds and other crops. Weed warning by guidance followed by detection is the next step in precision weed management using robotics. There might be different strategies to eradicate weeds through robots, like mechanical, electrical, and chemical. Robotic weed control systems have been developed and tested by various weed scientists. A self-governing robotic weed control system for sugar beet has been developed where there is a selective rotary hoe for weed removal, a machine vision system for within-row weed identification, and a machine vision guidance system. They mentioned that the robotic feature was capable of following the row on its own and selectively eradicating weeds. Their tests results showed that 99% of the sugar beets remained unaffected, while 41–53% of the weeds were eliminated successfully by the robot. Lamm et al. (2002) developed a real-time robotic weed management system on cotton fields. The system could move at a constant speed of 0.45 m/s. They mentioned that the entire robotic system could differentiate between weed and cotton plants while applying chemical spray only to the target weed. Blasco et al. (2002) developed a similar robotic weed management system for lettuce plants. Their system incorporated two systems with machine vision; one was for detecting the weeds in the images from the field, and the other provided trajectory information. They mentioned that the system could identify weeds by size, ignoring weeds present in the vicinity of lettuce plants. The system successfully identified 84% of the weeds present in the field.

16.7 CROSS-TALK BETWEEN QUALITY OF CONVENTIONAL AND GMO CEREAL GRAINS

The global population is exploding at an enormous rate of 80 million per year, and the ever-growing demand for nature and natural resources is getting out of hand. Developed nations have already looking for an alternative and come a long way forward in developing transgenic cereal grains, but developing countries are still hesitating to consider them an option due to various food security concerns (Hurburgh, 1999). The nutritional content in GMO crops has been a concern anyway; developing countries were reluctant to compromise on nutritional quality of the cereal grains for their huge populations, and their point of view is respected all over the world due to the unconventional and controversial nature of GMO crops. However, the world needs to move forward in a positive direction to meet the goals of sustainable development. Some scientists and researchers in the developing world are already coming out in public supporting GMO crops, including grains; simple logic emerging in this field is "How can a GMO grain be safe in the USA but suddenly become hazardous in Asia?" This line of thought is getting a lot of support from the general public as well as concerned citizens, including academicians, journalists, public figures, scholars, and students. Various critical findings have assessed the dietary value of transgenic grains, and the results were surprising as well. The nutritional quality of many transgenic grains was found to be superior to their traditional counterparts. According to a study conducted by Huang et al. (2004) in Illinois (USA), maize proteins are engineered through the use of antisense zein and expressing sense genes to increase nutritional quality. In this process through conventional breeding a mutant opaque-2(o2) was obtained, which is composed of superior amino acid essential for the production of Quality Protein Maize (QPM). Giri and Laxmi (2000) reported on transgenic rice production with the help of various beneficial genes, which are agronomically very useful. They highlighted various crucial things like the introduction of foreign genes to rice, which is achieved with the help of the successful implementation of many gene transfer protocols. A breakthrough experiment performed by Casas et al. (1993) reported the importance of the microprojectile bombardment technique for the production of superior-quality sorghum with high dietary value as compared to conventional sorghum. Improvement of sorghum is an important aspect from the perspective of food quality improvement and considered a current since it is used as a staple food by a large number of people in the Asian sub-continent as well as the African region. It is already an established fact that sorghum lacks the important amino acid lysine (K), and as a result, it has a very low protein content that can be an alarming situation for these people using it as a staple food. Even scientists across the globe credit this fact for the growing malnutrition problem among children in the region. An experiment described by Zhao et al. (2000) reveals the importance of *Agrobacterium*-mediated gene transfer technology in the engineering of T-DNA for the construction of a vector by incorporating more than ten lysine residues; the final product was found satisfactory and improved as compared to conventional sorghum available in the Asian and African market.

Another groundbreaking scientific find in the field of transgenic cereal grain was put forward by the historic "Golden rice" experiment. This initiative firmly established the importance of transgenic grains in the modern world. The normal rice varieties available in almost all parts of the globe have very low vitamin A content, resulting in a huge number of diseases associated with vitamin A deficiencies, including severe night blindness. Since rice is a staple food of more than two-thirds of the world, it was absolutely necessary to find a solution to this serious problem. Scientists finally decided to incorporate the genes responsible for β-carotene, which will eventually lead to high vitamin A content in the rice. This experiment was known as the "Golden rice" experiment due to the golden color of β-carotene, and it also appeared to be the final color of the transformed rice. At this point in time, scientists have finally confirmed that the genetically engineered rice variety contains up to 35 μg β-carotene per gram of rice, which is a very encouraging sign for future endeavors like this in the field of cereal grains (Tang et al., 2009).

Wheat is among the most important cereal grains and is utilized as a staple food by more than 35% of the population of the world, distributed throughout 60 countries, and it accounts for 10–20% of the total daily intake of calories in these regions. Even with such huge importance, very little interest has been given to generating or developing wheat crops for future generations. Only after the concept of sustainable development was introduced did the question of improvement of wheat varieties gain interest on the global stage, and as a result, wheat has become one of the last cereal grains to be transformed successfully. A very interesting experiment conducted by Tamás et al. (2009) in the prestigious Agricultural Research Institute of the Hungarian Academy of Sciences explained the role of a foreign gene in the upgrading of the nutritional quality of the conventional wheat plants. In the experiment, the authors isolated an albumin gene (AmA1) from *Amaranthus hypochondriacus* and incorporated it into an experimental bread wheat plant (*Triticum aestivum*). The final product showed results that exceeded the expectation of the authors, since the newly transformed product was not only rich in essential amino acids, but some functional parameters associated with the AmA1 protein also showed a significant improvement from the initial product.

CONCLUSION

There is rarely any doubt that the 21st century will be remembered for technology, biotechnology, electronics, and medical sciences, although the basic needs of the human population will remain the same. Due to the current COVID-19 pandemic, the development of developing countries was slowed significantly, which means there is an urgent need for preserving our food stocks for the future with more efficiency and precision. Cereal grains are here to stay; we are not in any situation to let them go from our dishes any time soon. Therefore, existing technologies along with new inventions should be made available to grassroots researchers as well as local farmers to get the most out of our food reserves.

REFERENCES

Adeyemi, O., Grove, I., Peets, S. and Norton, T., 2017. Advanced monitoring and management systems for improving sustainability in precision irrigation. *Sustainability*, 9(3): 353.

Ainsworth, E.A., Rogers, A. and Leakey, A.D., 2008. Targets for crop biotechnology in a future high-CO2 and high-O3 world. *Plant Physiology*, 147(1): 13–19.

Aktar, M.W., Sengupta, D. and Chowdhury, A., 2009. Impact of pesticides use in agriculture: Their benefits and hazards. *Interdisciplinary Toxicology*, 2(1): 1.

Ali, S., Bharwana, S.A., Rizwan, M., Farid, M., Kanwal, S., Ali, Q., Ibrahim, M., Gill, R.A. and Khan, M.D., 2015. Fulvic acid mediates chromium (Cr) tolerance in wheat (*Triticum aestivum* L.) through lowering of Cr uptake and improved antioxidant defense system. *Environmental Science and Pollution Research*, 22(14): 10601–10609.

Aron Maftei, N., Ramos-Villarroel, A.Y., Nicolau, A.I., Martín-Belloso, O. and Soliva-Fortuny, R., 2014. Pulsed light inactivation of naturally occurring moulds on wheat grain. *Journal of the Science of Food and Agriculture*, 94(4): 721–726.

Awais, M., Li, W., Cheema, M.M., Hussain, S., Shaheen, A., Aslam, B., Liu, C. and Ali, A., 2021. Assessment of optimal flying height and timing using high-resolution unmanned aerial vehicle images in precision agriculture. *International Journal of Environmental Science and Technology*, 19(4): 1–18.

Bajwa, A.A., Mahajan, G. and Chauhan, B.S., 2015. Nonconventional weed management strategies for modern agriculture. *Weed Science*, 63(4): 723–747.

Banik, P., Midya, A., Sarkar, B.K. and Ghose, S.S., 2006. Wheat and chickpea intercropping systems in an additive series experiment: Advantages and weed smothering. *European Journal of Agronomy*, 24: 325–332.

Blackshaw, R.E., Larney, F.J., Lindwall, C.W., Watson, P.R. and Derksen, D.A., 2001. Tillage intensity and crop rotation affect weed community dynamics in a winter wheat cropping system. *Canadian Journal of Plant Science*, 81(4): 805–813.

Blasco, J., Aleixos, N., Roger, J.M., Rabatel, G. and Moltó, E., 2002. AE—Automation and emerging technologies: Robotic weed control using machine vision. *Biosystems Engineering*, 83: 149–157.

Brodie, G., Ryan, C. and Lancaster, C., 2012. Microwave technologies as part of an integrated weed management strategy: A review. *International Journal of Agronomy*, 1–14.

Brown, R.B. and Noble, S.D., 2005. Site-specific weed management: Sensing requirements—what do we need to see?. *Weed Science*, 53: 252–258.

Carmona, D.M., Menalled, F.D. and Landis, D.A., 1999. *Gryllus pennsylvanicus* (Orthoptera: Gryllidae): Laboratory weed seed predation and within field activity-density. *Journal of Economic Entomology*, 92(4): 825–829.

Carter, J.G., 2011. Climate change adaptation in European cities. *Current Opinion in Environmental Sustainability*, 3(3): 193–198.

Casas, A.M., Kononowicz, A.K., Zehr, U.B., Tomes, D.T., Axtell, J.D., Butler, L.G., Bressan, R.A. and Hasegawa, P.M., 1993. Transgenic sorghum plants via microprojectile bombardment. *Proceedings of the National Academy of Sciences*, 90(23): 11212–11216.

Chatterjee, A. and Abraham, J., 2018. Microbial contamination, prevention, and early detection in food industry. In *Microbial Contamination and Food Degradation*. Academic Press.

Chauhan, B.S., 2020. Grand challenges in weed management. *Frontiers in Agronomy*, 1: 1–4.

Chauhan, B.S. and Johnson, D.E., 2010. The role of seed ecology in improving weed management strategies in the tropics. *Advances in Agronomy*, 105: 221–262.

Cheema, Z.A. and Khaliq, A., 2000. Use of sorghum allelopathic properties to control weeds in irrigated wheat in a semi-arid region of Punjab. *Agriculture, Ecosystems & Environment*, 79: 105–112.

Chernyad'ev, I.I., 2005. Effect of water stress on the photosynthetic apparatus of plants and the protective role of cytokinins: A review. *Applied Biochemistry and Microbiology*, 41(2): 115–128.

Christensen, S., Søgaard, H., T., Kudsk, P., Nørremark, M., Lund, I., Nadimi, E.S. and Jørgensen, R., 2009. Sitespecific weed control technologies. *Weed Research*, 49: 233–241.

Farooq, M., Bajwa, A.A., Cheema, S.A. and Cheema, Z.A., 2013. Application of allelopathy in crop production. *International Journal of Agriculture and Biology*, 15: 1367–1378.

Foschia, M., Peressini, D., Sensidoni, A. and Brennan, C.S., 2013. The effects of dietary fibre addition on the quality of common cereal products. *Journal of Cereal Science*, 58(2): 216–227.

Gayán, E., Condón, S. and Álvarez, I., 2014. Biological aspects in food preservation by ultraviolet light: a review. *Food and Bioprocess Technology*, 7(1): 1–20.

Gharde, Y., Singh, P.K., Dubey, R.P. and Gupta, P.K., 2018. Assessment of yield and economic losses in agriculture due to weeds in India. *Crop Protection*, 107: 12–18.

Giri, C.C. and Laxmi, G.V., 2000. Production of transgenic rice with agronomically useful genes: an assessment. *Biotechnology Advances*, 18(8): 653–683.

Godfrey, D., Hawkesford, M.J., Powers, S.J., Millar, S. and Shewry, P.R., 2010. Effects of crop nutrition on wheat grain composition and end use quality. *Journal of Agricultural and Food Chemistry*, 58(5): 3012–3021.

Hasan, M., Ahmad-Hamdani, M.S., Rosli, A.M. and Hamdan, H., 2021. Bioherbicides: An eco-friendly tool for sustainable weed management. *Plants*, 10: 1212.

Högy, P. and Fangmeier, A., 2008. Effects of elevated atmospheric CO2 on grain quality of wheat. *Journal of Cereal Science*, 48(3): 580–591.

Honfoga, B.G. and Van Den Boom, G.J.M., 2003. Food-consumption patterns in central West Africa, 1961 to 2000, and challenges to combating malnutrition. *Food and Nutrition Bulletin*, 24(2): 167–182.

Horikoshi, S. and Serpone, N., 2019. Microwave flow chemistry as a methodology in organic syntheses, enzymatic reactions, and nanoparticle syntheses. *The Chemical Record*, 19(1): 118–139.

Huang, S., Adams, W.R., Zhou, Q., Malloy, K.P., Voyles, D.A., Anthony, J., Kriz, A.L. and Luethy, M.H., 2004. Improving nutritional quality of maize proteins by expressing sense and antisense zein genes. *Journal of Agricultural and Food Chemistry*, 52(7): 1958–1964.

Hurburgh, C.R., 1999. *The GMO Controversy and Grain Handling for 2000*. Agricultural and Biosystems Engineering. Iowa State University.

Jamil, M., Cheema, Z.A., Mushtaq, M.N., Farooq, M. and Cheema, M.A., 2009. Alternative control of wild oat and canary grass in wheat fields by allelopathic plant water extracts. *Agronomy for Sustainable Development*, 29: 475–482.

Kaur, K.D., Jha, A., Sabikhi, L. and Singh, A.K., 2014. Significance of coarse cereals in health and nutrition: A review. *Journal of Food Science and Technology*, 51(8): 1429–1441.

Khaliq, A., Matloob, A., Irshad, M.S., Tanveer, A. and Zamir, M.S.I., 2010. Organic weed management in maize (*Zea mays* L.) through integration of allelopathic crop residues. *Pakistan Journal of Weed Science Research*, 16(4): 409–420.

Khaneghah, A.M., Moosavi, M.H., Oliveira, C.A., Vanin, F. and Sant'Ana, A.S., 2020. Electron beam irradiation to reduce the mycotoxin and microbial contaminations of cereal-based products: An overview. *Food and Chemical Toxicology*, 111557.

Kimball, B.A., Morris, C.F., Pinter Jr, P.J., Wall, G.W., Hunsaker, D.J., Adamsen, F.J., LaMorte, R.L., Leavitt, S.W., Thompson, T.L., Matthias, A.D. and Brooks, T.J., 2001. Elevated CO2, drought and soil nitrogen effects on wheat grain quality. *New Phytologist*, 150(2): 295–303.

Koehler, P. and Wieser, H., 2013. Chemistry of cereal grains. In *Handbook on Sourdough Biotechnology,* G. Marco and G. Michael (Eds.). Springer.

Kordas, L., Pusz, W., Czapka, T. and Kacprzyk, R., 2015. The effect of low-temperature plasma on fungus colonization of winter wheat grain and seed quality. *Polish Journal of Environmental Studies,* 24(1).

Kumar, D. and Kalita, P., 2017. Reducing postharvest losses during storage of grain crops to strengthen food security in developing countries. *Foods,* 6(1): 8.

Laca, A., Mousia, Z., Díaz, M., Webb, C. and Pandiella, S.S., 2006. Distribution of microbial contamination within cereal grains. *Journal of Food Engineering,* 72(4): 332–338.

Lamm, R.D., Slaughter, D.C. and Giles, D.K., 2002. Precision weed control system for cotton. *Transactions of the ASAE,* 45: 231.

Loladze, I., 2002. Rising atmospheric CO2 and human nutrition: Toward globally imbalanced plant stoichiometry? *Trends in Ecology & Evolution,* 17(10): 457–461.

Los, A., Ziuzina, D. and Bourke, P., 2018. Current and future technologies for microbiological decontamination of cereal grains. *Journal of Food Science,* 83(6):1484–1493.

Mathiassen, S.K., Bak, T., Christensen, S. and Kudsk, P., 2006. The effect of laser treatment as a weed control method. *Biosystems Engineering,* 95: 497–505.

Narayanamoorthy, A., 2003. Averting water crisis by drip method of irrigation: A study of two water-intensive crops. *Indian Journal of Agricultural Economics,* 58(3): 427–437.

Nkebiwe, P.M., Weinmann, M., Bar-Tal, A. and Müller, T., 2016. Fertilizer placement to improve crop nutrient acquisition and yield: A review and meta-analysis. *Field Crops Research,* 196: 389–401.

Noorka, I.R. and Da Silva, J.A.T., 2012. Mechanistic insight of water stress induced aggregation in wheat (*Triticum aestivum* L.) Quality: The protein paradigm shift. *Notulae Scientia Biologicae,* 4(4): 32–38.

Nösberger, J., Long, S.P., Norby, R.J., Stitt, M., Hendrey, G.R. and Blum, H. eds., 2006. *Managed Ecosystems and CO2: Case Studies, Processes, and Perspectives* (Vol. 187). Springer Science & Business Media.

Ortiz-Bobea, A., Ault, T.R., Carrillo, C.M., Chambers, R.G. and Lobell, D.B., 2021. Anthropogenic climate change has slowed global agricultural productivity growth. *Nature Climate Change,* 11(4): 306–312.

Peters, R.D., Sturz, A.V., Carter, M.R. and Sanderson, J.B., 2003. Developing disease-suppressive soils through crop rotation and tillage management practices. *Soil and Tillage Research,* 72: 181–192.

Pradhan, G., Meena, R.S., Kumar, S., Jhariya, M.K., Khan, N., Shukla, U.N., Singh, A.K. and Sheoran, S., 2022. Legumes for eco-friendly weed management in agroecosystem. In *Advances in Legumes for Sustainable Intensification,* R. Meena and S. Kumar (Eds.) Academic Press.

Praharaj, C.S., Singh, U., Singh, S.S. and Kumar, N., 2017. Micro-irrigation in rainfed pigeonpea—Upscaling productivity under Eastern Gangetic Plains with suitable land configuration, population management and supplementary fertigation at critical stages. *Current Science,* 112(1): 95–107.

Rehman, A., Cheema, Z.A., Khaliq, A., Arshad, M. and Mohsan, S., 2010. Application of sorghum, sunflower and rice water extract combinations helps in reducing herbicide dose for weed management in rice. *International Journal of Agriculture And Biology,* 12(6): 901–906.

Reicosky, D.C. and Allmaras, R.R., 2003. Advances in tillage research in North American cropping systems. *Journal of Crop Production,* 8(1–2): 75–125.

Reid, T.E., Kavamura, V.N., Abadie, M., Torres-Ballesteros, A., Pawlett, M., Clark, I.M., Harris, J. and Mauchline, T.H., 2021. Inorganic chemical fertilizer application to wheat reduces the abundance of putative plant growth-promoting rhizobacteria. *Frontiers in Microbiology,* 12: 458.

Renton, M., Lawes, R., Metcalf, T. and Robertson, M., 2015. Considering long-term ecological effects on future land-use options when making tactical break-crop decisions in cropping systems. *Crop and Pasture Science,* 66: 610–621.

Sabillón, L., Stratton, J., Rose, D.J., Flores, R.A. and Bianchini, A., 2016. Reduction in microbial load of wheat by tempering with organic acid and saline solutions. *Cereal Chemistry,* 93(6): 638–646.

Sawicka, B. and Egbuna, C., 2020. Pests of agricultural crops and control measures. In *Natural Remedies for Pest, Disease and Weed Control.* Academic Press.

Shahbaz, M. and Ashraf, M., 2013. Improving salinity tolerance in cereals. *Critical Reviews in Plant Sciences,* 32(4): 237–249.

Shiferaw, B., Smale, M., Braun, H.J., Duveiller, E., Reynolds, M. and Muricho, G., 2013. Crops that feed the world 10. Past successes and future challenges to the role played by wheat in global food security. *Food Security,* 5(3): 291–317.

Sun, F., Liu, J., Liu, X., Wang, Y., Li, K., Chang, J., Yang, G. and He, G. 2017. Effect of the phytate and hydrogen peroxide chemical modifications on the physicochemical and functional properties of wheat starch. *Food Research International*, 100: 180–192.

Sweeney, A.E., Renner, K.A., Laboski, C. and Davis, A., 2008. Effect of fertilizer nitrogen on weed emergence and growth. *Weed Science*, 56: 714–721.

Tamás, C., Kisgyörgy, B.N., Rakszegi, M., Wilkinson, M.D., Yang, M.S., Láng, L., Tamás, L. and Bedő, Z., 2009. Transgenic approach to improve wheat (Triticum aestivum L.) nutritional quality. *Plant Cell Reports*, 28(7): 1085–1094.

Tamošiūnė, I., Gelvonauskienė, D., Haimi, P., Mildažienė, V., Koga, K., Shiratani, M. and Baniulis, D., 2020. Cold plasma treatment of sunflower seeds modulates plant-associated microbiome and stimulates root and lateral organ growth. *Frontiers in Plant Science*, 11: 1347.

Tang, G., Qin, J., Dolnikowski, G.G., Russell, R.M. and Grusak, M.A., 2009. Golden rice is an effective source of vitamin A. *The American Journal of Clinical Nutrition*, 89(6): 1776–1783.

Temba, M.C., Njobeh, P.B. and Kayitesi, E., 2017. Storage stability of maize-groundnut composite flours and an assessment of aflatoxin B1 and ochratoxin A contamination in flours and porridges. *Food Control*, 71: 178–186.

Vadivambal, R., Jayas, D.S. and White, N.D., 2008. Mortality of stored-grain insects exposed to microwave energy. *Transactions of the ASABE*, 51(2): 641–647.

Varma, S. and Namara, R.E., 2006. Promoting micro irrigation technologies that reduce poverty. *Water Policy Briefing*, 23: 1–6.

Verdeguer, M., Sánchez-Moreiras, A.M. and Araniti, F., 2020. Phytotoxic effects and mechanism of action of essential oils and terpenoids. *Plants*, 9: 1571.

Wang, Y. and Frei, M., 2011. Stressed food—the impact of abiotic environmental stresses on crop quality. *Agriculture, Ecosystems & Environment*, 141(3–4): 271–286.

Wazir, I., Sadiq, M., Baloch, M.S., Awan, I.U., Khan, E.A., Shah, I.H., Nadim, M.A., Khakwani, A.A. and Bakhsh, I., 2011. Application of bio-herbicide alternatives for chemical weed control in rice. *Pakistan Journal of Weed Science Research*, 17: 245–252.

Yu, X., Chu, M., Chu, C., Du, Y., Shi, J., Liu, X., Liu, Y., Zhang, H., Zhang, Z. and Yan, N., 2020. Wild rice (Zizania spp.): A review of its nutritional constituents, phytochemicals, antioxidant activities, and health-promoting effects. *Food Chemistry*, 331: 127293.

Zahoranová, A., Henselová, M., Hudecová, D., Kaliňáková, B., Kováčik, D., Medvecká, V. and Černák, M., 2016. Effect of cold atmospheric pressure plasma on the wheat seedlings vigor and on the inactivation of microorganisms on the seeds surface. *Plasma Chemistry and Plasma Processing*, 36(2): 397–414.

Zegura, B., 2016. An overview of the mechanisms of microcystin-LR genotoxicity and potential carcinogenicity. *Mini Reviews in Medicinal Chemistry*, 16(13): 1042–1062.

Zhao, Z.Y., Cai, T., Tagliani, L., Miller, M., Wang, N., Pang, H., Rudert, M., Schroeder, S., Hondred, D., Seltzer, J. and Pierce, D., 2000. Agrobacterium-mediated sorghum transformation. *Plant Molecular Biology*, 44(6): 789–798.

Zimdahl, R.L. (Ed.), 2013. *Fundamentals of Weed Science*. Academic Press, pp. 295–344.

Index

Page numbers in *italics* indicate figures and in **bold** indicate tables on the corresponding pages.

For Product Safety Concerns and Information please contact our EU
representative GPSR@taylorandfrancis.com
Taylor & Francis Verlag GmbH, Kaufingerstraße 24, 80331 München, Germany

www.ingramcontent.com/pod-product-compliance
Lightning Source LLC
Chambersburg PA
CBHW080908220326
41598CB00034B/5510

9 781032 171593